Ethical futures: bioscience and food hc

Adam Spencer
July 2009.

Ethical futures:

bioscience and food horizons

EurSafe 2009

Nottingham, United Kingdom

2-4 July 2009

edited by:

Kate Millar

Pru Hobson West

Brigitte Nerlich

ISBN 978-90-8686-115-6

Photo cover: Eileen Bates-Walters

First published, 2009

EurSafe 2009 Committees

Organising Committee

Kate Millar
Pru Hobson-West
Brigitte Nerlich
Sujatha Raman
Sandy Tomkins
Emma Hooley

Scientific Committee

Raymond Anthony (USA)
Frans W.A. Brom (The Netherlands)
Johan De Tavernier (Belgium)
Ellen-Marie Forsberg (Norway)
Pru Hobson-West (UK)
Matthias Kaiser (Norway)
Karsten Klint Jensen (Denmark)
Kate Millar (UK)
Brigitte Nerlich (UK)
Susanne Padel (UK)

The EurSafe 2009 Congress is organised by:
- Centre for Applied Bioethics, University of Nottingham
- Institute for Science and Society, University of Nottingham
- Division of Animal Science, University of Nottingham

The organisers wish to gratefully acknowledge the support of:
- University of Nottingham
- City of Nottingham
- BBSRC Sustainable Bioenergy Centre (BSBEC)
- Wageningen Academic Publishers

Table of contents

Theme 3 – The ethics of bioenergy: research and technology development

Theme 4 – Animal futures: welfare and ethics

Theme 5 – Innovation in animal ethics teaching

Theme 6 – Sustainability in food production

Theme 7 – Concepts of risk, trust and safety in the food system

Theme 8 – Global food security

Theme 9 – Participatory methods and ethical analysis in biotechnology appraisal

Theme 10 – Foundational issues in agriculture and food ethics

Theme 11 – Ethical matrix development

Theme 12 – Ethical dimensions of organic food and farming

Theme 13 – Other contributions

Preface

Ethical futures: bio

Preface

Ethical futures: bioscience and food horizons

In an ever changing interconnected world, the agriculture and food system faces constant challenge in many forms, from issues such as the impacts of climate change, uncertainty surrounding the use of novel technologies, changing trading conditions, regional food insecurity, and the emergence of new zoonotic diseases. Alongside these challenges professionals working in the food system are faced with opportunities to improve food production and distribution, through the use of new farming techniques as well as the application of novel technologies. In terms of overall management and priority setting, what results is a complex food system that presents both public and private decision-makers with numerous choices and challenges.

As decision-makers attempt to balance these threats and opportunities in order to secure more sustainable production systems, the key question that arises is: What do we envisage as the future for agriculture and food production? In other words, what horizons do we see? Faced with an array of important choices, what possible futures can we imagine and what are the implications of current decisions we make? With numerous voices advocating different, and sometimes conflicting, approaches, from organic farming, wider use of GMOs through to *in vitro* meat production, this discussion of futures raises significant ethical questions.

With the additional pressure faced by decision-makers who are already presented with an array of notable ethical dilemmas, it has been argued that due to the increasingly global nature of these issues, the choices that society makes now in relation to the agriculture and food system are likely to have a long standing impact on the global community in a way that was less so in previous historical periods.

The premise of the underlying theme of this Congress is to explore some of the ethical dilemmas and challenges that the agriculture and food system may face over the coming decades, and to explore the role that ethical analysis may play in informing the decisions that will shape our shared future. This volume contains the extended abstracts presented at the 8th Congress of the European Society for Agricultural and Food Ethics and includes papers exploring a diverse set of themes. These include the role of novel technologies, the potential issues raised by the use of biofuels, the ethics of future animal production systems, concepts of global food security, and other contributions exploring our wider priorities in terms of food governance and educational aspects.

The European Society for Agricultural and Food Ethics (EurSafe) is an independent organisation, which provides a network for those who have an interest in the ethical issues involved in agriculture and food supply (www.eursafe.org). The Society is an interdisciplinary and cross-cultural organisation, which bridges academia and the world of practice. The Society aims to: (1) encourage international academic research and education on the ethical issues involved in agriculture and food supply; and (2) promote international scientific and public debate on the ethical issues involved in agriculture and food supply. One of the main activities of the society is to organise networking activities such as the EurSafe Congress. This year's Congress takes place at an important time, not only in terms of the need to reflect on the future of agricultural and food production and the ethical choices that we face, but also as this congress represents the Society's 10th anniversary.

The contributions found in this volume bring together presentations given by a diverse group of authors from academics, public sector professionals through to representatives from non-governmental organisations (NGOs). This diversity exemplifies the philosophy of the EurSafe society that aims to bring together different groups to discuss the ethics of agriculture and food production. This approach welcomes a wide variety of theoretical, practical and cultural perspectives. Many of the contributions represent work that applies a cross-disciplinary approach. In addition, the extended abstracts not only report on research outcomes, but also represent authors' early philosophical reflections, reports on developing research activities as well as some initial analyses of policy issues.

The contributions in this book, the extended abstracts, have been peer-reviewed by the scientific committee before being accepted for presentation at the Congress. The editors are extremely grateful to the many colleagues who freely provided their expertise during the reviewing process and editing, particularly in light of the time constraints that result from producing a volume that will be distributed at the Congress. As a result it should be noted that due to the tight deadlines not all contributions presented at the congress are included here. It is hoped that this volume serves as an interesting collection and acts as a source of stimulation that will contribute to wider debate and reflection on the future of the agriculture and food system. In addition to this aim, a number of the enclosed contributions highlight tools that can aid decision-making and help to clarify some of the ethical dimensions of emerging issues relating to agriculture and food production.

Kate Millar, Pru Hobson-West and Brigitte Nerlich

Keynote papers

Social justice in sustainable food initiatives: an inquiry

T.C. MacMillan, P. Steedman and S. Ripoll
Food Ethics Council, 39-41 Surrey Street, Brighton, BN1 3PB, United Kingdom;
tom@foodethicscouncil.org

Abstract

When Wal-Mart announced a wide programme of environmental commitments in October 2005, it signalled that sustainable development had fully entered the mainstream in the food sector and unleashed a flurry of activity by other major retailers and producers. The same period, in the UK at least, has witnessed a surge in policy activity to promote sustainable food. This included a major report by the government's Cabinet Office. Our contention is that, as sustainable development has risen rapidly up the mainstream food business and policy agenda, the concept has been transformed. In particular, the emphasis on social justice that features prominently in the Brundtland Commission's seminal definition has diminished. The focus is instead on resource efficiency. Meanwhile, environmental standards are in some cases becoming part of the 'cost-price squeeze' facing suppliers and, reportedly, contributing to workplace exploitation. In this paper we describe an Inquiry into 'Social Justice in Food and Farming' launched by the Food Ethics Council in 2009. We analyse the place of social justice in sustainable food initiatives in the UK and explain the methodology behind the Inquiry. We shall also report interim findings from the project.

Keywords: fairness, injustice, policy, sustainable development, methodology

Introduction

The Food Ethics Council has commissioned an Inquiry into Food and Social Justice. The Inquiry is happening in the UK but it will recognise that UK's supply chains and policy influence are international. This paper sets out the thinking behind the Inquiry and the approach it is likely to take.

What has social justice got to do with sustainable food?

Government, retailers, manufacturers and farmers have made commitments to sustainable food. Social justice is a key pillar of sustainable development, whether in food, farming or beyond. The classic definition of sustainable development – development that 'meets the needs of the present without compromising the ability of future generations to meet their own needs' – is above all about achieving a fair deal for the world's people, now and into the future (WCED, 1987). Looking after the environment is necessary to building a just society, but it is not sufficient.

This focus on social justice continues to underpin formal commitments to sustainable development. For example, the UK's current sustainable development framework not only has 'ensuring a strong, healthy and just society' as one of its two main objectives, but also one of the aims of 'living within environmental limits' (the other main objective) is to ensure that natural resources 'are unimpaired and remain so for future generations' (HMG, 2005). The framework sees the economy as a means to these ends, rather than a goal in itself, 'providing prosperity and opportunity for all', and allocating the costs of using resources fairly.

So social justice has everything to do with sustainable food. If we take our sustainable development commitments seriously in the food sector, fairness should be top of the agenda whether we are talking about trade issues, technology, the environment or health. Indeed, we cannot talk credibly about sustainable food without putting fairness at centre stage.

What would a fair food system look like?

Would a fair society be an equal society? It would doubtless be more equal than a world in which nearly a billion people are hungry, but fairness and equality are not the same thing. Inequalities in the incidence of diet-related disease might be unjust, for example, but they are not necessarily so: if I get diabetes and you do not, I might just be unlucky. And some differences might not matter: it is hardly unfair if you like boiled potatoes and I like them mashed.

So, to understand what a fair food system would look like, we need to work out which inequalities matter – clearly some are more important than others – and also whether they are unfair, as opposed to just unfortunate. Fairness is about what caused something to happen, not just about the end result. It follows that even a situation that looks equal can be unfair – an employee might rise through the ranks of an organisation despite discrimination, for example.

This means that a fair food system should:

- Provide equally for everyone's basic needs. Human needs include food, health care and gainful employment. Governments, international agencies and research organisations routinely measure which basic needs are being met, where and for whom.
- Be free from discrimination. Significant inequalities in outcome between social groups – for example in health or income – may reflect systematic inequalities of opportunity, or the domination of one group of people by another. But discrimination and domination are unfair even if they do not show up in statistics.
- Ensure people are free to lead lives they value. A level playing field is not always as fair as it seems – people are different, and some will need positive help to be in the game. A fair society helps people shape their own lives, for example by supporting them to take part in decisions that affect them.

How are we doing?

Modern food systems have managed to eradicate famine from most regions in the world. Over the last 40 years, increases in grain production have helped reduce the number of hungry people even as the earth's population has doubled to nearly 6 billion. Safety standards in developed countries have minimised the risk of food-borne diseases. Almost 3 out of 4 people in developing countries live in rural areas, and make a living, directly or indirectly, on agriculture. Food is a profitable business in the UK: we spend £172 billion a year on food and drink. Food represents the UK's largest remaining manufacturing industry, and the sector as a whole employs 3.7 million people.

There is, however, much work to do to ensure our food and farming systems are fair. They are resource hungry, wasteful, polluting and carbon intensive, creating risks and eroding opportunities for future generations. Environmental degradation goes hand in hand with increased human insecurity. While one billion people worldwide do not have enough to eat, another billion are obese. Job opportunities in food and agriculture are plummeting and many of those available are badly paid, insecure and at higher risk of workplace injury. Small farms and businesses are closing down or being relegated to low-profit marginal markets.

The UK food system's global reach enables us to 'off-shore' many of the social and environmental cost of production to countries with lower standards of worker, animal and environmental protection. The costs and rewards of contemporary food systems are unequally distributed. Casual workers, migrants, vulnerable women and excluded racial and ethnic groups are all too often at the sharp end.

Are we trying hard enough?

In some specific areas of policy, business and public interest in food, social justice is high on the agenda. Considerable effort has been invested in trying to understand and tackle inequalities in diet-related disease, for example. Unfair differences in consumption are the focus of work on food poverty, 'food deserts' and global hunger. In production, the Gangmasters Licensing Authority has been set up to eradicate the abuse of workers in the UK, and strong sales of fair trade products in the downturn indicate the depth of public commitment to achieving fair terms of trade for producers internationally. Support for small shops and 'buying local' is motivated by solidarity and support for the underdog, not just environmental concerns.

For all these significant pockets of work to promote a fairer food system, however, social justice barely figures in the UK's flagship initiatives to promote a sustainable food system. A key example is Food Matters – a major Cabinet Office report that sets out the framework for a UK food strategy (Strategy Unit, 2008). The report was welcomed by the Food Ethics Council and is in many ways pioneering, bringing together a broad sweep of health and environmental issues and offering the nearest thing Britain has had to a coherent food policy for decades. It highlights income-related inequalities in diet and health, and tensions between environmental and social objectives around issues such as fish consumption and climate change. It even commits in principle to fair prices for consumers and producers.

Yet even Food Matters does not address social justice – the links between these issues – because it does not examine the causes of inequalities or the factors that give rise to tensions between different social and environmental interests. For example, it does not consider the UK's position in international trade negotiations, although trade rules shape the capacity of UK businesses to meet their environmental and social commitments, and have a major bearing on social footprint of our food system internationally. Similarly, health inequalities arising from low-incomes are tackled by targeting support at low-income groups: that raising incomes could also contribute is not considered, even though the farming, food manufacturing, food retail and food service together account for a significant proportion of all jobs in the UK.

The silence of Food Matters on social justice is indicative of a wider neglect within UK initiatives to promote sustainable food. In contrast to the formal definitions of sustainable development outlined above, which start with social justice, most current policy and business programmes start with environmental issues. They focus on sustainability as a technical challenge of resource efficiency, lean production and cutting waste – of keeping on doing what we do now for longer – rather than a call to meet the unanswered human needs of today and to leave open the options for future generations. There are exceptions, of course – the Co-operative Group's approach to climate change and trade is one example (Co-operative Group, 2008) – but this is the overall picture.

Studies by City University and the European Commission confirm this view (L. Sharpe, personal communication). Whereas interviews find that the economic – understood as profitability – and the environmental pillars of sustainable development are very well elaborated by stakeholders in the food sector and in policy, understandings of social justice are foggy. Practitioners do not have a clear vision of what social justice entails, and implement unconnected and ad-hoc schemes on different issues such as nutrition and community programmes.

As recession bites, social issues are rising up the agenda. The concern is that falling incomes and high prices for some foods will severely exacerbate food poverty and hunger. Yet even now the urgency of this problem puts the focus on catering for the poor, rather than on tackling poverty. That is the difference between charity and social justice.

Why have an Inquiry?

The Food Ethics Council has commissioned an Inquiry into food and social justice. The aim of the Inquiry is to put social justice at the heart of efforts to promote sustainable food and farming. Often social justice is already there in the headline or in formal definitions of sustainable development, but it falls out of view when it comes to the nuts and bolts. We want the Inquiry to address the reasons for this.

The basic problem is that social justice seems too difficult. It is difficult as a concept – tricky enough that philosophers have wrestled for centuries with what means. It can also be politically difficult because it challenges us to test whether the rules of the game are fair.

We want this Inquiry to address both these obstacles:

- We will try to understand better what social justice means in practice in food and farming, and we have invited others to join us in this learning process. How fair is our food system?
- We want the Inquiry to prompt stakeholders to strengthen their commitments to social justice, both in principle and in practice. How can we make our food system fairer?

Holding a formal Inquiry, as opposed to a research project, will help to achieve these aims. The process provides opportunities to involve a wide range of stakeholders, both as members of the Committee of Inquiry and as witnesses. It is a deliberative process which is important because, like most ethical issues, the questions of values and priorities that arise in considering social justice are better resolved in conversation than on paper. Furthermore the process itself – the evidence sessions and the final report – provides exciting opportunities to highlight issues of social justice to a wide audience.

We will count the Inquiry as a success if, because of it, social justice moves centre stage in discussions of sustainable food and farming.

What will the Inquiry look for?

The Committee of Inquiry will agree and publish a call for evidence in May 2009. It follows from how we define social justice in earlier sections and from the rationale for the Inquiry that it is likely to seek evidence under two main headings:

1. How fair is our food system?

- Where is it providing equally for people's needs and where is it failing to do so?
  Evidence of this could include, at local, regional, national, global scales: statistics on nutrition, disease incidence, workplace health and employment relating to food; qualitative research on food poverty; and discussions of how to evaluate and prioritise different objectives.
- Where does our food system discriminate?
  Evidence of this could include: statistical analysis of the significance of differences between different groups according to factors such as gender, ethnicity or age; analysis of inequalities in factors that affect opportunity such as levels of resource use or access to markets by small-scale producers; analysis of power, for example using 'value chain analysis'; first hand accounts of discrimination, for example labour exploitation; and discussions of how to distinguish between difference and discrimination.

- Where does our food system enhance people's freedom to shape their own lives?
  Evidence of this could include: studies of how much control consumers or producers feel they have over their lives; examples of initiatives to involve people in policy, business or community decisions that will affect them; and discussions of how actions today can enhance the opportunities facing future generations.

2. How can we make our food system fairer?

- Where do controversies or dilemmas highlight institutions or rules that get in the way of promoting social justice?
  Evidence of these could include case studies of tensions between: different groups of people; this generation and future generations; and economic, social, environmental or animal welfare objectives. Case studies should include the following elements: (1) description of unjust outcomes; (2) tensions between social groups or objectives; (3) drivers of change; and (4) levers for change.
- What factors cause discrimination?
  Evidence of these could include: social and cultural analysis of issues such as gender discrimination and racism as they relate to food and farming; political and economic analysis of factors that affect the distribution of power and opportunity in the food system, for example retail concentration and deregulation of financial markets; discussions of whether such factors are unique to food and farming.
- Where are there specific opportunities to promote social justice?
  Evidence of these could include: recommendations for government, business or society that would resolve tensions, eliminate discrimination or promote participation in decision-making; examples of opportunities in areas of policy such as finance, competition, trade, labour, innovation and international development; analysis of the capacity of different institutions to influence systematic change; case studies of measures that have promoted social justice in the food and farming sectors, in the UK or other countries; and examples of measures that have promoted social justice in other fields and offer lessons for food and farming.

We expect to receive a large volume of evidence. Simply organising the evidence we receive under headings such as these and summarising the Committee of Inquiry's deliberations would be likely to make a significant contribution to understanding social justice in the food system and identifying ways to promote it. The Committee may even choose to use an established framework or set of concepts to structure its deliberations and present its findings. Possible frameworks for deliberation, analysis and recommendations include concepts of rights, power and risk.

A rights framework may be particularly worthy of consideration. The concept of human dignity that underpins social justice is also central to human rights. The use of a rights framework would provide a reference point for discussions about human needs and freedoms, it would help to assign responsibilities, and it would demand institutional mechanisms to ensure the accountability and the possibility for redress in the case of human rights violations. All countries have signed binding legal commitments to human rights, which include a number relevant to social justice in food and farming such as labour rights, and rights to food, water and development.

How will the inquiry work?

The Inquiry will be led by a Committee of sixteen members. Half are drawn from the Food Ethics Council. Half are leading figures from sectors and communities with a stake in food and farming. The Inquiry's final report will highlight areas of consensus and difference among the members.

The Committee plans to meet five times. At the initial meeting, in May, they will agree their Terms of Reference and publish a call for evidence. Over the summer, the secretariat will gather written evidence and identify witnesses to take part in oral evidence sessions. Three such hearings will take place during the autumn. The Committee will then prepare a final report, which will be published in the spring of 2010, together with the evidence submitted to the Inquiry.

References

Co-operative Group (2008). Sustainability report 2007/8. Co-operative Group, Manchester.

Strategy Unit (2008). Food matters: towards a strategy for the 21st century. Cabinet Office, London.

HM Government (2005). Securing the future: delivering UK sustainable development strategy. TSO, London.

World Commission on Environment and Development (1987). Our common future. Annex to UN General Assembly document A/42/427, Development and International Co-operation: Environment, August 2, 1987.

Animal futures: the changing face of animal ethics

P. Sandøe and S.B. Christiansen
Danish Centre for Bioethics and Risk Assessment, University of Copenhagen, Rolighedsvej 25, DK-1958 Frederiksberg C, Denmark; pes@life.ku.dk

Abstract

The aim of this paper is to ask what moral views about the relationship between humans and animals will predominate in the future. The paper describes the emergence of new attitudes to animals in the western world over the past two centuries. This development is seen as a *widening* rather than a *wholesale change* of the moral agenda. The main headlines on that agenda are: that animals are there for us to use, anti-cruelty, animal welfare, animals as friends, respect for animals as part of nature, and animal rights. Debates or controversies concerning our interaction with and use of animals reflect the dilemmas and contradictions to which these points give rise. The future of animals will be shaped through these debates and controversies.

Keywords: animal protection, animal ethics, cruelty, animal welfare

Introduction

Human-animal interaction in the western world has changed dramatically over the last 200 years or so, and remarkable changes have followed in the attitudes that humans have towards animals. Suggestions as to how the norms governing human relationships with animals will develop in the future need to take these developments into account.

The aim of this paper is therefore to examine the emergence of new attitudes to animals in the west over the past two centuries with an eye to identifying moral trends; and to ask what kinds of moral view of the human-animal relationship will predominate in the future.

One way to describe developments in animal ethics is as a gradual change of agenda: the idea would be that in the beginning animals were not at all recognized as worthy of moral concern, and now they are recognized as individuals worth protecting in their own right. This picture certainly possesses a kernel of truth. Thus, at least in the western world, many more people today than in the past consider animals as worthy of protection and this is increasingly reflected in legislation.

However, it may be more appropriate to describe the story as one involving *widening* rather than *wholesale change* of the agenda. New norms and ideals regarding animal treatment have entered the agenda in today's societies, but the old norms are still there. So in that respect the situation regarding animal futures is complicated, torn by dilemmas, inconsistencies and conflicts of interest.

In the following we begin by describing developments in the early nineteenth century, when movements and legislation aimed at *preventing cruelty* to animals first appeared. Next, the emergence of the idea of *animal welfare* after the Second World War is presented. The focus here changes from protecting animals from meaningless cruelty to shielding them from the adverse effects associated with intensive animal production and animal experimentation. Then, new developments in attitudes to pets and wildlife are introduced. These have in various ways given rise to the idea that animals deserve not only protection but also *respect*. Finally, a potential and highly complex moral agenda for animal futures is outlined.

Traditional ways of using animals and the emergence of anti-cruelty legislation

Within the mainstream of Western culture animals have traditionally been viewed as means to fulfil human needs. Until the nineteenth century animals in the Western world were legally protected only in their capacity as items of private property. Bans on mistreatment served to protect the rightful owner of the animals from having his property vandalized. Legally speaking, the animals themselves had no protection.

Things began to change in the nineteenth century. This was a reflection of more general ethical and political changes that had taken place in the eighteenth century – a century in which grand ideas of human rights and liberal democracy gained momentum. It was no longer accepted that the ruling classes could treat the lower classes in the way they treated their property. Together with revolutions in France and USA, the idea developed that all humans are equal, and that the role of the state is to protect the rights of all its citizens.

Whereas in the case of humans the focus is on political rights that allow people to pursue their own happiness, with animals (as with some weak or marginalised groups of humans) it does not seem to make sense to allow them to sort out things by themselves. Rather, in the various movements 'for' animals that developed around the beginning of the nineteenth century the aim was to place limits on what humans were allowed to do with, or to, animals in their care. The aim was animal protection rather than animal rights.

Of course, these developments were not driven by ideas alone. It also mattered that with growing urbanisation large parts of the population no longer lived in the countryside and so no longer took part in traditional rural pursuits. And it mattered that there was a general increase in average levels of wealth in many countries. Clearly, people who have enough to eat and do not have to strive daily to subsist are in a better position to discuss how animals are being treated.

All these conditions were in place in early nineteenth-century England, where the world's first law for the protection of animals was passed. Getting the law through both chambers of the parliament was a huge struggle for the two key figures in this reform, Richard Martin MP (Member of Parliament) (1754-1834) and his collaborator Lord Erskine (1750-1823). They were up against strong opposing interests, and a political climate in which many people found concern for animals effeminate and ludicrous (notice that at that time women had no role in political life). The formulation of the bill that finally passed through the British parliament in July 1822 was therefore, in many respects, a political compromise. The bill said:

> that if any person or persons having the charge, care or custody of any horse, cow, ox, heifer, steer, sheep or other cattle, the property of any other person or persons, shall wantonly beat, abuse or ill-treat any such animal, such individuals shall be brought before a Justice of the Peace or other magistrate (Ryder, 1989: 86).

There are three striking limitations here: (1) only some kinds of animals are covered; (2) only things done by people who do not own the animals are covered; and (3) only what is described as *wanton* cruelty is covered.

On the first point, it is remarkable that a number of species are not mentioned at all – for example dogs, cats, pigs and poultry. One reason for this is that, at the time, there was a custom of arranging fights between animals, e.g. cock fights and dog fights. These forms of 'sport' could be extremely cruel.

Another reason that not all animal species are covered is that there clearly is a hierarchy of animals – a moral ordering that has been called the sociozoological scale (Arluke and Sanders, 1996). The point of the scale is, that people rate animals as morally more or less important, and therefore more or less worth protecting, according to a number of factors. These include how useful the animal is, how closely one collaborates with the individual animal, how cute and cuddly the animal is, how harmful the animal can be, and how 'demonic' it is perceived to be.

The sociozoological scale is in many ways based on traditions and prejudices and its use as a basis for animal protection can be criticized on both scientific and ethical grounds. The point being made here is just that the scale is part of social reality. This reality is, among other things, reflected in the legislation that has been introduced to protect animals.

The second striking point about the 1822 bill mentioned above was that it only protects animals against things done by people other than the owner. This, of course, partly reflects a political reality, since those in power were typically the owners of land and livestock; by making sure that these people were not affected by the law it was easier to get it through both chambers of the parliament. However, there is another, more respectable reason, and this is related to the third of the mentioned limits in the scope of the 1822 law, namely that it only protects animals against 'wanton' cruelty.

The bill's advocates assumed that the animal owner wants to protect and make good use of his property. To a great extent, the way to do this is by treating the animals well. Bad animal treatment is thus seen as something that is irrational, or pointless, which can only be done by someone who does not share the owner's interest in protecting the value that the animal represents.

Negligence towards animals was, of course, not uncommon in the past, and there would have been cases of obvious conflict between the interests of the animals and the interests of the owners. The use of animals for blood sports is an obvious example of this.

However, in general people in the past had to treat their animals decently to get the most out of them; and in many ways it can be said that people and their animals lived under the same conditions in mutual dependence. This remains the case today in some third world countries.

So in the early nineteenth century the need to protect animals was equated with the need to protect animals against pointless cruelty. This equation underpinned most legislation aimed at protecting animals until at least the 1950s.

The concern to protect animals from cruelty seems to be derived from at least two moral sources. One is the utilitarian idea that suffering is bad in itself and that suffering does not become less bad just because it is experienced by a being of another species. This idea was most forcefully expressed by Jeremy Bentham in the late eighteenth century. The other source is the idea that our moral character matters, and that cruelty to animals is an expression of a bad moral character. This idea was clearly stated by Immanuel Kant and received strong popular expression in William Hogarth's *The four stages of cruelty* (1751).

The sufferings of animals used in *purposeful* ways, particularly in farm animal production, did not become major issue until the 1950s. This may partly be explained, as suggested above, by the assumption that to get the best out of the animals owners had to treat them well. But in part it may also be explained by focusing on moral character – for the use of animals to do useful things does not seem to be a sign of a bad character.

However, since the Second World War there has been a dramatic shift in the way animal use in food production is popularly perceived. With the emergence of intensive animal production the welfare of farm animals became an issue, and the moral character of those participating in the industrialisation of animal production was increasingly perceived as dubious.

New ways of using animals and the emergence of animal welfare initiatives

Since the 1950s new developments have taken place in the western world. On the one hand, human wealth has reached unprecedented heights; and on the other hand, the way animals are bred and raised has been the subject of considerable intensification. At the same time other ways of using animals for the sake of human well-being have developed on a large scale: for example, the use of animals as tools in biomedical research and the use of animals as pets.

The intensification of animal production in western countries was initiated by public policies in place before, during and after the Second World War. These promoted more abundant, cheaper food. As a result, animal production became much more efficient, as measured by the cost of producing each egg, or kilogram of meat, or litre of milk. The pressure for efficiency subsequently became market-driven, with competition between producers and between retailers to sell food as cheaply as possible, and thereby acquired its own momentum. In many ways this can be viewed as a success story. Thus, consumers in these countries are able to buy animal products at prices that are low relative to those charged in the past. In Northern Europe it was typical, in the immediate aftermath of the Second World War, for people to spend between 25% and 33% of their income on food. Now roughly 10-15% is usual. Again, by reducing the need for labour and by increasing farm sizes, farmers and farm workers have been able to maintain an income that matches what is common in the rest of society.

A number of conflicts have arisen between productivity and the interests of the animals, and the animals have paid the price. They now typically get less space per individual than they did previously, and many live in barren environments unable to exercise their normal range of behaviour, while genetic selection has been accompanied by increased problems with production-related diseases.

Particularly in Europe this development has given rise to a new kind of legislation aimed at protecting animals against the most extreme consequences of intensive animal production. The point of this legislation is to prevent farmers from doing what is, economically speaking, the most rational thing to do. For example, in intensive egg production it is economically sound to keep hens in small cages with very high stocking densities. In places without animal welfare regulation it is not uncommon to keep hens with as little space as 300 cm^2 per animal. The point is that if there is no regulation or other mechanism in place, egg producers will be obliged by market forces to keep their hens this way. The alternative is to produce at a higher cost than their competitors, and this is not feasible in the long run.

In Europe, the main response to such problems was initially through legislation. However, with the growth of international markets it has turned out that national legislation is vulnerable because of competition from producers from countries with less stringent laws.

An alternative approach is to find ways to allow the consumer's preferences regarding animal welfare to make themselves felt on the market. This can be done in two ways. The first, more direct, way is to label animal products so consumers can pay a premium for higher standards of animal welfare. This approach is in place in various production schemes. However, these schemes represent only a very small part of the market. The second, indirect approach is to work with fast food chains and retailers. To increase consumer confidence in their products, such outlets can (as some already have) define certain minimum standards of animal welfare that must be fulfilled by the producers from which they buy their meat, eggs, milk, and so on.

The main point to be made here is that the rationale behind animal protection has changed dramatically. In industrialised countries it is no longer the case that humans and animals live in a kind of symbiosis where animal welfare can be protected by a combination of self-interest and legislation aimed at preventing 'wanton cruelty'. People in these parts of the world are becoming more and more wealthy. And this wealth is partly achieved at a cost to the welfare of the animals which deliver products which (in both relative and absolute terms) become cheaper and cheaper. Therefore there has emerged a perceived need for legislation and other initiatives which place limits on the use of animals for purposes to which most people agree. Such initiatives fall under the heading of 'animal welfare' and may be distinguished from initiatives that are 'anti-cruelty'.

Companion animals, fascination with wild animals and the animal rights movement

Throughout human history the main role of animals in human life has been as utility animals. Animals have been needed as tools to provide transportation, clothing, food and – more recently – biological knowledge. Throughout history, however, humans have also kept animals as companions; and this kind of animal use has exploded in recent times. In many Western countries it is very common for families to have cats, dogs and other 'family animals', which do not appear to serve any other purpose than being good company. These animals are generally seen, not as tools, but as family members.

This represents a different kind of animal use, where an integral part of the role played by the animal is to serve as a companion. The main purpose of this variety of animal ownership is to have a life with the animal in which the animal is happy and thrives. This means, among other things, that the animals in question are typically no longer killed unless they are suffering.

Severe animal welfare problems may, however, still arise. Although, it may be assumed that being on top of the sociozoological scale and having a close relation with humans is a guarantee that the animal's welfare is assured, this is not necessarily the case. Despite the fact that the animal's owner wants the best for the animal, he or she may, out of ignorance and unintentionally, treat the animal in ways that jeopardize its health or welfare – for example by treating it as a human being rather than an animal of the species in question with its particular needs.

A parallel development has taken place in connection with wild animals. Whereas in the past wild animals were used as hunting animals, or as game, or were killed as pests or vermin, they are now increasingly objects of fascination and grave concern. The media and various forms of entertainment have, to a large degree, driven this new perspective. Today entire television channels are devoted to wildlife programmes. Zoos and various kinds of wildlife parks are flourishing.

Huge efforts are made all over the world to protect wild animals and their habitats. This is, of course, closely related to a general concern about man's destruction of nature and the environment. As a result a number of species, such as great apes, big cats, wolves, bears and birds of prey, which used to be hunted for e.g. meat, trophies or as pest, are now increasingly protected.

What these new trends have in common is opposition to the view that animals are here for us to use. Rather, companion animals are viewed as friends and wild animals as beings with a right of their own to exist. It is therefore no surprise that, since the 1970s, a number of new ideas and movements have developed which focus on the abolition of various forms of animal use.

The most radical of these movements, the animal rights movement, opposes most common forms of animal use. Supporters of the animal rights movement make up a relatively small group. However, the ideas of the movement, in a more or less pure form, have a much wider uptake. For example, many people are sceptical about fur production. They seem to be sceptical not just because they have concerns about animal welfare; rather, they seem to object to the idea that animals may be killed for the production of a luxury such as fur.

Other movements focus on wild animals as parts of nature and urge us to protect these animals together with other elements of wild nature. Here the concern is not for the individual animal and its rights. Rather it is for species or populations of animals. Radical holders of this view want human beings to stop from interfering with wild nature completely.

It would be quite wrong to suggest that these new ideas dominate the scene. Rather they exist side by side with the previous ideas, animals as property, anti-cruelty and animal welfare. So there is wide moral agenda set for the futures of animals.

Animal futures: where are we heading?

In some parts of the world basic anti-cruelty legislation is still missing. And on a global scale discussions about whether animals are for us to use without any restrictions will continue. In most western countries, however, the banning of cruelty to animals is no longer an issue, and minimum requirements regarding animal welfare are slowly being put in place.

The main debate in the western world seems to be between those who think it is acceptable to use animals as long as their welfare is looked after properly and those who are *in principle* against specific forms of animal use. This debate could significantly influence the keeping of various domestic animals and the management of wildlife.

An important factor affecting animal futures is the globalisation of the market for animal products and services of all sorts. Particularly for farm animal products, the global market is becoming more and more open. Therefore, increasingly, competition on the world market limits the effectiveness of local initiatives to improve farm animal welfare through legislation.

On the other hand, the global market may also serve, as already mentioned, as a vehicle for spreading and enforcing animal welfare initiatives. Thus in the food market large multinational companies involved in food processing, retail and catering are becoming more powerful. These companies have brands to protect and may be vulnerable to campaigns by NGOs aiming for better animal protection. Therefore international food companies may in the future become more engaged in animal protection.

Another important trend is the globalisation of the media and the entertainment industry. Stories about animals and their maltreatment seem to have a very wide appeal across cultures; and through these channels new ideas about how to treat animals may be very rapidly disseminated around the globe. Therefore distinctly western ideas of animal ethics may soon become global ideas.

Turning to the use of animals in research, this takes place in a scientific setting, or arena, which is becoming more and more international. The vehicle for this process is the international infrastructure of sciences, the backbone of which is made up of international scientific societies, international journals and international organisations. Here one may expect current developments to continue – namely, an increased focus on shared minimum standards for good scientific conduct, including standards on the housing of research animals and the implementation of the so-called three Rs (i.e. to reduce, replace and refine animal use). There is no reason here, in a foreseeable future, to anticipate a significant movement to stop animal use.

When it comes to wild animals, already there are, as mentioned above, strong international initiatives in place to protect biodiversity. In combination with tourism and other ways in which the protection of wild nature can become economically beneficial for local communities, strong international trends favour the protection of wild animals. However, there is a clear tension between two approaches to this: one emphasizing wise use of wild animals, the other leaning towards no use.

All the perspectives on animals presented in this paper – that animals are there for us to use, anti-cruelty, animal welfare, animals as friends, respect for animals as part of nature and animal rights – exist side by side in various combinations in today's world. Debates or controversies concerning our interaction with, or use of, animals reflect the dilemmas and contradictions to which these ideas give rise. The future of animals will be shaped through these debates and controversies.

References

Arluke, A. and Sanders, C.R. (1996). Regarding Animals. Temple University Press, Philadelphia, PA.

Ryder, R.D. (1989) Animal Revolution. Basil Blackwell, Oxford.

Sandøe, P. and Christiansen S.B. (2008). Ethics of animal use. Blackwell, Oxford.

Agrifood nanotechnology: is this anything?

Paul B. Thompson
W.K. Kellogg Professor of Agricultural, Food and Community Ethics, Department of Philosophy, Michigan State University, East Lansing, MI, USA; thomp649@msu.edu

Abstract

When does a new technology pose novel philosophical questions? The social and ethical implications of technology that exploit nanoscale properties of materials have been the focus of extraordinary attention. In both Europe and the United States, large commitments of public funds for the development of basic scientific capabilities have been accompanied by unprecedented levels of funding for research on the social and ethical questions that might arise in connection with nanotechnology. Two reasons for this commitment to social and ethical research were cited in debates before the U.S. Congress prior to creation of the National Nanotechnology Initiative (NNI). First, the Ethical, Legal, and Social Issues initiative of the Human Genome Project was cited as a success for having identified and provided a forum for debate over key issues that might have otherwise gone unrecognized. Second, agricultural biotechnology was described as a failure for having omitted attention to these issues. Congress was sufficiently persuaded by these arguments to direct U.S. agencies supporting nanotechnology work to have a social and ethical component. At the same time, some have questioned whether this attention to social and ethical issues is warranted by truly new aspects of nanotechnology, and have suggested that the elevated level of research on social and ethical issues is itself a function of 'hype' and the unprecedented amounts of research funding that have become available. This paper focuses on one sector in which nanotechnology is being applied: agricultural and food technology (henceforth 'agrifood'). The agrifood sector is not the 'hot' area for nanotechnology. Exciting and philosophically controversial examples exist in medical diagnostics and in the potential for human enhancement. However, there are applications of nanotechnology in the agrifood sector, and these examples provide an interesting case study because there have already been several reports by civil society organizations that highlight agrifood applications in calling for regulation, moratoria and public opposition to nanotechnology.

Keywords: nanotechnology, agriculture biotechnology, ethics, agrifood

Nanotechnology

It should be noted at the outset that nanotechnology is an extraordinarily broad concept. The literature is replete with lengthy debates over how the term should be defined. To the extent that there is a preponderance of opinion on this question, it revolves around the exploitation of nanoscale properties and processes in the creation of new tools and techniques. Defining the nanoscale is itself much less contentious: It is usually considered to be the range between 2 and 100 nanometers, a nanometer being one millionth of a meter. When ordinary substances or materials are produced in particles at the nanoscale, the ratio of surface to mass is significantly larger, so that physical and chemical properties associated with surface structure are dramatically more effective. This can result in different physical, chemical and biological reactivity. Nanoparticles of gold, for example, appear red in natural light. The low-level anti-microbial properties of silver are magnified greatly when silver particles are produced at the nanoscale.

Surface reactivity at the nanoscale has opened new vistas to industrial chemistry. Many substances can be shaped at the nanoscale through a process akin to crystal formation. The shiny metallic surface that can be seen on the inside of chip bags is a nano-thin layer applied to a plastic sheet in this way. (In order to be nanotechnology, it only needs to be nanoscale in one dimension.) The function here is to create a more effective barrier that both preserves freshness and retards microbial contamination. Other materials are being investigated for nano-thin barriers in packaging. Nano-encapsulation is a related nanotechnology in which the nano-thin barrier surrounds a particle that itself may be quite small (though not generally at the nanoscale). Encapsulation of chemicals allows them to be protected from heat, moisture, sunlight or contact with some other chemical until the encapsulating material is dissolved or broken down by some physical or chemical process. Encapsulation currently has applications in agriculture, where it is used in pesticides and fertilizers. Nano-encapsulation will introduce a new class of encapsulating substances. Encapsulation is also used in the food industry: Think 'Melts in your mouth, not in your hand!' As in agriculture, nano-encapsulation offers new possibilities for the food chemistry of flavors and added ingredients.

The nanoscale can also affect conductivity and magnetic properties of some substances, and their sheer size allows for novel ways to exploit the physics of spin, momentum and gravity. Nozzles create nano-sprays that allow water or other liquids to be absorbed more quickly or evenly, and the reverse is a nano-filter that is more effective at removing contaminants. The physics of nanotechnology has also opened up new horizons in electronics and computing. Of particular importance is the potential for linking processes such as ordinary chemical or protein bonding to devices that generate a tiny electronic signal. This has created a new class of automated biosensors that have the potential to dramatically reduce the time and cost of testing procedures. Biosensors may be i-pod size devices that can detect the presence of thousands of chemicals, or they may thin sheets smaller than a postage stamp that can be incorporated into a package to detect the presence of e-coli or salmonella.

Size and surface reactivity has also been exploited through techniques that create wholly novel chemical structures, especially from carbon. Carbon nanotubes are the most celebrated of a growing class of manufactured nanoparticles. The potential uses and risks of manufactured nanoparticles are only beginning to be understood and appreciated. Carbon nanotubes are both very light and very strong, and are already being utilized as strengthening agents in a variety of products and materials. Research on the potential toxic effect of carbon nanotubes has also shown that there is a great deal that we do not know about their effects in animal or plant tissue, or in the environment more generally.

Objecting to nanotechnology

It is very clear that (a) nanoparticles may be risky in ways that larger particles of the same substance are not; (b) some manufactured nanoparticles have been shown to toxic properties; and (c) toxicological methods for testing manufactured nanoparticles are not well defined, especially due to the lack of uniformity in the shape of manufactured nanoparticles. As such, there is a fully appropriate and burgeoning literature on the potential health and environmental risks associated with nanoparticles and especially manufacture nanoparticles such as carbon nanotubes. It is, at this writing, unclear what regulatory posture will be taken by agencies in Europe, Japan and North America.

It is also clear nanotechnologies of all sorts will become embroiled in power relationships. It is less clear exactly *how* they will be embroiled. Biosensors, for example, could have a decided leveling effect, especially if they make it feasible for people to provide definitive evidence of exposure to potentially damaging substances. Such a technology would tip the balance in favor of currently vulnerable parties. Nonetheless, nanotechnology will have an effect on power relationships because all technology has an effect on power relationships. Because powerful actors are often able to exploit their existing power to advantage, it is quite reasonable to be skeptical about the social impact of nanotechnologies deployed by powerful food system actors. Resistance to these impacts will, if we have learned anything from biotechnology, play out in complex ways, some of which will involve the cultivation of a broader climate of opposition. There are at least some conceptions of contemporary politics that could justify potentially misleading campaigns against nanotechnology as a means to cultivating this climate of opposition.

However, it is worth taking some pains to avoid self-deception. Even if there are socially and politically justifiable grounds for resisting agrifood nanotechnologies – and I believe it remains to be seen whether there are – an ethical analysis of *those* grounds should not conflate them with claims about health and environmental risk. That is, an activist might be able to justify something like fear-mongering on strategic political grounds, without regard to whether there actually are reasons for the activist to be materially concerned about health and environmental risks. There is thus still an independent philosophical question to be raised about risk, even if one holds open the possibility that raising questions about the risks of nanotechnology can be justified on the basis of socio-economic, rather than biological or chemical, concerns.

A content analysis of these reports reveals a number of interesting patterns that bear on the relationship between technological and philosophical novelty. First, the reports do in fact identify issues relating to toxicological methods and risk assessment that can reasonably be characterized as problems in the philosophy of the sciences underlying these fields. While not wholly new as philosophical issues, these problems intersect with political questions about regulation and uncertainty that have only become widely appreciated in recent times. Second, the social and ethical ills noted in these reports are generally typical of those associated with all forms of industrial agriculture, suggesting that there are *not* novel issues or even mechanisms uniquely tied to agrifood nanotechnologies.

Finally, the reports connect novel risks to traditional problems by exploiting ambiguities either in the definition of nanotechnology or the scope of agrifood. Risks tend to be associated with nanotechnologies (such as nanoparticles) that in fact have no specific agrifood application, though they may have uses in other consumer products such as cosmetics. As such, it is not unreasonable to interpret these reports as utilizing uncertainties associated with nanotechnology as a general phenomenon to mobilize support for public opposition to further industrialization of the agrifood sector.

Conclusion

My analysis implies that risks associated with novel materials such as carbon nanotubes or natural materials that do not typically occur at nano scales can reasonably inferred to be distinct from risks associated with will known materials that are only newly *described* as having reactivity at the nanoscale. While there are clearly some nanotechnologies that will require new approaches to risk assessment and risk management, there are other technological applications that would be more appropriately characterized as extensions of conventional engineering and chemistry. The rationale for this claim in the agrifood sector is that encapsulation, nozzles, filters and barrier technologies each represent approaches that have applications pre-dating the emergence of nanotechnology as a distinct type, and involve functions that do not rely on exploitation of nanoscale properties.

Although I believe my rationale ultimately supports the assumption with pragmatic considerations and broad analogies, I acknowledge that this may not prove convincing to skeptics. One cannot exclude the possibility that the shiny coatings on the inside of chip bags have some unknown hazard associated with them. Whether we, at the end, think that nanotechnology is something to be concerned about may have more to do with our background narrative for food production, processing and distribution. For some, foods are or were until recently the benign and commodious bounty of unspoiled nature. For me, most of the foods we eat (and have eaten since my youth) are fairly recent products of trial and error experimentation by farmers and scientists, and the same applies to the way they are grown. We have learned that some of these experiments have consequences that were never imagined by the experimenters. It is past time to approach experimentation with a more critical eye. But we must also realize that our current status quo in agrifood is perilously unstable, with many hidden risks and injustices. To resist change is to perpetuate a system with known hazards, vulnerabilities and social inequities. And that is something I cannot endorse.

Theme 1 – Governance of the food system: ethical challenges

Overcoming the moral psychology of denial: how to communicate more effectively about food ethics

Raymond Anthony
Department of Philosophy, University of Alaska Anchorage, USA; ranthon1@uaa.alaska.edu

Abstract

Besides creating incentives for the public to change their behavior and adopting effective policies for removing impediments, it is also critical to understand what might be immobilizing the public when it comes to making morally responsible food choices. Here, an area of food ethics that has received little attention is that of overcoming moral denial in an information-saturated society. That is, while many of us can accept the evidence that our farming practices and consumption habits may not be environmentally sustainable or good for animals and our farmers, we find it difficult to accept our responsibility for such enormity. This papers sheds light on how to overcome diffusion of moral responsibility, procrastination and bystanding in the food system. Here, the focus will be on how we should employ more effective communication when it comes to food ethics discourse. Merely delineating or countering with information, emphasizing the severe consequences of inaction and calling for immediate action is insufficient to overcome moral denial. Overcoming moral denial involves creating ethical narratives that break down and stand up to dominant forms of discourse about food. What is critical is reframing and placing narratives about food in the context of deeply held citizenship commitments and principled-driven alternatives. The case of animal agriculture is used as a backdrop to help discuss how virtues of stewardship, accountability and care may dislodge the social and conceptual hegemonies of the present day industrial-agribusiness narrative. It also serves as a way to cast our moral imagination to seek animal and farmer friendly practices and inspire collective determination and build social consensus.

Keywords: food ethics, animal ethics, agricultural ethics, ethical consumerism

Introduction

Ethical consumerism as it relates to food does not only involve what we eat but *how* we do so. In liberal democracies, it is critical that consumers also participate in the governance structure of the food system in a meaningful way in order that their values and concerns are respected and addressed, respectively. A significant impediment, what I term 'truancy of accountability,' needs to be identified in some detail and overcome in order that we may more effectively address concerns related to the 'how' dimension of ethical consumerism and moral governance regarding food. Many of us live with a kind of moral psychology of denial when it comes to our responsibilities to the food system and this in turn constrains our actual behavior when it comes to being *good* actors about food.

My aim in this essay is to address one aspect of the systematic malaise in ethical consumerism about food. Here, I will identify some of the philosophical stumbling blocks that perpetuate truancy of accountability. As I tease out the contributing elements that lead to the current moral psychology of denial about food, I will use examples from animal agriculture to help lubricate my discussion. I will end by making a few brief observations about how to communicate more effectively in this current climate of truancy/denial.

The moral psychology of denial and our food

Truancy of accountability can be characterized as a disposition of omission, a kind of complacency if you will, that is associated with being willfully disconnected or intentionally isolated from the goings-on in our food system. Here, 'truant consumers' fail to recognize their shared responsibility to others in the food system or choose to turn a blind eye or deaf ear to their role in the organizational structure of the global food system (Due to space, I am unable to elaborate on this here). The moral psychology of denial contributes to truancy of accountability in the following ways:
1. Conditioned helplessness in the face of the enormity of the food system and its issues.
2. Diffusion of responsibility: the 'bystander effect'.

Stanley Cohen (2001) argues that in an 'information-saturated society,' the capacity to deny a level of awareness can be internalized as business as usual on the social front. In 'States of Denial: Knowing About Atrocities and Suffering', Cohen contends that 'far from being pushed into accepting reality, people have to be dragged out of reality'. In his view, we are confronted by a fundamental paradox of denial – a simultaneous state of 'knowing and not-knowing' where denial betrays the fact that we have knowledge of the existence of problems and its moral implications. In applying Cohen's concepts to truancy of accountability in the food system, we see certain parallels with psychological processes that are highly pertinent to the ethics of what we eat. The fact that many of us remain passive 'bystanders' is not due to the lack of 'knowledge' about the state of our food system. That is, despite knowing that all is not well with the state of our agriculture and food system, we do not act accordingly to the implications of what we know.

1. Conditioned helplessness in the face of enormity

One form of widespread denial may be reflected in a kind of helplessness that is connected to the perception that the scale and nature of the problem are so unprecedented that we lack the moral and social mechanisms for addressing and processing them. Here, industrialized animal agriculture, for example, can be likened to a hegemonic icon or entrenched structural paradigm. As Singer and Mason write (see also Rollin, 1995; Thompson, 2001), 'Factory farms have taken over and no one had noticed' (2006: vii). Industrial forms of farming *IS* the face of conventional agriculture and must have always been and thus, must be. Many of us having grown up with one major paradigm may not be able to envision real alternatives to Concentrated Agricultural Feed Operations (CAFOs). Thus, this would explain the refusal by many of us to contemplate the moral status and viability of alternatives, i.e., family farms, for example (whose existence appears merely fabled or quaint). The pervasiveness of the current system in our collective consciousness is perhaps the most powerful evidence of our denial. We cannot even fathom that there could be a moral dimension with identifiable transgressors (let alone that *we* could be the transgressors) and victims.

In the case of animal agriculture again, the language of 'factory farms', 'industrialized agriculture', 'animal machines', and 'animal protein' are themselves a form of denial through conceptual 'masking devices'. These devices serve to objectify and obscure the noninstrumental moral nature of the other. For consumers, ownership of the social and moral problems facing the food system is deflected on and entrusted to producers, government institutions and industry agents, who may not be empowered to act in any meaningful way themselves.

Conditioned helplessness can take numerous forms. There are masking devices that are employed as 'coping strategies' that (a) help to assuage or moderate discord we might feel and others which fuel (b) detachment so that we don't feel. In both instances, truancy of accountability is met with modest recourse and leaves us off the hook from being more active in the food system. Briefly, we will look at these strategies in turn (as they relate to aspects of animal agriculture).

a. *Discord moderating strategies* when employed by consumers, although easing the awkwardness of coming to terms with the animals in animal agriculture, only perpetuate the myth that animals are but 'absent referents' (Adams, 2002) or are not morally considerable in themselves. If consumers are generally unwilling to consider the welfare of farmed animals they would not have to think about them as subjects and thus, as mere objects, (Serpell, 1986), consumers need not be held responsible for their fate. These strategies at best, occasion a thin sense of responsibility to the food system, and include (Te Velde, 2002):

i. Strategies to ameliorate conflict

Here, consumers selectively attend to those aspects of welfare they can influence and which reinforce their view of the acceptability of farming animals under the current scheme. They may discount the presence or severity of welfare impediments too. Their participation in the ethical governance aspect of the food system need only be minimal then. For example, if they pay more at the register so that laying-hens have freer range and gestational sows can turn around as a consequence of their having complained, consumers believe that they have discharged their duties adequately.

ii. Strategies that dismiss the capacities of animals

Here, consumers deny that there exist welfare problems by underestimating the capabilities of animals. The upshot of this impoverished view of the capabilities of animals is that they are provided with restricted opportunities. Consumers may believe that since animals can cope in the current system, they miss nothing, that is, 'animals on factory farms have never known any other life' (Pollan, 2002).

iii. Strategies that feign concern

Here, a myopic view of welfare is taken. Participation in ethical governance involves ensuring that this minimal view is met. As Fraser *et al.* (1997) suggest, welfare has three essential components. Animal welfare includes psychological and biological felicity or wellness and an animal's opportunity to express species-characteristic natural tendencies. Good welfare also involves giving farmed animals adequate and proper attention. If welfare is subservient to economics and farming is driven not by attending to the needs of animals, but to production efficiency priorities, then consumers and producers alike may feel some exoneration when they trade off one welfare impediment for another provision.

b. Masking devices that breed *detachment* reflect a myopic view that farming remains the vocation and source of income of a select few and that consumers are physically removed and have little or no emotional attachments to farmed animals and for that matter farming in general. Here, consumers intentionally choose to remain 'functionally ignorant' about what actually goes on conventional farms because this falls outside their realm of expertise or moral jurisdiction (Te Velde, 2002: 217-218).

2. Diffusion of responsibility: the bystander effect

Another major form of denial is diffusion of responsibility. Cohen argues that violent crimes occur in the presence of on-lookers due to a 'passive bystander effect.' Omission or abdication of personal responsibility to stop the injustice is due to the expectation that others will act. In the food system, we have been conditioned to rely on the other millions of agents (in the US), and thus, we – any individual moral agent or her nuclear community – can become incapacitated. In the case of a fractured food system, we become both bystander and perpetrator; a likely internal conflict that can only exacerbate our denial.

i. Transference of blame

A variety of diffusion of responsibility is *blame shifting or transference*. When the consumer passes the 'buck' onto others in the case of animal agriculture, for example, she defers to producers, retailers, government agents and other industry agents to 'offer meat that is produced in an animal-friendly way' (Te Velde, 2002: 214). Government and industry agents alone then become charged with the task of ensuring good animal welfare (as well as safe and quality assured food). But by shifting the burden of responsibility to another party, the consumer endorses a very thin version of ethical governance by evading culpability and disavows any influence she has over the food system.

ii. Concealment

Tied closely to dismissing animals as morally considerable beings, consumers can deflect their ambivalence over consuming animal flesh and products by going along with what everyone else thinks about the value of animals. Employing 'cuisine disguises' to conceal the true nature of the origin of their food, this strategy protects consumers from having to consider the moral status of animals in the first place because 'everyone eats meat.' There is near consensus that they are acting appropriately, since animals are 'meant to be eaten' (Adams 2000).

iii. Emotional distance

Emotional distance is another way in which diffusion of responsibility surfaces. Some consumers may contend that since they do not care for farmed animals directly, their sympathies would be misplaced.

iv. Misrepresentation

Misrepresentation points to an oversimplified picture of agriculture that distorts the plight of animals, unevenly distributes the burden of responsibility to industry agents, overemphasizes the role of government, mislabels farmers as uncaring, and portrays the current gains as good enough.

Communication amidst the current climate of denial

In the limited space for this extended abstract, I offer a few modest ways to frame further discourse about food ethics in order to meet the challenges of the moral psychology of denial and overcome truancy in the food system (and help to promote the consumer end of ethical governance):

- Situate food issues in the context of laudable values, such as stewardship, strengthening community, integrity, care, and justice.
- Characterize initiatives to address food security, animal welfare and environmental sustainability in terms of being responsible, 'forward-thinking,' planning for the future, and embracing 'smart' and balanced or wholesome alternatives.
- Use images, narratives, analogies or metaphors that are vivid and provide a 'conceptual hook' to help consumers personalize their connection to the food system, and make sense of information so that they can reason appropriately.
- Link personal responsibility with prevention of near term concerns and effectiveness in resolving existing ones within manageable timeliness and at affordable personal expense.
- Stress existing solutions and how consumers can act on them responsibly and effectively with visibly beneficial outcomes.
- Provide meaningful feedback once guidance and comment are solicited, i.e., evaluate the effectiveness of communication initiatives periodically and encourage consumers to be part of this feedback loop.

References

Adams, C. (2000). The Sexual Politics of Meat: A Feminist-Vegetarian Critical Theory Tenth Anniversary Edition. New York: Continuum Publishers.
Anthony, R. (2004). Risk Communication, Value Judgments and Animal Welfare: Revisiting Britain's Foot and Mouth Crisis Journal of Agricultural and Environmental Ethics Special Supplement 17: 363-383.
Arendt, H. (1987). Collective Responsibility. In: J.W. Bernauer (ed.) Amor Mundi: Explorations in the Faith and Thought of Hannah Arendt. Boston: Martinus Nijoff Publishers.
Berry, W. (2002). How We Grow Food Reflects our Virtues and Values, in Gregory Pence (ed.) The Ethics of Food: A Reader for the 21st Century. Lanham, Massachusetts: Rowman and Littlefield Publishers.
Cohen, S. (2001). States of Denial: Knowing about Atrocities and Suffering Oxford: Polity Press.
Farm Animal Welfare Council. January 2002. Foot and Mouth Disease 2001 and Animal Welfare: Lessons for the Future. London: DEFRA Publication.
FAO (2008). Capacity Building to Implement Good Animal Welfare Practices. http://www.fao.org/ag/againfo/home/en/news_archive/2009_animalwelfare.html. accessed 19 March 2009.
Fraser, D., Weary, D., Pajor, E. and Milligan, B.N. (1997). Scientific Conceptions of Animal Welfare that Reflect Ethical Concerns Animal Welfare 6: 187-205.
Lakoff, G. (2004). Don't Think of an Elephant: Know Your Values and Frame the Debate. Canada: Chelsea Green Publishing Company.
Lean, G. (2001). The Plague that Never Was The Independent on Sunday. 4 March 2001.
McDonald, M. (2001). Canadian Governance of Health Research Involving Human Subjects: Is Anybody Minding the Store? Health Law Journal 9: 1-21.
Pollan, M. (2002). The Unnatural Idea of Animal Rights The New York Times Magazine (Section 6), 58-64 and 100-110.
Pogge, T. (1992). Institutional Approach to Humanitarian Intervention Public Affairs Quarterly 6(1): 89-103.
Regan T. (1983). The Case for Animal Rights. Berkeley, CA: California University Press.
Rollin, B.E. (2008). Animal Ethics and the Law: Agricultural Animals and Animal Law, Michigan Law Review: First Impressions, 106(5): 143-146.
Rollin, B.E. (1995). Farm Animal Welfare: Social, Bioethical, and Research Issues. Ames, IA, Iowa State University Press.
Serpell, J. (1996). In the Company of Animals: A Study of the Human-Animal Relationships. Cambridge: Cambridge University Press.
Singer, P. and Mason, J. (2006). The Ethics of What We Eat: Why our Food Choices Matter. US: Rodale Inc.-Holtzbrinck Publishers.
Singer, P. (1990). Animal Liberation Revised Edition. New York: Avon Books.
Te Velde, H., Aarts, N. and Van Woerkum, C. (2002). Dealing with Ambivalence: Farmer's and Consumers' Perceptions of Animal Welfare in Livestock Breeding Journal of Agriculture and Environmental Ethics 15(2): 203-219.
Thompson, P.B. (2001). Reshaping Conventional Agriculture: A North American Perspective Journal of Agricultural and Environmental Ethics 14(2): 217-229.
Thompson, P.B. (1995). Spirit of the Soil: Agricultural and Environmental Ethics. London: Routledge.
Woods, A. (2001). Kill or Cure? The Guardian. 28 February 2001.
Worster, D. (1984). Good Farming and the Public Good, In Meeting the Expectations of the Land: Essays in Sustainable Agriculture, In: Jackson, W., Berry, W. and Coleman, B. (eds.). 31-41. San Francisco: North Point Press.

Consequences of the EC framework of health claims made on food: are we protecting the consumer by means of human trials?

L. Escajedo San Epifanio
Constitutional Law, Inter-university Chair in Law and the Human Genome, University of the Basque Country, University of Deusto, Spain; leire.escajedo@ehu.es

Abstract

With regards to initiatives adopted in other countries, the EU regulation on functional foods was delayed for well over a decade. A complex sharing of competences and the variety of cultural and legal conceptions that coexist around food are factors which help explain the potential source of conflict in the food sector. Another is the fact that this is a strategic sector for many countries. The EU appealed to a peculiar interpretation of its competence of *minimum harmonisation*. The White Paper on Food Safety (2000) reflected the objective of harmonising a high level of consumer safety and protection with the backing of a scientific authority of acknowledged prestige and independence (the EFSA). Within this regulatory framework, in 2006, a regulation on *health claims made on foods* was adopted. It is noteworthy that the Union draws a distinction and places these products outside the scope of drugs. However, in order to publicise a health claim on a food it is required a scientific substantiation that includes, among others, data from studies in humans 'addressing the relationship between consumption of the food and the claimed effect is required for substantiation of a health claim'. This requirement, very high in terms of comparative law, looks for a high protection for the consumer but it has a sad counter effect. The regulation which protects the rights of humans who participate in clinical trials has been centred on medicines, and it is practically non-existent with reference to food studies on humans. Some Member State authorities and Research Ethic Committees are reacting to fill this gap in the law. This is the case, for example, of the Spanish Clinical Research Ethic Committees, within the Law on Biomedical Research (2007). Nevertheless, in the medium-term, it is necessary to provide a more adequate solution.

Keywords: food labelling, health claims, functional food, clinical trials

Functional foods and policies of the European Union

In the 1950's, a probiotic drink named 'Yakult' began to be sold in Japan. This is considered to be the beginning of the *modern functional food market*, though the Japanese scientific community did not define the concept of functional food until the early 1980's. Broadly speaking, this expression is used to refer to a set of foods characterised by having a specific health benefit for their consumers, besides their conventional nutritional contribution. In the medium and long term, they can improve the collective health, thus reducing pharmaceutical and health costs.

Nonetheless, this potential benefit for one's health is not the cornerstone of governmental measures that are being adopted in different countries on functional food. The key to these policies is to decide whether to allow publicising these health properties or not and how this should be done. A refusal would impede food producers from obtaining economic compensation for their investment in the research and development of these types of food. On the other hand, an overly broad authorisation can compromise the rights of consumers or even their health. The search for the equilibrium point in this disjunctive started in many countries in the 1990's. However, in the European Union, the regulation on functional foods did not take place until the Regulation EC nº 1924/2006.

Furthermore, although it took almost ten years to reach a consensus, many refer to it as a 'hasty' regulation. Its process of implantation was supposed to be finished at the end of January 2010, but it has been recently delayed. There are too many obstacles to overcome. Perhaps this context helps us understand why the level of demands for *health claims* in food has been so great. In many cases, this level is so high that it requires the undergoing of clinical trials with food. As the focus of the regulation is on protecting the consumer, the European authorities have not yet understood what this entails. A new protagonist has been left without protection: the human being who participates in this research. The Member States are now dealing with this important lack.

Dealing with obstacles to free commerce in the EU food market

Many of the peculiar traits of the agro-food market of the EU are present in the European policy on functional foods. In the last two decades, within Europe and across the world, a growing phenomenon of integration has taken place in the market for goods and services. The agro-food market has been a source of constant conflict at European and global levels. This has a simple explanation, given its economic and social importance, the agro-food sector constitutes one of the most regulated sectors within Member States. Additionally, it is very closely linked to the traditions of each country, to its different legal and cultural conceptions. Comparatively speaking, integration of the agro-food sector raises more conflicts and of a greater variety than those of any other productive sectors.

In accordance with the *principle of mutual recognition*, any product from a Member State should be admitted throughout the Union. The only exceptions permitted by the Treaty must be due to issues dealing with public health or protection of the consumers (art. 30 ECJ). States still keep important competences in these two matters and this provides them with legitimacy in order to establish some limitations to free trade. This precaution makes reference not only to food, but to any other product. However, the community institutions, aware of the sensitivity of this sector, began to accept as legitimate the placing of more obstacles to the free trade of food than that placed on other products.

In the 1990's, the controversies surrounding new food, especially transgenic and the food crises of the new dioxins and mad cow disease placed this market in a delicate position. The absence of an agile, centralised food authority with capacity to quickly execute its decisions was very much noticed. With the passage of time, there was a demonstration that the mistrust between countries, and the moratoriums and other practices which acted as obstacles to decision making did not have sufficient legal basis. But by then, the harm had already been done. The key to a common market and the only aspect that is indispensable, is that there must be a consumer who wants to use the market. If there are no consumers who trust in the common food market, then practically there is no market (Escajedo, 2008a).

In order to repair the damaged trust, the community institutions resorted to a capacity recognised by the Treaties. In order to ease the integration of the market, the Union can promote an approximation of those legal, regulatory or administrative dispositions of the Member States that directly affect the functioning of the market. This entails a *harmonisation of minimums* of those matters, that belonging to the authority of the States, this directly influences the policies of market integration.

This possibility was used in order to adopt the *White Paper on Food Safety* (2000). This book is presented as a compromise whereby the protection of the consumers and food safety are going to be raised throughout the Union. This is reinforced by a scientific entity whose authority is recognised by Member States, but is independent from them and from European authorities (EFSA). Although as already mentioned, the competences of public health and food safety belong mostly to the States; however the de facto aim is that the level is so high that no State can appeal to the need for greater food safety or greater protection of the consumers. If a State observes that a level can be improved, then the step to be followed is not to place an obstacle, but rather to inform the Union authorities in order that improvement can be extended throughout the market.

The 'minimum' harmonisation of the food market and its effects on functional food

This interpretation of the legal-constitutional basis provides an explanation as to why Regulation EC no. 1924/2006 especially highlights in its narrative recitals the purpose of guaranteeing a 'high level of protection for consumers'. From a socio-political perspective, Margaret Ashwell (2002) reminds us that functional food is not a defined set of food; but rather, food from which we can *express a special feature*. The need to express that possible beneficial influence is posed in a moment in which the majority of western countries have prohibited *the attributing of medicinal properties to food*. This was in accordance with institutions such as the FAO or the WHO. Also, European regulation had been prohibiting 'claims to preventive, therapeutic or healing properties' in food. Following the trail of earlier regulations (Council Directive 79/112/EEC), Directive 2000/13/EC upholds that prohibition and adds that any claim on a product (food or not) must be 'truthfully, sufficient and not misleading'.

Regulation EC no. 1924/2006 sets out to establish the conditions under which nutritional statements and health claims can be set out for food. Nutritional statements are those which make reference to the content of food, such as *low fat* or *rich in fiber*. Currently authorities are trying to define some comparative limits (the nutritional profiles). Besides functional properties, a statement can be made regarding a food which reduces the risk of bearing certain diseases or that positively contributes to children's development or health. This is what is known as health claims. The Regulation establishes two different processes in order to solicit the authorisation of a health claim for a food. On the one hand, it regulates the claims referring to the reduction of disease risk and children's development and health; and secondly the *remaining* health claims regarding a food. The regulation of this second group of foods, with contributions from food industry and the food authorities of the EU countries, resulted in a drafted list of claims. Once this list is finalised, industry will be able to use these health claims when two conditions are met: that these are substantiated in generally accepted scientific tests and that these are well understood by the average consumer.

A more complex path can be foreseen for health claims that *allege a reduction in the risk of contracting a disease* or those which make *reference to children's development and health*. These claims require the ad hoc authorisation of the European Commission after a decision by the EFSA. EC Regulation no. 353/2008 which establishes implementing rules for this procedure warns that these health claims must be based on a solid scientific basis and provides details as to what should be understood by this. The EFSA has the obligation to verify the soundness of this claim, and although its decision is not binding, the Commission must provide reasons if it goes against that decision.

Among the principles for the scientific substantiation, it states that 'Data from studies in humans addressing the relationship between consumption of the food and the claimed effect is required for substantiation of a health claim' (Regulation no. 353/2008). This need to prove in humans the relation between the consumption of a food and the required effect necessarily entails clinical trials with food, which in some cases is little different from those undertaken with medicines or health products.

The lack of protection for humans who participate in clinical food trials

This requirement has a sad counter effect. The regulation which protects the rights of humans who participate in clinical trials has been centred on medicinal products, and it is practically non-existent with reference to human food trials. Besides, our legal system has struggled a great deal at times to define medicinal products from other products which do not deserve such label. Even when from a subjective point of view, both subjects require a similar protection, that legal barrier prevents the subjects of food clinical trials to benefit from the thrust that has been provided by the European Union to fundamental rights in this field.

Since as early as the 1970's, the European Communities have placed their attention on biomedical research, especially on the needs of the pharmaceutical industry and the regulation of the market for medicines. Both the possible breach of fundamental rights as well as the different ethical sensibility of the Member States has been taken into account in community policies and with the tools used in this matter, in accordance with what has been the growing concern in different European countries.

These and other efforts, although not properly *harmonising* the ethical and legal treatment of Biomedicine as the EU does not have competence for this, are in fact achieving an important coordination of positions. The promotion of biomedical research and the guarantee of high standards in quality and efficiency of the medicinal products produced in the EU have brought about a clarification of the regulations and an effort to protect the fundamental rights of those involved in biomedical research. This result has been achieved by *inviting* States to include measures such as the protection of subjects in trials within the legal regulation created to transpose some Directives.

Some Community Directives approved in the time period ranging from the Charter of Nice (2000) to the Treaty of Lisbon have, in a meaningful manner, drawn closer the protection that Member States provided to fundamental rights. This is despite the EU's lacks of competence in this matter. Along these lines, it should be noted that Directive 2001/20/EC sets out the laws, regulations and administrative provisions of the Member States relating to the implementation of good clinical practice in the conduct of clinical trials on medicinal products for human use. Directive 2004/23/EC defines standards of quality and safety for the donation, procurement, testing, processing, preservation, storage and distribution of human tissues and cells. Finally Directive 2005/28/EC lays down principles and detailed guidelines for good clinical practice with regards investigational medicinal products for human use, as well as the requirements for authorisation of the manufacturing or importation of such products. Directive 2001/20/EC deals with research on humans carried out to establish the effects of the consumption of one or several medicinal products, or in other words clinical trials with medicinal products. Under this Directive, all community medicinal products must comply with standards of quality, safety and efficacy.

However, its provisions also make reference to the need for the subject of the trial to provide *consent after being duly informed* on the research to be performed. Furthermore, it establishes special provisions when dealing with especially vulnerable subjects (such as minors or adults without full capacity to consent). The right to informed consent is a right linked to the autonomy of the person and therefore an essential content of a fundamental right. The Directive deals with its lack of competence by limiting itself to invite the 'national legislation' to take measures. But this invitation has had a very noteworthy response by the Member States.

Directive 2004/24/EC, regulates the obtaining and use for different purposes of separate parts of the human body (cells, tissues, etc.) and takes into account the informed consent for its obtaining; the rights of the subject over its samples and the protection of data (DNA) contained in the sample or associated to it. Finally, Directive 2005/28/EC incorporates into the legal regulation of the community those principles recognised in the Helsinki Declaration (1996 version), whereby a text of the World Medical Association, which was of reference but had no legal character, has became part of the derived Law. In these cases the Union could not claim its direct competence, not even to approximate the legislation, as its capacity for minimum harmonisation is limited by its treaties to matters which directly involve the functioning of the market. In this case, it practically appeals to its moral authority, as it only highlights which issues should be regulated by the States to adequately protect the rights of the subjects involved.

Regulations on food health claims have not yet followed the trail of the aforementioned drafted list of claims. This is probably due to the fact that the greater part of the content of these regulations is not associated with what has been the regulation of biomedical research. However, as it is easy to foresee that the member States also bear the risk of not realising this lack of protection, a reminder of this would have been very desirable. The EFSA has recently warned on this issue in its *Guidance for applicants* and some of the Member States are now dealing with this lack. In the following section I look at how Clinical Research Ethics Committees in Spain have responded to this issue.

The temporary response by the Clinical Research Ethics Committees in Spain: a need for satisfactory measures in the medium term

As in other countries, Clinical Research Ethics Committees (CRECs) came into being in Spain with the aim of guaranteeing the methodological, ethical and legal adequacy of clinical trials of medical and pharmaceutical products. The first regulation to this effect was Royal Decree 944/1978, although the definite consolidation of these institutions was reached with the Law 25/1990 on Medicinal Products (Romeo Casabona *et al.*, 2006) Article 64 of this Law expressly provided that 'no clinical trial may be performed without the previous report by a Clinical Research Ethics Committee, independent and duly accredited'.

At present, there are almost two hundred committees of this type throughout all of Spain (Escajedo, 2008b). Their task is to verify the compliance with the principles and guidelines detailed in the good clinical practices and the authorisation requirements for the fabrication or import of medicinal products in research for human use. They were created for the purpose of performing an ethical control before the undertaking of research, but also to guarantee that the results of the clinical trials supported the quality, safety and efficacy of the medical products. In this manner, *they verify the methodological, ethical and legal* adequacy of research on human beings and consequently, the acceptability of the results obtained in these. The entry into force of the health claim regulations has occurred at a time of transition.

The Law on Biomedical Research (2007) established that 'the authorisation and development of any research project on human beings or their biological material' requires a favourable decision from a new category of committees, the *Research Ethics Committees*. As a result, many of the guarantees for the subjects who participated in trials with medical products have been extended to subjects who donate their biological samples for experimentation and to those who participate in other types of trials. Different circumstances have delayed the effective creation of these new committees but have not stopped the application of the measures of protection at present. In accordance with the provisions of the third transitory disposition of this Law, the *CRECs*, created under the legislation of medicinal products, 'can assume their competences' and this is how it is being applied in practice.

Several of the CREC that currently operate in the institutions which carry out biomedical research are voluntarily assuming, among other practices, the protection of subjects who participate in clinical trials with food. In this manner, they comply with the task which UNESCO (2005) attributes them with, in their guide of committees: to be *platforms to balance the good of science, human rights and the public's interest.* However, this is only a short term solution. It is desirable that the final development of the Law on Biomedical Research, via the next *Royal Decree on Research Ethics Committees* takes into account human beings who participate in trials with food. Even better would be if the EU itself, in accordance with the path that it has followed regarding medicinal products, also promotes an awareness among Member States of the need to protect these people.

Acknowledgements

This work has been funded by a Grant to Research Groups by the Basque University System, ref. IT-360-07, Department of Education, Universities and Research of the Basque Government.

References

Ashwell, M. (2002). Concepts of Functional Food, ILSI, 2002.

Council Directive 79/112/EEC of 18 December 1978 on the approximation of the laws of the Member States relating to the labelling, presentation and advertising of foodstuffs for sale to the ultimate consumer.

Council Directive 2000/13/EC on labelling, presentation and advertising of foodstuffs to the final consumer.

Escajedo, L. (2008a). The constitutional basis of the Consumer Protection within the EU, Revista de Derecho Político, nº 70, 2008, 223ss.

Escajedo, L. (2008b). La libertad fundamental de investigación y los nuevos controles éticos a la actividad biomédica, AA.VV., Los avances del Derecho ante los Avances de la Medicina, Thomson- Aranzadi, 2008: 1187-1202.

Romeo Casabona, C.M., Emaldi Cirión, A., Escajedo, L. *et al.* (2006). La ética y el Derecho ante la Biomedicina, Deusto, 2006.

UNESCO (2005). UNESCO, Guide No. 1. Establishing Bioethics Committees. United Nations Educational, Scientific and Cultural Organization.

Developing an ethical aquaculture food index for international food trade

A.S. Haugen[1] *and M. Kaiser*[2]
[1]*Telemark University College, Department of Environmental and Health Studies, Bø, Norway; arnes@entri.no*
[2]*University of Bergen, Centre for the Study of the Sciences and Humanities, Bergen, Norway*

Abstract

Aquaculture products make up the fastest growing food sector world-wide. Production and trade is truly international, and markets are diverse. Expectations from European consumers are comprehensive, covering not only product quality, but also issues like traceability and sustainable production methods. So-called political consumerism provides for a diversification of market segments where fair trade and ethical issues play a role. In its trade with aquaculture producing countries, mainly countries in Asia, Europe faces a challenge in spelling out what these expectations amount to and how to measure them. In particular the expectations relating to ethical issues in aquaculture production and trade are far from clear. How to specify common standards is a challenge. In this paper the authors present the initial work plan to develop an ethical aquaculture food index (EAFI) as part of a larger EC funded project on Sustaining Ethical Aquaculture Trade (SEAT; coordinated by the University of Stirling). An initial overview over existing ethical guidelines and standards, such as the FAO Guidelines for sustainable fisheries and aquaculture, will provide the background for working out possible candidates and constraints for such an EAFI. Participatory stakeholder processes in Europe and four Asian countries will probe deeper into a realistic set of possible index standards. Who the relevant stakeholders are is in itself a challenging question. To some extent the ethical matrix will serve as a tool to facilitate these processes. The pros and cons of voluntary versus mandatory ethical certification schemes need to be assessed, and it is far from clear which is a realistic and promising choice. The project aims at a more holistic, evidence-based perspective to the fundamental research question: what constitutes 'fair and sustainable' trade? Post harvest impacts, socio-economic dependencies (especially of smaller value chain actors) and consumer understanding are key areas which can augment the principle standards and certification schemes currently available for aquatic products, many with singular emphasis on biologically-based resource management assessments of sustainability. The final result will be tested for policy relevance through a Delphi process.

Keywords: fair trade, food ethics, ethical index, ethical matrix, stakeholders

Background of the study

Aquatic food products are becoming a major part of people's diet worldwide. As Daniel Cressey observes, the 'only way to meet the increasing demand for fish is through aquaculture' (Cressey, 2009: 398). The aquaculture industry offers more and more species for the menus in restaurants around the world, a 'secret that most diners are blissfully unaware of: farmed fish are everywhere' (*ibid.* 398). In the globalized economy, aquaculture products are often produced in far away countries before they reach the consumer. Asia is a particularly strong producer with 86% of the world's fishers and fish farmers being located in Asia (FAO, 2009: 23). Products are exported to mainly the markets of the European Union, Japan and the United States. Aquatic products are the most internationally traded food commodity. Production and trade is truly international, and markets are diverse. The EU is now the World's largest seafood market. In 2006, almost half of the world's food fish came from aquaculture. The recent FAO study

The State of World Fisheries and Aquaculture 2008 (FAO, 2009) forecasts that by 2020 aquaculture will contribute with 60% of food fish.

At the same time there is no sign that consumer scepticism will diminish. Retailers and supermarkets are attentive to these concerns, as are NGOs who in part increase them. The critical tone towards aquaculture has been dominant in the media, even among the more serious papers, for quite a long time. The German *Die Zeit* stated in an article in 2001 that aquaculture 'is nothing but the transformation of cheap protein in a form, that given the wrong impression – namely that it is real salmon – flogs it overpriced to the consumer' (Luyken 2001; our translation). The FAO report states thus that 'concerns about human health and the social and environmental impacts of fisheries and aquaculture show no sign of abating' (FAO, 2009: 95). The critical trend may rather be on the increase as various forms of so-called political consumerism appear to gain a foothold in the richer parts of the world. The more traditional forms of boycotting certain products (as e.g. Shell experienced as a consumer response to the Brent Spar affair) are now among others supplemented by what Micheletti calls 'positive political consumerism' (Micheletti, 2006: 24), i.e. through labelling schemes that are beyond compliance and do more than governmental regulation may require. Eco-, organic- and fair trade labels are examples of this. For the aquaculture industry this would mean to attend to a wide range of concerns that could be summarized under the label of 'ethics'. Aspects of this would be socio-economic conditions of production, environmental impacts, ecological footprint and long-term sustainability (cf. Kaiser, 2006), nutritional value, energy consumption, food-miles and climate effects, animal welfare and equitable trade.

As an example of the ambivalence in the attitudes among the public, there is an interesting debate in the Norwegian media. The exports of Norwegian (hitherto wild-caught) cod to the European market experienced a decline in recent years due to the strong competition from farmed *Tilapia* and *Pangasius*, both mainly imported from Asian countries. This has resulted in the Norwegian seafood industry questioning whether Norwegian tax money should be used for development aid which in the end is used to outcompete Norwegian products. An interesting response form Norwegian seafood companies is to focus on the use of certification systems, such as the Marine Stewardship Council (MSC), to explore for the market the difference between wild caught cod and farmed *Tilapia* and *Pangasius* (Dagens Næringsliv, 2009). No such critical response has yet been put forward in Norway in relation to Norwegian farmed salmon.

The rapid growth of Asia-Europe aquaculture trade places a particular challenge to Asian producers who mostly operate under very dissimilar conditions from their European counterparts. Existing labelling schemes have typically imposed narrow and localized interests, and have often been unrealistic for Asian producers. The project 'SEAT – Sustaining ethical aquaculture trade', coordinated by David Little from the University of Stirling and financed by the European Commission is set to meet this challenge (see Little *et al.*, 2009, this volume). As partners in this project, our task will be particularly directed towards the development of an Ethical Aquaculture Food Index (EAFI). In the remainder of the paper we shall outline preliminary ideas and challenges as we understand them at this point.

Initial workplan

The initial work plan to develop an ethical aquaculture food index (EAFI) is a combination of several research steps. Most of them are intimately connected to work in other work packages of the SEAT project and there will be a flow of information between them.

1. Preliminary awareness-raising among project participants on the role and function of ethical considerations in the project.

2. Critical review of current ethical standards for aquaculture, related food production and delivery, including: eco-labelling, nutritional, food safety, animal welfare, fair-trade, organic, internal retail standards.
3. Identify costs, benefits and risks against livelihood contexts associated with compliance to voluntary and mandatory standards for different stakeholders in the food chain, from producers to retailers to consumers.
4. Assess interaction and areas of overlap between voluntary and mandatory ethical certification schemes. This step ultimately aims at reducing the certification costs for the producer and thus may influence positively their overall compliance.
5. Assess governance structures of existing certification schemes. Awareness, communication and transparency are important parameters of governance here.
6. Review of relevant surveys of European consumer needs, consumer understanding and prospective interpretation of existing and emergent food and seafood standards. There is a fragmentation of consumers with a plurality of preferences and values. Similarly, food scares of various kinds have shown differentiated effects in different countries and socio-political cultures.
7. Review of relevant ethical principles and adaptation of ethical decision-making tools incorporating participatory approaches.
8. Identification and description of relevant stakeholders, both primary and secondary, will be made for the purpose of inclusion in participatory ethical tools.
9. Implementation of participatory ethical tools incorporating expert advice and public discourse. The ethical matrix approach (Kaiser and Forsberg, 2000; Kaiser *et al.,* 2007; Mepham, 1996; Mepham *et al.,* 2006) will be used to structure participatory processes in project countries.
10. Develop the concept and practicalities of an EAFI. Post harvest impacts, socio-economic dependencies and consumer understanding are key areas which can augment the principle standards and certification schemes currently available for aquatic products, many with singular emphasis on biologically-based resource management assessments of sustainability. Part of the quality assessment of the proposed EAFI will be an ethical Delphi, conducted with selected experts (Millar *et al.,* 2007).

Ethical considerations

A core challenge when developing the EAFI is to decide who should be regarded as stakeholders and whether this is about humans only or whether it includes nonhumans as well. The question is whether and to what extent the biota or ecological entities such as animals, plants, species, communities and ecosystems might or should be regarded as stakeholders in the deliberative ethical process. This means basically to identify parties that are perceived to have a moral standing or status that makes them worthy to care about when making ethical assessments about values at stake. If there are such nonhuman stakeholders, this will have consequences for the representation among stakeholders in the participatory deliberative processes that this project is going to embark on. Since such parties for obvious reason cannot represent themselves, they need to be represented by proxy. So even if nonhuman stakeholders are recognized as potential stakeholders, i.e. assigned some moral standing, the question of who is to represent them is not settled. It is not obvious that NGOs which in some Western countries often take the role of speaking on behalf of the animals, should also be given that role when the context is Asian aquatic food production.

Zimmerman (1998: 3) says that philosophical theories of environmental ethics represent efforts to examine critically the notion that nature has inherent worth and efforts to inquire into the possibility that humans have moral duties to animals, plants and ecosystems. This means to sort out the extent to which ecological entities might have instrumental or non-instrumental value. The moral duties will stem from the extent of non-instrumental valuations and the kinds of instrumental valuations of ecological entities. Non-instrumental valuations are here used for situations when ecological entities are perceived to have a worth as ends in themselves. Instrumental valuations on the other hand refer to situations where ecological entities are valued as means to other ends, such as for economic, nutritional, aesthetic or recreational needs.

The first question in relation to the EAFI development is to find whether some ecological entities should be valued non-instrumentally as ends in themselves, and thus as stakeholders with a moral status that make them worthy as parties to take into consideration when making ethical judgements. The second question focuses on the stakeholders' long-term needs and preferences, whether for humans only or also for ecological entities. These might either be purely instrumental values such as food, water, shelter, comfort and economic income, or cultural values such as adventure, recreation and aesthetics. The two questions constitute then the basis for identifying which moral duties humans might have in relation to the sustainability both of human and nonhuman stakeholders.

For the EAFI development, a basic challenge will be to sort out what might be the practical consequences of the different positions the stakeholders might have in relation to the moral status of ecological entities, and how these should be handled. Some of the practical challenges faced in this connection might be illustrated by the following three questions asked in an investigation of ethical issues related to salmon farming in Norway and environmental interests (Haugen 2008):

1. Cultural landscapes versus salmon farming impacts.
   On what ethical grounds might it be fair that the negative impacts of agriculture on original or primary terrestrial ecosystems, which has resulted in the formation of cultural landscapes, are accepted, while at the same time salmon farming and its negative impacts on aquatic ecosystems are not?
2. Wild salmon versus the European minnow (*Phoxinus phoxinus*) species.
   On what ethical grounds might it be fair that a government committee, like the Norwegian Wild Salmon Committee, is appointed in an effort to secure the wild salmon's future existence, while at the same time the minnow, an invasive species in Norwegian wildlife, is perceived as a threat that should be combated?
3. Animal welfare versus salmon game fishing.
   On what ethical grounds might it be fair to be concerned with the animal welfare of domesticated animals, such as farmed salmon, while at the same time salmon game fishing is accepted? (Haugen, 2008: 459)

Ethical tools at work

We shall approach the task of developing an EAFI through the in-depth study of four particular cases, i.e. cultured species (tilapia, pangasius catfish, peneid shrimp, macrobrachium prawns), and look at their production in four different Asian countries (China, Vietnam, Thailand, Bangladesh; cf. Table 1 in Little *et al.*, 2009, this volume). The mix of countries and species gives opportunities to assess key sustainability issues.

Given the cultural and socio-economic diversity between producing and consuming countries, a particular challenge will also be to not let the best stand in the way for the good, i.e. not to impose top-down standards merely anchored in the affluence and cultural values of Europe.

However, there will be a need for reflection on the use of the ethical tool we intend to employ. The ethical matrix is well studied and tested within various settings and in relation to various technologies and ethical choices (see e.g. Mepham, 1996; Kaiser and Forsberg, 2000; Kaiser *et al.*, 2007; Mepham *et al.*, 2006). There are still some debates about some aspects of the matrix, both philosophical and practical, but by and large it represents a workable tool to analyse concrete ethical issues. However, when crossing cultural borders as in this project, one should not take for granted that even the very basic concepts and principles communicate easily to stakeholders educated in a very different cultural framework. The typical principles used in an ethical matrix (welfare or beneficence / non-malificence, dignity or autonomy, justice or fairness) refer back to traditions of ethical thinking anchored in the Western hemisphere. It is still an open question how such concepts and ideas fare in inter-cultural dialogues, specifically between Asia and Europe.

However, in regard to the principles, prospects may not be all that dim. Perhaps one can assume that there is something like a global reference frame for ethical thinking, perhaps expressed differently in different cultures, but essentially unified. Darryl Macer, who has spent much of his academic life in Japan and is now working for UNESCO Bangkok, argues for precisely this point. The basic principles, originally inspired by the principles of medical ethics (Beauchamp and Childress, 1994), are for Macer expressions of a 'love of life' which he sees expressed in Asian ethical thinking. 'Love is the desire to do good and the need to avoid doing harm. It includes love of others as oneself, the respecting of autonomy. It also includes the idea of justice, loving others and sharing what we have – distributive justice. These cover all the ideas and concepts of bioethics, and are found in ancient writings around the world – both as descriptions of behaviour and as prescriptions that others have made on the desirable standards of society. This heritage can be seen in all cultures, religions, and in ancient writings from around the world.' (Macer, 2004, 1).

Assuming that Macer has a valid point would mean that there is a good chance to pursue the project with Asian stakeholders utilizing the ethical matrix – possibly in a slightly modified form – in the participating countries. In this sense the project will also be a test case for the usability of the ethical matrix in an inter-cultural environment.

References

Beauchamps, T.L. and Childress, J.F. (1994). Principles of Biomedical Ethics, 4th ed. (1st 1979) Oxford: Oxford University Press.

Cressey, D. (2009). Future fish, Nature, vol. 458, 398-400.

Dagens Næringsliv (2009). (Norwegian newspaper) Three articles appeared 20.03.09, 21.03.09 and 27.03.09, Oslo, Norway.

FAO (2009). The State of World Fisheries and Aquaculture 2008, available at: http://www.fao.org/docrep/011/i0250e/i0250e00.HTM, accessed 4 April 2009.

Haugen A.S. (2008). Values behind biodiversity and moral duties – Development of a conceptual model for ethical environmental accounting using salmon farming as a case study, PhD Thesis, University of Oslo, Norway. 481 p.

Kaiser, M. (2006). Turning cheap fish into expensive fish? The ethical examination of an argument about feed conversion rates, in: M. Kaiser and M.E. Lien (eds.), Ethics and the Politics of Food, Wageningen Academic Publishers: Wageningen 2006.

Little, D.C., F.J. Murray, T.C. Telfer, J.A. Young, L.G. Ross, B. Hill, A. Dalsgard, P. Van den Brink, J. Guinée, R. Kleijn, R. Mungkung, Y. Yi; J. Min, L. Liping, L. Huanan, L. Yuan, Y. Derun, N.T. Phuong, T.N. Hai, P T Liem, V N Ut, V.T. Tung, T.V. Viet, K Satapornvanit, T. Pongthanapanich, M.A.Wahab, A.K.M. Nowsad Alam, M.M. Haque, MA Salam, F. Corsin, D. Pemsl, E. Allison, M.C.M Beveridge, I. Karunasagar, R. Subasinghe, M. Kaiser, A. Sveinson Haugen, S. Ponte (2009). Sustaining ethical trade in farmed aquatic products between Asia and the EU, this volume.

Luyken, R. (2001). Aus kleinen Fischen grosse zaubern, Die Zeit Online, 06. September 2001, available at: http://www.zeit.de/2001/37/Aus_kleinen_Fischen_grosse_zaubern, accessed 4 April 2009.

Macer, D. (2004). The Behaviorome Project: Moving forward with ideas, available at: http://www.eubios.info/abc5bk.htm, accessed at 4 April 2009.

Micheletti, M. (2006). Political consumerism: Why the market is an arena for politics. In: M. Kaiser and M.E. Lien (eds.) Ethics and the politics of food, Wageningen Academic Publishers: Wageningen.

Millar, K., E. Thorstensen, S. Tomkins, B.Mepham and M. Kaiser (2007). Developing the Ethical Delphi, Journal of Agricultural and Environmental Ethics, vol. 20, 53-63.
Zimmerman, M.E. (1998). General Introduction, In: Environmental Philosophy: From Animal Rights to Radical Ecology, Vol. 2, Ch. 1, M.E. Zimmerman (ed.), Prentice Hall, New Jersey, USA.

Exploring a mode of agri-food governance: international food standard setting and expertise

R.P. Lee
Centre for Rural Economy, School of Agriculture, Food and Rural Development, Newcastle University, Newcastle upon Tyne NE1 7RU, United Kingdom; r.p.lee@ncl.ac.uk

Abstract

The paper examines a particular mode of agri-food governance: international food standard setting. Although, from a social science perspective, standard setting is an activity often overlooked due to the technical and scientific content of the process, it is imbued with moral and ethical dimensions. Standards set norms and conventions and in doing so help to configure the behaviour of many actors involved in the governance of the agri-food system. Moreover, standards have assumed a heightened significance in regulating the increasingly globalised trade in food products. The focus in this paper is upon how standard setting is conducted. Firstly, the potential contribution of the concept of epistemic communities to the interpretation of international food standard setting is discussed. Secondly, the paper details the policy framework in which international food standard setting takes place. Thirdly, the role of scientific expertise to the standard setting process in the Codex Alimentarius Commission – the intergovernmental organisation in which public food standards are set – is explored through a case-study of the Codex Nutrition Committee. In this case-study the construction, articulation and presentation of knowledge claims is a crucial element of standard setting, challenging the authoritative knowledge claims deemed crucial to the establishment of an epistemic community. The paper concludes by reflecting upon the political and ethical implications of the growing importance of technical expertise to the conduct of agri-food governance.

Keywords: agri-food governance, expertise, knowledge claims, Codex Alimentarius, dietary fibre

Introduction

This paper examines a particular mode of agri-food governance: international food standard-setting. As core components of regulatory systems governing the agri-food system, international food standards are produced through the activities of scientific experts, government regulators, transnational agri-food companies and intergovernmental organisations. As Busch (2000) has noted, standards set norms and conventions and in doing so help to configure the behaviour of many actors involved in the governance of the agri-food system. At a time when regulation is an increasingly important tool of governance, the conduct and significance of standard setting within the agri-food system remains under-researched by social scientists. In order to address this the paper examines the influence of transnational networks of experts – termed epistemic communities – in the setting of international food standards through a case-study of the on-going attempt to agree an international definition of dietary fibre in the Codex Alimentarius Commission (the Codex). Using the concept of epistemic communities, Haas (1992) suggests that scientific experts in a particular domain can become active agents in the regulatory process, beyond the position of isolated communicators of scientific opinion.

The role of expertise in global governance: epistemic communities

The importance of self-organising networks of expertise to international policy co-ordination – such as the formulation of international conventions, agreements and standards – has been identified by some international relations scholars using the concept of epistemic communities. Frequently, international policy co-ordination occurs around scientific and technical issues, such as food safety, environmental pollutants or communicable diseases. These issues require the input of specialists in order to provide evidence on the scientific problem. The concept of epistemic communities has been employed to develop various analyses of how international agreements are formulated in particular policy domains through the contribution of these expert groups. Haas (1992: 3) defined epistemic communities as '...a network of professionals with recognised expertise and competence in a particular domain and an authoritative claim to policy-relevant knowledge within the domain or issue-area'. He also set out the four characteristics which epistemic communities exhibit. They are: a shared set of normative and principled beliefs; shared causal beliefs; a shared notion of validity; and a common policy enterprise. The first three characteristics are principally internal to the community. Normative and principled beliefs give the epistemic community a social rationale – they become certain that the correct application of their understanding will be 'for the best'. Causal beliefs are central to the epistemic community concept and are produced by claims to truth being challenged within the community until an agreed understanding about 'how this problem is manifest' is reached. Finally, the ways in which causal beliefs are assessed is the validation of knowledge, in other words the criteria for assessing truth claims. From these internal mechanisms, an epistemic community becomes actively engaged in a common policy enterprise. This is the focus of practice for the epistemic community – it is where their communal expertise is directed.

Uncertainty over scientific and technological questions poses particular problems for international policy-making. In such situations, the formation of an epistemic community with an authoritative claim to knowledge can be influential in producing consensus around their particular judgment. The concept does make an important contribution in revealing that knowledge – in particular scientific and technical knowledge – has a formative role in the production of international regulation. In this regard, the concept can be usefully employed to explore the process of setting international food standards. However, by suggesting that the claims to knowledge are held by epistemic communities are authoritative – that is they are can provide a 'knock-down argument' (Pinch and Bijker, 1984) – the concept appears to be less robust in areas of contention or controversy. In an account of an epistemic community of cetologists operating in the area of the regulation of commercial whaling, Petersen (1992) suggests that epistemic communities can be open to fracture if disagreement occurs over scientific methods. The emphasis of Petersen (1992) is upon how scientists mobilise, or rather how they fail to mobilise, against intergovernmental decision-making which does not take account of scientific evidence. It is not clear how networks of experts respond in situations which are both scientifically and politically contentious.

In the remainder of the paper, the concept of epistemic communities will be used to examine the standard setting process in the Codex Nutrition Committee. In particular, the ability of scientific experts to form an influential group based upon authoritative knowledge will be considered. It is suggested that, in this case-study, the contested and negotiated process of standard setting means that an epistemic community cannot be established. Instead the process is characterised by the production, articulation and contestation of knowledge claims from various sources.

A mode of agri-food governance: international standard setting

International food standards impact upon the structure of the agri-food system, but the technical discussions involved in their production, and the international fora in which these discussions take place, means that the origin and negotiation of standards is frequently obscured. In such regulatory contexts, the concept of epistemic communities suggests that transnational networks of scientific experts can become highly influential. The strength of this assertion is investigated using the case of the Codex Nutrition Committee.

In the Codex, standard setting is guided by procedures. In these procedures, government delegates, who negotiate standards within Codex committees, are defined as risk managers who have responsibility for risk management (Codex, 2007a). Risk management, so defined, constitutes the political domain of standard-setting in Codex, alongside the recently agreed activity of risk assessment policy (which is discussed below). Those who provide scientific advice are termed risk assessors and have responsibility for risk assessment. The division between risk management and risk assessment in the risk analysis framework used by Codex (another component, risk communication, is not discussed here) has a procedural function in assigning standard setting tasks to particular groups of experts. Following Codex procedures, scientific activity and discussions in the Codex ought to be restricted to the domain of risk assessment. Primary responsibility for risk assessment is said to rest with expert groups, which deal with requests for scientific evidence made by risk managers working within Codex committees. A number of expert groups exist on a permanent basis. These groups work to provide Codex committees with evidence when requested, covering topics such as food additives and pesticide residues. However, not all Codex committees have a corresponding, permanent expert group. In such cases, Codex committees seeking the input of risk assessors can make a request to FAO/WHO for advice. If a Codex risk management committee has a corresponding expert group, it should at least be clear where the request should be directed to. For those Codex risk management committees without a corresponding expert group, the relationship between the committee and the FAO/WHO, as risk assessors, becomes more uncertain as there is no formal organisational relationship between these committees and FAO/WHO. Risk management activities conducted in Codex committees should be undertaken on the basis of scientific advice from FAO/WHO. If this advice is not available from FAO/WHO, but advice is still required, then other expert groups may become more influential in the standard setting process. These expert groups can be international advisory bodies, such as the European Food Safety Authority (EFSA), international professional organisations or national scientific advisory boards and institutions.

Standard setting in the Codex Nutrition Committee

The following case-study derives from observations, interviews and document analysis undertaken during 2007 and 2008. Observations of the conduct of meetings were undertaken as a public observer. Forty interviews were conducted with scientists, government delegates, food industry and consumer representatives over the course of the research. The operation of Codex occurs, in part, through the distribution of written comments and papers, and analysis of these and related documents was completed. By way of introduction to the case-study, the Nutrition Committee is a Codex committee which deals with standards pertaining to the nutritional and dietary aspects of internationally traded food and has an area of competency stretching across different Codex committees. In terms of risk assessment, there is no standing expert group which provides advice to the Nutrition Committee. As there is no standing expert body for the Nutrition Committee, it is unclear where scientific advice on nutrition should be produced. In the words of one government delegate 'who is our risk assessor?' (Interview, November 2007).

Since 2004 the Nutrition Committee has undertaken work to define principles by which risk analysis for nutrition should be conducted. The principles state that: 'Nutritional risk analysis comprises three components: risk assessment, risk management and risk communication. Particular emphasis is given to an initial step of Problem Formulation as a key preliminary risk management activity.' (Codex, 2008). Problem formulation is deemed important in that 'it fosters interactions between risk managers and risk assessors to help ensure common understanding of the problem and the purpose of the risk assessment.' (Codex, 2008). This is an explicit recognition that risk assessment should not take place in isolation from risk management. According to Millstone (2009), since 2003 the Codex has led a move towards a 'co-evolutionary' model of standard setting, a key feature of this which is the representation of specific and situated risk assessments. In the case of the nutritional risk analysis, the inclusion of problem formulation was done with explicit recognition of the introduction of risk assessment policy in the Codex procedures.

The nutritional risk analysis principles recognise that interactions should occur between risk managers and risk assessors. It does not, however, resolve the question of who should provide scientific advice to the Nutrition Committee. At the 2007 meeting of the Nutrition Committee, the question of scientific advice for nutrition was raised during discussion of these principles. The FAO representative at this meeting emphasised that FAO/WHO Joint Expert Groups give independent advice and should be probably be the only source of advice for Codex committees (Personal note, November, 2007). The comments of the FAO representative came in response to the assertion in the nutritional risk analysis principles that despite the FAO/WHO being a primary source of scientific advice, other sources may be used. The FAO representative took issue with the suggestion that other sources of advice could be chosen by the Nutrition Committee. Further, they stated that if a request for scientific advice is made by a Codex committee then the FAO/WHO are obliged to meet this request (Personal note, November 2007). The ability of the FAO/WHO to deliver scientific advice is crucial to standard setting in the Codex if the parent organisations are designated as the primary source of scientific advice. This is problematic. For example, the EC takes its scientific advice on food safety and nutrition primarily from the European Food Safety Authority (EFSA), not FAO/WHO Joint Expert Groups, when setting EC legislation. Likewise, other member governments of Codex take scientific advice from other national and international scientific expert bodies outside of the UN system.

Curiously, the lack of a standing committee for scientific advice on nutrition has not resulted in an absence of scientific input from the FAO/WHO in Codex standard setting. In some instances, FAO/WHO have proactively provided scientific advice which has not been requested by risk managers. A longstanding source of contention within the Nutrition Committee has been the agreement of a definition for dietary fibre. In November 2008 a draft definition was finally agreed in the Nutrition Committee and this definition has been sent for adoption by the Codex Commission. The contention over dietary fibre has, primarily, concerned the types of food components which can be classified as dietary fibre and so contribute to the fibre levels measured in food products. Naturally, some food industry groups (such as the International Dairy Federation and the International Life Sciences Institute) favoured broader definitions of dietary fibre, returning higher values for dietary fibre in many products (so providing a platform for nutritional claims). In contrast, in 2007 the FAO/WHO presented evidence in the form of a Scientific Update, which asserted that term dietary fibre should only apply to 'intrinsic plant cell wall polysaccharides.' (Cummings and Stephen, 2007: S13). In essence this definition limits dietary fibre to those polysaccharides found in fruits, vegetables and wholegrains.

The contribution of the FAO/WHO took place without the explicit request of risk managers and, importantly, represented a significant departure from the draft definition which was being negotiated in the Nutrition Committee. According to a delegate to the Nutrition Committee, this was a classic situation of scientific advice being produced in a competitive relationship with other sources of advice (Interview, November 2007). Despite the controversial nature of the discussions over the definition of dietary fibre, a compromise solution was eventually reached (or 'fudged' in the words of one commentator (Personal communication, January 2009)). The agreed definition of dietary fibre represents a compromise solution in three main ways. Firstly, national governments are free to determine whether to include shorter chain polysaccharides (such as oligosaccharides) as dietary fibre. Secondly, the definition allows for the inclusion of synthetic and recovered carbohydrate polymers as dietary fibre (which were excluded in the FAO/WHO definition), as long as these components can be demonstrated to have a physiological effect of benefit to health (which is not easy to prove). Thirdly, national governments are free to determine the required nutrient content level for claims ('source of' or 'high in') made for dietary fibre in liquid foods. However, the question of the most appropriate methods of analysis for measuring these components as been left for further discussion. Therefore, the agreement reached in 2008 is, in a number of ways, open to interpretation. Therefore the nutrient content claims for dietary fibre on food products will vary depending upon the approach taken by national governments.

The agreed definition for dietary fibre in Codex moved beyond that stated by the FAO/WHO Scientific Update in 2007, by incorporating oligosaccharides and synthetic carbohydrates polymers. Risk assessment in this case could not be judged to have comprised an authoritative claim to knowledge. The papers comprising the Scientific Update were produced by a group of scientists with recognised expertise in dietary fibre and were published in an established academic journal, the European Journal of Clinical Nutrition. However, the Scientific Update attracted critical comments from some government delegations. The delegation of the EC suggested that the presentation made by on behalf of the FAO/ WHO expert committee was: 'A suggestion which takes us back where we started many years ago.' (Personal note, November 2007). Thus, the iterative process of risk analysis was taking place. The FAO/ WHO had intervened in the Codex standard setting procedure (risk management) using scientific advice (risk assessment) presented by a risk assessor to risk managers (EC delegation) who questioned the basis (risk assessment) of his argument (risk management). The same risk assessor was asked by a government delegate at a subsequent meeting whether he thought he had compromised his scientific integrity 'because I was a risk assessor and they were risk managers and the two should never meet' (Interview, December 2008).

Conclusions

The contention over defining dietary fibre in the Codex demonstrates that the establishment of authoritative claims to knowledge is especially difficult in standard-setting. The division between science and politics is mobilised rhetorically in order to carry out what Gieryn (1983) calls boundary work; an activity which designates who is inside a knowledgeable group and who is outside. However, in the case of dietary fibre the attempt to exclude risk assessors from the group of risk managers was unsuccessful, as the contention over the science of dietary fibre was closely entangled with the political negotiation of the definition and the implications of this for nutrition claims on food products. Even though the procedures of Codex have, or are being, adapted to recognise that 'problem formulation' or 'risk assessment policy' is an important element of standard setting, in this case-study scientific experts are able to transcend any rhetorical divisions between risk assessment and risk management. The provision of scientific advice to the Nutrition Committee has proved contentious, with competing interests advocating different sources of advice. Contestation over scientific advice emerges from the lack of recognised authoritative claims to knowledge. A relationship exists between the contested nature of knowledge claims in such circumstances and the significance of regulatory competition within international food standard-setting. In the case of dietary fibre, not only is the dispute one between scientific experts who disagree on appropriate evidence and analysis, but also between regulatory systems which are brought together under the intergovernmental fora of the Codex. As a result, the assertion that unified, scientific knowledge can be established and simply applied to standard setting activities is to seriously misconstrue the iterations which take place between science and politics. In such conditions there is little chance that a network of professionals can establish an uncontested and authoritative position in the sense used by Haas (1992).

References

Busch, L. (2000). The moral economy of grades and standards, Journal of Rural Studies, 16:3, 273-283.

Cummings, J.H. and Stephen, A.M. (2007). Carbohydrate terminology and classification, European Journal of Clinical Nutrition, 61 (Supplement 1), S5-S18.

Haas, P.M. (1992). Introduction: epistemic communities and international policy co-ordination. International Organisation, 46 (1), 1-35.

Millstone, E. (2009). Science, risk and governance: radical rhetorics and the realities of reform in food safety governance, Research Policy, in press.

Petersen, M.J. (1992). Whalers, cetologists, environmentalists, and the international management of whaling, International Organisation, 46 (1), 147-186.

Pinch, T.J. and Bijker, W.E. (1984). The social construction of facts and artefacts: or how the sociology of science and the sociology of technology might benefit each other, Social Studies of Science, 14, 399-441.

An ethical reflection on the thoughts of Asbjørn Eide on the right to food

J.L. Omukaga
Katholieke Universiteit Leuven, Naamsestraat 40, 3000 Leuven, Belgium; jlikori@yahoo.co.uk

Abstract

Expert opinion today confirms an overall adequate world food security situation. The world is capable of feeding its population. A current priority concern is to explore ways and means of distributing the world's food resources. Instrumentalisation of the human right to food is a conspicuous activity within these discussions. This paper is an ethical reflection on the work of Asbjørn Eide on the human right to food. It discusses the role of 'cosmopolitan' ethics as a motivation for the implementation of the right to food among the concerned partners in the fight against hunger.

Keywords: right to food, ethical view

Introduction

The human right to food and to be free from hunger is an international law with a sound basis on the Universal Declaration of Human Rights (UDHR) and its subsequent elaborations. Tribute has been paid to the human rights expert, Asbjørn Eide, as the pioneer champion of the human rights approach to the problem of hunger.[1] He led a team of experts in the process of clarifying and instrumentalizing the right to food to the level where the international community adopted it as law binding on all the member States of the United Nations. The right to food law is among the landmark contributions in the struggle against hunger in the last two decades. It comes within the initiative of the World Food Summit, 1996, and the Millennium Development Goals, 2000, to reduce hunger by half by 2015. While the latter two initiatives set targets to be pursued, Asbjørn Eide's struggle facilitated the entrenchment of the right to food into the international legal framework. This paper, above all, is an ethical reflection on Asbjørn Eide's contribution to the right to food as an approach to the problem of hunger in the background of the WFS and MDG targets. It shall unfold in three sections: The first section is a focus on the current hunger situation in the Great Lakes Region of Africa as the background and point of reference. The second section shall then highlight the key points of concern in the thoughts of Asbjørn Eide on the right to food. The third section is the important ethical reflection on this approach. In the conclusion, this reflection shall plead for cosmopolitan ethics to be a core motivation for the implementation of the right to food.

Current hunger situation: focus on the Great Lakes Region of Africa

Five years after the official adoption of the right to food, 22-27 November 2004, and as we close in on the target time, 2015, the world hunger is even a bigger cause of worry. At the definitive World Food Summit of 1996, the number of world's undernourished stood at 800 million, then described by the world leaders as 'intolerable and unacceptable'. Today this is estimated to the round figure of 850 million (FAO report, 2008). Though progress towards the WFS target is fairly on course, that is, the reduction of general

[1] Cf. Torkel Opsahl and Jan Helgesen, 'Asbjørn Eide: A Tribute' in *Broadening the Frontiers of Human Rights*, ed. Donna Gomien (Oxford: Oxford University, 1993). This was a tribute to Asbjorn Eide from scholars upon on the his birthday anniversary 1992. See especially the introduction to the recent edited articles on Asbjørn Eide's work, Morten Bergsmo, ed., *Human Rights and Criminal Justice for the Downtrodden: Essays in Honour of Asbjørn Eide* (Leiden: Marinus Nijhoff, 2003).

hunger prevalence, the situation is complicated when one considers the reduction of the actual numbers of the undernourished persons targeted by the Millennium Development Goals. Experts attribute this to the irregular trends in population growth rates in different parts of the world. For example, while many regions of the world are projected to register a substantial reduction in the number of the undernourished owing to the expected overall decreasing trend in population growth rates, the population growth rate in Sub-Sahara Africa shall remain beyond 2% by 2030. It will still remain beyond unit percentage (1.4%) by the close of 2050 compared to less than 0.5% in all other regions of the world. The region will exhibit a very gradual reduction in overall undernourishment towards 2015. To demonstrate the impact of this trend in the Sub-Sahara region, experts point at its devastating effects in the Great Lakes Region and the horn of Africa. From Table 1 below, statistics indicate very high population of the undernourished and high prevalence of undernourishment.

With the exception of Uganda, more than one third of the population of each of the other five countries was undernourished in 2005. The DR Congo was the worst hit, followed by Burundi and Tanzania in the second and third positions respectively. At the moment, Kenya is on the spotlight with estimated 3.5-4.5 million people feared starving with reported sporadic deaths. With this trend in population growth, not to mention other worse effects such as war, disease, drought, floods, poverty and lack of relevant policy frameworks, both the current and the future situation of undernourishment in this region remains worrying.

Highlights of Asbjørn Eide's thought on the right to food

In his pursuit of the human rights approach, Asbjørn Eide employed the normativist-activist method. As a lawyer, he believes strongly in the relevance and power of norms to change the lives of people. His involvement in the human rights struggle did not only imply ideological contribution to important agreements and commitments of the United Nations in upholding the rights of their citizens, but his practical involvements in meetings, workshops, discussions and crucial interventions also reveal his passionate desire to protect the advancement of human rights. This disposition allows him to make brilliant contributions to human rights law as well as promoting human rights policy in appealing for more commitment from the side of the States.

Secondly, Asbjørn Eide joined other scholars in clarifying the content, subject and object of the right to food as well as the mechanism to promote compliance (Alston, 1984). But his expert ingenuity stands out in the development of the economic, social and cultural rights (ESCR). His work on a framework

Table 1. Undernutrition in the countries of the Great Lakes Region of Africa by 2005 (FAO, 2006).

Country	Population (million)	Undernourished (million)	Proportion (%)	Food supply (kcal/person/day)		
				1990-1992	1995-1997	2002-2004
Burundi	6.8	4.5	66	1,900	1,700	1,660
Kenya	32.0	9.9	31	1,980	2,060	2,150
Tanzania	37.0	16.4	44	2,050	1,880	1,960
Rwanda	8.4	2.8	33	1,950	1,830	2,110
Uganda	25.8	4.8	19	2,270	2,220	2,370
DR Congo	52.8	39.0	74	2,170	1,770	1,590
Total	162.8	77.4	47.5	12,320	11,460	11,840

for state obligations for ESCR is today adopted as the reference for discussions on such obligation, 'it realized a fruitful effort to design a broader composite framework for identifying more precisely state obligations in relation to specific components of the right to food,' (Oshaug and Eide, 2003). Asbjørn Eide identified a dynamism within the human rights system which essentially consists of the interrelationship between freedom and demands. While freedom sets limits to the exercise of state authority, demands require action by the State, which necessitates the existence of state administration machinery equipped for this task. This dynamism ultimately begets the corresponding obligations of the States to facilitate the implementation of human rights. Accordingly, he proposed a three-level framework within which the State would fulfill its obligation in implementing the right to food: The State would intervene at the levels of respecting, protecting and fulfilling the right to food. The expert contribution of Asbjørn Eide in this regard was an enormous effort that established a lasting balance between the right to food and all other rights in the system. Likewise, the question of freedom and demand at this level introduces the moral perspective of interpretation of the right to food.

Thirdly, Asbjørn Eide's proposal for the Right to Food Matrix pioneered a framework for working out obligations for implementing the right to adequate food in the context of household food and livelihood security. It helped to work out context-specific meaning of different levels of food security, link it to appropriate level of state involvement, and concretize the appropriate state obligations to realize the right to food. While the State's duty involved passive non-interference for those already able to feed themselves, it directly provided for the deprived and destitute. After tireless research and consultative effort, the content of the right to food shaped up and constituted the General comment 12 article paragraph. 8. Thus:

> The right to adequate food is realized when every man, woman and child alone or in a community with others, have physical and economic access at all times to adequate food or means for its procurement. The core content of the Right to Adequate Food implies the availability of food in quantity and quality sufficient to satisfy the dietary needs of individuals free from adverse substances, and acceptable within a given culture, (and) the accessibility of such food in ways that are sustainable and that do not interfere with the enjoyment of other human rights.

Fourthly, in pursuit of the implementation of this holistic approach to an ESCR, Asbjørn Eide played a key role in drafting the voluntary guidelines recommended to support the progressive realization of the right to adequate food in the context of national food security. These guidelines 'represent the first attempt by governments to interpret an economic right and to recommend action to be undertaken for its realization.' They are a valuable support both for the States and non-State in their efforts to elaborate and implement the strategies on the right to food spelled out in the General comment 12. With these highlights, the struggle of Asbjørn Eide to instrumentalize the right to food is nothing short of a thorough scholarly commitment. His holistic method, interpretational ability, collaborative acumen, passionate diplomacy and his devotion to its implementation and protection, accorded the right to food an exemplary force among the Economic, Social and Cultural Rights. When all seemed done, Asbjørn Eide remarked: 'An impressive set of legal standards is available. Regrettably, much of it is soft law... their impact depends primarily on their recognition and the degree to which they can be brought to influence political choices and legal behavior. This however, can be much influenced by academics, including research institutions and think-tanks, by NGOs as agents of civil society and by national institutions for human rights.' In the spirit of this invitation we here wish to highlight the moral force implicit in the human right to food as analyzed by Asbjørn Eide.

An ethical reflection on the right to food

Three basic ethical concerns can be observed from Asbjørn Eide's comprehensive contribution to the right to food. These include the regard accorded to the place of individual freedom and dignity, the issue of individual moral responsibility and above all the dynamics of the global ethics of food enshrined in the prescription of the right to food matrix. As regards the place of individual freedom and dignity, Asbjørn Eide confirms it in many ways as the core expression of what is at stake in the hunger narrative. With hunger, the victim's physical appearance is disfigured and wasted, his/her personality in society is proportionately lowered and his/her spiritual/psychological ability is dented sometimes without hope of full recovery thereafter. The individual human victim suffers indignity. Throughout his interpretation Asbjørn Eide strikes a noticeable balance between the responsive state legal structures and the place of the individual person. His primary concern in identifying freedom and demand as forces that regulate the State's involvement at implementation, secured the basic respect for the individual in the entire process. 'While freedom sets limits to the exercise of state authority, demands require action by the State which necessitates the existence of state administration machinery equipped for this task. The individual is engaged as a moral agent but his/her freedom is moderated by the state legal structures. Ultimately, the individual's threatened moral significance enjoys positive regard over the legal and social structures at his/her service.'

On the question of responsibility, it is observable that despite the leading role of the Nation-State in implementing the right to food, Asbjørn Eide secures the responsibility of the individual from the point of view of correlativity between duties and rights. The individual, having duties to other individuals and to the community to which he/she belongs, is under a responsibility to strive for the promotion and observance of the recognized rights. In this sense, individual obligations involve, as a minimum, a duty not to overconsume and not to waste food. To this extent, Asbjørn Eide brought to focus the elaborate moral doctrine of correlativity. A moral consideration imparts an indissoluble link between every human duty and correlative right: 'any rights claim implies correlative duties.'

Having secured the rightful place for the individual human person, and on the basis of correlativity between individual duties and rights, the right to food narrative goes on to espouse an elaborate global ethics of food. Correlativity first begins and operates within an individual person, but eventually this link also manifests its social dimension in the interaction between individuals in society. On the individual level, mankind's actions proceed in freedom, that is, from his/her own initiative, conviction and sense of responsibility, and not under coercion of any sort. This basic bond eventually extends human rights and duties, to its social dimension by recalling a human person as a social being by his/her very nature. The result is a constant interaction between people in which both rights and duties are mutually exchanged leading to a well-ordered society: 'to one man's right there corresponds a duty in all other persons: a duty, namely of acknowledging and respecting the right in question.' The good order in society is promoted when individual members recognize and perform respective rights and duties accordingly. In this respect, Asbjørn Eide agrees with many ethicists that human rights 'represent universal values and constitute an ethical imperative to safeguard the dignity of every human being, providing fundamental norms of outcomes and processes of action to this end.' In the situation of gross inequalities in the abilities to address the hunger problem on regional level such as exhibited in the Sub-Sahara region of Africa, and, even worse, in the Great Lakes Region, the current debate on the ethics of nationality and cosmopolitan ethics, becomes very relevant. While the ethics of nationality posits very credible arguments that highlight the peculiarity of the moral significance of our nationalistic social attachments, the concerns of the right to food resonate more with the view posited by cosmopolitan ethics. Contra nationalists argue that 'There is nothing morally significant about duties to compatriots that is greater in weight than to non-compatriots given the contingent fact of our communal membership.' More specifically Martha C. Nussbaum adds: 'our highest obligations are to our fellow human beings, whether or not we are fellow citizens.' The cosmopolitan ethics affirms that our duty to the whole of humanity at the same time presumes our special moral connections to our family, friends or neighbors. With regard to the right to food, there is a bigger obligation to bring together all the aspects of moral significance in humanity around the world.

The two aspects upon which the right to food matrix was constructed, that is, the household food security model and the hierarchy of state obligations, provide a formidable infrastructure for cosmopolitan ethics. According to Asbjørn Eide: 'Achieving food security at household level means ensuring that sufficient food is available throughout the territory, that supplies are relatively stable, and that everyone within that territory in need of food has the capacity to obtain it in order to lead a healthy and productive life.' Party States are further obliged to take maximum steps to progressively achieve full realization of the right to food through international cooperation, and to ensure an equitable distribution of world food supplies in relation to need. In this arrangement the international community has the obligation to cooperate both with the State parties and the individual persons, especially in situations where the State is seen to fail to meet its overbearing obligation to its citizens. The entire matrix therefore is a practical example of operational cosmopolitan ethics whose concern recognizes the moral significance of the other no matter the divide. Ultimately, the cosmopolitan ethical attitude motivates partnerships that brings on board both State and non-State actors. This partnership is both necessary and crucial in advancing war against hunger and more so the motivation behind it.

Conclusion

A Focus on the moral content of the right to food as structured by Asbjørn Eide, yielded a discussion on the attention accorded to the value of the individual person, as a victim and participant, the strength of the right to food on the issue of moral responsibility and the view of the right to food matrix as a basis for a practical locus for global ethics of food. While the legal interpretation of the right to food highlights the central role of the State to respect, protect, and fulfill the right to food, its ethical aspects highlight the priority of the approach, namely: the human person both as a victim of hunger and a partner in the search for the solution. Consequently, the progress in the implementation of the right to food must unfold within the framework of reciprocal partnership between the individual person and the social structures that build up to the entire global community. This crucial reciprocity should, in turn, be inspired by cosmopolitan ethics which confers moral significance to the individual and the State as much as it does to the international community.

Reference

Alston, P. (1984). International Law and the Right to Food. In: Eide, A., Eide, W.B., Goonatilake, S., Gussow, J. and Omawale (eds.) Food as a Human Right. United Nations University press, Tokyo, 162-174 pp.

De Haen, H. (2005). Foreword. In: Eide, W.B. and Kracht, U. (eds.) Food and Human Rights in Development. Intersentia publishers, Oxford, Vol. I, xxiii-xxiv pp.

Eide, A. (2007). Freedom from Hunger as a Basic Human Right: Principles and Implementation. In: Pinstrup-Andersen, P. (ed.) Ethics, Hunger and Globalization. Springer publishers, Dordrecht, The Netherlands, Vol. 12, 93-110 pp.

Eide, A. (1984). The International Human Rights System. In: Eide, A., Eide, W.B., Goonatilake, S., Gussow, J. and Omawale (eds.) Food as a Human Right. United Nations University press, Tokyo, Japan, 152-161 pp.

Eide, A. (2005). The Importance of Economic and Social Rights in the Age of Economic Globalisation. In: Eide, W.B. and Kracht, U. (eds.) Food and Human Rights in Development. Intersentia publishers, Antwerpen, Volume I, 3-37 pp.

Eide, W.B. (2005). From Food Security to the Right to Food. In: Eide, W.B. and Kracht, U. (eds.) Food and Human Rights in Development. Intersentia publishers, Antwerpen, Volume I, 67-94 pp.

Eide, W.B. and Kracht, U. (2005). The Right to Adequate Food in Human Rights Instruments. In: Eide, W.B. and Kracht, U. (eds.) Food and Human Rights in Development. Intersentia publishers, Antwerpen, Belgium,Volume I, 99-116 pp.

Miller, D. (2008). The Ethics of Nationality. In: Brook, T. (ed.) The Global Justice Reader. Blackwell publishers, Oxford, 284-305 pp.

Nussbaum, M.C. (2008). Patriotism and Cosmopolitanism. In: Brook, T. (ed.) The Global Justice Reader. Blackwell publishers, Oxford, 306-314 pp.

Office of the High Commissioner for Human Rights (2008). International Covenant on Economic, Social and Cultural Rights. http://www.unhchr.ch/html/menu3/b/a_cescr.htm. Accessed 6 March 2009.

Opsahl, T. and Helgesen, J. (1993). Asbjørn Eide: A Tribute. In: Gomien, D. (ed.) Broadening the Frontiers of Human Rights. Oxford University press, Oxford, 1-12 pp.

Oshaug, A. and Eide, W.B. (2003). The Long Process of Giving Content to an Economic, Social and Cultural Right: Twenty-Five Years with the Case of the Right to Adequate Food. In: Bergsmo, M. (ed.) Human Rights and Criminal Justice for the Downtrodden: Essays in Honour of Asbjørn Eide. Brill Academic Publishers, Leiden, 325-367pp.

World Food Summit (1996). Rome Declaration on World Food Security. http://www.fao.org/docrep/003/w3613e/w3613e00.HT. Accessed 5 March 2009.

The European Common Agricultural Policy, obesity, and organic farming

Dita Wickins-Drazilova
Department of Philosophy, Lancaster University, Furness College, Lancaster, LA1 4YG, United Kingdom; d.wickins-drazilova@lancaster.ac.uk

Abstract

The current rise of prevalence of obesity and weight-related diseases is a serious public-health problem. Various anti-obesity measures that are being introduced are contradicted by a powerful obesogenic intervention: the Common Agricultural Policy (CAP). This undermines healthier choices by supporting overproduction of high-calorie foods. The CAP should be reformed to be more focused on health and quality. As organic farming is an existing system of production that aims towards better quality of food, five arguments are presented on how organic agriculture might be a beneficial anti-obesity measure. Firstly, organically produced food is richer in nutritional value; secondly, it offers a less obesogenic option, as chemical pollution in foods and environments may be linked to weight gain; thirdly, organic farming creates a more diverse countryside which is inviting for walking and other recreational activities outdoors; fourthly, it increases employment, and there is a strong association between unemployment and weight gain. Finally, the EU could support availability of allotments. This would be a useful anti-obesity measure as people could grow their own fresh fruit and vegetables, as well as get a regular healthy exercise.

Keywords: CAP, obesity, public health, organic farming, allotments

Obesity-related health problems

Obesity is defined as a body mass index (BMI) over 30 and is classified as a disease by the World Health Organization. It is generally held that obesity carries various health risks to individuals and is associated with higher incidence of cardiovascular diseases, diabetes and other health problems. It is also associated with reduced life expectancy and is regarded as one of main preventable causes of death in the developed world. Obesity is therefore recognised as a leading public-health problem in Europe. It has implications for national health systems, and, as such, imposes costs on others.

On the other hand there are authors and organization that disagree with these commonly held claims. For example, J. Eric Oliver argues that various evidence shows that obesity is not in itself a disease or a cause of diseases, but rather a symptom of health problems, and most of the statistics showing a rapid rise of obesity incidence do not take into account that the categorisation of obesity has been changed several times over time (Oliver, 2006). But no matter whether obesity is a disease in itself, and/or a cause or consequence of serious diseases, even Oliver doesn't dispute the fact that the weight of an average European as well as the incidence of weight-related health diseases are rising.

A good deal of political discussion and many media articles assume that individuals are mostly responsible for their weight and weight-related health problems. In another paper my colleague Garrath Williams and I have argued that people who become obese should not be considered responsible or as imposing costs on others (Wickins-Drazilova and Williams, 2009), because the choices we make depend very much on the range of opportunities open to us and the costs of those choices. All sorts of organisations, including the state, continually influence the choices we make and alter the opportunities that are open to us. We may desire to adopt a healthy lifestyle, but we are constantly in the grip of strong but subtle currents guiding us to unhealthy behaviours. The main features of an 'obesogenic' society are that there is an abundance of cheap and easily accessible high-calorie food, and that sedentary activities predominate over manual labour and physical exercise.

Many states have recently introduced anti-obesity interventions to tackle the problem of rising obesity, such as providing support to offer healthier meals in schools. However, the governments give a contradictory message by supporting interventions that undermine healthier choices. An example of a powerful intervention that contributes toward an 'obesogenic' environment is the current system of the European Union agricultural subsidies.

The Common agricultural policy from a public-health perspective

The European Union intervenes extensively in the price of food, principally via the Common Agricultural Policy (CAP). The EU spends almost half of its annual budget on agricultural subsidies. The subsidies were introduced in many European countries after the Second World War in light of the experience of war and post-war food shortages. Its purpose is to ensure that Europe is self-sufficient in food production so that people don't ever starve again.

However, it has been pointed out that, over the years, CAP subsidies have led to overproduction, oversupply and overeating. Liselotte Schäfer Elinder argues that the current system of agricultural subsidies is perverse in terms of health. The EU system of subsidizing and supporting agriculture encourages massive overproduction, which artificially lowers the prices of foods. Schäfer Elinder claims that this makes the CAP policy counter-productive from the point of view of obesity. Although the 2003 CAP reforms have been aimed to break the link between subsidies and amount of production, they have had only an insignificant effect. The solution would be to phase out the market support of agricultural producers (Schäfer Elinder, 2005).

The UK Faculty of Public Health agrees with Schäfer Elinder in the sense that health needs to be considered when food policy is planned, as the current CAP leads to overproduction of 'unhealthy' foods. But the report doesn't call for scrapping the whole system of subsidies, but advocates gradual reforms leading to a 'healthy' CAP. Some of the changes the report suggests are lowering financial support of beef and dairy products, converting produce of fat into fuel and lubricants, and increasing financial support of fruit and vegetable production. A persuasive argument that the report points to is a figure showing that, while the WHO dietary recommendations state that fruit and vegetables should form a quarter of our diet, the CAP support of this healthy food group is lower than 10%; and while meat and diary products should also form a quarter of a healthy diet, over 50% of the CAP budget is spent on subsidizing this production (Faculty of Public Health, 2007).

The way the CAP distributes its support to various food groups needs to be changed from a health and obesity perspective. However, it still remains questionable what food groups deserve more subsidies. While fat is usually blamed for the rise of obesity, others have pointed out that weight gain can be blamed on increased caloric intake consumption caused by subsiding grain production (Pollan, 2006). Other authors have pointed out that it may not be food but sugary drinks have lead to the rapid rise of weight (Newby, 2007).

Leaving aside the consequences that changes suggested by Schäfer Elinder for farmers and rural communities: how would consumers, and their health, benefit? It is true that for example in the UK food prices have dropped in last 50 years in comparison to average income (National Statistics, 2009). But there is no guarantee that scrapping the system of subsidies would raise the price of food and lead to people eating less and/or healthier. It might lead to consumption of less meat and dairy products, but it would not necessarily lead to healthy eating of foodstuffs such as relatively expensive fruit and vegetables. Such reforms would probably not lead to healthier eating as people would tend to buy more food high in calories and low in nutrition. These changes would impact most on lower socio-economic groups, as for them food still takes significant proportion of their income.

The future of organic farming

With clear evidence still missing on what foods actually contribute to rise of obesity (Newby, 2007), and the risk of the economically vulnerable being pushed even more towards cheaper processed foods and nutritionally empty calories, what can be done? Even fruit and vegetables, which are clearly under-subsidized by the current CAP, aren't in themselves healthy if they are full of pollutants and lack most of their nutritional value. Although European farmers have to follow certain quality standards, and recent reforms tried to decouple subsidies from the amount of production, the CAP is still predominantly focused on quantity. Farmers are subsidised, as well as paid on the market, by weight, not the quality of their produce. However, as explained above, the European agricultural policy needs to be more focused on quality and lessening over-production. The EU already supports a system of farming that guarantees higher quality of food produce that fits within this profile: organic farming.

Organic farming is about the interconnection of the health of soils, plants, animals and humans (IFOAM, 2009). The principles of organic farming include crop rotation, and strict limits on chemical synthetic pesticide and fertiliser use, livestock antibiotics, food additives and GMOs, as well as stricter rules on animal welfare, thus maintaining quality of soil and water supplies, and protecting environments, biodiversity and rural communities.

Organic farming has developed greatly in Europe in the last 20 years. Recent statistics show that almost 4% of agricultural land in the EU is used for organic farming, with over 13% of agricultural land used for organic farming in Austria and almost 10% in Italy and Latvia (FiBL RIOA, 2009). Debates on the future of organic farming are focused on whether this form of agricultural production should stay a marginal alternative to conventional food production, or whether organic farming should be supported to become a mainstream or even the prevailing form of agriculture in Europe (Padel *et al.,* 1999).

The future of organic farming could be supported by the claim that it could be a useful anti-obesity measure. Therefore it is useful to consider the suggestion of a CAP reform towards larger support of organic farming. In five arguments I will explore how organic production might be beneficial to public health and reduction of weight-related diseases.

Obesity-related reasons for supporting organic farming

Firstly, organic farming allows farmers to produce smaller quantities of a better quality. Although there is still some doubt over the health benefits of organic food, it can be shown that industrial fertilizers increase productivity but decrease the amount of nutrients such as iron, zinc or calcium (Halweil, 2009). In general, due to better quality of soil, organically grown plants are richer in minerals, vitamins and antioxidants. The nutritional 'erosion' of food quality means that many people are undernourished and obese at the same time (*ibid.*). And although feeding animals growth hormones is now forbidden in the EU, meat from factory-farmed animals is often poor in essential nutrients compared to meat from free-range animals (Pollan, 2008). This is due to the fact that most animals in conventional agriculture are fed on grains while they need a more diverse diet (ibid). Bruce Ames suggests that nutritional deficiency might be what contributes to obesity. Our nutritionally poor diet makes our bodies 'hungry' in the effort to obtain essential nutrients (Ames, 2006). Eating more organic food that is higher in nutrients could tackle the rise of weight-related diseases.

Secondly, organic farming creates fewer pollutants than conventional farming, and pollution has been linked to obesity. Agriculture production has a large share in polluting the air, soil and water, as well as demanding fossil fuels. Our bodies are increasingly exposed to industrial chemicals: in food as well as in the environment, and a link between environmental chemicals (including compounds found in pesticides) and the increase of body weight has been documented (Newbolt *et al.*, 2008). Evidence also shows that antihistamines, taken to treat allergies, which can be linked to pollution, are also linked with weight gain (Aronne, 2002). Organic farming, on the other hand, bans the use of synthetic pesticides and fertilisers and prefers free-range farming of animals. It also requires less petrol and diesel. Thus it can be beneficial for all, as well as for the health of farmers as their work environment contains less chemical pollutants such as pesticides.

Thirdly, organic farming means better environmental land management, resulting in a more diverse and uncultivated countryside, which is more inviting for people to spend and enjoy time outside. Regulations such as keeping more animal outside enable leaving more land for walking and recreating outdoors. Organic farming also creates more diverse countryside, that is attractive for hiking, cycling and tourism, and indeed many organic farms benefit from eco-tourism. Although there is no requirement on organic farmers to set aside more lands than conventional farming, the system of wide and varied crop rotations allows soil recovery time. Unfortunately the recent changes in the 2008 CAP reform have gone the other way as the set-aside program has been scrapped.

Fourthly, organic agriculture employs more mechanical and physical methods as well as stricter animal welfare standards, which means that organic farming labour requirements are higher. A survey found that organic farming provides 32% more jobs per farm than equivalent non-organic farms. (UK Soil Association, 2009). So moving towards larger proportion of organic farming would improve employment and regional development. This measure could be beneficial from the perspective of weight-related diseases as obesity has been associated with unemployment and low socio-economic status in general. Although the connections involved may be complex, there are strong associations between occupation and weight gain (Ball and Crawford, 2005). So although low socio-economic groups cannot in general currently afford organic food, they can benefit from organic farming. The UK Soil Association study estimated that: 'If all farming in the UK became organic over 93,000 new jobs directly employed on farms would be created' (UK Soil Association, 2009).

The fifth suggestion would be in line with the original CAP mission of improving and maintaining food self-sufficiency in Europe. As well as subsidising farms and markets, the CAP could support people in growing their own food. Farms and local councils could be financially supported to rent allotments to people as well as organisations such as schools. Growing their own produce could mean an improvement of the diet, as people would be encouraged to grow, eat and appreciate more fresh and locally produced fruit and vegetables. That would lead to greater awareness among consumers of where food comes from. Gardening would also encourage fitness as it is a healthy exercise, as according to the UK government many people fail to exercise for the recommended minimum of 30 minutes a day, five days a week (NHS, 2009). Half-an hour of gardening on most days, which is approximately what is needed to keep an allotment, is regarded as a moderate exercise that burns over 150 calories (ibid). Thus increasing the availability of allotments could be a hugely useful intervention for tackling weight-related diseases, and this could be especially beneficial for low socio-economic groups, who cannot usually afford to buy fresh fruit and vegetables and do not often have big gardens to grow them.

Conclusion

The current CAP is counter-productive from a health and anti-obesity perspective. The CAP seems to be difficult to reform, and part of the problem is that CAP is funded totally from the EU budget, while other policies cost much more, but are funded out of national budgets. Truly 'healthy' changes to the CAP need to be introduced to help current health policies and anti-obesity interventions to be effective.

One of the solutions might be stronger support of organic farming. Organic agriculture isn't only about the production of healthier food: organic farming creates a less obesogenic environment. I am not claiming that tackling the rise of obesity and weight-related diseases should be the main reason for increased support of organic farming. Nor am I claiming that conventional farming is the main contributor to the current rise of people's weight. However, more support of organic farming needs to be discussed as a possible intervention that might reverse the current trend of increasing incidence of weight-related diseases. The EU could aim towards making organic farming as common as conventional means, thus making it cheaper than it is at the moment. The EU could also launch support campaigns for allotments, increasing the availability of healthy fresh food as well as encouraging healthy exercise.

References

Ames, B. (2006). Low micronutrient intake may accelerate the degenerative diseases of aging through allocation of scarce micronutrients by triage. Proceedings of the National Academy of Sciences: 103.

Aronne, L.J. (2002). Drug-induced weight gain: non-CNS medications. In: Aronne, L.J. (ed.) A Practical Guide to Drug-induced Weight Gain. McGraw-Hill: Minneapolis.

Ball, K. and Crawford, D. (2005). Socioeconomic status and weight change in adults: a review. Social Science and Medicine: 60.

The Faculty of Public Health and Birt, C. (2007). A CAP on health? The impact of the EU CAP on public health. Available at: http://www.fphm.org.uk/resources/AtoZ/r_CAP.pdf. Accessed 21 March 2009.

FiBL Research Institute of Organic Agriculture (2009). Organic and in-conversion agricultural land and farms in the European Union. Data results from 31.12.2007. Available at: http://www.organic-europe.net/europe_eu/statistics-europe.htm. Accessed 21 March 2009.

Halweil, B. (2009). Still no free lunch: nutrient levels in U.S. food supply eroded by pursuit of high yields. Worldwatch report from September 2007. Available at: http://organic.insightd.net/science.nutri.php?action=view&report_id=115. Accessed 21 March 2009.

IFOAM (2009). The principles of organic agriculture. Available at: http://www.ifoam.org/about_ifoam/principles/index.html. Accessed 21 March 2009.

National Statistics (2009). Family Spending 2007. Available at: www.statistics.gov.uk/cci/nugget.asp?id=1921. Accessed 21 March 2009.

Newby, P.K. (2007). Are dietary intakes and eating behaviours related to childhood obesity? A comprehensive review of the evidence. The Journal of Law, Medicine & Ethics 35: 35-60.

NHS (2009). Healthy living for everyone: 10-minute workouts. Available at: http://www.nhs.uk/Livewell/loseweight/Pages/Tenminuteworkouts.aspx. Accessed 21 March 2009.

Oliver, E.J. (2006). Fat politics: the real story behind America's obesity epidemic. Oxford University Press.

Padel, S., Lampkin, N. and Foster, C. (1999). Influence of policy support on the development of organic farming in the European Union. International Planning Studies: 4.

Pollan, M. (2006). The omnivore's dilemma: the search for a perfect meal and a fast-food world. Bloomsburry. London.

Pollan, M. (2008). In defence of food: An eater's manifesto. New York: The Penguin Press.

Schäfer Elinder, L. (2005). Obesity, hunger, and agriculture: the damaging role of subsidies. British Medical Journal 331: 1333-1336.

UK Soil Association (2009) Employment on organic farms. Available at: http://www.soilassociation.org/Web/SA/SAWeb.nsf/848d689047cb466780256a6b00298980/988dc846f1df6e508025716f00392620!OpenDocument. Accessed 21 March 2009.

Wickins-Drazilova, D. and Williams, G. (2009). Ethics and public policy. In: Epidemiology of obesity in children and adolescents: prevalence and aetiology. Springer series: Epidemiology and public health (in press).

Theme 2 – Climate change and the future of farming

The price of responsibility: ethics in a time of climate change

M. Gjerris and C. Gamborg
Danish Centre for Bioethics and Risk Assessment, University of Copenhagen, Rolighedsvej 25, 1958 Frederiksberg C, Denmark; mgj@life.ku.dk

Abstract

With the basic human ability to influence the lives of others follows an inevitable ethical responsibility. Quite often we choose to forget this, just following the norms of society or our own preferences. In many situations, this works fairly well. However, norms may be challenged and sometimes we might be obliged to set aside our own interests. Furthermore we might end up in situations where we acknowledge the ethical responsibility, but where we are in doubt about how to respond, since the elements of the situation are uncertain or we have conflicting values. Climate change is just such a situation. It obviously raises technical issues such as: how we can maintain our current transportation patterns and at the same time cut down on CO_2 emissions? But behind such a question and the possible answers to it, lie ethical choices based on our fundamental values. This paper examines two cases: Biofuels and GM plants. Both are often suggested as ways of coping with the inescapable impacts of climate change. The paper argues that part of our responsibilities is to set forth and discuss the broader societal visions and values implicit in these technologies before rushing to implement them after a brief search for any new specific ethical issues. The ethical discussions related to climate change are rather a question of understanding familiar problems in a new context, where climate change and its consequences accentuate the need for addressing some of the perennial ethical questions.

Keywords: biofuels, GM-plants, responsibility, visions, values

Introduction

Ethics involves complex and interwoven issues, because the problems, concerns and challenges that force us to think along ethical lines are intricate and connected with each other. This is one of the most obvious, yet overlooked truths in the age of applied ethics where every new human endeavour or every new technology seems to give birth to a new branch of specialized ethical reflection. Hunting for the special, unique and singular ethical issues that arise within this, and only this area, ethicists and sociologists gather at the feet of scientists to understand the technicalities and figure out what makes this area so special. More often than not the result is disappointing and once again the conclusion that the ethical issues found are the same as everywhere else: 'How to be good' and 'what is good'? Questions that are then left unattended because we are already moving on in search for the next issue that might contain unique problems thus legitimizing ethical reflection in the eyes of scientists and policy-makers.

Ethics has to some degree become a specialized branch of policy-making, twisted and turned as a discipline to meet the demands of those who have a vested interest in the developments of science and technology whether they applaud it or resent it. And it has become so specialized that there is more than a tendency of forgetting that often the most crucial ethical questions are those that can be found in all areas of human existence. And that ethics is not something you are done with at some point, but a continuous effort to understand our basic values in light of our current challenges (Gjerris, 2009).

If we forget this and only focus on what is 'new' and what is 'special' we miss the chance to see the interconnectedness of our problems and we fail to understand that our decisions within specialized areas of science and technology have to reflect broad societal values and not just the technically rational, but limited values that can enter a cost-benefit analysis. In this paper we will argue how and why it is problematic to forget that ethics is a complex and interwoven issue by focusing on two of the most frequently suggested ways of mitigating the consequences of climate changes: biofuels and GM plants. We point out that part of our ethical responsibility is to clarify and discuss the broader societal visions and values embedded in our decisions, just as we need to clarify the unique ethical challenges these technologies might hold before rushing in to them.

As Shue (2001) points out, a lot of the philosophical writing on climate change has focused on fundamental issues of distributive justice. Such issues are needless to say important, but other issues should be addressed as well. There is currently an increasing amount of literature on the ethics of climate change (e.g. Garvey, 2008; Jamieson, 2007), drilling out or adding to, the issues raised in the Buenos Aires Declaration on the Ethical Dimensions of Climate Change in 2004, and especially in the United Nations so-called white paper on the ethical dimensions of climate change (UN, 2006). However, before looking at some of these issues, we need to address what the challenges of climate change are.

The challenges of climate change

The global climate is changing, that much is indisputable. That human CO_2 emissions play a central role in this is also hard to deny, although it is possible to find dissent on this point. But looking at the overwhelming tendencies in the research that the International Panel on Climate Change (IPCC) under the auspices of the United Nations bases their conclusions on, it is hard to deny that humans seem to play a crucial role (Parry *et al.,* 2007). What will happen and when it will happen seems to be more of an open question. Right now it looks as if many of the estimates of IPCC from their 2007 report are in fact are too modest. Things seem to be happening faster and sooner than expected, e.g. the melting of the ice masses at the poles. Whether this is because the estimates were always too careful or whether new and unexpected feed-back mechanisms are beginning to show, is not for us to say. We will just briefly mention some of the more important consequences of the climate changes that IPCC predicts we will experience in the future: more extreme weather types, rising sea levels, changes in temperature and climate leading to changed living conditions for wildlife and plants, changed conditions for agriculture, increased pressure on fresh water infrastructure and increased problems with invasive species.

All these challenges mean that there will be increased pressure on the Earth´s ability to sustain human life. Drastic changes in lifestyle and culture seem to lie ahead. Even if we choose to believe that human CO_2 emissions from transportation, industry and agriculture has nothing to do with what is happening, we will still have to adapt to a planet that will change its weather drastically in the years to come. To adapt to these changes and perhaps mitigate them, two strategies in particular keep popping up in the debate. To adapt agriculture to the changing environmental conditions it has been suggested to genetically modify plants to make them more robust and able to cope with more extreme weather conditions. To mitigate the changes by cutting down on CO_2 emissions one of the most discussed opportunities is replacing fossil fuels with biofuels. We use the term 'biofuels' here as any solid (e.g. wood), liquid (e.g. biodiesel) or gaseous fuel (e.g. biogas) derived from relatively recently dead biomass, which is distinct from fossil fuels derived from long dead biological material.

Technical or ethical questions?

Both GM crops and biofuels are issues that are full of technical and scientific questions. What traits to modify for, which genes to use, what risks will the new plants constitute to the environment? What is the energy efficiency of biofuel technology, will it lower CO_2-emissions? These are questions that call for technical answers and risk analysis. But behind these questions lie other questions about the goals that we pursue by these technologies and the ethical values that they bring into play. Are they really the best answers to the questions that we ask these days about how to ensure food is on the table in the future?

Questions like this are bound to be closely connected to the values that we live our lives by. Our point here is to say that we miss out on this aspect, if we immediately decide that we can only solve the challenges ahead by using technologies such as these and then begin discussing how to use them most efficiently.

Ethics understood as the extraction of values from our presumptions about the solutions we seek to our problems is an indispensable 'tool' here. It is by reflecting ethically that we get a chance to see how the technical solutions are connected with values about lifestyle, technology or nature. But these issues are, as we mentioned in the beginning, not new issues. These issues can be found within many technologies and are therefore not especially connected to these technologies. The manipulation of nature that takes place in GM-technology has its precedents in conventional breeding and the threat of biofuel plants to fuel the needs of the rich, instead of feeding the poor are already present in today`s agricultural systems.

Case A: GM plants

Could – or should – genetic modification of agricultural crops be part of the solution to mitigate the negative consequences of climate change? Or phrased another way, is the use of GM plants the right way to meet our responsibilities in mitigating the effects of global warming on agriculture? Whichever way the question is phrased, the ethical issues raised are about issues such as: manipulation of nature, increased industrialized farming, dependency issues related to centralization of development of new varieties in a few major multinational companies etc. There are also questions more specific to the technology such as the risks and benefits of the individual varieties. These are questions specific to this area. But just as important are the questions of socio-economic effects, impact on the human-nature relationship and consequences for the organization of agriculture. And these are much broader questions transcending the often very limited discussions of GM technology, since they are also related to other kinds of technological development.

Since the first GM plants were created in 1983 they have been advocated as an almost obvious technology with the potential of substantial benefits, not least in terms of helping to feed an increasing global population estimated to reach 8 billion in 2030, if used 'appropriately' to achieve production increments in agriculture. For an equal amount of time GM technology and its applications in food production in particular have been the target of considerable public scepticism and severe criticism, cf. some of the issues raised in the previous paragraph (Jørgensen, 2009). A considerable proportion of the European consumers are reportedly critical towards the GM technology (Gaskell *et al.,* 2006)

When it comes to climate change, proponents argue that with a complex forecast of drought, cold spells, salt effects and diseases, enhanced crops suited to different (and changing) environments can be produced more quickly and efficiently through genetic modification with environmental, nutritional and economic benefits. That said, proponents of the use of the GM technology caution that so far the technology is not efficient enough – but it can be developed, especially if EU approval procedures are being speeded up (Holm, 2009).

On the other hand, it is argued by opponents, if we are to take seriously the challenge of contributing to a more environmentally sustainable and sociably just world, then we have to ask if GM technology is in fact the best answer to our present challenges, before we rush to lift the alleged 'red tape' surrounding GM crops approvals in the EU. Opponents do not necessarily rule out GM crops wholesale (depending on the type of objection of course) but call for careful consideration of a number of issues. One example is the question of Intellectual Property Rights (IPR). Should the GM crops based on locally adapted species be public domain or the intellectual property of a few multinational GM producers? (Jørgensen, 2009). Will the GM varieties actually result in increased yields, as there is an issue of poverty to address – in other words can the farmers using the GM crops also afford to pay for irrigation and fertilizers?

(Zerbe, 2004) Will proper risk assessments procedures be established in the developing countries, e.g. to allow for other, local production GM free strategies? (Cohen, 2005).

As it should be clear climate change provides a new context for some of the perennial ethical questions, such as: What is the right relation to nature? How far does our responsibility for the needs of other people extend? Who can rightfully own what? Questions that we cannot begin to answer before we have clarified our deeply held values, queried our assumptions and challenged our beliefs through thorough ethical (and political) discussions.

Case B: biofuels

Twenty or thirty years ago biofuels were considered a necessary and important field of research but not one which gave rise to controversies the way biofuels do today. Experiments were done with different species, the yield of various forms of biomass was measured in an attempt to find cost effective ways to produce biofuels. Moreover, environmental effects were assessed, e.g. looking at potential nutrient leakage from the cultivation system. In other words, research into biofuels was framed in terms of sustainability along the dimensions of ecology and economy, but much less on the social dimension. Today, almost the opposite seems to be the case.

Biofuels raise many questions of a technical character, some of which have been answered, but the big challenge seems to lie within the societal and ethical questions: Who should bear the costs and burdens of mitigating global warming and who should get the benefits – from an intra- and intergenerational perspective? (Felby, 2009) The view of FAO (2008) is that some of the key elements of concern in relation to the development of sustainable biofuels are food production and CO_2 emission reduction. First, however, a brief background on biofuels.

Biofuels are heralded worldwide as ways to reduce the reliance on fossil fuels, increase energy based on renewable resources and a way of mitigating climate change. But biofuels are also increasingly the target of severe criticism. It is claimed to cause hunger, use up much needed agricultural subsidies and to be inefficient in net CO_2 reduction (Bach, 2009). Biofuels are unique in the sense that it is the only type of renewable energy which can be used for heating, electricity and transport. In relation to the last category, transport, it is estimated by the International Energy Agency (IEA) that biofuels should be able to reduce CO_2 emissions by 30 to 50 percent compared with fossil fuels in cars and further down the line also in ships and aeroplanes. Bioethanol accounts for approximately 20% of all liquid biofuels within the EU, albeit part of this production is also based on imported biomass, from e.g. Brazil. These developments raise several ethical issues. 'First-generation biofuels' are made with conventional technology from feedstocks, such as wheat, which could instead enter the animal or human food chain. Second-generation biofuels use advances in conversion technology and are based on a variety of non-food crops such as straw, wood and dedicated bioenergy crops such as perennial grasses. As the biomass for second generation biofuels are a byproduct of e.g. food production, this production will not divert any additional land.

Hunger

A much debated topic is whether biofuels add to worldwide hunger (by using food crops as feedstock) or if they actually help in alleviating poverty and thereby hunger? The answers to these questions depend upon how the question is more exactly phrased, and the nature of the accompanying assumptions. Although it is difficult to identify the exact reasons for the steep increases in worldwide food prices between the autumn of 2007 and the summer of 2008, at least part of it seemed to be connected to the use of corn for bioethanol (Bach, 2009).

Subsidies and distribution of wealth

Will biofuels be a factor in terms of increasing employment and supporting economic growth? Again, some examples point to positive aspects: the rural poor in the third world getting energy and income at the same time by growing biomass for biodiesel. On the other hand, biomass is seldom refined in the poor countries, and intensive biomass production is said to have the same negative socio-economic effects in terms of unequal land distribution and power as sugar or oil palm production. And if production takes place in the more wealthy countries, issues about subsidies crop up; OECD (2007) has recently calculated that up to 50 percent of production costs are subsidies.

Reduction of CO_2 emissions

Biofuels are said to be able to cause a reduction in CO_2 emissions between 30 and 50 percent as compared with fossil fuels (Bach, 2009). However, allegedly taking up new land to produce the necessary biomass will mean an increase in emission of CO_2 of two to nine times more than the biofuels can decrease CO_2 emissions over a thirty year period through substitution with fossil fuels (Felby, 2009).

Many of the estimates that the above mentioned analysis rest on are subject to ongoing scientific debate. There are no clear answers. This goes to show how the decisions we are currently making often rely on uncertain knowledge. But the examples also illustrate the underlying ethical questions which need to be addressed when undertaking technologically based efforts to change energy production patterns to address climate change.

Conclusions

The global climate is changing. More extreme weather is coming and living conditions for plants, animals and humans will change profoundly the next 50-100 years. Climate change follows the pattern of other areas of human interest and technology and the societal, political and ethical issues related to climate change are thus also being discussed. Examples of this development are two of the most suggested mitigating technologies around: biofuels and GM crops. These technologies are both discussed from a technical and an ethical perspective.

The problem is that the increased specialization of ethics that has taken place since applied ethics became popular in the 1970s often prevents the deeper and more fundamental ethical issues from being discussed, since these issues are not unique for the technology in question. Thus what is common to all areas ends up being homeless in a discussion of climate where we always seek what is special for the different areas that provoke ethical discussions. This is problematic from an intellectual point of view as there is a need to discuss fundamental issues within new contexts if we are to deepen our understanding of them. But it is also a societal and political problem since the decisions we make nowadays about how to react to the coming climatic changes need to be rooted in foundations which are as transparent as possible. This is necessary since the decisions will have wide-reaching effects for human culture now and in the future. If we do not focus on the basic ethical questions and our responsibilities once again before rushing into technical solutions having only discussed the 'new' ethical issues, we risk taking decisions that do not reflect values of our societies – and the question then becomes, whether they will be effective at all or if they will eventually lead to a more fragmented society.

References

Bach, F.B. (2009). Biobrændstoffer: Sult, subsidier og manglende CO_2-effekt. In: Gjerris, M. *et al.* (eds.) Jorden brænder. Copenhagen, Alfa, pp. 161-164.

Cohen, J.I. (2005). Poorer nations turn to publicly developed GM crops. Nature Biotechnology 23, pp. 27-33.

Felby, C. (2009). Biobrændsler – Planter til mad og energi. In: Gjerris, M. *et al.* (eds.) Jorden brænder. Copenhagen, Alfa, pp. 155-160.

Food and Agricultural Organisation of the United Nations (FAO) (2008). Bioenergy, food security and sustainability – Towards an International Framework. Food and Agricultural Organisation of the United Nations. http://www.fao.org/.

Garvey, J. (2008), The ethics of climate change. Right and wrong in a warming world. London: Continuum.

Gaskell, G. *et al.* (2006). Europeans and Biotechnology in 2005: Patterns and Trends: Eurobarometer 64.3. Brussels, European Commission.

Gjerris, M. (2009). This is not a hammer – on ethics and technology. In: Bedau and Parke, E. (eds.) Our Future with Protocells: The Social and Ethical Implications of the Creation of Living Technology. MIT Press.

Holm, P.B. (2009). GMO: En løsning på ændrede klimaforhold. In: Gjerris, M. *et al.* (eds.) Jorden brænder. Copenhagen, Alfa, pp. 167-173.

Jamieson, D. (2007). The moral and political challenges of climate change. In: Moser, S.C. (ed.) Creating a Climate for Change. Cambridge: Cambridge University Press, pp. 475-484.

Jørgensen, R.B. (2009). GMO. Den rette måde at løfte ansvaret på? In: Gjerris, M. *et al.* (eds.) Jorden brænder. Copenhagen, Alfa, pp. 174-180.

Parry, M.L., Canziani, O.F., Palutikof, J.P., Van der Linden, P.J. and Hanson, C.E. (2007). Climate change 2007 Cambridge University Press, Cambridge.

Shue, H. (2001). Climate. In: Jamieson, D. (ed.) A companion to environmental philosophy. Oxford: Blackwell, pp. 449-459.

The Royal Society (2008). Sustainable Biofuels: Prospects and Challenges. The Royal Society. http://royalsociety.org/document.asp?latest=1&id=7366.

Zerbe, N. (2004). Feeding the famine? American food aid and the GMO debate in Southern Africa. Food Policy 29, pp. 593-608.

Climate change and the future of farming: challenges from urban agriculture

A.E.J. McGill
International Union of Food Science and Technology (IUFoST): Future for Food, 89 Melvins Road, Riddells Creek, Victoria 3431, Australia; albert.mcgill@futureforfood.com

Abstract

Climate change, population growth and the global economic downturn have had a negative impact on food security. A need to double the food supply by 2050, loss of existing arable land and increases in transportation costs have acted as stimuli for the development of urban agriculture. Some case studies are described and the related positive and negative impacts are discussed. System defects are highlighted and remedies proposed.

Keywords: vertical farms, food security, urban health

Introduction

Farming is under significant pressures to change. This applies whether considering the industrialised western approach or the more traditional methods used in developing countries. Changing weather patterns have demonstrated extreme variations producing droughts, floods, storms and increases in ambient temperatures (Lobell *et al.*, 2008). The consequent impacts of lower yields, higher costs and loss of species diversity have negatively affected food production. Further effects from the competition of biofuels and continuing losses of arable land to development have damaged food security. Lack of universal acceptance of alternatives such as genetic modification (GM) technologies has not ameliorated the problems (Brown and Funk, 2008). The changing world population balance has made this an increasing problem for urban dwellers (Table 1).

The combination of the effects of climate change, increasing scarcity and cost of fossil fuels, and the global economic recession have had varied impacts on different socio-economic urban groups. A wide choice of foods remains available to higher status groups less affected by price rises. Little and more limited choices, predominately from fast-food outlets, are the lot of those of lower status (London Food Strategy, 2006). Urban agriculture may provide a significant part of the solution as the results of a number of case studies indicate.

Table 1. Changing world population balance (FAO, 2008).

Urban population %	2007	2015	2030
World	50	53	60
Developed countries	75	76	81
Developing countries	44	48	56

Case studies and development programmes

Deltapark, Rotterdam, was one of the earlier concepts that would have housed 300,000 pigs, 1.2 million chickens, tens of thousands of fish and a giant vegetable growing area under one roof of a building comprising 200 hectares (Anon., 2006). This has yet to materialise.

The London Food Strategy (LFS) reported previously (McGill, 2006), is a food system fully integrated with the broader development programmes for that city, and is now at the implementation stage (LDA, 2007).

Daichi-o-Mamoru-Kai, Tokyo (McGill, 2006) remains active and successful in delivering produce directly from farm to customer within the city.

Dongtan Eco-City is an integrated city development southeast of Shanghai on Chongming Island, has a projected area of 1,200 hectares and is intended to hold a population of 80,000 by 2020. It will provide its own energy, food crops and treat its own waste. It should be completed by 2010 (Lim, 2008).

Eco-City Melbourne, Australia, is an environmentally friendly suburb of the city designed by the Victorian Eco-Innovation Lab (VEIL) at Melbourne University funded by the State government. It has similarities to the Dongtan project and includes urban agriculture through a vertical farming approach. This is a concept exhibited to the public in February 2009, but yet to be initiated (Dowling and Craig, 2009).

Organizations such as the Resource Centres on Urban Agriculture and Food Security (RUAF) have had active programmes for some time, for example, the Cities Farming for Future (CFF) (Van Veenhuizen, 2006) and their early workshop 'Growing Cities growing food' in 1999, which reported on 16 city case studies. Another active design programme run by Dr Despommier at Columbia University, USA, that emphasises the advantages of vertical farming (Chamberlain, 2007). Similar schemes were reported by Vogel (2008) and included the use of river based sites.

Discussion

The demands of expanding urban communities for not only enough food but also for a choice in its production and handling methods, encourages initiatives in developing urban agriculture and incorporating some of its elements into 'green' building design. On a large scale these have been proposed as eco-cities and some have reached the implementation stage. Such developments will pose challenges for both the urban dweller and the traditional farmer, with consequences for animal welfare. Although all of the projects cited have considered the agricultural/horticultural production processes quite well, almost none have explained how produce from greenhouses, vertical gardens, roof cultivations and basement livestock will be handled for further storage, processing or distribution. The existing food supply chain has well established links and hubs organised mainly by large retail conglomerates, none of whom seem to have been engaged in these projects during development. The present food supply chain has established systems to ensure the essentials of food safety in association with government regulations and compliance. The diversity of production proposed in most projects would require considerable tuning to reach even the present standards of quality that most consumers take for granted. The opportunities for the urban populations to become more intimately engaged with their own food production, processing and handling are strongly welcomed and could add significantly to both the social and economic dimensions of city living. Availability of a greater diversity of fresh produce would be of benefit to the diets and health, particularly of those in lower socio-economic groups. Only the London project (London Development Agency, 2005) has well documented evidence of dealing with the concerns raised here. These opportunities will come at the expense of traditional farming and rural communities which will add significantly to the drift away from farming and the loss of whole communities.

Conclusions

The holistic nature of the projects cited leads to them being grouped as eco-cities and being seen as a solution to many environmental problems of energy, water, transport and waste disposal management. With almost no exceptions, the projects seen to so far fail to consider all aspects of the food supply chain, particularly with respect to produce quality and safety. Greater involvement of food industry professionals will be required if the improvements to diet and resulting benefits to health of whole populations are to be realised.

References

Anonymous (2006). Delta dawn? Pigs, fishponds and crops – all housed in skyscrapers. Livestockhorizons 2(3): 4-5.

Brown, M.E. and Funk, C.C. (2008). Food Security Under Climate Change. Science 319: 580-581.

Chamberlain, L. (2007). Skyfarming. New York Magazine, 2 April.

Dowling, J. and Craig, N. (2009). Eco suburb plan unveiled for city. The Australian, 25 February.

FAO (2008). World Food Security and the Challenges of Climate Change and Bioenergy. 3-5 June, Rome, Italy. http://www.fao.org/foodclimate. Accessed 27 March 2009.

London Development Agency (2007). Healthy and Sustainable Food for London. Implementation Plan. http://www.lda.gov.uk. Accessed 4 January 2009.

London Food Strategy (2006). London Sustainable Food Hub. http://www.lda.gov.uk. Accessed 31 March 2007.

Lim, C. (2008). Dongtan Eco-City. FuturArc New Architecture 10: 40-43.

Lobell, D.B., Burke, M.B., Tebaldi, C., Mastrandrea, M.D., Falcon, W.P. and Naylor, R.L. (2008). Prioritizing Climate Change Adaptation Needs for Food Security in 2030. Science 319: 607-610.

McGill, A.E.J. (2006). Urban challenges and solutions for ethical eating. In: Kaiser, M. and Lien, M. (eds.) Ethics and the Politics of Food, Wageningen Academic Publishers, Wageningen, the Netherlands, pp. 368-375.

Van Veenhuizen, R. (ed.) (2006). Cities Farming for the Future – Urban Agriculture for Green and Productive Cities. http://www.ruaf.org/ Accessed 27 March 2009.

Vogel, G. (2008). Upending the traditional farm. Science 319: 752-753.

Animal agriculture and climate change: ethical perspectives

A. Nordgren
Centre for Applied Ethics, Linköping University, SE-581 83 Linköping, Sweden; anders.nordgren@liu.se

Abstract

There is a growing consensus among climatologists that humans – by emissions of greenhouse gases – cause global warming, and a report by the Food and Agriculture Organization of the United Nations points out the livestock sector as an important contributing factor. We might expect that mitigation of global warming to the extent it is caused by this sector would be of high political priority. In fact, this issue is almost completely neglected by politicians. Several organisations have responded to this report, however. In this research project, we investigate – from an ethical perspective – different views on how to mitigate climate change to the extent it is caused by animal agriculture. The project consists of three parts. The first part is an ethical analysis of reports, statements and academic articles. Four main views can be found in this material: intensive high tech animal agriculture, extensive organic animal agriculture, laboratory-grown meat (*in vitro* meat, 'meat without animals') and vegetarianism. We clarify and discuss critically the values and ways of resolving value conflicts implicated by these four views. In the second part of the project, we carry out interviews with representatives of key stakeholders in Sweden such as farmer associations, environmental organisations, political parties, vegetarians and proponents of *in vitro* meat. We investigate their views on the four options, their value commitments and approaches to value conflicts. The results of the first two parts constitute the basis for the third part in which we put forward policy proposals within the 'weak anthropocentrism' of the Swedish legal framework.

Keywords: meat production, global warming, values, value conflicts

Scientific background

This paper discusses the initial findings of a Swedish ethics project on how to mitigate climate change to the extent it is caused by animal agriculture. It begins by presenting the scientific background to this study.

As pointed out in the 2007 report by the Intergovernmental Panel on Climate Change (IPCC), it is becoming increasing clear that humans cause – or at least contribute substantially to – global warming. The extent of global warming is not clear, however. The Panel presents several scenarios representing different degrees of severity of the effects. The main cause is the emissions of carbon dioxide, mainly due to fossil fuel use (IPCC, 2007).

However, a report by the Food and Agriculture Organization of the United Nations (FAO) maintains that the contribution of the livestock sector to global warming is larger than previously estimated. The emissions of greenhouse gases from this sector constitute 18 per cent and are thereby larger than those from the transport sector, including aviation (Steinfeld *et al.*, 2006).

In its estimations, FAO focuses on the whole production chain from feedcrop to beef, transport not included. 9 per cent of the global emissions of carbon dioxide are related to changes in land use, especially deforestation. When forests are harvested or burned in order to give way for grazing land and feedcrop production – this is mainly the case in Latin America – large amounts of carbon dioxide are emitted into the atmosphere from vegetation and soil. The report claims that nearly three-quarters of the deforestation is caused by extensive beef farming. The rest comes from soy growing for animal feed. 37 per cent of the emissions of methane (which has 23 times higher warming potential than carbon dioxide) derive from enteric fermentation in ruminants (cattle, sheep). Moreover, 65 per cent of the emissions of nitrous oxide (which has 296 times higher warming potential than carbon dioxide) come from the livestock sector, mainly from manure (Steinfeld *et al.*, 2006). An illustrative example (from Japan) of the impact of animal agriculture on climate change is that production of 1 kg of beef leads to a warming potential equivalent to 36.4 kg of carbon dioxide (Ogino *et al.*, 2007). This corresponds to the amount of carbon dioxide emitted by a car every 250 km (Fanelli, 2007).

Although the problem is global, Sweden – the focus of this study – is also causally responsible, indirectly and directly. In Sweden, land use is not a problem. However, Sweden contributes indirectly to the emissions of carbon dioxide by purchasing feedcrops such as soy from Latin America (if we take FAO's deforestation perspective into account). Sweden also contributes directly by greenhouse gas emissions from its own animal production, although the size of this contribution is unclear.

Values and value conflicts

Given the serious and urgent character of global warming and the scientific facts about the contribution of animal agriculture to this warming, one might expect that mitigation of the global warming to the extent it is caused by animal agriculture would be of high priority on the political agenda. However, this issue is almost completely neglected by politicians. A clear example is the proposal of the Commission of the European Communities on how to reduce the greenhouse gas emissions up to 2020 (Commission of the European Communities, 2008). Another example is the recent official report on Swedish climate policy. Although it briefly mentions the FAO report, it does not propose any concrete measures for mitigation of climate change to the extent it is caused by animal agriculture (SOU, 2008).

Why is this problem of meat production neglected? In part, because it is new knowledge, and it takes time to digest and implement. More important is probably that unresolved value conflicts are involved. There is a history of strong support of meat production by the EU and the Swedish government. Due to these previous value commitments, it might be difficult for many politicians to accept that this sector contributes to global warming and to present measures to mitigate its impact. This is basically a conflict between the value of the established livestock sector and the value of the welfare of future human generations (which would motivate mitigation measures). An even more fundamental reason is perhaps that meat consumption is a sensitive area, closely related to the lifestyle and basic values of many people. Politicians might be afraid of steering people's lifestyle in a particular direction. This is basically a conflict between the value of individual autonomy (and privacy) and the value of the welfare of future human generations (motivating steering measures).

Four approaches to mitigation

However, values are not only of key importance in the neglect of the impact of the livestock sector on climate change. They are also essential in different views on how to mitigate it. A preliminary investigation indicates that there are at least four main views in this regard:

1. Intensive high tech animal agriculture
   FAO argues that extensive animal agriculture – mainly cattle grazing on pasture – is responsible for two-thirds of the greenhouse gas emissions, while intensive animal agriculture accounts for just one third. FAO therefore proposes a more intensive production, i.e. increased productivity in livestock production (in the narrow sense) and in feedcrop production. Moreover, the report describes closed cattle stalls in which the manure is collected and the methane is used for biofuel production. FAO also suggests that the emissions of methane and nitrous oxide can be reduced by improved animal diets. In addition, FAO proposes increased production of monogastrics such as pig and poultry

rather than ruminants, since the former have much lower emissions of methane and nitrous oxide (Steinfeld *et al.*, 2006).

2. Extensive organic animal agriculture
   Several organisations criticize FAO and claim that organic animal agriculture is a better way of mitigating climate change than intensive production. It is argued that organic animal agriculture has the potential to reduce greenhouse gases by a diet that is lower in protein and higher in fibre and by limiting the number of animals to the land available for manure application. In addition, it consumes less energy, mainly because it does not use chemical fertilizers (ISIS, 2007; IFOAM, 2004; HSUS, 2007). More specifically, a Swedish study has shown that organic beef production, in which cattle is raised on grass rather than on concentrated feed, may emit 40 per cent less greenhouse gases and consume 85 per cent less energy (Cederberg and Stadig, 2003; cf. Ogino *et al.*, 2007).
3. Laboratory-grown meat
   Proponents of *in vitro* meat have recently argued that this 'meat without animals' could possibly be a long term alternative to animal agriculture and lead to a reduction of green house gas emissions. Farm animal stem cells are obtained without killing any animals. These are then induced to differentiate into muscle cells that are cultured and finally organised into three-dimensional muscle structures (The In Vitro Meat Consortium, 2009).
4. Vegetarianism
   Vegetarians maintain that stopping animal production completely would be the best way to mitigate climate change as far as it is caused by this production. This goal would be achieved by consumers changing their lifestyle and no longer eating meat (International Vegetarian Union, 2009). A less radical solution would be to eat less meat.

These four views implicate different values and norms as well as different ways of resolving value conflicts. We have seen two examples of value conflicts above, namely the conflict between the value of the established livestock sector and the welfare of future human generations, and the conflict between the value of individual autonomy (and privacy) and the welfare of future human generations. But there are also other possible value conflicts such as: the welfare of the present human generation vs. the welfare of future human generations (involving values like sustainability, solidarity and intergenerational justice), and the welfare of developed countries vs. the welfare of developing countries (involving values like burden-sharing in the mitigation process and justice between rich and poor countries).

All four views have implications for how to resolve value conflicts like these. However, of particular importance to our study is the value conflict between the welfare of the present human generation and the welfare of animals in the livestock sector, and the conflict between the welfare of future human generations and the welfare of animals in this sector.

Objectives

With this preliminary investigation in mind, the objectives of the project are:
1. to clarify values, norms and ways of resolving value conflicts implicated by the four main views on how to mitigate climate change to the extent it is caused by animal agriculture: intensive high tech animal agriculture, extensive organic animal agriculture, laboratory-grown meat and vegetarianism;
2. to clarify values, norms and ways of resolving value conflicts implicated by the views of key stakeholders in Sweden – e.g., The Federation of Swedish Farmers, The Swedish Board of Agriculture and political parties – on how to mitigate climate change to the extent it is caused by animal agriculture;
3. to discuss critically the ethical arguments for and against the four main views and the views of the key stakeholders in Sweden;
4. to put forward policy proposals – within the Swedish legal framework of 'weak anthropocentrism' – on how to mitigate climate change to the extent it is caused by animal agriculture.

Method

The project combines empirical investigation and philosophical analysis. The former consists of a content analysis of reports, statements and academic articles as well as an interview study, the latter of a clarification of values (of different types), norms and ways of resolving value conflicts found in this material as well as a critical investigation of the tenability of the arguments.

The research consists of three parts:

1. An ethical analysis of reports, statements and academic articles on animal agriculture and climate change.
   First, we follow the development in the agricultural sciences very closely regarding scientific estimations of the impact of various types of animal agriculture on climate change. This is an important basis for our ethical analysis. Second, we investigate the four main options mentioned above with regard to explicit and implicit values and norms as well as different ways of handling value conflicts. In addition, four further parameters will be investigated.
   - *Potential for mitigation of climate change.* Vegetarianism and laboratory-grown meat have both obviously high potential compared to intensive and extensive animal production, because according to these former options no animals are to be used and consequently no greenhouse gases will be emitted from animal production. But what is the relative mitigation potential of intensive and extensive animal production?
   - *Handling of the increasing global demand for meat.* FAO estimates that meat consumption will double by the year 2050. This is mainly due to the increasing meat consumption in the growing middle class in many developing countries (Steinfeld *et al.* 2006). Should this increasing demand be met by producing more meat or by influencing people to eat less meat?
   - *Feasibility.* To what extent can the various options be expected to be effective means of mitigating climate change, given its global and urgent nature?
   - *Impact on animal welfare.* An important issue is how climate change can be mitigated without lowering animal welfare. Vegetarianism and laboratory-grown meat are both extremely animal-friendly compared to intensive and extensive animal agriculture, since they do not involve animals whatsoever. But what are the relative strengths of intensive and extensive animal agriculture regarding animal welfare?
2. Interviews with representatives of key stakeholders in Sweden on animal agriculture and climate change.
   We will interview representatives of key stakeholders such as The Federation of Swedish Farmers, The Swedish Board of Agriculture, organic farming associations, environmental organisations, political parties, vegetarians and proponents of *in vitro* meat. We are interested in their views on how to mitigate climate change to the extent it is caused by animal agriculture. There are three main reasons why we want to carry out this interview study:
   - A focus on Sweden is important, since this nation has a reputation of being one of the leading nations in environmental work and could be expected to be a leading nation also in the work of mitigating climate change to the extent it is caused by animal agriculture.
   - According to our preliminary investigation, there are almost no reports or written statements from Swedish key stakeholders on how to mitigate climate change to the extent it is caused by animal agriculture (see, however, LRF, 2009), therefore we need to interview representatives of these stakeholders in order to obtain knowledge of their values and ethical views.
   - In order to be realistic, our policy proposals for the Swedish government have to be based on knowledge of the Swedish context, and the key stakeholders' values and views are of special importance in this regard.

   The interview questions will be constructed on the basis of the analysis carried out in part (1) of the project. We will ask about their value commitments, their views on the four options and their evaluation of the four options in terms of the four parameters mentioned above. In particular, we will ask about their views on various concrete steering measures regarding meat production and meat consumption.

3. Policy proposals for the Swedish government
   On the basis of the results of part (1) and (2) of the project, we will put forward well-balanced policy proposals on how to mitigate climate change in animal agriculture. Our policy proposals will be based on the 'weak anthropocentrism' of the Swedish environmental and animal welfare legislation, rather than biocentrism or ecocentrism (cf. Des Jardin, 2005). This ethical approach is 'anthropocentric', because humans are allowed to use animals for their own benefit and because the welfare of future human generations is of central importance. It is 'weak', because animal welfare is to be taken seriously and because this sets certain limits to animal use: not all uses are acceptable and a balancing has to be carried out (cf. Nordgren, in press). Of key importance is the concept of sustainable development. We understand this concept in terms of the classical statement by the Brundtland Commission that sustainable development is to meet the 'needs of the present without compromising the ability of future generations to meet their own needs' (World Commission on Environment and Development, 1987: 85), but with the important addition: 'and without unnecessarily compromising animal welfare.'

Acknowledgements

This project is funded by the Swedish Research Council. Participants are Anders Nordgren (PI) and Henrik Lerner at Linköping University, and Bo Algers and Stefan Gunnarsson at the Swedish University of Agricultural Sciences.

References

Cederberg, C. and Stadig, M. (2003). System expansion and allocation in life cycle assessment of mild and beef production. International Journal of Life Cycle Assessment 8: 350-356.

Commission of the European Communities (2008). Proposal for a Decision of the European Parliament and of the Council on the effort of Member States to reduce their greenhouse gas emissions to meet the Community's greenhouse gas emission reduction commitment up to 2020. Available at: http://ec.europa.eu/environment/climat/pdf/draft_proposal_effort_sharing.pdf. Accessed 19 March 2009.

Des Jardin, J.R. (2005). Environmental ethics: An introduction to environmental philosophy. Wadsworth Publishing Company, Belmont, USA.

Fanelli, D. (2007). Meat is murder on the environment. Available at: http://environment.newscientist.com. Accessed 19 March 2009.

Humane Society of the United States (HSUS)(2007). An HSUS Report: The impact of animal agriculture on global warming and climate change. Available at: www.hsus.org/web-files/PDF/farm/hsus-the-impact-of-animal-agriculture-on-global-warming-and-climate-change.pdf. Accessed 19 March 2009.

Institute of Science in Society (ISIS) (2007). Mitigating climate change through organic agriculture and localized food systems. Available at: www.i-sis.org.uk/mitigatingClimateChange.php. Accessed 19 March 2009.

Intergovernmental Panel on Climate Change (IPCC) (2007). Climate Change 2007: Synthesis Report. Available at: www.ipcc.ch. Accessed 19 March 2009.

International Federation of Organic Agriculture Movements (IFOAM) (2004). The role of organic agriculture in mitigating climate change. Available at: www.ifoam.org/press/positions/pdfs/Role_of_OA_migitating_climate_change.pdf. Accessed 19 March 2009.

International Vegetarian Union (2009). Meat eating and global warming (links). Available at: http://www.ivu.org/members/globalwarming.html. Accessed 19 March 2009.

Lantbrukarnas Riksförbund (LRF, 'The Federation of Swedish Farmers')(2009). Fakta om kött och klimat. Available at: www.lrf.se/vi-arbetar-med-/notkottsproduktion-/fakta-om-kott-och-klimat--/. Accessed 19 March 2009.

Nordgren, A. (in press). For our children: The ethics of animal experimentation in the age of genetic engineering. Rodopi, Amsterdam, The Netherlands.

Ogino, A., Orito, H., Shimada, K. and Hirooka, H. (2007). Evaluating environmental impacts of the Japanese beef cow-calf system by the life cycle assessment method. Animal Science Journal 78: 424-432.

Statens offentliga utredningar (2008). Svensk klimatpolitik ('Swedish climate policy'). SOU 2008:24. Available at: www.regeringen.se/content/1/c6/09/96/94/8393cd02.pdf. Accessed 19 March 2009.

Steinfeld, H., Gerber, P., Wassenaar, T., Castel, V., Rosales, M. and De Haan, C. (2006). Livestock's long shadow: environmental issues and options. Food and Agricultural Organization of the United Organization (FAO), Rome. Available at: www.fao.org/docrep/010/a0701e/a0701e00.htm. Accessed 19 March 2009.

The In Vitro Meat Consortium (2009). Why in vitro meat? Available at: http://invitromeat.org/content/view/12/55/. Accessed 19 March 2009.

World Commission on Environment and Development (1987). Our common future. Oxford University Press, New York, USA.

Animal ethics in a time of climate changes

H. Röcklinsberg
Dept. of Ethics, Centre for Theology and Religious Studies, Lund University, Sweden; Helena.Rocklinsberg@teol.lu.se

Abstract

According to recent studies such as the scientific background of the ICPP climate report, the question is not *whether* but *when* changes in weather type, temperature and sea level will occur. Therefore, preparation and adjustments in a range of fields are needed. Regarding farming, the role of animals is discussed, as well as the options for securing their future welfare. Different scenarios are possible. Three possible lines of changed praxis are described, and the paper explores what ethical rationales could underpin each scenario. According to the first scenario, increased efficiency in farming is recommended. This solution combines efficient breeds and the 'least polluting' species. In ethical reasoning this view would mainly have utilitarian support. In the second scenario, husbandry decreases but is restricted to species that could be used for food *and* in pharmaceutical industry, e.g. bioreactors. This scenario includes an ethical rationale related to 'complex usage' combined with utility. A third scenario would be to adopt an organic standard, optimizing or harmonizing animal numbers and species to the available land, while concentrating on vegetable and crop production. In this third scenario, 'sustainability' and an ethical contract view are vital. Standpoints regarding these ethical elements influence arguments about the role of animals in a warmer future, and are thus highly relevant to overall decisions on acceptable new praxis.

Keywords: animal ethics, climate change, food insecurity, intensive farming, bioreactor, organic farming, species-specific behaviour

Introduction

In January 2009 Rajendra Pachauri, chair of the United Nations Intergovernmental Panel on Climate Change, IPCC, himself a vegetarian, fuelled the climate debate by stating that reduced meat consumption would be as necessary as reduced transport, since meat production itself causes about 20% of global greenhouse gas emissions (FAO, 2009). Moreover, food security, as well as food safety will be issues of increased importance due to future climate changes. According to the Food and Agriculture Organisation (FAO) of the United Nations (FAO, 2009), rural communities, particularly those living in already fragile environments, face an immediate and ever-growing risk of increased crop failure, loss of livestock, and reduced availability of marine, aquaculture and forest products.' Since about one third of the total greenhouse gas emissions can be traced back to the agricultural and forestry sectors, essential solutions are to be found in these areas, not least since they are closely related to land use. For instance about 70% of the world's agricultural land is used for livestock production, including grazing and growing crops for feed. Increased sustainability in systems for land and manure management as well as water use is needed to mitigate against climate changes and meet new demands for food security. Food products of animal origin have a higher climate impact than vegetables (livestock constitute nearly 80% of all emissions from agriculture), (FAO, 2007: 112), hence land and farm management practices need to be established that mitigate negative climate changes. On a theoretical level, an effective and quick solution would be to refrain from all animal production, using land for vegetables with low level of greenhouse gas emissions. Such a vegan world would easily meet the interests of certain groups, and possibly be the most animal friendly solution following the logic of 'no life no suffering'.

However seen on a more realistic level, animal products will continue to be part of agriculture. Not only because of the demand for meat and milk, but also because the benefits for the production cycle, not least in a sustainable management system refraining from artificial fertilizers, and for biodiversity in graze land.

Given that animals will continue to be used in farming, what ethical considerations for human-animal-relationship arise in relation to climate changes? Related to this one has to consider what husbandry systems will be promoted as acceptable adaptation to climate changes, and what impact these could have on animal welfare. In this paper I will briefly discuss what role animals may play in agriculture adapted to climate changes within three different possible scenarios, and highlight what ethical considerations would arise and need to be considered. Although linked to FAO's suggested new practices for mitigating global warming the following scenarios are freely formulated as possible aims for a society concerned with climate change. Due to the ongoing work of formulating international comprehensive climate policies, it is difficult to find actual political decisions taken on farming in any society. Hence three different ethical rationales or interpretations of what is a responsible and sustainable management of animal husbandry are presented through the scenarios.

Scenario 1: increased intensity and fewer species

The first scenario takes its point of departure in the FAO statement that 'by far the largest share of emissions come from more extensive systems, where poor livestock holders often extract marginal livelihoods from dwindling resources and lack the funds to invest in change. Change is a matter of priority and vision, of making short-term expenses (for compensation or creation of alternatives) for long term benefits.' (FAO, 2007: 114). Hence intensification is seen as one option for decreased emissions, given that it is combined with sustainable land management and careful choice of species. It is well known that methane gas from ruminants constitutes a large part of total methane emissions, and one option is to radically reduce ruminant livestock in favour of increased production of broiler, fish and other aquaculture. Although aquaculture and fisheries are themselves widely threatened by climate change in some geographical areas, intensification in other and land based fish production can be further developed to cover a larger part of the world's food consumption of animal protein.

What would this scenario mean for animal welfare, and more generally, in terms of animal ethics? Intensification does not by necessity imply lower welfare for each single animal. In general it does however imply higher density as well as striving for lower cost and higher yield per animal. A utilitarian perspective is well suited to underpin such a husbandry system. An overall positive outcome can be achieved on an environmental level since these species cause lower emissions than large ruminants, as well as by better land management by reducing overgrazing. It would also imply more land for crop production, and hence create options to contribute to reduced food insecurity.

A potential difficulty caused by mitigating climate changes through intensive farming of certain species relates to how to handle a situation of food insecurity and demand for a minimal level of animal welfare. In a utilitarian animal ethics equal interests are given equal consideration, which means that in a case of acute food scarcity, human need for food is more important than animal interest in a certain cage size. Both needs are regarded as basic needs, the latter however of less significance than essential food supply. In a normal situation however, utilitarian animal ethics argues that human basic needs of food can be met without animal products from husbandry systems compromising with animal welfare.

A further gain in utilitarian terms regarding intensive aquaculture and broiler production above the aforementioned advantages is the increased number of animals. Taken that these animals have more positive than negative experiences the bare fact that they are alive adds to the overall positive result. This is especially so since there will be a higher number of poultry and fish individuals than of ruminants.

Moreover, even if a number of the individuals concerned have low welfare, the overall positive outcome outweighs their negative experiences.[2]

From another ethical standpoint however, it would be difficult to accept increased intensity in broiler and fish farming solely by its positive outcomes on a general level. Adherents to the view Sandøe and Christiansen (2008) call 'respect for nature' (e.g. by Holmes Rolston) would argue the individual animal's possibility to perform species-specific behaviour is too restricted, e.g. in intensive fish farming where fish do not have the possibility to move over long distances. A related argument against intensive husbandry of any kind is given by animal rights theories, i.e. deontological animal ethics, which argues that each animal should be respected in itself, and has an intrinsic value. In an intensive broiler system the wellbeing of each hen might be very much compromised, i.e. the single hen is not respected and regarded as important in itself. Hence none of these views would accept clustering of individuals in an ethical balancing process. Contrary to the respect for nature view however, animal rights theories also argue for the right to life, and dismiss all instrumentalisation of animals. Therefore any farming system that aims to use animals for the sake of human preferences is regarded as unethical.[3]

In summary it seems the utilitarian perspective is the most suitable to meet food insecurity through intensive aquaculture and broiler production. Balancing usage of new technology and sustainable land use on the one hand with the basic needs of animals and the number of animals on the other, could result in an argument in favour of intensive aquaculture and broiler production.

Scenario 2: double role of animals

In a second scenario the main concern is to meet as many human needs as possible while using animals. The aim is to strive for efficiency on as many levels as possible. According to this scenario agriculture's contribution in mitigating against climate change would be to reduce the number of production animals. The animals should, however, be as efficient and useful as possible. The main focus then lies on breeding animals with double roles, such as sheep or chickens as so called bioreactors. A transgenic sheep may produce proteins for human pharmaceuticals in its milk, and a genetically modified chicken may have certain valuable proteins or antibodies in its eggs.

Due to climate change human health will be under threat in new ways, such as new pests and diseases caused by viruses developing in higher temperature or humid climate. One efficient way to create new medicines is to produce them in animals. These animals are of course much more expensive than ordinary sheep and layers and cannot be kept in as large numbers as today's farming. Hence we should only breed and keep bioreactors, which could not only help cure humans, but also serve as food.

Here an ethical argumentation similar to the one connected to the first scenario is appropriate. A utilitarian approach could be used to defend such a combined husbandry and pharmaceutical system. But the focus lies even more so on the most efficient use of each animal. When the climate is warmer and each animal adds to global warming by its mere existence the most useful way of using animals, without compromising too much of their welfare, is called for. It could be supposed that the welfare of each individual animal is better looked after when it is much more expensive, and its health is important also for human health. Thus, in utilitarian terms this scenario comes close to a win-win-situation.

---

[2] Peter Singer is the most well-known utilitarian defender of respect for animal preferences. *Animal Liberation*, 2nd edn. Thorsons, London.

[3] Tom Regan is the most well known animal right's defender. *The Case for Animal Rights*. Routledge, London.

In comparison, a 'respect for nature' view would have difficulties accepting genetic modification of animals, regarding it as unnatural. Also an animal rights person would be sceptical, since the modification serves human interests, not the animal itself. Furthermore, the argument of better care that is derived from a higher value would be met with the view that the animal is not regarded more valuable in itself, but as a result of its instrumental role. As soon as it is not a useful bioreactor, it would be turned into slaughter for meat. This is, on the other hand, exactly what convinces a utilitarian – the usefulness that is derived from the animal's double role.

Scenario 3: organic farming

According to a third scenario the best way to mitigate the negative changes of climate is global conversion to organic farming of all farmland. Organic farming has a higher resilience to changes in temperature, water supply and other factors of environmental stress thanks to, among other things, well developed recycling systems and higher biodiversity. These aspects show the importance of animal husbandry; (closed) cyclic systems build on use of the farm's own resources, and biodiversity on a high number of different species interacting on the farm for farm sustainability and higher soil fertility. Husbandry would need to be adjusted to land capacity since organic farm animals are kept to a large extent outdoors. Performing species-specific behaviour reduces stress levels, enables positive experiences, and causes these animals to contribute to farm management (e.g. pigs preparing fields) and improves biodiversity (e.g. in graze land by keeping chicken and ruminants together). In addition to positive environmental effects, organic farming is closely linked to other values such as social responsibility, fairness and human and animal health. The third scenario aims at creating a holistic and sustainable farming system that as such develops a win-win exchange for humans, animals and environment.

Hence one central ethical rationale behind this scenario is the understanding of sustainability in two dimensions; environmental and social (which can be seen as a solid basis for economic sustainability). Furthermore it can be related to an ethical contract view arguing that domesticated animals are useful for humans, and that domestication has been useful for animals. Humans have the right to use them for some purposes, but parallel to a contract, also have direct obligations to the animals such as respecting their innate behavioural needs in regard to environment, feed, social and sexual behaviour and may not inflict pain or cause stress (Lund *et al.*, 2004). In Aristotelian terms this would be to respect the essence and the telos of the animal by enable it to flourish. In line with this, Martha Nussbaum elaborates on a 'contractarian capabilities' approach to animal ethics, which considers animals as having basic entitlements, being entitled to justice in terms of fair treatment (Nussbaum, 2007). Showing this kind of respect built on contractarian thinking by implying a strict reduction of production animals adjusted to land capacity, fits well with an adaptation to climate changes through conversion to organic farming.

One obvious difficulty with contractarian ethical models for human-animal relationships is the fact that animals cannot partake in an agreement in the form of a contract, or any other theoretical act of reciprocity. They may however have representatives, or the humans in charge may 'think from the position of the individual animal' before action is taken in any direction. This is not easy, and different humans may come to different conclusions. In order to counterbalance such relativism, respect for species specific behaviour becomes a core role as an important translator of and nuanced way to mirror animal welfare.

Summary and conclusion

The impact of livestock on climate is undisputable. Much less obvious is how to handle this impact, not least since food insecurity is one consequence of global warming. The three scenarios sketched show optional ideals to strive for when mitigating climate changes. Different ethical aspects were discussed, and different ethical theories were used to elaborate the possible rationales. It has not been the aim of this ethical inquiry to argue in favour of any of the three scenarios, nor for any of the ethical theories. It is important to note however, that the more animal integrity is regarded as ethically relevant, the less intensive the farming system. Thus responses to the impact of livestock production on climate change, and the choice between the three scenarios seems to depend on this issue, and how it is interpreted, rather than on the choice of ethical theory.

References

FAO (2009). Climate change and Food security. Statement on www.fao.org/climatechange, March 18, 2009.

Ilea, R.C. (2009). Intensive Livestock Farming: Global Trends, Increased Environmental Concerns, and Ethical Solutions. Journal of Agricultural and Environmental Ethics 22 (2) Springer, pp. 153-167.

Lund, V., Anthony, R. and Röcklinsberg, H. (2004). The ethical contract as a tool in organic animal husbandry. Journal of Agricultural and Environmental Ethics 17(1), 23-49.

Nussbaum, M.C. (2007). Frontiers of justice. Disability, Nationality, Species Membership. Belknap, Harvard University Press, Cambridge.

Regan, T. (1988). The Case for Animal Rights. Routledge, London.

Sandöe, P. and Christiansen, S.B. (2008). Ethics of animal use. Blackwell publishing, Oxford.

Singer, P. (1991). Animal Liberation, 2nd edn. Thorsons, London.

United Nations Intergovernmental Panel on Climate Change (ICPP) (2007). Livestock's long shadow: Environmental issues and options.

Theme 3 – The ethics of bioenergy: research and technology development

Technological innovation and social responsibility: challenges to Bavarian Agriculture and the provision of bioenergy

Ch. Dürnberger[1], B. Formowitz[2], H. Grimm[1] and A. Uhl[2]
[1]*Institut TTN (Institute Technology, Theology and Natural Sciences), D-80335 Munich, Germany; herwig.grimm@elkb.de*
[2]*Technologie und Förderzentrum Straubing (Technology and Support Centre, D-94315 Straubing, Germany; Anne.Uhl@tfz.bayern.de*

Abstract

The objective of the interdisciplinary project described in this paper was to explore whether the production of biomass for energy can be ethically sound and live up to sustainability criteria. In a first step, an Ethical Delphi was carried out with an interdisciplinary body of 27 experts, who answered three questionnaires on bioenergy. Generally the experts agreed that biomass production for energy can substantially contribute to meeting future energy demand. Nevertheless, they did not consider bioenergy as a universal remedy for energy scarcity, but rather as one element in a comprehensive portfolio of energy sources to ensure a sustainable energy policy on the local, national and international level. In the next step of the project the results of the Ethical Delphi will be correlated with paradigmatic and normative concepts that influence the debate in order to create an 'Ethical Matrix' tailored to bioenergy. Detailed case studies will finally deliver guidance for ethically sound bioenergy in Bavaria. This article presents the first results of the Ethical Delphi, method descriptions and a working hypothesis that presents a real option for an ethically sound, decentralised bioenergy production in Bavaria.

Keywords: energy from biomass, local production, sustainability, Ethical Delphi

Introduction

Energy shortage and climate change have made the search for alternative energy sources one of the critical challenges of our times. Among the possible alternatives, energy from biomass has inspired a particularly controversial ethical debate revolving around agriculture's social responsibility to feed people and the need for ethically sound ways to produce bioenergy (MNP, 2008; Mitchell, 2009). Particular attention has been paid to the exploration of these issues for decentralized production of bioenergy. An interdisciplinary project of the Institute Technology Theology Science (TTN) and the Technology and Support Centre (TFZ; Southern Germany) aims at identifying ethically relevant criteria for contrasting and comparing alternative energy supply scenarios. This will allow reflection on the value of influential concepts such as 'local production', 'agricultural development', or 'climate friendliness' of energy from biomass, and contribute to clarifying the controversial debate. In a first step, a 3-phase 'Ethical Delphi' (ED) was carried out that allowed controversial issues to be structured according to expert opinion. A body of experts, including natural scientists, representatives of the humanities, decision-makers, and agricultural practitioners, were invited to share their views on the issue and to investigate decision-making processes. In a second step, the results of the ED will be correlated with paradigmatic and normative concepts that influence the debate in order to create an 'Ethical Matrix' (Mepham, 1996, 2006) tailored to energy from biomass. This Matrix will make the communication between stakeholders, scientists and consumers more transparent through visualizing the fields of tension. It will also provide foresight of future conflicts, pointing out the ethically relevant issues. Detailed case studies will finally deliver guidance for ethically sound bioenergy from biomass in Bavaria. Overall, the project consists of three parts which are the ED, the 'Ethical Matrix' and a 'citizen conference'. This article points out first results of the applied method (ED) and formulates the working hypothesis that local production presents a real option for an ethically sound biofuel production in Bavaria within limits.

Methodology: the ethical Delphi

The first phase of the project was conducted according to the methodology of the 'Delphi,' originally designed to combine the knowledge and abilities of a diverse group of experts. This method has been further developed for moral issues and described by Millar *et al.* (2006). An ED is an iterative participatory process between experts for exchanging views and arguments on ethical issues. Anonymity of the participants is central to the process. The method is structured around the notion of a virtual committee where the exchange of ideas is conducted remotely through a series of opinion exchanges. The ED within the project was carried out in three steps, each of which investigated experts' opinions on particular aspects of the bioenergy debate. It was held from August 2008 to March 2009. Its aim was not only to gain insight into expert opinion but bring scientific facts and societal and moral issues together. The experts' assessment provides a solid basis to identify the most important and controversial topics and to frame the subsequent research accordingly.

Expert identification and recruitment

Since the recruitment of experts is of crucial importance to the process of the ED, significant effort was expended for bringing together a balanced group of experts. Therefore, the experts were identified according to the following four key criteria (Millar *et al.*, 2006): Experts should (a) reflect a wide range of institutional interests and perspectives; (b) present a diversity of viewpoints; (c) represent a wide knowledge base; (d) hold a range of worldviews and ethical positions. Following these criteria, a manageable group of 41 experts were identified and invited to participate. Twenty-seven of them replied positively (66%). Since the project focuses on the German – respectively Bavarian – context, only German speaking experts were invited from Germany, Austria and Switzerland. The group of participants consisted of 22% natural scientists (e.g. meteorologist, agricultural scientists), 22% humanities and social scientists (e.g. theologians, ethicists), 22% representatives of interest groups (e.g. farmer associations, consumer groups), 19% entrepreneurs, and 15% others (e.g. politicians). All experts have experience in the field of bioenergy or related subjects.

Stages of the ethical Delphi process

The experts where introduced to the three-stage process and asked to answer online-questionnaires that were sent by e-mail. In the first round they answered ten general questions and indicated their degree of expertise on the issues. The first questionnaire aimed to gather general standpoints, perspectives and interests, while providing the opportunity to comment on the questions and add important missing aspects. The questions addressed the following topics related to the biomass production for energetic use: potentials and risks, politics and economy, fields of social conflicts, ethical aspects, relevant actors, technological potentials, sustainability, research, and alternatives. To give an example, question number six (Q6) asked for technological potentials as follows: *What technologies can substantially contribute to the provision of energy from biomass?* The correlating answers named technologies such as: *hydrated vegetable oil, biomass-to-liquid, ethanol, biogas, etc.* One question also asked for additions and topics not covered by the questionnaire.

These qualitative responses were clustered in order to identify relevant categories. Subsequently, hypotheses were formulated that correlated with each of these categories. E.g. category environmental protection: *The biomass production for energy significantly contributes to environmental protection.* Or for the category mobility and transportation: *The future of bioenergy is its use in biofuels for transportation.* This process resulted in 58 hypotheses plus related sub-hypothesis.

In the second round of the ED, the experts were presented 58 hypotheses and were asked for their degree of agreement within the spectrum of 1 to 5 (1 = agree; 5 = disagree). If the expert did not feel qualified enough to give an answer or not comfortable with the hypothesis, the opportunity was provided to abstain from judgement and to comment on the hypothesis. This round allowed a *quantitative assessment* of the experts' opinion on particular topics, with respect to the backgrounds of the experts. In a concluding round, the experts were sent the results, giving them the arithmetic mean, and percentage of agreement. They were asked to comment on the results and give a general feedback on the entire ED. Particularly the results of the second round are of interest for the debate and will direct the future project.

Results of the ethical Delphi

Potentials

Starting with the potentials, the experts believed that bioenergy had a positive effect on agriculture (arithmetic mean 1.9). Representatives of the humanities tended to hold a more pessimistic view on the issue than the average participant (2.5) and especially the entrepreneurs (1.7). However, the experts agreed on the hypothesis that the future of bioenergy lies in its stationary use (1.7) and not in fuel for transport (3.7). Outliers in the latter case were the practitioners (1.8). Also on the hypothesis of a direct relation between food scarcity and bioenergy, a difference between the opinions of practitioners and scientists was apparent. Practitioners tended to disagree on the hypothesis that bioenergy production pushes food scarcity (4.3), whereas the scientists saw a connection (2.3).

Risks

The question, whether bioenergy production has a negative effect on biodiversity, was answered diversely. Only the representatives of the humanities, working in the field, consented that bioenergy threatens biodiversity. When estimating the risks, we found strong disagreement of representatives of the humanities and practitioners. However, perceptions of potentials and risks were strongly dependent on region, infrastructure, species and technology. Therefore, it is of great importance to judge individual cases only.

Politics and economy

A clear tendency to consider political regulation of bioenergy production as a necessity for sustainability was prevalent among stakeholders. Particularly experts from the humanities agreed on the need for political regulation (1.0). Correspondingly, when asked whether future development should be left to the free market, experts strongly disagreed (4.4). Regarding subsidies as a governance tool, 80% of experts agreed on the hypothesis that only technologies with a positive impact on climate should be supported. In this category experts used their option to give feedback, indicating that political and economic issues should be addressed on a case by case basis due to the large number of unknowns.

Societal aspects

Amongst others, one topic of concern was the use of genetically modified plants. 66.7% of the experts agree that GMOs should not be used to produce energy from biomass. The hypothesis that the use of GMOs can never be sustainable polarised the group. Again, the practitioners thought differently. They did not agree that the cultivation of genetically modified plants can never be sustainable (4.0). This opinion is far from the average among all participants (2.9). Asked for the most relevant dimensions of the positive development of bioenergy, experts consider the social and ecological dimension of sustainability very important (Figure 1).

Ethical aspects

Compared to other topics, experts had rather similar opinions on ethical aspects. For instance, 83% agreed to the need to search for and develop bioenergy technologies because of our moral responsibility for future generations. Also the hypothesis that 'food safety has priority' gained strong agreement and brought very clear results with 96% agreeing. Furthermore, the experts concurred that too little attention has been paid to international justice (1.7). 58% thought that the current energy consumption in the industrial states is immoral. 88% agreed on the hypothesis that bioenergy should only be produced according to international sustainability standards. However, regarding justification of energy needs to interfere with individual freedom, e.g. birth regulation (arithmetic mean 4.1), or restriction of mobility (3.2), opinions ranged more widely (Figure 2).

Relevant actors

Figure 3 illustrates the experts' opinion on whether specific actors play a significant role and whether they should be considered more seriously in the debate.

Technology and practice

Regional production of biomass was considered a necessary precondition for sustainable bioenergy (70%). Interestingly, regional production was considered be the most promising way not only within

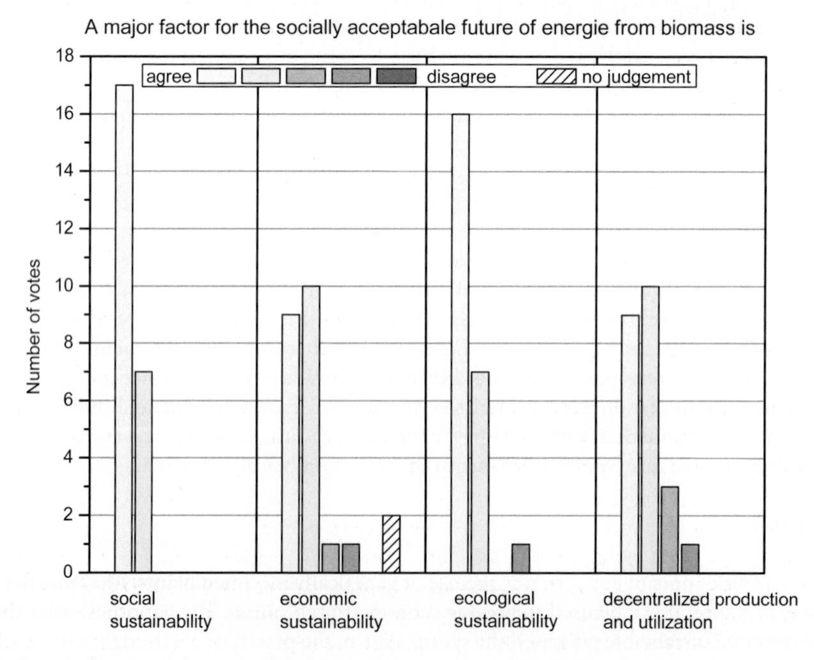

Figure 1. Criteria for societal sustainability.

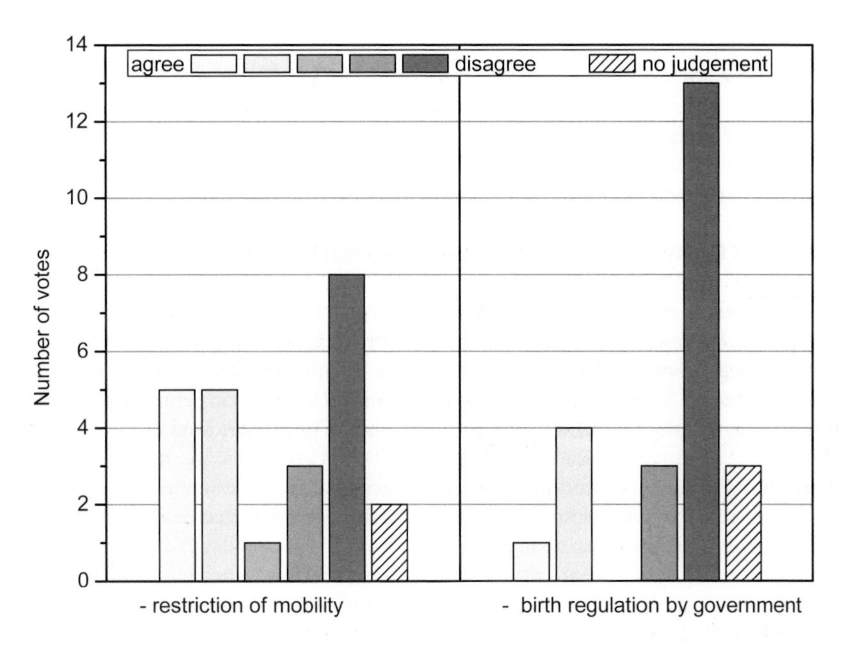

Figure 2. Energy problem and limitation of personal freedom

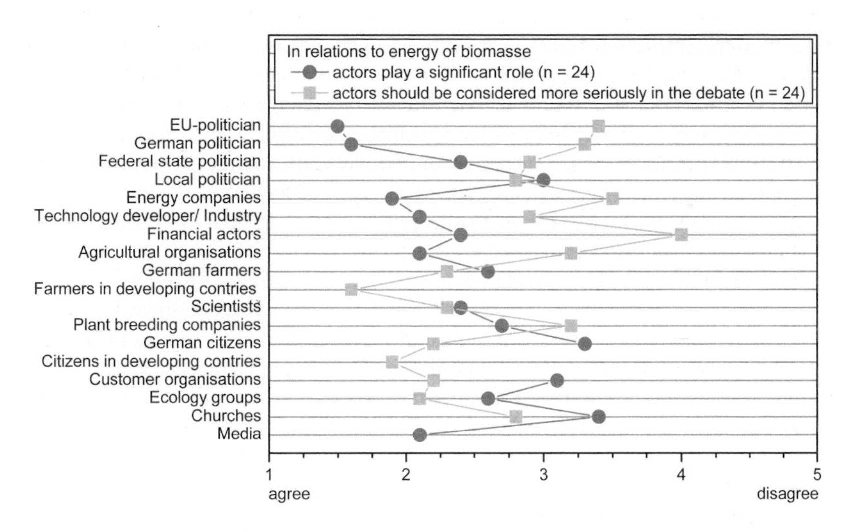

Figure 3. Relevant actors and their exertion of influence.

Bavaria, but also on the national and European level. Experts were asked to assess alternative sources of energy like wind and solar power that were considered more promising than bioenergy. Only nuclear energy was clearly assessed negatively. Experts agreed that only an 'energy mix' (91%) can tackle the energy problem. All (100%) agreed that a sensible use of energy is the precondition of any successful energy policy.

Research

The development of international sustainability criteria and corresponding control mechanisms were considered a research topic of utmost importance (78%). Two thirds (68%) of the experts believed that research should not pursue the genetic modification of plants for biomass production.

Ethically sound biomass production: a Bavarian perspective

As a main outcome, the presented results indicate that energy from biomass will be one part of the future energy portfolio if economic and ethical sustainability criteria can be met. The project's hypothesis is that Bavaria – known for predominantly small-scale agriculture – can serve as an example to demonstrate how the production of energy from biomass can be ethically sound whilst being efficient and economically successful (knowing that the market for biomass is presently highly subsidised).

Considering economic criteria, short transportation and regional use of raw materials and closed material cycles could be of key importance. Furthermore, decentralised markets that depend on local prices and given situations promise a higher net farm product, e.g. by using maize silage to produce biogas instead of animal feed, the total agricultural value added can be increased. Thus enhanced job security and economic benefit could strengthen rural areas. A lot of projects about energy from agricultural feedstock have been financed in Bavaria and fiscal advantages given for e.g. rapeseed oil fuel, and electricity and heat from biogas plants (StMUGV, 2007). Unsurprisingly, in 2007 one third of the total energy manufactured from agricultural commodities in Germany was produced in Bavaria (Bernhard, 2007).

Regarding ethical criteria, rather abstract concepts like 'sustainability', 'justice' or 'environmentally friendly' are commonly discussed. These concepts have to be further developed and the vague notions filled with content. Building on the results of the ED and utilizing the results of former research and experiences, technologies will be compared and estimated in terms of their social, economic and environmental impact with a focus on the Bavarian context. This will be the next step of the project.

Recent developments in rapeseed oil fuel might already testify to the potential for bioenergy in Bavaria. Between 2004 and 2007 the number of small-scale oil mills in Bavaria increased from 93 to 246 (Uhl *et al.*, 2007). Rapeseed oil fuel that replaces fossil fuels is considered to reduce CO_2-emissions, contribute to climate protection and reduce dependence on the petroleum industry (Widmann and Remmele, 2008). Moreover, it can be used as cold pressed oil without being further processed, and the by-products like presscake can be used as animal feed locally. Taking into account these aspects, rapeseed oil fuel can be related to ethical criteria and be compared to other technologies.

Therefore, in the next step, a tool for ethical assessment of different biomass energy technologies will be developed, following the concept of an Ethical Matrix (Mepham, 1996; Mepham *et al.*, 2006). This will allow an assessment of current bioenergy policy in Bavaria and development of criteria for ethically sound production of energy from biomass.

Conclusions

In general, the experts agreed that bioenergy can substantially contribute to meeting future energy demand. Nevertheless, they did not consider bioenergy a universal remedy for energy scarcity but only as one source of energy amongst others. Their opinions clearly indicated that a mix of renewable energy sources and a sensible and efficient use of energy are essential to tackle the energy problem. However, consideration of ethical aspects will be crucial for any future technology for production and use of energy from biomass. Furthermore, the experts uniformly thought that biomass heating plants were a reasonable technology for an agriculture structured in small units. With these results in mind, an Ethical Matrix will be developed with a focus on the ethical soundness of the local production of energy from biomass. This matrix will help evaluate and compare different technologies for biomass production for energy and give guidance to their future development in Bavaria.

References

Bernhard, O. (2007). Bayern ist Klimakompetenzregion. Bayerisches Staatsministerium für Umwelt, Gesundheit und Verbraucherschutz, Pressemitteilung Nr. 318/2007.

Mepham, B., Kaiser, M., Thorstensen, E., Tomkins, S. and Millar, K. (2006). Ethical Matrix. Manual. LEI, The Hague, The Netherlands.

Mepham, B. (1996). Ethical analysis of food biotechnologies: An evaluative framework. In: Mepham, B. (ed.) Food ethics. Routledge, London, England, pp. 101-119.

Milieu en Natuur Planbureau (MNP) (2008). Local and global consequences of the EU renewable directive for biofuels – Testing the sustainability criteria. Available at: http://www.mnp.nl/bibliotheek/rapporten/500143001.pdf. Accessed 16 March 2008.

Millar, K., Tomkins, S., Thorstensen, E., Mepham, B. and Kaiser, M. (2006). Ethical Delphi. Manual. LEI, The Hague, The Netherlands.

Mitchell, D. (2009). A note on rising food prices. Policy research working paper of the World Bank/Development Prospects Group. Available at: http://www-wds.worldbank.org/external/default/WDSContentServer/IW3P/IB/2008/07/28/000020439_20080728103002/Rendered/PDF/WP4682.pdf. Accessed 16 March 2008.

Bayerisches Staatsministerium für Umwelt, Gesundheit und Verbraucherschutz (StMUGV) (2007). Klimaprogramm Bayern 2020. Available at: www.umweltministerium.bayern.de. Date of access: 12.03.2009.

Uhl, A., Haas, R. and Remmele, E. (2007). Befragung von Betreibern dezentraler Ölsaatenverarbeitungsanlagen. Im Auftrag der Union zur Förderung von Oel- und Proteinpflanzen e. V. Berichte aus dem TFZ, Nr. 15. Straubing: Technologie- und Förderzentrum im Kompetenzzentrum für Nachwachsende Rohstoffe, 68 pp.

Widmann, B. and Remmele, E. (2008). Biokraftstoffe. Fragen und Antworten. Straubing: Technologie- und Förderzentrum, Germany, 16pp.

Keeping warm in an ethical way – is it acceptable to use food crops as fuel?

C. Gamborg[1], K.H. Madsen[2] and P. Sandøe[1]
[1] University of Copenhagen, Danish Centre for Bioethics and Risk Assessment, Rolighedsvej 25, DK-1958 Frederiksberg C, Denmark; chg@life.ku.dk
[2] AgroTech, Institute for Agri Technology and Food Innovation, Udkærsvej 15, DK-8200 Aarhus N, Denmark

Abstract

Since 1990 crop based products which may be used as food are not allowed in the Danish heating supply system – for 'ethical' reasons. This ethically founded resistance seems to flourish in other parts of the world as well: you cannot burn food when there is hunger in the world. At the same time, setting agricultural land aside or growing non-food crops on a minor part of the fields has been seen by many as a fairly uncontroversial solution to periodical surplus production of food within the European Union (EU). This is despite the fact that the net result is the same: less food is produced, i.e. a smaller part of the production area is used for producing crops for food. The main issue to be addressed in this paper is to understand what lies behind this difference in acceptance. In the paper the background and arguments used in the debate are described and discussed to see whether ethically relevant differences between the two ways of reducing surplus agricultural production can be discerned. The paper argues that one obvious way of explaining the difference is by means of the so-called act- and omission-doctrine found in discussions on medical ethics.

Keywords: agriculture, act, omission, bioethanol, biofuels

Introduction

In the past fifty years, focus in farming has been on increasing yields through intensive production schemes. However, these schemes have periodically resulted in excesses – too much food being produced and too much land in production. It is expensive to store or export the surplus food produced to other world markets, and it may also disrupt world trade patterns. In 1988 the European Union (EU) introduced set-aside as a political measure to help reduce periodical large and costly surplus production food under the guaranteed price system of the Common Agricultural Policy in Europe. The obligatory set-aside system has only very recently been abolished (January 2009). Set-aside in general refers to the practice of leaving part of the farm uncultivated or put to uses other than food or feed production for a period of time. Another reason for setting aside land is to improve soil fertility and reduce pest and weed problems after prolonged periods with intensive agriculture. On part of the set-aside area, non-food crops could be grown; some for producing bioenergy in various forms.

Setting agricultural land aside or growing non-food crops on a minor part of the fields, traditionally less than 10% of an area, has been seen by many as a fairly uncontroversial solution to periodical surplus production of food within the EU. Another way to make use of the productive capacity of agricultural land is to change the end use. Thus, instead of using crops such as wheat for *food*, they can be used to produce bio-based energy.

However, since 1990 crop based products which may be used as food are not allowed in the Danish heating supply system – for 'ethical' reasons. This ethically founded resistance seems to flourish in other parts of the world as well: you cannot burn food when there is hunger in the world.

Judged merely by looking at the end result, the consequences for potential food supply are the same: A smaller part of the potential production on agricultural land ends up as food, regardless of whether the crops are used for energy purposes or the land capable of producing the food crops is set aside. Thus, it may come as a surprise that wheat and other crops grown for human consumption are not allowed to be used in district heating plants as fuel.

The main issue to be addressed in this paper is to understand what lies behind this difference in acceptance. We will examine the use of a distinction often found in discussions on responsibilities in medical ethics; the so-called act- and omission-doctrine, and ask whether in the present context this distinction can be claimed to bear moral weight. Thus we shall ask if it is possible to use the distinction to explain the difference in stance towards saying no, for ethical reasons, to the use of food crops for energy while at the same not objecting to the use of set-aside to reduce production. Furthermore, we will discuss what actually constitutes an act or an omission when it comes to using food for energy or setting land aside. Finally, we have to ask whether the two cases – burning food for energy and setting land aside – actually have the same outcomes. For a start we describe the background and arguments used in the debate.

No use of food crops for energy?

Biomass of various kinds is being used worldwide to produce biofuels. There is an international debate about biofuels – a debate which in various forms has been going on for the best of twenty-five years now – focusing on different topics over the years such as self-sufficiency, energy efficiency, fuel costs, types of biomass and at the latest climate change mitigation and sustainability (FAO, 2008; The Royal Society, 2008; Tolleson, 2008). We are not going to enter this comprehensive discussion here. Our focus is specifically on the use of food and fodder crops and if the use of these for energy stands out as a particular ethical problem compared with alternatives where no crops are produced in the fields or where non-food crops are grown. Several European countries, including Germany and Sweden, have blended biofuels into commercially sold transport fuels for years, and are producing significant amounts of liquid biofuels such as bioethanol based on wheat and other food crops or biodiesel from oilseed rape and other oil producing plants – but in Denmark, this is currently only to a very minor extent. Several bioethanol plants are however, being planned or are under construction. Danish biodiesel has been produced since 1990 but only for export markets. Part of the reason for this delay can be traced back to a debate concerning biofuels based on food crops.

Nearly twenty years ago, the use of wheat for energy purposes was, in Denmark, only relevant in relation to district heating. At that time, in the beginning of the 1990s, the Danish government issued regulation concerning future energy planning to the municipalities. The government order, which is still valid, in effect bans the use of products which could be used for food purposes, including fish and rapeseed oil, surplus butter, wheat and the like, as fuel in the collective heating supply, in heating plants larger than 0.25 MW capacity (Ministry of Transport and Energy, 2005: §13). The reason behind the ban was at that time a rather widespread view that it is unethical to burn food for energy when there is hunger in the world. Instead of using surplus food crops, a lot of district heating plants had to use other types of biomass, such as straw and wood chips – as well as domestic household waste. To a lesser extent, willow from dedicated short rotation coppices was used for energy purposes.

This ethically founded opposition to traditional food crops' use in the energy supply chain is not only a Danish phenomenon. It seems to be present in other countries and contexts as well. Recently, the well-known American environmentalist Lester Brown, founder of the WorldWatch Institute and since 2001 founder and president of Earth Policy Institute which explicitly aims to provide a vision and a road map for achieving an environmentally sustainable economy, expressed apprehension about rising food prices. He ascertained that one of the dire consequences would be an increasing amount of people starving worldwide: 'The grain required to fill a 25-gallon SUV gas tank with ethanol will feed one person for a year. The grain it takes to fill the tank every two weeks over a year will feed 26 people.' (Brown, 2006: 1). He claimed that cars, and not people, would claim most of the increase in the world's grain consumption – 'leaving supermarkets and service stations to compete for grain' (*ibid.*). Rising grain prices could become a threat for the 2 billion poorest people in the world; in other words you cannot burn food when there is hunger in the world. In a fierce debate on the British newspaper The Independent's website, one of several comments to the feature article on 'the burning question' ran as follows: 'It is absurd to burn food for fuel unless it is one's intention to starve the world. What have our governments in mind for us? It does not appear they care about renewable resources to eliminate both hunger and soot' (The Independent, 2008). Although factual counter arguments have been raised, the main question we shall set out concerns the underlying ethical issues.

Ethically based arguments

As previously stated, the main issue to examine here is whether it is possible to say that it is unethical to burn food crops as long as there is hunger in the world and also to consider it ethically acceptable to choose a strategy where food crops are not being produced because of the set-aside policy. It is this *double* question we are concerned with.

The issue of setting aside agricultural land is not academic. Thus until very recently, approx. 170,000 hectares in Denmark have been set aside, meaning that no food was produced on that area. If it is only about consequences of feeding a hungry world, then – in principle – there should be no difference between omitting to grow food by setting agricultural land aside for other purposes and the use of food crops for energy purposes.

Acts and omissions

The question is whether the difference, if there is one, between acting and omitting to act can be described or defined in a way that bears general moral weight? Commonsense would inform many to draw an ethically relevant line between an act and an omission (of an act). According to the act-omission doctrine, it is morally more problematic if you actively do an act which expectedly will have a negative consequence than it is to omit doing an act where the omission would have the same result. Hence, there is said to be a moral difference between e.g. hindering and failing to help or lying and not telling the truth even though they may have the same outcome.

Think for example of the difference between sending poisoned food to a person in a poor country, hereby killing another person, and omitting to make a contribution to foreign aid organisations resulting in a person dying in a poor country from starvation, a person which – all things equal – otherwise could have been saved. In the first instance, an atrocious and morally reprehensible act, a crime is committed, for which the person responsible should be punished whereas many people would probably regard the second instance as a mild sin of omission, if a sin at all.

In the case of using food crops to produce energy, we could then equate an overt act with the transformation of food (crops) to energy, and omission would be to set land aside. Hence, the reason why the use of grain for energy – either directly as feedstock in heating plants or as a basis for liquid fuel – is seen as ethically controversial is that by doing this you carry out an overt *act* of destroying food. Setting land aside means you abstain from producing on the land, you make an *omission* in terms of growing food crops. To act is to do something, while an omission is a failure to act in circumstances where one has the ability and opportunity to act.

If we refer to larger bodies of ethical theory, there is a complex philosophical debate to be had about the moral significance between act and omission. A consequentialist would dismiss the importance of the distinction – i.e. there is no intrinsic moral difference between act and omission. If an act or an omission has the same total outcome, they are either as good or as bad as each other (Oderberg, 2000). On the other hand, according to non-consequentialist ethics, outcomes are by no means all that matter, we should also consider (positive and negative) duties, and as such the acts-omission distinction is more of a corner-stone of non-consequentialist ethics.

It should be noted that the act-omission distinction should not be seen as a way to distinguish between what we are obliged not to do and what we are allowed to do. In addition, in both cases, there has to be something more than mere accidental action of failure to act. The main thrust of the act-omission distinction seems to be that cases of action or positive commission demand reasons which are morally weightier than those justifying an omission. For example, not lying is held to more morally basic than telling the truth.

Another issue of discussion is the way the doctrine is put forward. One criticism is that what appears to be a morally relevant distinction between acts and omission often really is a distinction between deliberate immoral action and actions which non-intentionally cause bad effect. If you walk past a person drowning, it can be hard to argue that there is a difference between killing the person and omitting to save the drowning person. In other words, 'By doing nothing' you can actually do something. Deliberate absence of (bodily) movement can constitute acting negligently.

If we turn to the case of burning food and using set-aside policies to produce less food, it is a real question whether setting land aside should actually count as an act or an omission (in the terminology of the act-omission doctrine). Within the current framework of EU subsidies, set- aside is almost considered as an active measure to maintain land in good agricultural condition, and not as land lying fallow in the long-established sense of the word. For example, there had to be crop cover to prevent nitrogen leaching, it was a requirement that set aside areas were cut down every two years, and there were rules and regulations concerning plant cover characteristics (EC, 2004). In this case, it seems hard to argue that setting aside land is an omission rather than just another way of seeding the land with different crops – in other words an act. Since the motive for setting land aside and the motive for producing food crops to be burned are the same, there seems to be no reason for distinguishing between setting land aside and the use of produced food crops for energy purposes. In other words, it does not seem possible in terms of motives to say that it is unethical to burn food crops while it is ethically acceptable to choose a strategy where food crops are not being produced through the set-aside policy.

Surplus food for the starving?

Even if the ethical distinction between act and omission is accepted (i.e. we find the distinction to bear moral weight and we can sort out what constitutes an act/omission) there is still a problem: If it is regarded as unethical to utilise food crops for energy purposes as long as people starve in other parts of the world, the logical implication would be that the food crops entering the energy supply would actually be able benefit the starving.

However, this is not necessarily the case. In spite of the fact that sufficient amounts of food to feed all are normally being produced on a global scale, time and again food aid is needed in several places. The EU has a periodic surplus of food crops, which is also used for foreign aid. Foreign aid experience from the EU, International Red Cross and the United Nations World Food Programme suggests that such relief can lead to an increase in external dependency by unintentionally hampering local food producers' ability to sell their goods at reasonable prices. Hence, foreign food aid should only be used as a last resort and not as a more or less permanent instrument to reduce surplus production. The ultimate aim of a food aid programme, according to the UN, should be to take away the need for food aid. Thus, in most cases, food will not be 'taken off the plate' through the production of food crops for energy purposes in Denmark, or in other European countries. Instead, it has been suggested by the WTO (World Trade Organisation) that members of WTO should, whenever possible, should donate monetary funds for purchase of food in the aid recipient country.

Conclusion

So, is it possible to say that it is unethical to burn food crops to gain energy while at the same time saying it is okay to abstain from producing food crops?

We have illustrated that a much used distinction between act and omission may explain why some consider burning food crops as ethically unacceptable (being an overt act), whereas setting agricultural land aside (an omission) – in principle having the same net consequence (less food produced) – is considered ethically acceptable. However, even if this distinction in some cases bears moral weight, it is debatable if setting land aside actually would count for an omission in a relevant sense. Finally we have discussed whether sending surplus food production from the rich parts of the world to poor countries will help poor people in the long run. Thus the current Danish aid policy is based on the idea of helping the poor countries to feed themselves (and not giving surplus production other than in emergency situations).

Our conclusion therefore is that it is difficult to see how one can say that it is unethical to burn food crops to gain energy while at the same time saying it is okay to abstain from producing food crops. This is of course only a small part of the discussion about how to keep warm in an ethical way.

References

Brown, L.R. (2006). Supermarkets and service stations now competing for grain. Eco-Economy Update July 13, 2006-5 Earth Policy Institute. 2 pp.

European Commission (EC) (2004). Commission Regulation (EC) No 795/2004. of 21 April 2004 laying down detailed rules for the implementation of the single payment scheme provided for in Council Regulation (EC) No 1782/2003.

Food and Agricultural Organisation of the United Nations (FAO) (2008). Bioenergy, food security and sustainability – towards an international framework. Food and Agricultural Organisation of the United Nations. Available at: http://www.fao.org/fileadmin/user_upload/foodclimate/HLCdocs/HLC08-inf-3-E.pdf. Accessed 2 March 2009.

Ministry of Transport and Energy (2005). BEK nr 1295 af 13/12/2005 Bekendtgørelse om godkendelse af projekter for kollektive varmeforsyningsanlæg. Available at: https://www.retsinformation.dk/Forms/R0710.aspx?id=22516. Accessed 5 March 2009.

Oderberg, D.S. (2000). Moral theory: A non-consequentialist approach. 2nd edition. Blackwell: London.

The Independent (2008). Blog entry Tuesday 15 April 2008. Available at: http://blogs.independent.co.uk/openhouse/2008/04/have-your-say-5.html. Accessed 9 March 2009.

The Royal Society (2008). Sustainable biofuels: Prospects and challenges. The Royal Society, London.

Tolleson, J. (2008). Not your father's biofuels. Nature 451: 880-883.

Use of Italian national food survey to estimate CO_2 emissions related to 'out of season' consumption: kiwifruits as a case study

L. Mistura, M. Ferrari, S. Sette and C. Leclercq
National Research Institute on Food and Nutrition, via Ardeatina 546-0178 Rome, Italy; mistura@inran.it

Abstract

Kiwifruits are available on the Italian market in all seasons but the kiwifruits produced in Italy are available mainly from November to April, whereas in other months these fruits are shipped from other continents (e.g. New Zealand, Australia, Chile and Argentina). The last Italian food consumption survey (INRAN-SCAI 2005-06) was used to estimate the CO_2 emissions related to this 'out of season' consumption of kiwifruit. The survey was conducted from October 2005 to December 2006 on 3,323 subjects. Food consumption was assessed on three consecutive days through individual dietary records. Overall, 14% of people taking part in the survey ate kiwifruit during the survey period. The data used to determine kiwifruit consumption over the year was skewed towards the months in which they are grown in Italy with a per capita intake of 11.4 g/day (18% of consumers) versus 6.5 g (12% of consumers) in May to October. Extrapolated to represent the Italian population (59 million citizens), it was estimated that 183,346 tonnes of kiwifruits are consumed in Italy each year, 10,630 tonnes of local origin and 79,716 tonnes from abroad. These kiwifruits may travel distances from abroad between a range of 12,000 km (Argentina) to 18,000 km (New Zealand). Based on an estimated emission of 15.0 g CO_2/tonne/km for freight transport by small container vessels, the estimated CO_2 emissions for kiwifruit that is consumed 'out of season' ranges from 14,349 to 21,523 tonnes per year. If Italian consumers chose to limit their kiwifruit consumption to the months in which they are locally produced a very significant reduction of CO_2 emission could be obtained. There is a need for information campaigns directed to consumers so that (1) they become more aware of the damages caused by greenhouse gas emissions, (2) they understand the link between their food consumption patterns and greenhouse gas emissions, (3) they modify their food choices accordingly.

Keywords: food consumption, kiwifruit, greenhouse gas, seasonality

Introduction

In recent years, consumers are increasingly concerned about their personal impact on CO_2 emissions, in relation to food consumption, home energy consumption and transportation (Weber and Matthews, 2008). Among these three sources of emission, food probably represents the easiest opportunity for consumers to lower their CO_2 emission through modified personal choices. In the wide range of food categories, fruit is a fitting example to show how CO_2 emissions could be reduced if consumers had food habits close to those of their grandparents, i.e. eating 'in season' fruit.

In Italy there is a great diversity of fresh fruit and some fruit is available mainly in a limited period corresponding to the local harvesting period. This is the case for peaches, apricots, cherries etc. Other fruit items such as kiwifruits, bananas, pineapples, apples, pears and lemons are available all year long.

Kiwifruit, *Actinidia deliciosa,* have only recently become a fairly common and highly appreciated fruit in Italy. Bananas and pineapples are exotic fruit available in Italy exclusively as imported products, while pears and apples are available all year long due to the use of preservation techniques. Lemons are also available all year and their main use as fresh fruit is as condiments. Kiwifruits are available on the

Italian market in all seasons, but the kiwifruits produced in Italy are available mainly from November to April whereas in other months kiwifruits are mainly shipped from countries outside of Europe: e.g. New Zealand, Australia, Chile and Argentina (National Institute of Statistics, 2009).

Methods

The last Italian food consumption survey (INRAN-SCAI 2005-06) (Leclercq *et al.*, 2009) was used to estimate the CO_2 emissions related to the 'out of season' consumption of kiwifruits. The survey was conducted using a random sample of 1,300 Italian households from the four main geographical areas of Italy (North-West, North-East, Centre and South and Islands). The survey was carried out from October 2005 to December 2006 including 3,323 subjects aged 0.1 to 97.7 years. Food consumption was self-recorded by subjects (or by caregivers in the case of children) for three consecutive days through individual dietary records.

The percentage of individuals consuming each item was calculated. Any subject who consumed at least one item during the survey period was classified as consumer.

The amount of CO_2 emissions related to kiwifruit consumption was calculated by multiplying the estimated amount of imported kiwifruits (in tonnes) by the average distance between harbours of departure and those of arrival (in km) and by the CO_2 emissions per km per kg for freight transport of small container (Department for Environment, Food and Rural Affairs, 2008). Statistical analysis was performed using SAS for windows (release 8.01, SAS Institute Inc., Cary, NC, USA).

Results and discussion

Overall, 91% of subjects enrolled in the survey consumed fresh fruit during the 3 days of survey and no relevant trend in percentage of consumers was observed during the year (Figure 1).

On the other hand, the distribution of the consumption of the individual fruit items which are available all year long shows a small decrease in the percentage of consumers during the summer period. This is because the variety of fruit available in this period is wider than in other periods of the year and consumers tend to replace some of the 'out of season' fruit with the 'in season' ones (Table 1). For the specific case of kiwifruit, 14% of subjects ate kiwifruits during the survey (Table1).

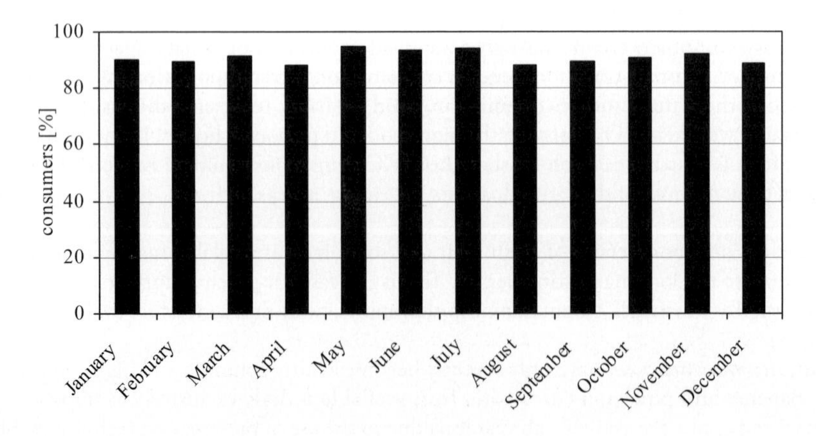

Figure 1. Percentage of consumers of fresh fruit by months of consumption.

Table 1. Percentage of consumers of specific fruit items by months of consumption.

Months	Population sample	Kiwifruit (%)	Lemons (%)	Pineapple (%)	Banana (%)	Pear (%)	Apple (%)	Cherries (%)	Peaches (%)
January	318	22.3	19.8	3.1	34.0	28.9	65.1	0.0	0.0
February	488	15.2	17.6	0.8	35.0	28.9	68.0	0.0	0.2
March	189	20.6	16.9	2.6	43.9	30.7	64.6	0.5	2.1
April	301	16.9	19.3	4.3	42.2	31.9	70.1	0.0	0.7
May	319	21.6	16.0	6.6	50.8	26.3	73.0	11.0	1.6
June	315	12.7	24.4	4.8	36.8	28.9	48.3	24.1	47.0
July	434	9.4	8.8	1.8	29.7	25.8	36.6	6.2	58.3
August	85	0.0	18.8	2.4	35.3	29.4	42.4	0.0	51.8
September	106	7.5	34.0	2.8	32.1	32.1	61.3	0.0	29.2
October	248	7.7	25.4	4.0	35.5	41.5	64.9	0.0	10.9
November	420	12.6	15.5	1.9	36.7	34.3	65.2	0.0	0.7
December	100	12.0	18.0	2.0	32.0	30.0	58.0	0.0	0.0
All year	3,323	14.4	18.1	3.0	37.1	30.4	60.5	4.2	15.6

The trend of kiwifruit consumption over the year was unbalanced towards the months in which kiwifruits are harvested in Italy with a per capita intake of 11.4 g/day (18% of consumers) in November to April versus 6.5 g (12% of consumers) in May to October. When extrapolated to the total Italian population (59 million citizens), it was estimated that 183,346 tonnes of kiwifruit are consumed in Italy in one year, with 103,630 tonnes of local origin and 79,716 tonnes from abroad. The distance to transport kiwifruits from abroad ranges from 12,000 km (Argentina) to 18,000 km (New Zealand). Based on estimated emissions of 15.0 g CO_2 per tonne/km for freight transport of small container vessels, the estimated amount of CO_2 emissions for kiwifruits consumed 'out of the season' ranges from 21,523 to 14,349 tonnes per year.

In conclusion using kiwifruit as a simple case study, national food consumption surveys could be powerful tools to assess the CO_2 emissions related to 'food miles'. The applied methodology can also be used to identify CO_2 reduction options for consumers of other products.

If, based on their sensitivity to the damages caused by greenhouse gas emissions, Italian consumers chose to shift their kiwifruit consumption to the months in which they are locally produced, a very significant reduction of CO_2 emission could be obtained. There is a need for information campaigns directed to consumers so that 1) they become more aware of the damages caused by greenhouse gas emissions, 2) they understand the link between their food consumption patterns and greenhouse gas emissions, 3) they modify their food choices accordingly.

References

Department for Environment, Food and Rural Affairs (2008). 2008 Guidelines to Defra's GHG Conversion Factors: Methodology Paper for Transport Emission Factors. Available at: http://www.defra.gov.uk/environment/business/envrp/conversion-factors.htm#transport. Accessed 5 January 2009.

Leclercq, C., Arcella, D., Piccinelli, R., Sette, S., Le Donne, C. and Turrini, A. (2009). The Italian National Food Consumption Survey INRAN-SCAI 2005-2006. Main results in terms of food consumption. Public Health Nutrition (in press, doi:10.1017/s1368980009005035).

National Institute of Statistics (2009). Coeweb – Foreign trade statistics. Available at: http://www.coeweb.istat.it/. Accessed January 2009.

Weber, C.L. and Matthews, H.S. (2008). Food-miles and the relative climate impacts of food choices in the United States. Environmental Science & Technology 42: 3508-3513.

Theme 4 – Animal futures: welfare and ethics

Culling of day-old chicks: opening the debates of Moria?

S. Aerts[1,2], R. Boonen[2], V. Bruggeman[3], J. De Tavernier[2] and E. Decuypere[2,4]
[1]*Animal Welfare and Ethology Group, Katholieke Hogeschool Sint-Lieven, Hospitaalstraat 23, 9100 Sint-Niklaas, Belgium; Stef.Aerts@kahosl.be*
[2]*Boerenbond Chair Agriculture & Society, Centre for Science, Technology and Ethics, Katholieke Universiteit Leuven, Kasteelpark Arenberg 1 – bus 2456, 3001 Leuven, Belgium*
[3]*Katholieke Universiteit Leuven, Schapenstraat 34 – bus 5100, 3000 Leuven, Belgium*
[4]*Biosystems Department, Katholieke Universiteit Leuven, Kasteelpark Arenberg 30 – bus 2456, 3001 Leuven, Belgium*

Abstract

Intensive selection in layer poultry has resulted in a situation in which the male sex is no longer useable in a meat production context. This results in a massive culling of day-old male chicks in hatcheries (approximately 4.2 billion/year, worldwide). The main techniques used in the European Union are maceration and gassing. Recently, significant societal debates have been raised on this issue, especially in The Netherlands, and in the United Kingdom. This will probably spread across Europe. For a number of reasons, this is a highly polemic practice. Not only is it little known, it also involves a visually disturbing action on a small and furry animal. The ethical questions that arise on this issue may not be the same as the societal ones. There are two distinct fields of ethical issues to be considered: (1) issues directly relevant to the animals; i.e. animal welfare problems associated with the sexing and/or killing of the chicks, and (2) issues relevant to humans; i.e. the killing as such, including the age of the chicks, its integrity, its naturalness, and the sustainability of the practice. Moreover, some of the alternatives may lead to genetic modification in animals, and this may in turn provoke new debates. Animal welfare during sexing can be assured in lines where plumage colour or plumage growth are sex-related traits. Taking the complexity of the procedure and the time to death as the most important risk factors, maceration seems to be the ethically best culling technique available, but may be societally less acceptable. Pre- or early-incubation sexing techniques – currently still commercially unavailable – would avoid all production of male chicks, thereby preventing any welfare problems, while at the same time being economically and ecologically more profitable. The most far-reaching criticism against this practice, underlying most mentioned above, is that it represents a profound instrumentalisation of animals. Producing and then killing day-old animals because they have no more use than being feed sources seems a serious violation of the animal's *telos*. It is apparent that even the alternatives resulting in no male chicks being produced or born are prone to the same criticism. The entire discussion hints at the question of whether the killing of animals as such is morally defensible. The issues entailed therein are mostly the same, but generally more pronounced. In that view, the day-old chicks case can be regarded as a *pars-pro-toto* for animal production as a whole.

Keywords: gassing, maceration, spent animals, instrumentalisation, hatchery

Introduction

Within poultry production, there are great differences in productivity between the two sexes of the same production type (and line). Raising male animals of the highly specialised layer breeds (or hybrids) has become an economical impossibility. Right after the chicks are released from the hatcher, the male chicks are therefore selected manually and are killed immediately by electrocution, maceration, or CO_2 gassing; only the latter two are allowed in the EU.

Since 1925 the selection procedure is executed using morphological differences between the two sexes. At that time, the only difference that could be used was a small difference of the cloaca (Martin, 1994). If pressure is exercised on the abdomen of the chick, the cloaca protrudes slightly, making it possible to assess the morphology. It is clear that this is a very detailed procedure that can only be done with the greatest concentration. Few people are able to do this at a commercial scale. Alternatives have been developed through classic selection. Sex-specific single-gene traits for feathering speed or feather colour have been introduced that make it possible to visually differentiate, which is certainly less prone to error.

The yearly production of these 'useless' male chicks amounts to 280 million in Europe, 230 million in the US and at least 4 billion worldwide (relating to the 5.7 billion laying hens; Evans, 2007).

A first set of alternatives to the killing of these chicks is to identify the sex of the chick as early as possible in its ontology (i.e. *in-ovo*), during incubation or even before. One may also try to change the embryo's sex by interfering in the sex determination or sex differentiation processes. Both options would potentially decrease some of the problems and critiques that are faced when killing male chicks.

The ethical issues

It is clear that the killing of such a high number of animals cannot be ethically neutral, and it merits an in-depth discussion of the why and how.

A Dutch preliminary analysis of the case (Woelders *et al.*, 2007) identified six elements in an ethical framework surrounding the killing of day-old chicks. Five of those we consider relevant to our analysis, but we would like to summarise these in three broad terms: animal welfare, age at killing, and instrumentalisation. The latter would then contain the notion that the animal is becoming a 'by-product' of egg production, that its integrity is harmed, and that the system is not 'natural' anymore. To these we add sustainability issues and (possible) GM issues.

Animal welfare

During the sexing process, the chicks' welfare can be harmed during the sexing itself, and during the killing. Sexing is normally performed while the chicks are transported from the hatcher to the transport crates. Welfare problems can arise when each animal has to be manipulated individually as is the case with cloacal sexing (which additionally involves abdominal pressure).

Although these should in principle only be moderately stressful (little pain, moderate fear), it is clear that feather sexing involves even less manipulation, which is less complex and only necessary for half of the animals. Judging by animal welfare, feather sexing is the better option. But in general, sexing seems to add little pain or fear to that which is the result of the transport between incubators and transport containers.

The killing itself is clearly a much greater source of potential welfare problems. When the maceration or the gassing is not executed correctly (e.g. too many animals at once, gas concentration too low), serious injuries and a lot of pain are probable results. The correct maintenance and operation of the material is a *conditio sine qua non* for any decent level of welfare. We first ask whether welfare is harmed during normal operation, and we will estimate the risk of problems arising.

Maceration is technically a relatively simple system, and due to the high speed of the maceration blades the animals are killed within a fraction of a second. It seems therefore unlikely that there is any conscious feeling of pain. As the introduction into the machine is not very different from the introduction into the transport crate (a short drop), it seems that there is no real welfare problem with maceration. Arguably, the risk of problems arising is low, as long as there is no deliberate diversion of normal operation parameters (e.g. introduction speed).

Gassing on the other hand is slightly more complex, and the time to death is certainly longer. Fear as well as pain are relevant here; fear can be induced by the potentially aversive properties of the gas used (Raj *et al.*, 2006) and by the possible accumulation of animals if they are introduced too rapidly; pain will be primarily the result of the latter. Thus, although there need not be welfare problems during normal operation, the risk seems to be higher with gassing.

We will not discuss any of the other 'non-official' methods used in practice. Suffocation in bags, cooling in fridges or freezers, and the like; these do not even merit consideration.

Age at killing

A very prominent aspect of this issue is the atypically young age at which the animals are killed. Few other animal production types involve the killing of animals immediately after birth (*in casu* hatching), possibly with the exception of the male offspring of dairy cows, in certain parts of the world.

As Woelders *et al.* (2007) already stated it is difficult to coherently argue why killing day-old chicks would be an ethical problem, but not the slaughter of broiler chickens at an age of five weeks. And, also in veal, lamb, and pork production animals are killed long before they reach their 'natural' maximum age.

Instrumentalisation

The sense that the killing of day-old male chicks is the ultimate example of the limitless instrumentalisation of animals in modern animal production seems to be the basis of most of the opposition against the killing of male chicks. Killing animals so shortly after their birth because they are of no further use than (in the best cases) to serve as food for other animals seems to be a grave infringement of a male chicken's *telos*.

That the animal is only a by-product of egg production is something that can hardly be contested. This is also true in other 'animal production' fields, e.g. dairy production. This does not necessarily mean it is wrong. Whether the integrity of a chicken is more harmed by being killed within the first 48 hours of its life, or by being kept for 72 weeks for egg production and then being killed, is doubtful. The same line of thought is applicable to the 'naturalness' argument.

Again, it seems difficult to coherently argue why all this would be a problem when an animal is killed and used for pet food, but not when used for human food, and why this would be a problem with chicks from hatcheries, but not with fish from fisheries.

Sustainability

On a more practical level, one could also argue that there are sustainability issues connected to the killing of day-old chicks. Without analysing these in great detail, it is clear that the social component of sustainability is relevant, but also that significant economical and ecological progress is possible.

Indeed, significant amounts of inputs are used to produce eggs and chicks that are not useful for the goal they are produced for (egg production). Half of the parent stock, and its associated feed, energy, etc., and half of the hatchery capacity, and its associated energy consumption, are 'useless'.

A short analysis of alternatives

In-ovo *sex determination*

Any method that relies on a sex discriminating factor that can be assessed during the embryonic or foetal stages, needs to be able to sample and analyse this factor automatically, rapidly, and as simply as possible (e.g. through a colour reaction). Not only is it necessary that the procedure can be used on massive amounts of eggs, but each result needs to be traced back to the individual egg, that is then sorted out (Phelps, 2003). A prototype of such a system is developed by Embrex and Life-Sensor, in which the oestrogen concentration in the allantoic fluid is measured at day 17 of the incubation (of 21 days in total). This is a method that is – at present – only feasible late in the incubation. This means that the foetus is in principle fully developed.

One may question whether there is a fundamental difference between the killing of a grown foetus and the killing of a day-old chick. It may only 'hide' the practice for the general public, but that is a visual difference at best. The ethical issues remain. If there are welfare problems during killing, these will also be applicable to these un-hatched chicks. Neither will it fundamentally solve the age issue, nor the instrumentalisation question (especially when it is done late in the incubation); it only shifts it to a still younger age. It will only be less visible – and thereby more acceptable to society?

Earlier sex determination could be based on molecular DNA markers situated on the sex chromosomes. Standard laboratory techniques such as PCR could be used to establish the sex of the embryo. Although these techniques are highly reliable, they are at this time still too expensive and too specialised to be useful in full-scale industrial processes. If these techniques can be developed further so that they are able to determine the sex of the embryo at day 10 (or even earlier), this would not only be an improvement towards the societal questions, it would also deal with some of the ethical issues such as the welfare issues (if it can be done before the nervous system is fully developed).

If the sex could be determined before incubation, even more positive effects can be expected. Not only would the incubation costs decrease substantially (presumably halved), the egg mass would still be available for industrial use or consumption. It may even have technical advantages over early-incubation assessments (Klein *et al.*, 2003). Unfortunately, it is complicated to exactly locate the area to sample, making the technique complex, slow, and therefore expensive. Significant technical advances are still necessary before the massive ethical advantages of such method can be exploited.

In general we see that *in-ovo* sexing has bigger advantages the earlier the sex can be determined. Some of the 'theoretical' issues (welfare, age, maybe even instrumentalisation) could be circumvented and the 'practical' (sustainability) profits will be larger..

Influencing sex determination and sex reversal

There are some indications that in quail it is possible to influence the primary sex ratio of the offspring by altering the hormone balance of the parental stock (Pike and Petric, 2006). If this could be confirmed – and brought to commercial scale – in chickens, this could lead to fewer male chicks in layer breeds. Economically, this can be interesting, but it does not solve the problem as it only lowers the number of males, but the remaining number still needs to be killed. Additionally, hormone treatments for agricultural animals are seldom accepted by the public.

There have been experiments trying to surgically, chemically or biologically reverse the sex of chickens (Kagami, 2003). Although some limited success has been reached, no male-to-female reversal has been reported. Even if this can ever brought to success, it remains to be seen whether the advantages outweigh the disadvantages. We believe this will not be the case.

Genetic modification

There are some possibilities to use genetic modification (GM) techniques to avoid having to kill male chicks. One could e.g. introduce a lethal gene at the Z chromosome (male chicks are ZZ) and compensate this at the W chromosome (females are ZW). Only female eggs would then hatch. One could also introduce a fluorescent protein at Z and use this to divert male eggs early in the incubation (or even before).

It is clear that this GM solution needs to be embedded in a broader debate on GM animals. At this time, it still seems a no-go-zone; although clear advantages can be seen with reference to the ethical issues raised above (welfare, age, sustainability).

Nevertheless, it seems that alternatives that result in less (or no) male offspring – GM or not – are prone to the instrumentalisation critique. Preventing the conception or the birth of half of the offspring seems no less instrumentalising than killing them.

Conclusion

Moving from post-hatch sex determination to pre-hatch has significant advantages, not only societally, but also ethically if it can be done before the full maturation of the central nervous system. Unfortunately, this does not avoid all ethical questions, as is shown in the paragraphs on age and instrumentalisation of animals.

In the present situation, in which commercial pre-hatch sex determination is not available, we believe that (1) there is a need for stricter enforcement of the regulatory framework, and (2) that maceration is the better option over gassing.

The ethical analysis presented indicates that the killing of day-old male chicks does not result in problems other than those that are inherent to the animal production system as a whole. The fundamental opposition raised against the killing and the instrumentalisation of animals are inherent to animal production (conventional or other), and are certainly not limited to the egg production chain. They may be more apparent, maybe more extreme and visually more disturbing, but these are ethically irrelevant.

In conclusion, it is difficult to condemn the killing of day-old male chicks as wrong, without questioning animal production as such. This is a slippery slope along which most people would not be prepared to go. Nonetheless, it is a (higher-order) debate one should not shy away from. In the end, it may well prove that the male chick case is a *pars-pro-toto* for animal production as a whole. Maybe, through the discussion about this case, we can 'open the gates of Moria' (Tolkien, 1954) and see whether the human race really 'dug too deep'.

References

Evans, T. (2007). Growth slows in global demand for poultry products. Poultry International 46(8): 12-16.

Kagami, H. (2003). Sex reversal in chicken. World's Poultry Science Journal 59: 15-18.

Klein, S., Bauleir, V., Rokitta, M., Marx, G. Thielebein, J. and Ellendorff, F. (2003). Sexing the freshly laid egg: development of embryos after manipulation; analytical approach and localization of the blastoderm in the intact egg. World's Poultry Science Journal 59: 39-45.

Martin, R.D. (1994). The Specialist Chick Sexer. Melbourne, Bernal Publishing.

Phelps, P., Bhutaka, A., Bryan, S., Chalker, A., Ferrel, B., Neumann, S., Ricks, C., Tran, H. and Butt, T. (2003). Automated identification of male layer chicks prior to hatch. World's Poultry Science Journal 59: 33-38.

Pike, T.W. and Petric, M. (2006). Experimental evidence that corticosterone affects offspring sex ratios in quail. Proceedings of the Royal Society B 273: 1093-1098.

Raj, A.B.M., Sandilands, V. and Sparks, N.H.C. (2006). Review of gaseous methods of killing poultry on-farm for disease control purposes. The Veterinary Record 159: 229-235.

Tolkien, J.R.R. (1954). The Fellowship of the Ring (Lord of the Rings, part one) George Allen and Unwin, London.

Woelders, H., Brom, F.W.A and Hopster, H. (2007). Alternatieven voor doding ééndagskuikens: technologische perspectieven en ethische consequenties. In opdracht van het Ministerie van Landbouw, Natuur en Voedselkwaliteit. Animal Sciences Group van Wageningen Universiteit & Research Centrum, Lelystad, 29 pp.

Ethical principles for the use of animals in Austrian legislation

R. Binder[1], *H. Grimm*[2] *and E. Schmid*[3]
[1] *Institute for Animal Husbandry and Animal Welfare, University of Veterinary Medicine Vienna; Member of the Animal Welfare Council, Veterinärplatz 1, A-1210 Vienna, Austria*
[2] *Institute TTN Munich (Theology, Technology, and Natural Sciences), Marsstr. 19, D-80335 Munich, Germany; Herwig.Grimm@elkb.de*
[3] *Official Veterinarian, Animal Protection Ombudsman; Member of the Animal Welfare Council, Römerstr. 11, A-6900 Bregenz, Austria*

Abstract

In 2008, the Dutch Ministry of Agriculture, Nature and Food Quality commissioned a comparative study into the inclusion of animal ethics in animal welfare policy-making in four European countries. One of those countries was Austria. A small group of experts met in a workshop to investigate research questions provided by the Dutch project in order to outline the current Austrian situation. The results of the working group were presented in a detailed report. It provided information on how ethical principles are represented in the Federal Austrian Animal Protection Act (APA) and what measures have been taken to live up to them. This article gives an overview of the key findings. After a short introduction to the history and the scope of the APA (1), the ethical background of the APA is briefly outlined (2). The introductory part is followed by answers to the four research questions given by the project's organizers. The first addresses moral principles in animal ethics which are represented in the APA (3). The consecutive section deals with governmental instruments that have been installed to implement the APA and incorporated principles (4). The APA contains a number of measures that are listed and explained. Since not all of them are effectively working, doubts and critical attitudes are outlined where appropriate. Finally, the article focuses on problems and obstacles that hinder the efficient moral and legal consideration of animal welfare in practice (5). The presentation of all results brought together in the report is beyond the scope of this article. Therefore, only specialties such as the concept *Mitgeschöpflichkeit* (fellow creatures) in its secular, non-theological, meaning and specific Austrian governmental instruments are discussed.

Keywords: use of animals, animal protection act, animal welfare implementation, governmental instruments

Introduction and historical background

Since 1st of January 2005, the Federal Austrian Animal Protection Act (*Bundesgesetz über den Schutz der Tiere – Tierschutzgesetz (TSchG)*, BGBl. I Nr. 118/2004 as amended by BGBl I Nr. 35/2008; in the following APA) has replaced the animal welfare acts of the nine Austrian provinces (*Laender*). Until then, animal welfare was the responsibility of the *Laender*. As a consequence, animal protection was regulated in 10 different animal protection acts and 38 derivative regulations in Austria. In the 1990s first attempts to harmonize the confusing situation failed. On the initiative of Austrian Non-Governmental Organisations (NGOs), a referendum 'A right for animals' was held and supported by 459,096 people in 1996. In 2003 the Austrian government decided to draft a federal animal protection act (*Bundes-Tierschutzgesetz*) which was adopted by the Austrian parliament in the spring of 2004 (online version: www.vu-wien.ac.at/vetrecht).

Due to the Austrian constitution the APA does not cover the following areas:
- hunting and fishing (competence of the *laender*);
- animal transportation (animal transportation act);

- animal experimentation (animal experimentation act).

Ethical background

The most important aspect of the APA is its ethical – and primarily pathocentric – approach. Thus animals are legally protected because of their moral standing irrespectively of the utility for humans. This position is usually termed 'respect for the intrinsic/inherent value of animals'. Therefore the objective of the APA has to be interpreted in a non-anthropocentric manner since it integrates animals into the community of legally protected entities for morally relevant reasons. The protection of life and well-being are considered to be cornerstones of the APA. In terms of the moral basis, animal ethicists in the 20th century made the point clear that we owe moral respect to certain animals (see Singer, 1979; 1985; Regan 1997[1985]; Rollin, 1998[1989], etc.). This intention of the APA is expressed by the term 'fellow creature' (*Mitgeschöpf*) in § 1, indicating that animals are to be respected for their own worth and that man carries a moral responsibility to protect and consider their interests for moral (non-anthropocentric) reasons: 'the Federal Act aims at the protection of life and well-being of animals based on man's special responsibility for the animal as a fellow-creature.' Consequently, any suffering and/or injury is in need of justification, as formulated in § 5 of the APA: 'It is prohibited to inflict unjustified pain, suffering or injury on an animal or expose it to serious fear.' These paragraphs (§§ 1 and 5) reflect the two guiding ethical principles of the APA which are: the protection of animals' well-being and protection of their lives.

Another central aspect of the APA is the consideration of scientific knowledge as the obligatory basis for generating further regulations (decrees) as well as interpreting existing rules. The transformation of abstract principles into concrete obligations is not always easy. However, any infringement of the two principles mentioned is in need of justification and the APA has to ensure that those principles are respected in the various fields of the usage of animals.

Moral principles in animal ethics and their representation in the APA

The working group was asked to answer the following research question: in analogy with the three professional normative principles in laboratory animal science (3 Rs: reduction, refinement, replacement) or research/treatment on patients (beneficence, non-maleficence, justice, autonomy) are there any comparable principles that could structure in a similar sense the practice of animal agriculture, fishery, wild life management, circus animals? Consequently, principles that could structure the responsibility of moral agents (persons) for moral patients (animals) were sketched. The working group identified a number of principles. However, not all of them are integrated in the APA but are nevertheless influential concepts discussed in animal ethics. Hence, they are mentioned as pertinent principles which could function as sources to evaluate human use of animals. Whereas the first group gathers rather abstract concepts (a), the second group addresses practice oriented principles and concepts (b).

a. *Telos*, dignity of living creatures; fellow creatures; integrity; beneficence; *Artgerechtheit*; fairness; proportionality (*Verhältnismäßigkeit*); protection of life; etc.
b. Justifying reason; five freedoms; *Tiergerechtheit*; *nil nocere*; ethical matrix (including the relevant principles; see Mepham, 2000); etc.

Since most of these concepts are familiar to the reader, only two shall be discussed briefly because they do not easily translate into English. These two principles are *Artgerechtheit* and *Tiergerechtheit*. *Artgerechtheit* can be best understood as a concept close to Rollin's formulation of an animal's *telos*. Animals should be allowed to express 'the unique, evolutionarily determined, genetically encoded, environmentally shaped set of needs and interests which characterize the animal in question – the 'pigness' of the pig, the 'dogness' of the dog, and so on' (Rollin, 1998 [1989]: 146). Therefore this concept goes beyond suffering and pain and hence beyond the pathocentric approach. *Tiergerechtheit* on the other hand focuses on the needs of the *individual animal*, as they are formulated in the five freedoms. *Tiergerechtheit* comprises the concept of *Artgerechtheit* since one aspect of an animal's well being is to express its species specific behaviour. Although this principle is not easily implemented and is abstract in nature, it was integrated in the second group (b) because it integrates the principle *nil nocere* that can be put in practice and is more easily to be measured.

The next question was how these principles are reflected in the APA. To answer this question, two tables were developed that relate the principles outlined above to paragraphs in the APA. Comments and explanations of how the particular principle has to be understood and implemented in the APA were given. It appeared that only two principles were *not* really explicitly covered; first of all the concept of the *telos*. However, as mentioned above, the term is well considered in the concept of *Artgerechheit* that is implicitly addressed in § 13/3 which says that the ability to adapt to keeping conditions must not be overstrained. Secondly, the term 'dignity' does not appear in the APA. Despite this fact, the group agreed that the main idea of the inherent worth of animals that is addressed by the term 'dignity' is well considered in the term *Mitgeschöpflichkeit* (fellow creature) that expresses the inherent value of creatures.

Governance instruments

The working group was asked for types of governance instruments and structures that have been established to guarantee that these ethical principles are applied and implemented in various (non-lab) animal practices such as farming of cows, pigs, rabbits. Also the last question, how these instruments work in practice is addressed in this section and explanatory reasons are given why these instruments perform well or badly.

*Regulations (*Tierhaltungsverordnungen*)*

The Federal Minister of Health has to put forward regulations regarding kept animals that specify the abstract framework provided by the APA. In these regulations minimal standards for keeping conditions for farmed animals and companion animals are formulated. In the case of farmed animals, the Federal Minister of Agriculture has to consent to these regulations. Specific decrees have been issued on circuses, zoos, and shelters.

The problem with these regulations is that they quite often erode the intentions of the APA when applied to specific areas. Especially in the area of farmed animals we find a great number of articles that obviously contradict the fundamental principles of the APA. This applies e.g. to the cases of castrating male piglets and the dehorning of calves without anaesthesia, or the minimum requirements for the keeping of rabbits.

Strengthening of animal protection (§ 2)

The federal, regional and municipal authorities are obliged to deepen the public understanding and awareness for animal protection and, to the extent possible within their budgets, to promote and support animal friendly keeping systems, research on animal welfare as well as other matters of animal protection.

An example for the implementation of this regulation is the recently launched project called 'Tierschutz macht Schule'. Its aim is to train teachers and integrate animal protection in the timetables of pupils. In order to keep the workload for teachers as small as possible, they are provided with information material and guided and trained in how to deal with the animal protection issue within a politically correct framework (www.tierschutzmachtschule.at).

Animal Protection Ombudsman (§ 41)

Each *Land* has to appoint an Animal Protection Ombudsman (APO), whose duty is to represent the interests of animal protection. The APO has the status of a party in administrative proceedings. He/she is authorized to survey all files of the proceeding and to request any relevant information. The authorities have to assist the APO in exercising his/her duties.

A critique related to the APOs, who are predominantly vets, is that they are not trained in law and can therefore often not efficiently deal with legal issues. Further, the question of the irreconciliability of the APOs' responsibilities remains unsolved (e.g. some APOs are also official vets).

Animal Protection Council (§ 42)

An Animal Protection Council (APC) has been installed, advising the Federal Minister for Health regarding animal protection issues. Its main duties are to comment on drafted regulations; develop guidelines necessary for uniform execution of the subject; reply to inquiries and wording of recommendations resulting in the course of implementation of the APA; evaluate the execution of the APA as well as working out proposals towards improvement of implementation; and work out an annual report on the activities of the APC.

The main criticism of the APC relates to its decision making procedures. Since decisions are taken by majority, the composition of the APC is central. Presently, 11 out of 31 members explicitly represent animal welfare interests (two representatives of NGOs, nine APOs). Consequently, animal protection claims can be overruled by a majority quite easily.

Animal protection report (§ 42)

After consulting the Animal Protection Council, the Federal Minister for Health submits an Animal Protection Report (APR) to the National Council every two years.

The APR has been delivered once for 2005/06. Due to missing indicators and/or agreed parameters and criteria, the APR develops no innovative force to improve the situation. One further reason for the thoroughly uncritical nature of this report is the fact that it is a governmental report.

Authorisation (§ 23)

The following kinds of animal keeping require authorization: zoo (§ 26), circus (§ 27), using animals for events (§ 28), animal shelters (§ 29), keeping animals within business activities (§ 31), ritual slaughter (§ 31/5).

Reporting

The keeping of wild animals with special requirements regarding husbandry conditions is only allowed on the basis of a report to the authority regarding the animals kept. Such reports shall contain the name and address of the keepers, the number and maximum number of animals kept, the place where they are kept and further information enabling the authority to judge the matter.

Check of keeping facilities (§ 18/6)

An instrument addressed in the APA but not at all implemented yet is a compulsory administrative approval procedure for new types of serially manufactured stable systems and new types of technical equipment for keeping animals.

Execution and supervision (§ 35)

The execution and supervision of the APA appears to be a critical point. A main problem is the insufficient awareness of violations of the APA. The protection of animals as a moral and legal duty seems not to be manifest in practice and in the conscience of many vets and jurists who are responsible for the execution and supervision of the APA.

Qualifications of animal keepers and animal care staff (§ 12, 14)

The APA mentions that animals shall be looked after by a sufficient number of staff possessing the appropriate ability, knowledge and professional skills. Although criteria for the qualification of animal keepers are formulated, they only demand a rather basic qualification. This reflects the low degree of conscience for the moral and legal duty to protect animals.

Selbstevaluierung (self-evaluation system)

In order to bring the legal requirements of the APA into practice, a self-evaluation system for farmers has been developed. Its aim is to enable every farmer to check his/her animal housing system in a self-evaluation procedure (Ofner *et al.*, 2007). The farmers are provided with checklists and manuals that include all requirements of the APA. The checklists are formulated in easily understandable questions (yes-no-answers). They are published on the homepage of the Ministry of Health. The self-evaluation is mandatory for farms within the AHS (Animal Health Service) and integrated into the risk assessment for cross compliance controls. Checklists and manuals were developed for cattle, pigs, poultry, sheep and goats (checklists online 2009).

Problems and vital obstacles

General reasons for malfunctioning

In contrast to the German and Swiss legal systems animal protection is not yet enshrined within the Austrian constitution. Therefore, animal protection is automatically overridden by constitutionally protected values, such as freedom of religion and freedom of research. Thus an efficient execution of the APA is not possible, if animal protection collides with one of the constitutionally granted rights (Figure 1).

Verbandsklagerecht

Another reason for the deficient execution of animal protection law is the lacking authorization of NGOs to act as representatives on behalf of animals' interests in legal procedures (*Verbandsklagerecht*).

Lack of conscience

A main problem is the insufficient awareness of violations of the APA. This is true for many animal keepers as well as for a great number of vets and jurists. Therefore violations of the APA very often are

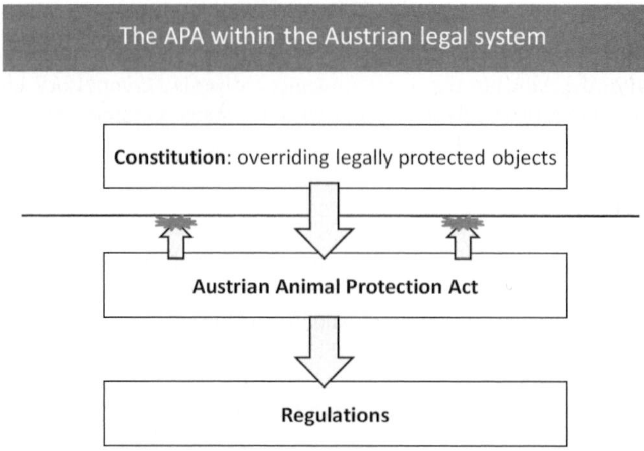

Figure 1. The APA within the Austrian legal system.

considered as peccadilloes. The protection of animals as a moral and legal duty is not manifest within the cultural awareness yet and has not become part of common sense morality.

Practical obstacles

Many animal farmers cannot live up to the minimum required in the APA without severe changes in practice. Consequently, there are strong pressures that work against the efficient implementation of the APA.

Explanatory reasons for inefficient execution of the APA

a. not enough personnel resources (e.g. official vets);
b. lack of knowledge of official vets (e.g. treatment of exotic animals; legal issues);
c. conflicts and bias of vets (e.g. service provider vs. animal protection);
d. missing manifestation of animal protection in the education of vets, farmers and jurists;
e. the complexity of animal protection issues.

Other explanatory reasons

Considering *pet animals* we face a great lack of knowledge with regard to exotic species; this causes a lot of welfare problems. In *farm animals* the main reason is the low status of the animal and missing education of farmers on animal welfare issues. Indirectly, one reason is the strategic focus on the 'nice' side of animal production in advertisements. This avoids dealing with the issue in a transparent way.

Conclusions

The federal Austrian animal protection act can be considered a step in the right direction since it addresses moral principles and measures to implement animal protection in relevant fields. However, implementation and execution in particular suffer from severe problems and often fail to further the issue positively. The main reasons can be seen in its subordinated rank within the legal system and the lack of awareness that the APA is a legal text with binding force for all citizens. Although the legal protection has been institutionalized, animal protection has not been satisfactorily incorporated into the commonly held morality and social practices. Due to the lack of awareness and understanding, the implementation of the APA is currently insufficient.

References

Checklists online (2009). Handbücher und Checklisten zur Selbstevaluierung Tierschutz. Available at: http://www.bmgfj.gv.at/cms/site/standard.html?channel=CH0804&doc=CMS1157545064200. Accessed 7 April 2009.

Mepham, B. (2000). A Framework for the Ethical Analysis of Novel Foods: The Ethical Matrix. Journal of Agricultural and Environmental Ethics 12: 165-176.

Ofner, E., Schmid, E., Schröck, E., Troxler, J. and Hausleinter, A. (2007). Self evaluation of animal welfare by the farmer: a report of application on Austrian cattle farms. Animal Welfare 16: 245-248.

Regan, T. (1997[1985]). Wie man Rechte für Tiere begründet (The Case of Animal Rights. In: Singer, P. (ed.) In Defence of Animals. Oxford). In: Krebs, A. (ed.): Naturethik. Grundtexte der gegenwärtigen tier- und ökoethischen Diskussion. Frankfurt a. M., pp. 33-47.

Rollin, B.E. (1998[1989]). The Unheeded Cry. Animal Consciousness, Animal Pain and Science (expanded ed.). Ames: Iowa State UP.

Singer, P. (1979). Practical Ethics. Cambridge University Press, Cambridge.

Singer, P. (ed.) (1985). In Defence of Animals. Blackwells, Oxford.

Text of APA and relating decrees: www.vu-wien.ac.at/vetrecht

Conserving poultry biodiversity: choosing is not always losing!

R. Boonen
Centre for Science, Technology and Ethics, Kasteelpark Arenberg 30 – bus 2456, B-3001 Leuven, Belgium; Ruben.Boonen@biw.kuleuven.be

Abstract

This paper considers the notion of biodiversity in chickens (*Gallus gallus*). Biodiversity consists of a genetic, an aesthetic and a culture-historical aspect. To ensure future food production, genetic diversity needs to be protected. It is generally thought that genetic diversity is very low for commercial poultry breeds, because the animals are – within their own breeds – phenotypically very similar and only two companies control worldwide layer and boiler selection. Studies show however that this is not necessarily true. The broiler male lines that were used in these studies show a higher heterozygocity than local chicken breeds and different subspecies of *Gallus gallus*. Studies also show that although genetic diversity within old breeds is rather low, they represent a rather large part in the founding-father genome-equivalent, which makes them interesting to maintain genetic diversity within the species. Because conservation of breeds is rather expensive, one has to choose which breeds are worth conserving and which are not. A problem here is the ill-defined term 'breed', because – in the end – a breed is whatever breeders say it is. Since biodiversity also relates to aesthetics and to culture/history, which cannot be expressed in an objective, numerical value, estimating the total value of a breed is not just a simple calculation. One has to take these difficulties into account while making a well-founded decision about which breed is worthy of conservation and which is not. Depending on the goal – which diversity has to be saved? – different groups of breeds are required. The author argues that there is no need to conserve all breeds, because some breeds are strongly related – genetically, historically or aesthetically – to others. Choosing is not always losing!

Keywords: breed, conservation, chicken

Introduction

For several decades, livestock husbandry underwent significant management transformations with regard to the breeding, taking care of and rearing of animals. But it was not only the environment which has changed; performance characteristics and with them the animals have also changed. Only a handful of domesticated breeds were selected for very specific production parameters and this selection process resulted in a few highly specialised single-purpose production breeds, which gradually replaced old local breeds. The most significant performance progress has been made in chickens: commercial stocks can lay over 300 eggs per hen per year or about 30 times as much as their ancestor *Gallus gallus* and broiler chickens have the most efficient feed conversion of all meat-producing animals. With only two companies performing broiler and layer selection globally (Hendrix Genetics and Erich Wesjohann Group) and the animals – within their own types – looking phenotypically very similar, chickens might have a very low biodiversity.

Since biodiversity encompasses more than genetic diversity, this paper starts with a discussion about several aspects of biodiversity. When talking about conservation of breeds in order to conserve existing biodiversity, it is important to know what is meant with a 'breed', a very ill-defined term. Because conservation schemes are rather expensive and difficult to set up, one has to choose which of the known chicken breeds have to be conserved and which ones not. In this paper, some decision tools with their pros and cons are presented in order to select those breeds which need to be preserved for specific conservation goals.

What is biodiversity?

Genetic diversity

The increasing world population and the changing climate conditions continue to challenge agriculture to produce enough food in different production contexts. In order to cope with known, but also with yet unknown problems, it might be necessary to use specific breeds which are already naturally selected for specific parameters, such as tropical breeds, known for their tolerance to heat or to specific diseases. From this point of view, the international community is interested in those extraordinary breeds. Most of them however are currently severely threatened, mostly through being crossed with, or being replaced by, imported high-producing breeds. Therefore, conservation schemes are needed to protect the already existing genetic diversity. It is easier to protect what we already have than to (re-)create what is lost.

Genetic diversity within a species includes both interbreed and intrabreed diversity. The first is the difference between two separate breeds and the latter is the diversity between the individuals among one specific breed. Both are equally important because a breed can only survive if there is enough intrabreed diversity available in order to adapt to changing conditions. If the effective population size is too small, an inbreeding depression might occur, leading to extinction of the breed, which also reduces the interbreed diversity.

Aesthetic diversity

Within one species, a large variety between breeds appear: type (the mutual proportions of the different parts of the body), colour and structure of hair or feathers, and ornaments like horns, and combs. Usually, the notion 'phenotype' is used to indicate this aspect of diversity, but aesthetic diversity is larger than the phenotype. Some breeds look the way they are because the creator(s) preferred certain characteristics which are not important for production, but for social or ritual purposes or even just because of personal taste. Talking about the aesthetic value of a breed incorporates this extra condition in phenotype. Although beauty is not important in the case of production animals, aesthetic diversity is of importance for breed conservation. Because of their peculiar appearance, some breeds are kept by amateur breeders. Such small scale interventions might help to ensure a certain future for rare breeds.

Cultural-historical diversity

In contrast with contemporary single-purpose breeds which are spread worldwide, most local breeds exist for several decades or even centuries in a particular region. Those breeds are typical for some landscapes, well-known for their excellent taste in local, traditional dishes and for playing a central role in folklore. Some of these breeds are silent witnesses on artworks such as paintings. Just like these paintings, the breeds could be considered as 'living heritage', having some historical value, in need of conservation for future generations. Especially in developing countries, animals play an important role in local cultures and habits, so if the breeds get lost, the culture looses an important aspect of its identity.

What is a breed?

Although the term 'breed' is a commonly used notion, and although several authors have tried to formulate an all-embracing definition, there is to date no single definition available in scientific literature. One often refers to common hereditary, uniform characteristics or a common geographical isolation or history, an artificial selection, etc. (Rodero and Herrera, 2000). In contrast to those who try to find a complete definition, other points of view abandon the scientific search for such an all-embracing definition, as can be seen in the following two examples. First, Hall (2004) interprets Lerner and Donald

(1966) as follows: 'a breed is whatever a government says it is'. Secondly, Lush (1994; Cited in Burditt *et al.*, 1995 and FAO, 2007: 340) postulates that:

> 'A breed is a group of domestic animals, termed such by common consent of the breeders, ... a term which arose among breeders of livestock, created one might say, for their own use, and no one is warranted in assigning to this word a scientific definition and in calling the breeders wrong when they deviate from the formulated definition. It is their word and the breeders common usage is what we must accept as the correct definition.'

Taking these quotes together, one might say that 'a breed is whatever breeders say it is' and in practice, this seems to be the only definition that fits. For example:

- The Rhode Island Red exists with two different combs: single and rose comb. By contrast, the only difference between a Belgian bantam and a Waas bantam is that the first has a single comb and the latter a rose comb.
- While Japanese bantams (Chabo) exist in more than 30 different colours, the only differences between a Brakel, a Zingem laying fowl, a Zwalmvalley fowl and a Zottegem are small differences in colour.
- The type of a Wyandotte in Germany is totally different than the original type.
- Araucanas are very different in many countries. A breed-specific characteristic in some countries are the ear tufts. This is caused by an autosomal, dominant, lethal gene, the so-called Et-gene. Some of the offspring don't have these typical tufts and are not considered as 'belonging to the breed', although they are genetically true Araucanas, except for one allele.

When talking about conservation of breeds, it is very important that conservationists are familiar with these unwritten rules in order to make well-founded decisions.

Chicken biodiversity

The Food and Agriculture Organisation (2007) estimates that there are about 17 billion chickens in the world. Looking at the volume of meat and eggs produced each year (FAO-STAT, 2006), there are about 8.17 billion production animals, parental stock not taken into account. Hence, less than half of the global chicken population belongs to non-commercial flocks, possibly a local chicken breed. Comparing Scherf (2000) and FAO (2007) shows that knowledge about the risk-status of all known breeds is increasing, but that data are still incomplete and a lot of breeds are still unknown. From the 1,273 breeds known by name (FAO, 2007), only 25% are without any risk to extinction, 40 breeds are already extinct and for more than 40% the status is still unknown.

Looking to scientific literature, it is remarkable that most studies about genetic biodiversity cannot be compared at all. In all of them, heterozygocity is used to show how big inter- and intrabreed diversity is, but very often different markers are used, which makes comparison almost impossible. On the other hand, these studies inform us about the present situation within commercial and traditional local stocks, although more structured research is needed.

For commercial lines, there are differences between several production types. In broiler lines, heterozygocities are measured from about 55% up to 80% in some male broiler lines (see for example Crooijmans *et al.*, 1996; Eding *et al.*, 2002; Hillel *et al.*, 2003). Laying hens are less heterozygote and values differ with the bloodline: circa 50% heterozygocity within brown layers, circa 40% within white layers and circa 30% within Leghorn-based white layers (Crooijmans *et al.*, 1996; Eding *et al.*, 2002; Hillel *et al.*, 2003; Ivgín and Bílgen, 2002; Sharma *et al.*, 2001). By comparison, the heterozigocity of the different subspecies of Red Jungle Fowl used in the same experiments is about 60 to 65%.

Within traditional local breeds, there is a very large range: from about 20% up to 60% (among others: Eding *et al.*, 2002; Hillel *et al.*, 2003; Moiseyeva *et al.*, 2003; Niu *et al.*, 2002; Vanhala *et al.*, 1998; Wimmers *et al.*, 2000). There are several explanations for this. If a breed is very rare, the degree of inbreeding is much higher, due to a smaller breeding stock, resulting in less heterozygocity. On the other hand, local landraces have a much higher heterozygocity, because they are spread over large regions, resulting in much more genetic diversity. Most European breeds have a rather low heterozygocity for several reasons. Since poultry exhibitions were – and still are – popular, local chicken populations were split up into different artificial but genetically very related breeds. Also inbreeding is very common within show breeders, because homogene breeding populations are needed to breed top-animals with almost identical looks.

Nevertheless, local breeds do make an important contribution to chicken biodiversity. Eding *et al.* (2003) compared commercial layer and broiler lines with old Dutch breeds and found that the loss of founding-father genome-equivalents (Lacy, 1989) would be about 40% if the old breeds disappeared. Local breeds also have a very large aesthetic and historical diversity. Unfortunately, in contrast with heterozygocity, these characteristics cannot be measured objectively, but this does not mean that they do not need to be taken into account.

Choosing is not always losing

Since conservation schemes are expensive and difficult to set up, one has to make a choice between all existing breeds. It is very important to look first at the goal of conservation: maintaining genetic diversity or saving cultural-historical value for the future, keeping in mind that those goals imply two very different groups of breeds. Depending on a specific goal, several procedures can be used to make sure the preferred breeds can be saved from extinction.

First of all, one might decide to look at the risk-status of the breed. A critical view is needed, because some breeds are strongly related and have to be considered as one breed-cluster. This will result in a totally different risk-status, since the population-size will increase. Moreover, the FAO (2007) wonders if a critical breed is still the same authentic breed, because high inbreeding levels and genetic drift might alter the genetic composition of the population.

The best decision tool is the maximum-utility-strategy (Bennewitz *et al.*, 2007), where all diversity parameters are taken into account and not only the genetic diversity. As mentioned earlier, it is not easy to make certain decisions when one is not familiar with the existing breeds within a species. Dialogue with breeders and breeding organisations is necessary to make a good estimate of the breed's value.

'Who is responsible for conservation?' and 'who will invest in these expensive conservation programs?' remain pressing questions. The cultural-historical and aesthetic parts of diversity can be understood as a local concern. Therefore every local authority, society and individual has to decide if they are prepared to make a contribution to this issue and how much this would be. On the other hand, Mendelsohn (2002) rightly sees genetic diversity as an assurance for future food supply. Therefore the biodiversity topic is of global interest and requires international support.

Conclusion

Biodiversity consists of 3 aspects: genetic diversity, aesthetic diversity and cultural-historical diversity. Genetic diversity is needed to ensure future food production. Unlike the commonly held belief, the limited number of selection companies and the similar phenotypes within commercial stocks does not mean that the genetic diversity of chickens is reduced. Nonetheless, old breeds are responsible for diversity which cannot be found in commercial stocks. Local breeds also have a large aesthetic and cultural-historical value, but this cannot be expressed in an objective, numerical value. Therefore, estimating the total value of a breed is thus not just a simple calculation. Another difficulty is the ill-defined term 'breed'. One might say that a breed is whatever breeders say it is, because there are several exceptions on nowadays-used scientific definitions. One has to take these difficulties into account, while making well-founded decisions about which breed is worthy of conservation and which is not. The kind of diversity which has to be saved depends on the goal one wants to achieve. Therefore, it is possible that different groups of breeds are needed. Some breeds are much more related to other breeds, having common genetic, cultural-historical or aesthetical aspects. In sum, not every breed will add something extra to biodiversity needs. Here, choosing is not always losing.

References

Bennewitz, J., Eding, H., Ruane, J. and Simianer, H. (2007). Selection of Breeds for Conservation. In: Oldenbroek, K. (ed.) Utilisation and Conservation of Farm Animal Genetic Resources. Wageningen Academic Publishers, Wageningen, the Netherlands, pp. 131-146.

Burditt, L., Desilva, U. and Fitch, J. (1995). Breeds of Livestock [Online]. Available at: http://www.ansi.okstate.edu/breeds. Accessed 20 March 2009.

Crooijmans, R.M.P.A., Groen, A.F., Van Kampen, A.J.A., Van der Beek, S., Van der Poel, J.J. and Groenen, M.A.M. (1996). Microsatellite Polymorphism in Commercial Broiler and Layer Lines Estimated Using Pooled Blood Samples. Poultry Science 75: 904-909.

Eding, H., Crooijmans, R.P.M.A., Groenen, M.A.M. and Meuwissen, T.H.E. (2002). Assessing the contribution of breeds to genetic diversity in conservation schemes. Genetics, Selection, Evolution 34: 613-633.

Food and Agriculture Organisation of the United Nations (FAO) (2007). The State of the World's Animal Genetic Resources for Food and Agriculture. Rischkowsky, B. & Pilling, D. (eds.). Food and Agriculture Organisation of the United Nations, Rome, Italy, 511p.

FAO-STAT (2006) Livestock primary and processed [Online]. Available at: http://faostat.fao.org/site/569/DesktopDefault.aspx?PageID=569. Accessed 6 May 2008.

Hall, S.J.G. (2004). Livestock Biodiversity: Genetic Resources for the Farming of the Future. Blackwell Science Ltd, United Kingdom, 269 pp.

Hillel, J., *et al.* (2003). Biodiversity of 52 chicken populations assessed by microsatellite typing of DNA pools. Genetics, Selection, Evolution 35: 533-557.

Ivgín, R. and Bílgen, G. (2002). Estimation of Genetic Distance in Meat and Layer Pure Lines Using Randomly Amplified Polymorphic DNA. Turkish Journal of Veterinary and Animal Sciences 26: 1117-1120.

Lacy, R.C. (1989). Analysis of founder representation in pedigrees: Founder equivalents and founder genome equivalents. Zoo Biology 8: 111-123.

Lerner, M. and Donald, H.P. (1966). Modern Developments in Animal Breeding. Academic Press, London and New York, 269 pp.

Mendelsohn, R. (2002). Analysis: The challenge of conserving indigenous domesticated animals. Ecological Economics 45: 501-510.

Moiseyeva, I.G., Romanov, M.N., Nikiforov, A.A., Sevastyanova, A.A. and Semyenova, S.K. (2003). Evolutionary relationships of Red Jungle Fowl and chicken breeds. Genetics Selection, Evolution 35: 403-423.

Niu, D., Fu, Y., Luo, J., Ruan, H., Yu, X., Chen, G. and Zhang, Y. (2002). The Origin and Genetic Diversity of Chinese Native Chicken Breeds. Biochemical Genetics 40: 163-174.

Rodero, E. and Herrera, M. (2000). El Concepto de Raza. Un Enfoque Epistemológico. Archivos de Zootecnia, 49: 5-16.

Scherf, B.D. (2000). World watch list for domestic animal diversity. FAO, Rome, Italy, 726 pp.

Sharma, D., Appa Rao, K.B.C., Singh, R.V. and Totey, S.M. (2001). Genetic diversity among chicken breeds estimated through randomly amplified polymorphic DNA. Animal Biotechnology 12(2): 111-120.

Vanhala, T., Tuiskula-Haavisto, M., Elo, K., Vilkki, J. and Mäki-Tanila, A. (1998). Evaluation of Genetic Variability and Genetic Distances Between Eight Chicken Lines Using Microsatellite Markers. Poultry Science 77: 783-790.

Wimmers, K., Ponsuksili, S., Hardge, T., Valle-Zarate, A., Malthur, P.K. and Horst, P. (2000). Genetic distinctness of African, Asian and South-American local chickens. Animal Genetics 31: 159-165.

Comparative analysis of the inclusion of animal ethics in European animal welfare policy-making

Tjard de Cock Buning
Athena Institute, VU University, De Boellelaan 1085, 1081 HV Amsterdam, the Netherlands; tjard.de.cock.buning@falw.vu.nl

Abstract

A comparative analysis of the inclusion of animal ethics in animal welfare policy-making in four European countries (Austria, the Netherlands, Norway and the United Kingdom) reveals several similarities and differences in national experiences. This analysis shows that considerable similarities exist with respect to the framing of ethical issues in animal welfare policy-making. All four study countries have adopted quite similar animal welfare legislation that is further specified in guidelines and codes of practice. Probably the most interesting difference between the four countries concerns the interpretation and relative weight of the notion of enhancing natural behaviour in animal welfare legislation. The UK explicitly opted to reformulate this notion of behaviour as set out by the Brambell Committee into normal behaviour, thus introducing current farming practices rather than nature as the benchmark for measuring welfare. In Norway, natural behaviour refers somewhat ambivalently to either behaviour that can be observed in a species' wild relatives or in the animals when placed in natural or semi-natural environments. Austria and The Netherlands opted to embrace the notion of natural behaviour as one of the main guiding principles in animal welfare legislation but struggle with the application in more specified guidelines and codes of practice.

Keywords: governance, harmonisation, animal welfare

Introduction

There are several forces working to transform independent national legislation on animal welfare to meet a kind of European wide framing. At the macro level the EU system strives for trade and economical reasons towards harmonisation of national legislations. At the same level the Green Parties in the European parliament use the harmonisation argument to lobby for higher levels of animal welfare. At the grass root level, animal welfare is an important topic among concerned citizens and consumers organized in various, and sometimes highly professionalised, animal welfare organisations. At the level of national governments and politically engaged ministers a politically 'fair' compromise is sought between the European and societal pressures. However, 'fair' appears to be a complex trade off between national interests of the agricultural industry, historically grown legislation and implementation options, and societal arousal.

The usual response of administration to these forces is necessarily limited to the four instruments they have in their possession: legislative, economical, educative and modern (networks and processes). From a comparative perspective the legislative instruments are the most interesting as all European member states are intensively monitored and evaluated by the EU on their legislation, while the other three instruments are highly dependent on the national resources available. Especially from an ethical perspective the related legal and political documents (notes to the parliament) offer a wealth of normative arguments underpinning specific animal welfare policy. Interestingly four countries recently presented new formulations of their national animal welfare acts: the Netherlands, UK, Austria and Norway. As a result of a grant from The Dutch Ministry of Agriculture, Nature and Food (LNV) we were able to investigate and compare the guiding visions, being explicit or implicit, beyond those reforms.

The analyses of the ethical assumptions and framing of the UK, Norway and Austria are presented in side papers. In this paper I will focus on the comparison of the countries (and taking on board the Netherlands) and reflect on the outcome.

Methodology: choosing a frame for comparison

We and our partners from UK, Austria and Norway, analyzed the new national proposals for animal welfare legislation according to the following 4 research questions:

1. In analogy with the three professional normative principles in laboratory animal science (3 Rs: reduction, refinement, replacement) or research/treatment on patients (beneficence, non-maleficence, justice, autonomy) are there any comparable principles that structure in similar sense the practice of animal agriculture, fishery, wild life management?
2. How are these principles positioned in national legislation on animal welfare with respect to (non-lab) animal use? Detailed references to articles in acts and governmental notes to the parliament, discussion in parliament (minutes indicating pros and cons).
3. What types of governance instruments and structures have been erected to guarantee that these ethical principles are applied and implemented?
4. How do these governance instruments function in practice?

The rich material of the national reports showed that the legalistic and political formulations could be interpreted, positioned and labelled according to the classification of ethical principles as defined by Mepham (2000) in his ethical matrix approach for the agricultural domain, which is inspired by and in analogy (though not identical) with the classification Beauchamp and Childress (2001) proposed for the biomedical domain. However, the category of 'justice' appeared to be empty in all legislative texts. However, synonyms like 'fair balance between...', 'proportionality' were mentioned but left to structures of committees and processes of deliberation without detailing. In Table 1 an overview is presented of the abstract ethical principles at the top and the more detailed operational translations to law and order, downward. Apart from a highly relevant philosophical discussion about differences between concepts such as intrinsic value, inherent worth and being a subject of life, it is interesting to see that an analogous (legal) policy concept is used in three countries to describe the basic premises of the legal reform. The UK is clearly an exemption in this set of 4 countries. Referring basically, also in parliamentary debates, to (an implicit) 'veterinarian' notion of welfare.

The category *norms* refers in this table to ethical/legal rules that define what is said to be allowed or not (prohibition and obligation) and *criteria* refer here to terms upon which an inspector is able to verbalise. At the bottom, a row *concepts* is added, listing new policy constructions being defined to facilitate implementation of the policy intentions, like putting in place an Animal Council in Austria, defining 'Integrated animal friendly husbandry systems' in the Netherlands.

The political-legalistic perspective

Almost all analysed texts animal welfare acts have been formulated at the level of a legal framework, e.g. clarifying the main goals and subordinated lines of issues that will be defined in the near future or reference is made to existing acts and rules. The related notes to parliament clarify the philosophy behind the structure of the new legislation. For those familiar with the national history of the animal welfare debate and legislation, these documents disclose the shifts in the ongoing process of political negotiation, i.e. what stays, what is changed in the reform and what might be subject to change in the future. The last two aspects reflect the impact of societal debates and show how the state intends to deal with these issues in terms of political-legalistic framing.

Table 1. Comparison between four countries under study (columns) and a 'hierarchy' of normative notions from guiding visions towards operational criteria for normative deviant behaviour.

	The Netherlands	Austria	Norway	UK
Guiding principles	Intrinsic value	Intrinsic value Mitgeschöpflichkeit (Fellow creature)	Intrinsic value	Animal welfare
Ethical principles				
Doing well (beneficence)	Duty of care regarding health and wellbeing	Well being and life	Treating animals well, with care, in a positive sense	Promotion of welfare
Non-maleficence	Protection against unnecessary suffering Respect the limits of adaptation	Protection against unjustified suffering and pain Respect the limits of adaptation	Protection against unnecessary suffering and pain	Prevention of harm
Autonomy	Animals are free to choose their own natural species specific behaviour Perspective of the animal is guiding	Reverence for life	Considering natural needs and instincts	Freedom to express normal behaviour
Norms	Act when animals are in need of help No interventions without medical reason Caring for wild and nature Transparency	Act when animals are in need for help No interventions without medical reason Do not torment No killing without good reasons Prohibition of unjustified pain, suffering or injury or expose to heavy fear	Healthy well functioning animals – physically and mentally	Positive welfare is encouraged and 'Unnecessary suffering' will be prosecuted
Criteria	Right food Right physical housing, right social housing No pain No suffering No injury Not sick Normal behaviour Suffering	Right physical housing Right social housing Proper care of sick animals No harm inflicted to the offspring Only animal species that can be kept without welfare problems No breeding programs inflicting negative welfare	During slaughter no pain, stress, anxiety and suffering	Right food Right physical housing Right social housing No pain No suffering No injury Not sick Normal behaviour Forbidden to: - mutilate - poison - animal fighting

Table 1. Continued.

	The Netherlands	Austria	Norway	UK
Criteria (continued)	Forbidden to: - mutilate - poison - animal fighting - long transportation No breeding programs inflicting negative welfare During slaughter no pain, stress, anxiety and suffering.			
Concepts	Integrated animal friendly husbandry systems Robustness Responsible consumer	Animal welfare council Animal ombudsman Public understanding for animal protection	Basic needs Quality of life	Unnecessary suffering Promotion of welfare

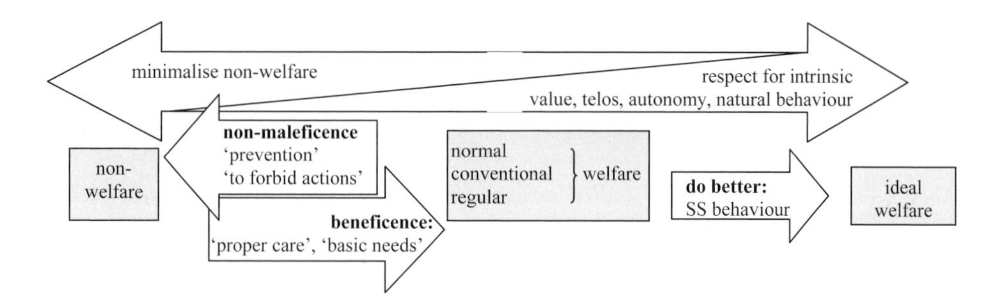

When we analyze the national texts on these aspects, it shows an overall consensus regarding a policy to guard the 'the Five Freedoms of Brambell': hunger and thirst; discomfort; pain, injury and disease; normal behaviour; fear and distress (Table 2) (Bennet *et al.*, 2004). However, all countries studied, except the UK, interpret the fourth freedom of '*normal* behaviour' as '*natural* behaviour' instead. By changing the scope from 'normal' to 'natural' they pay tribute to their national debates on 'good farming', in a normative sense, taking on board referential notions like 'naturalness' and 'quality of life'.

When analysing in more detail the assumptions and drives behind the animal welfare debate in relation to the actual structuring of the legislations, one might reframe the policy actions into three major domains of the animal welfare debate (each including ELSI aspects like public issues, welfare research and legal provisions): refrain from doing harm (non-maleficence), do good (beneficence) and do better (naturalness). The first two are nowadays mainstream discourses prescribing what should not be allowed and what is considered professional husbandry, in conventional farming. However, in Norway, Austria and Holland policy documents refer to a category of 'ought's', or conceptions of the good, that is in most countries not part of the conventional practice of husbandry. This is the domain of policy intentions, most often lobbied by animal welfare organizations, and as such granted in the political arena. At this level of governmental and political ethics the intention to comply with species specific behaviour or the *telos* of the animal are articulated in a specific and limited sense to implement positive changes in respect to conventional husbandry. For that reason this category of prescriptive maxims challenges farmers to increase 'conventional animal welfare' towards some conception of 'ideal welfare'. A conception that is, however, much debated. It is important to notice that both 'beneficence' and 'do better' are part of a 'positive welfare' debate (arrows heading to the right), but the implicit reference of the former is strongly driven by the notion of 'non welfare'.

Concluding remarks

This comparative survey of the four countries made clear to us that the echo's of four famous animal ethics (i.e. Singer, 1975; Regan, 1985; Rollin, 1981; Verhoog, 2007) reached the administration in the form of political intentions for further policy actions, except for the UK. At the same time these intentions come in direct conflict with the vested and politically acknowledged values of farmers as agricultural entrepreneurs, and the state running of an export market. At the level of state policy, the ministry of economic affairs and the ministry of agriculture constitute inter-institutionally a dilemma without a solution. Neither mono-ethical theories nor pluralistic ethical theories appear to be well equipped to guide policy makers towards rational, defendable and fair solutions. What seems to be necessary is the development of a new domain in *applied ethics*, that has its locus at the level of the state: ethical models of distributive justice.

Table 2. Development of governmental policy and instruments at three domains of the animal welfare debate.

Negative welfare debate				Positive welfare debate	
refrain from doing harm (non-maleficence)		do good (beneficence) (obligations)		do better (realize their telos) (attitude, political intentions)	
Not sick (B3)	NAUH	Proper food (B1)	NAUH		
Not injured (B3)	NAUH	Right social housing (B4)	NAUH		
No pain (B3)	nAUH	Right physical housing. (B2)	NAUH		
No suffering (B5)	nAU	Care for basic needs (B4)	NAUH	Physiological and behavioural needs	NA
No distress (B5)	H			According latest knowledge of animal welfare	NA
No abnormal behaviour (B4)	U	Guard the freedoms of Brandell (B1-5)	NAUH	Natural species specific behaviour(b4)	NAH
No anxiety and stress (B5); Not helpless (B4)	HN	Healthy and well functioning animals(B3-4)	N		
No hunger/thirst (B1)	nNA	Proper care by illness (B3)	NAH		
No unnecessary med. mutulations / interventions	OU	Provide help when in need (B3)	AH		
No harm by new procedures	N	Competence courses animal owners	NH		
No harm by new technologies	N	Slaughtering without pain, anxiety, and suffering(B5)	NH		
No harm in the offspring	A				
No breeding programs with negative welfare effects	A				
No animal fights (B5)	NAUH				
No bestialities	N				
No life bait (B5)	N				
No unhealthy food	H				
No access to waste/ manure	H				
Animals not as prize	NH				

Abbreviation: N=Norway, A= Austria, U= UK, H= the Netherlands. B=Brambell freedoms: 1- hunger and thirst; 2-discomfort; 3- pain, injury and disease; 4-normal behaviour; 5-fear and distress. Lower case n = Norway when the prohibition is explicitly related to a prima facie reflection, i.e. pain without cause.

References

Anonymous Austria (2005). Federal Austrian Animal Welfare Act (Bundesgesetz über den Schutz der Tiere).

Anonymous UK (2006). Animal Welfare Act.

Anonymous Norway (2008). Animal Welfare Act.

Anonymous The Netherlands (2009). Animal act.

Bennett, R.M., Broom, D.M., Henson, S.J., Blaney, R.J.P. and Harper, G. (2004). Assessment of the impact of government animal welfare policy on farm animal welfare in the UK, Animal Welfare 13: 1-11.

Beauchamp, T.L. and Childress, J.F. (2001). Principles of Biomedical Ethics. OUP, New York. pp. xx.

Mepham, B. (2000). A Framework for the Ethical Analysis of Novel Foods: The Ethical Matrix. Journal of Agricultural and Environmental Ethics 12, 165-176.

Regan, T. (1985). The Case for Animal Rights. University of California Press, California.

Rollin, B. (1981). Animal Rights and Human Morality. Prometheus Books, New York.

Singer, P. (1975). Animal Liberation: A New Ethics for Our Treatment of Animals. New York: Avon Books.

Verhoog, H. (2007). The tension between common sense and scientific perception of animals: recent developments in research on animal integrity. Wageningen Journal of Life Sciences 54(4): 361-373.

Principles of animal ethics in Norwegian legislation and governance

E.M. Forsberg
Work Research Institute, P.O. Box 6954 St. Olavs plass, 0130 Oslo, Norway; ellen-marie.forsberg@afi-wri.no

Abstract

Norway is revising its animal welfare legislation these days. In the current *Animal Protection Act* the basic animal ethics principles are to avoid suffering, to treat animals well and to consider their natural needs and instincts. In addition, a principle stating the need to balance our duties towards animals with the needs and interests of humans is expressed by the formulation 'unnecessary suffering'. These principles (only with slightly different formulations) are retained in the new act. The novelty of the new act is shown through its insistence on enhancing respect for animals. New is also the recognition of 'the animal's value of its own' (intrinsic value), which is stated in the Act's general provisions. Respect for animals and their intrinsic value is hard to follow up in practice, particularly at a time when available resources for inspection seem to have decreased. With the new legislation increased emphasis is therefore put on building good attitudes among animal keepers and the general public. The challenge is to provide adequate measures to achieve in practice the intentions of respect for animals expressed in the new act.

Keywords: animal ethics, ethical principles, animal welfare legislation

Introduction: historical developments

Cruelty against animals has been forbidden in Norway since 1842, but Norway did not have a separate act on the protection of animals until 1935. With this act Norway became one of the most progressive countries in this area. In the 1960's technological development in animal husbandry necessitated a revision of the law, and in 1974 the *Act on Animal Protection* came into force. Even though the law has been revised on several occasions, this is still the one regulating this area. The general provision of the Act is that: 'Animals are to be treated well and the animal's instincts and natural needs shall be considered so there is no risk that it suffers without cause.' In addition to the *Act on Animal Protection* there are a number of regulations (26 regulations per 2002/3), guidelines, etc.

The current legislation is under revision due to major changes over the last 30 years. Society's attitudes towards animals have changed, the structure of agricultural production has changed, new production species have been introduced (fish, ostrich, deer, etc.), knowledge of animal behaviour has increased, etc. Norway still wants to be in the forefront internationally with regard to animal welfare, and the new *Act on Animal Welfare* is to reflect this. The change of name from Animal *Protection* Act to Animal *Welfare* Act implies a stronger focus on the individual animal. A proposal for a new act was sent on hearing and a revised version was then submitted for discussion in parliament. In the end of March this year the responsible parliamentary committee submitted the final recommendation containing the text that will become the new act. This text included – as we shall see – some slight changes from the version presented to the parliament in the autumn of 2008. The new Act builds on the parliamentary white paper *Proposition to the Storting No 12 (2002-2003) Animal Welfare and Animal Husbandry*. In this white paper a concern is voiced that the current act's concept of 'unnecessary suffering' is not sufficiently precise. In the white paper an explicit ethical platform was developed:

- Animals have a value of their own. Handling of animals shall be carried out with care and respect for the species. This involves having extensive consideration for animals' natural needs and to actively prevent sickness, injuries and pain.
- Persons who are caring for animals shall have knowledge about the animal's behavioural needs and its needs concerning nutrition, social and physical environments. They shall take responsibility for the animal's basic needs, and treat them correctly in the event of sickness and injury.
- Animals shall be kept in environments which give them a good quality of life.
- Healthy functioning animals – physically and mentally – shall be a condition for all types of breeding.
- Before new technological solutions are used it must be proven probable that these solutions do not reduce animal welfare. New operational methods must have as little negative impact on animals as possible.

The hearing draft of the new act included most of the ethical platform from the white paper. In the hearing proposition it was stressed that the goal is to 'enhance the respect for animals by strengthening their position in society' (1.5.1). The paragraph stating the purpose of the act said: 'The purpose of this Act is to facilitate good animal welfare and respect for animals.' The third paragraph stated: 'Animals have an intrinsic value irrespective of the usable value they may have for man. Animals must be treated well and protected against danger of unnecessary stresses and strains. Consideration shall be given to the animal's physical and mental needs on the basis of the animal's distinctive character and its ability to have positive and negative experiences.' The law proposition presented to the parliament in November 2008 showed an important change in that § 3 now stated only that '[a]nimals are to be treated well and be protected against danger of unnecessary stresses and strains.' However, in the final version presented by the parliamentary committee the notion of intrinsic value was reintroduced. The new act must be considered a strengthening of the protection of animal welfare.

The basic structure of norms

The current act has been amended several times and is a mixture of very general principles and very specific directives. The body of regulations accompanying the act is also quite diverse, and the same concerns are formulated in slightly different ways in the different regulations. The new act will be more generic and the body of regulations will be reviewed and adjusted to be more internally coherent, so that the whole legislation gets more of a systematic structure. There is also an intention of reducing the number of regulations, for instance by combining the regulation of all production animals into one regulation with specific chapters on the specific species. The goal is to have a logical system starting with the act's purpose and general provisions stating the overarching principles, which are then translated into general principles in the Act and more specific directives in regulations.

In the 1974 *Animal Protection Act,* the paragraph of purpose states that: 'Animals are to be treated well and the animal's instincts and natural needs shall be considered so there is no risk that it suffers without cause.' In this paragraph we can identify three general ethical norms that structure legislation in Norway:
a. avoiding suffering;
b. treating animals well, with care, in a positive sense;
c. considering natural needs and instincts.

The proposed new act states that: 'Animals are to be treated well and be protected against danger of unnecessary stresses and strains.' In the new act 'avoiding suffering' has been replaced by 'protected against danger of stresses and strains', which is intended as a strengthening. The new act does not include a statement on natural needs and instincts in the introductory paragraphs, but includes a later paragraph under 'special provisions on the keeping of animals': 'The animal keeper is to make sure that animals are kept in environments that provide good welfare based on species specific and individual needs, including providing the possibility for stimulating activities, movement, rest and other naturally occurring behaviour. The living environment of the animals is to strengthen good health and contribute to safety and satisfaction.' Although formulations differ somewhat I believe it is fair to say that the basic principles of the old act are taken on into the new legislation.

In some sense it is artificial to treat the three principles as separate. For instance, the way the current act describes accommodation probably refers to all three: 'Persons who own, or have in their charge, livestock, pets, or animals kept in captivity in other ways, shall ensure that an animal has fully adequate accommodation, with enough room, suitably warm, with sufficient light and access to fresh air, etc., in accordance with the needs of the animal species in question.' (§ 4)

In addition to the three principles presented above, we may identify two other ethical concerns that are common to the old and new legislation. The use of the words 'risk' and 'danger' may seem to imply a principle of caution. It is not clear if this is identical with the precautionary principle. Many of the hearing letters to the new act stressed that the precautionary principle should apply in this area, but it is probably reasonable here to interpret the principle of caution as an instruction, in cases of doubt, to err to the advantage of the animal. Another concern that can be found in both the current and the proposed act is a clause that breeding shall not provoke common ethical reactions. This may be understood as a specific ethical clause, applying to a particular issue. It is the responsibility of the Food Safety Authority to make a judgement on this, and this judgement shall take into account the scientific evidence on welfare consequences. Specific issues related to this clause will also be of interest to the Council on Animal Ethics.[4]

Intrinsic value and respect for animals

An important development with the new act is the notion of intrinsic value and respect for animals. The new act states that '[t]he purpose of this Act is to facilitate good animal welfare and respect for animals' and it acknowledges the animal's 'value of its own' (intrinsic value). As we have seen this was included in the first hearing version, removed in the version presented to parliament (apparently because the Ministry of Justice and the police considered it hard to operationalise and therefore not apt for inclusion in a law) and then reincluded in the final version. It is stressed that intrinsic value only has a symbolic value in the act and has no practical consequences other than supporting other paragraphs. I will claim that the statement on the animal's intrinsic value and the act's purpose to facilitate respect for animals are referring to one and the same principle: respect for intrinsic value. It can be argued that the principle of respect for intrinsic value is simply a way to accord animal's legitimate moral standing, i.e. that they are morally considerable. This may imply that they have moral rights or at least confirm that some general ethical principles apply to animals. It is made clear, however, that the proposed act is not intended to accord animals rights. It is also stressed that the statement of respect for animals is intended to have practical implications. The hearing document explains: 'The act shall contribute to animals being recognised for their own characteristic properties and as having intrinsic value. The requirement to show respect for animals implies restrictions on use and killing of animals that exceed those that follow from pure welfare considerations.' (p. 17). Thus the principle is intended to accord moral protection over and above avoiding suffering and treating animals well. An example here may well be the protection of animals against bestiality (§ 14 in the new act). As this prohibition is unrelated to the possible suffering of the animal it seems implicitly to refer to the violation of the integrity of the animal.

[4] The Council on Animal Ethics is addressing the issue of breeding in 2009.

The principle of balancing and justification

The qualification 'unnecessary' or 'without cause' implies that one may justifiably violate the animal ethics principles if one has good reasons for it. The whole concept of 'justified suffering' has been debated: critics have objected that Norway has a law that defends the suffering of animals (Larsen, 2003). The Ministry has tried – unsuccessfully – to explain the concept of 'unnecessary suffering' in the commentary to the proposed Act: 'The Ministry is of the opinion that the law cannot protect animals against all burdens and stresses. The most important is that one seeks to hinder danger of burdens or stresses that are unnecessary or unacceptable, and that can be avoided.' (p. 23)

The existence of qualifications like 'unnecessary' means that the basic ethical principles identified above must be considered *prima facie*, i.e. that they hold unless they must yield to other ethical concerns that in the particular circumstance are held to have stronger force. Such moral principles will typically be principles that protect the interests of human beings and in practice it will be difficult to distinguish genuine moral principles from simple economic interests here. The balancing of the *prima facie* principles must be justified *all things considered*. What justified suffering is a matter of judgement and this judgement may easily be influenced by economic interests. Therefore there are many regulations that guide the use of judgement when determining what 'unnecessary suffering' is and this alleviates the use of judgement. Some actions are forbidden and a number of functional and technical requirements are given by regulations. Still, as all situations cannot be regulated, judgement must in many cases be used. This is for instance necessary for evaluating new technology developments in breeding, new production methods, etc.

It may be interesting to consider an example of such balancing in Norway. The Council for Animal Ethics advised in 1998 the government not to allow catch and release sports fishing for ethical reasons. Indeed, the activity does not seem to be necessary. The former Animal Health Authority also stated that this was in conflict with the Animal Protection Act. Still, in May 2008, the Ministry of Fisheries and Coastal affairs notified interested parties that there would be no sanctions against catch and release fishing. The Ministry stated that they will not use catch and release as a strategy for managing river fish stocks, but will allow private persons to practise this form of sports fishing. Critics say that the Ministry here was more interested in protecting the economic interests of land owners in the districts, and that they hesitated to make a decision that would be unpopular among a fairly substantial share of the population (voters).

The governance system

Above, I have presented some central ethical principles in Norwegian animal welfare legislation. We shall now take a brief look at how the principles and norms are followed up in practice.

Over the last ten years there have been a number of changes in the governance structure in Norwegian animal welfare legislation. Earlier animal protection was secured by the Norwegian Animal Health Authority and a network of local and regional inspectors that at the same time had commercial veterinarian practice. This allowed the inspectors to gain information about animal health in their area from their work in the field and from their customer contact, but it may also have made them less independent. In 2004 the Animal Health Authority was merged with other related authorities into the Norwegian Food Safety Authority (FSA). The animal welfare inspectors of the FSA are no longer veterinarians with a private practice and it has been claimed that they have more limited resources than before. The FSA claims that the fusion has not been to the disadvantage of animal welfare, and that there would likely not have been more resources available if the separate authority had been retained, but this is disputed. Being an animal welfare inspector is now a full time position and the number of inspectors is therefore reduced and each inspector's region has increased. Still, there has been an increase in the number of inspections carried out.

In addition to the inspectors there are also local animal welfare committees that inspect where there are no strict, technical requirements in legislation and where judgement therefore is crucial.[5] The reason for establishing such lay people committees is that the use of judgement (for instance on what is unnecessary suffering in a given case) should not be strictly expert based, but should be in touch with common sentiments in the population and among the stakeholders. The committees are to keep up to date on the keeping of animals in their district and are to perform unannounced inspections. If the situation is not satisfactory the committee is to guide the owner or they can issue injunctions. The FSA allocate funds to the animal welfare committees, functions as the secretariat of the committees, provide veterinarian and legal advice, and provide training of the committee members. The animal welfare committees and the FSA inspectors have only limited resources and have to prioritise their work in accordance with the general inspection strategy, which is risk based. This implies that they will focus on the situations where the risk of violation of the law is the greatest, at the expense of following up the principle of treating the animal well in the broader sense and the principle of allowing the animals to express natural behaviour. Only unnecessary suffering will be penalised and the police or the courts will dismiss the case if it is not thoroughly shown that the suffering is unreasonable in relation to the purpose. In 2007 the FSA and the animal welfare committees carried out in total 7441 inspections. Only in 71 cases was the situation reported to the police. A major weakness of the governance system is therefore that in spite of the fact that they are in contact with animal keepers, they seem to have only moderate impact on the situation in practice.

The use of animal welfare indicators, like behavioural indicators, is increasing. These indicators may include the presence of joy, curiosity, playfulness, etc. and absence of stereotypical behaviour. This may make inspection more time sensitive; that the animals seem to be well functioning at the time of the inspection does not guarantee that the welfare is good at other times. However, it is possible to indirectly inspect the ability to express natural behaviour, as well as inspecting whether the animals likely experience satisfaction, by inspecting their housing conditions, etc. Internal control systems may also to a larger extent be used with the new legislative regime. In the current Animal Protection legislation only the aquaculture industry is required to have internal control systems, but this may change. In that case it will be possible to inspect not only the situation at the specific time where the committee or the inspector visits the animal keeper, but also the animal keeper's routines for good animal keeping. In addition, mandatory courses on proper animal keeping are currently being developed for all areas of professional animal use, including transportation. More responsibility is therefore being placed on the individual animal keeper. Agricultural organisations have been complaining that the burden of documentation now imposed on people involved in animal husbandry is becoming so massive that it will soon be impossible to comply in practice.

Conclusion

The hearing document for the new act repeatedly stressed respect for the animal:

> 'The purpose of the act is to enhance respect for animals. This amounts to a difference with regard to what is expressed in the current animal protection act. The Act's purpose with regard to good animal welfare and respect for animals will to a large extent be related. Good animal welfare will be a good foundation for respect. Similarly, respect for animals will contribute to good welfare.' (p. 17)

[5] This is mostly the keeping of pets, but the committees may also inspect farms, etc.

The stress on respect implies a focus on attitudes. Of course it is difficult for an inspector to assess an animal keeper's respect for the animals. An important function of the law is therefore simply to strengthen good attitudes among the public and animal keepers and according animals intrinsic value is intended to contribute to achieving this goal. It seems there is a tendency away from state control towards individual responsibility. An illustration of this is the focus on whistle-blowing in § 5 in the new proposed Act. The local community is encouraged to engage more with animal keepers' actions, either by complaining to the animal keepers directly or by notifying the police, the animal welfare committees or the FSA. § 4 makes clear that everyone has a duty to help suffering animals.

In order for this tendency towards more individual responsibility and respect for animals to work, legislation must be followed up – not only with regulations – but with education of the public and animal keepers. Moreover, public discussion of what respect for animals means in practice must be stimulated. Animal welfare Non-governmental Organisations (NGOs) play an important role here, but The Norwegian Council for Animal Ethics should also be mobilised for this work. However, this requires providing the Council with adequate resources to take a sufficiently proactive position. It is important that the new legislation is followed up with adequate information and motivation measures so that the celebrated notions of respect for animals and intrinsic value do not end up as paper tigers.

References

Council for Animal Ethics (1998). Angling – 'catch and release'. Available at: http://org.umb.no/etikkutvalget/. Accessed 1 April 2009.

Larsen, G. (2004). Dyrevern eller samfunnsvern?. Unipub, Norway.

Animal Protection Act. Available at: http://www.lovdata.no/all/nl-19741220-073.html

Ot. prp. nr. 15. (2008/09). Om lov om dyrevelferd. http://www.regjeringen.no/nb/dep/lmd/dok/regpubl/otprp/2008-2009/otprp-nr-15-2008-2009-.html?id=537570&epslanguage=NO. Accessed 1 April 2009

St.meld. nr. 12. (2002-2003). Om dyrehold og dyrevelferd. Available at: http://www.regjeringen.no/nb/dep/lmd/dok/regpubl/stmeld/20022003/Stmeld-nr-12-2002-2003-.html?id=196533. In English: http://www.regjeringen.no/upload/kilde/lmd/bro/2002/0003/ddd/pdfv/246168-parliamentary_report_number_12_on_animal_husbandry_and_animal_welfare_recovered.pdf. Accessed 1 April 2009.

Innst. O. nr. 56. Innstilling fra næringskomiteen om lov om dyrevelferd. http://www.stortinget.no/Global/pdf/Innstillinger/Odelstinget/2008-2009/inno-200809-056.pdf Accessed 1 May 2009.

Ethics for vets: can ethics help to improve animal disease control?

S. Hartnack[1], H. Grimm[2], P. Kunzmann[3], M.G. Doherr[1] and S. Aerts[4]
[1]DCR-VPH, Vetsuisse Faculty, Univ. Bern, P.O. Box, 3001 Bern, Switzerland; sonja.hartnack@itn.unibe.ch
[2]Institute Technology-Theology-Natural Sciences, Marsstr. 19, 80335 Munich, Germany
[3]Ethics Center Jena, Zwaetzengasse 3, 07743 Jena, Germany
[4]Catholic University College Ghent, Belgium

Abstract

During major animal disease outbreaks in the EU controversial discussions arose about the ethical soundness of mass culling and vaccination. The ethical question 'what should I do, all things considered?' (Mepham, 2008), is clearly difficult to answer in this context especially for veterinarians who are not trained in ethical thinking. To clarify some of the ethical dilemmas which veterinarians face in their professional life, interdisciplinary workshops with ethicists and veterinarians from Switzerland, Germany and Austria were initiated. The discussions revealed that in an emergency situation like a contagious animal disease outbreak, official veterinarians are faced with various tasks and corresponding conflicting responsibilities. In some cases a clear lack of competence in specific fields, or responsibilities which reach beyond the professional competences of a veterinarian, were identified. The most prominent ethical dilemma was found in the attention paid to an efficient and fast disease control including culling interventions and animal welfare questions. In discussions on potential role conflicts of official veterinarians, three different and not necessarily congruent veterinary role models evolved, those of: (1) the general public (society); (2) the veterinary professional organizations and; (3) the self-perception of the individual veterinarian. Additionally, an unresolved question became evident: whereas the perspective of animal welfare ethics is focused on the individual animal (where transgressions are to be found on individuals) it remains open for discussion how this perspective could be extended to animal populations. Concerning the question about the (un-)justified killing of animals by veterinarians with regard to the 'dignity of creature' concept in the constitution of Switzerland, three dimensions became obvious. Whereas the first two dimensions are focused on the animal's life (future and past) or the *telos*, the third dimension incorporates the concept of dignity and the animal-human bond. With the aim of identifying and discussing the ethical dilemmas in a prevailing animal disease, Bluetongue disease, currently emerging in North-Western Europe, was chosen. In different workshops the ethical matrix developed by Ben Mepham and the animal disease intervention matrix have been applied to Bluetongue disease control scenarios, with an intervention emphasis on compulsory mass vaccination. This matrix approach helped workshop participants to identify, structure, weigh and partly resolve the dilemmas mentioned.

Keywords: veterinary medicine, animal disease intervention matrix, dignity of creature

Veterinarians in animal disease control

During major animal disease outbreaks in the EU controversial discussions arose about the ethical soundness of mass culling and vaccination. Compared to the amount of literature referring to ethical questions in animal experimentation, animal production systems or genetic engineering, literature about ethical deliberations in the context of animal disease control remain scarce. In particular, the highly contentious topic of mass culling as a means to stop the spread of contagious diseases mainly appears in case reports of outbreaks of avian influenza, food-and-mouth-disease or classical swine fever, but is rarely subject of a systematic ethical deliberation. Veterinarians, who are in charge of decision-making and the implementation of animal disease control measures, find themselves in tensions between the officially prescribed national or EU policy and a public which becomes more and more critical. The EU policy in this context still relies on the four pillars of animal disease control as they were implemented in the 19th century as the foundation of the official animal disease control policy: (a) early detection and notification of suspect cases; (b) containing the spread by transport restrictions; (c) culling and disposal of infected and suspect animals; (d) compensation of the owner for financial losses. This strategy, which has been successful in limiting the burden of disease or even eradicating a number of various diseases, led to the decision in the beginning of the 1990s to prohibit vaccination against food-and-mouth-disease, classical swine fever and avian influenza. The rationale behind this decision was the objective to build up healthy or naïve animal populations in the EU which never had had contact with infectious pathogens, neither in the form of field virus nor in the form of vaccine virus. Still it was considered that outbreaks due to the introduction of a pathogen were possible. But this event would have been immediately detected in a non-vaccinated population which would show distinct clinical symptoms compared to a population with a number of vaccinated animals where an infectious pathogen could possibly spread some time before being detected. To maintain the status of being free of a certain pathogen, culling of infected and susceptible animals was considered necessary. This approach was considered to be cost-effective and in line with animal health on a population basis. Subsequent to these decisions, legislation concerning international trade has been amended and countries with a certified 'freedom of disease' status may, accordingly to the WTO-SPS agreement, exert their right to impede the trade of animals and animal products.

In the meantime, the societal attitude towards animals, their role or function and the valuation of their moral status have changed (Kunzmann, 2007). The non-integration of these changes in – most of – the contingency plans for animal disease control poses a problem and may be one of the reasons for the ethical dilemmas faced by veterinarians in animal disease control. This became evident in the analysis of the crisis after a successful containment of a FMD outbreak in the Netherlands. In this case it was concluded that 'the ignorance of the societal function of animals and countryside was the cause of the crisis' (Van der Zijpp *et al.*, 2004).

With the aim of defining the relevant ethical components for veterinarians in animal disease control and to clarify ethical dilemmas, interdisciplinary workshops with ethicists and (official) veterinarians were initiated (Hartnack *et al.*, 2009). The following questions were tackled:

1. What are the constitutive aspects of ethical dilemmas faced by official veterinarians in animal disease control?
   - What are the tasks he or she is responsible for?
   - How do these tasks conflict in practice causing ethical dilemmas?
2. What are the different roles of veterinarians? Is this conceived not by the tasks at hand, but by the expectations towards veterinarians of different stakeholders like the public, the media or even the veterinarians themselves?

Concerning the various tasks and responsibilities, they can be subsumed in six different fields: disease control, animal welfare, organization, occupational health and safety of the control personnel, support of the animal holder and communicating information. The first three fields should be clearly within the competence of official veterinarians. Concerning occupational health and safety during the case of emergency culling, assistance from (human) public health authorities should be requested. Further, communication was clearly not considered to be a competence a veterinarian is trained for. From the statements of the participants it can be concluded that it is important for the individual veterinarian to be aware of his/her abilities and about potential lack of competence in order to seek adequate support. The most important ethical dilemma was found in the attention paid to an efficient and fast disease control, including culling interventions versus animal welfare issues. In the case of emergencies it could easily happen that what is legally required and implemented – securing the least maltreatment of animals possible -will be temporarily abandoned. Since it is assumed that animal disease outbreaks will continue to occur at least sporadically and cannot any longer be considered as to be always exceptional, it is imperative for the decision-makers in animal disease control to be aware about these existing ethical dilemmas and to prevent them as far as possible.

Coming to the second question of the veterinarians' various roles and responsibilities it became evident that the understanding 'what a veterinarian should be and do?' depends on three different and non-congruent role perceptions. First the societal view and expectation could quite often be described as to be a vet like 'James Herriot'. Second, the veterinary organisations represent the professional role perception, eager to defend the veterinarians on the job market but mainly focusing on the largest of their interest groups, i.e. the veterinary practitioners and their responsibility for helping the individual animal. And thirdly there is the self-perception of the individual veterinarian; probably nobody chose the veterinary profession with the aim of killing animals most efficiently, and the 'healers' role model' is certainly predominant among veterinary students. In the heat of an animal disease outbreak the discrepancies between these different roles could aggravate the ethical dilemmas and impede the finding of adequate solutions. During animal disease outbreaks, veterinarians have very limited capacities to care for the most animal friendly and humane measures, since they are obliged to carry out the directives. There is little potential (for them) to change the parameters that led to these conflicts in those times; however, veterinarians would be in a position to suggest changes in the actual system that might help to prevent such conflicts in the future.

Additionally, an unresolved question became evident: whereas the perspective of animal welfare ethics and animal protection law is focused on the individual animal (where transgressions are to be found on individuals) it remains open for discussion how this perspective could be extended to animal populations. In a joint statement by the Swiss Ethics Committee on Non Human Gene Technology (ECNH) and the Swiss Committee on Animal Experimentation (SCAE), 'concerning a more concrete definition of the dignity of creation', it has been proposed that protection of populations should be laid down in other legislation e.g. conservation of nature and preservation of national heritage, or environmental protection (ECNH, 2001). But so far the criteria or principles on which the decision to kill the animals in one holding or region, in favour of other animals or animal populations, is taken remain unclear.

Veterinarians and the killing of animals

With the aim of clarifying the ethical relevant aspects of the killing of animals for disease control, the question about the (un-)justified killing of animals was approached in a broader way (unpublished data). The overall perspective was to answer the question: 'As a veterinarian, where and when would I feel – with good reasons – uncomfortable with killing animals?' An inventory of the numerous occasions when animals are killed with the involvement of veterinarians in a large sense was made. Some of the occasions were consistently accepted (sick and suffering animal, with no chance of recovering) or rejected (male day-old chicks). Some the occasions were inconsistently assessed and further differentiation was needed, e.g. if alternatives were considered. An approach was undertaken to define the ethical component which leads to (un-)justification of the killing of animals from a veterinarian perspective. Despite existing animal protection laws it was still felt that in some occasions killing is legal, but there is an intuitive impression that it is not morally justifiable. One example is the disposal of male day-old chicks from laying hen breeds which are put into a shredder since they have no economic value. It is assumed that they will not suffer and so a pathocentric argument cannot be used. Here the concept of 'dignity of creature' as laid down in the Swiss constitution and the animal welfare legislation which came finally came into force in September 2008, may offer new possibilities to base the intuitive moral judgment on reasons. In this perspective three dimensions that could be used to explain the (non-) acceptability of killing of animals became obvious. For the individual life of the animal it states that animals have an intrinsic value, independent from human considerations- close to Regan's position that considers (higher) animals as 'subjects of a life'. Therefore the expected or possible of life matters, of which we deprive an animal by killing it. Secondly, the past or 'fulfilled' life is a criterion. How much of the animals potential of life has been fulfilled – which sheds light in the case of the male day-old chicken obviously deprived of the main period of their life. In bioethics it remains controversial whether the concept of a *telos* could be used to assess clearly infringements of the first two dimensions; whereby *telos* means that living beings have a certain 'goal' or 'meaning' in their lives which ought to be fulfilled as completely as possible.

The dignity concept or the 'dignifying' of animals also becomes important within the third dimension. The dignity of the animal is embedded in the human-animal relationship and is deployed here. It was not only understood as something like a protective cover around the animal or as an intrinsic property of the animal. Rather it can be understood as an active relation where human dignity emerges when respecting the proper life of the animal. 'Dignity emerges' in this relationship, to the degree the human actor as an appreciating, dignifying person respects the attributed value of the life and the wellbeing of an animal. A violation or disregard of the animal's dignity will not only affect the animal, but due to the mutual, reciprocal relationship, also the human being (including the veterinarian who kills the animal). It may mean that disregarding the 'dignity of animals' may harm and affect human dignity. Referring to deliberation required by the animal legislation, it was concluded that the less complete or fulfilled the first and the second dimension of an animal's life are perceived to be, the more important the reasons to kill an animal must be. During the discussion it became evident that veterinarians in emergency situations, due to unwanted, surplus or non-economically profitable animals, may not dispose of alternatives to the killing of the animal and that the justification is based on the emergency situation. Relating to this, a main point of critique was the limits and constraints of the animal production system that forces veterinarians to act immorally. This applies especially to culling, where a strong case can be made that the present agricultural system and the correlating market, with its tremendous demand for transport of animals, meat and food, carries an unavoidable and high risk of spreading animal disease. Given the moral conflicts indicated before, veterinarians should participate in initiatives to modulate some of the factors that lead to larger outbreaks and the necessity of culling.

A practical approach: use of the ethical matrix and the animal disease intervention matrix for the BTV-8 mass vaccination

Bluetongue disease caused by the Bluetongue disease virus serotype 8 is considered to be an emerging disease in Europe causing up to 10% case- fatality rates in cattle and 40% in sheep (Conraths *et al.*, 2009). After its first appearance north of the Alps in August 2006, BTV-8 continued to spread and outbreaks occurred almost simultaneously in Belgium, France, Germany and the Netherlands. Since spring 2008 a vaccination is available. Concerning bluetongue, legislation in the EU was amended (1266/2007/EC) with regard to the control, monitoring, surveillance and restrictions on movements, but it remains the responsibility of the individual member states to decide about vaccination. Switzerland decided for a compulsory mass vaccination for cattle, sheep and goats. In some areas this decision was controversially discussed and refused by some farmers. In a workshop with participants originating from the veterinarian or closely related fields, the ethical matrix (EM) developed by Ben Mepham (Mepham *et al.*, 2006) was discussed and filled out. The ethical matrix is a conceptual tool designed to help decision-makers to reach sound judgements or decisions about the ethical acceptability and/or optimal regulatory controls for existing or prospective technologies in the field of food and agriculture. Most of the participants did not know the EM before. Hence it was introduced and applied in moderated small group discussions. In a feedback round, the participants expressed their belief that the EM provided a useful approach to get an overview about the situation and to initiate fruitful discussions on ethical issues in the field of animal disease control. It also became evident that further considerations about the importance of the different principles and their associated values are needed in order to decide about appropriate animal disease strategies. To do this, it is necessary to reflect on the methodology and limitations of practice oriented approaches (Siep, 2008; Zichy, 2008).

In two different workshops, the animal disease intervention matrix (ADIM) was applied. The aim of these workshops – one primarily with veterinarians and the other mainly attended by experts from agriculture – was to compare and contrast various disease interventions applied in the case of BTV-8 on the basis of relevant indicators. These indicators include ethics, practicability demands, efficiency, etc. The ADIM was originally developed for assessing control strategies for potential H5N1 outbreaks in Belgium (Aerts, 2006). Some of the indicator questions of the original ADIM were clearly focussed on H5N1 and had to be amended in order to be applied to BTV-8 mass vaccination scenarios. The matrix aims to identify the ethically best animal disease intervention scenario. In order to segment the general problem into smaller parts, 15 different objectives that a good disease control scenario should achieve are considered. In the spreadsheet-based ADIM the impact of different disease control scenarios or methods on the objectives is assessed by answering indicator questions and assigning a score to them. Finally an overall score for every scenario is given, allowing users to compare (rank) the different scenarios with regards to their overall stakeholder 'acceptance'.

During these two workshops with veterinarians and experts from agriculture, it became evident that the ADIM offers the possibility to approach the complex issue of BTV-8 vaccination in a structured and transparent way and even allows respect for controversial points of view. Interestingly the two workshops with different stakeholders yielded similar results.

It was highly appreciated that this approach helped to structure the discussions and to prevent overlooking important aspects. It became evident that there is a need to reach a consensus between different stakeholders about the overall aim of an animal disease control strategy, preferably *before* a real outbreak takes place. In the case of BTV-8 it remains open if the disease control aim is to eradicate the virus or to mitigate its impact. This question, however, cannot be answered by the ADIM; it needs to be provided by the disease 'managers' (veterinary authorities) in collaboration with scientists.

Conclusions

Whereas in human medicine, a public health ethics, distinct from medical ethics, has recently evolved, no explicit veterinary public health ethics seems to be 'emerging' so far. The conclusions of the abovementioned workshops and the approach taken in the systems we discussed here at least suggest two things: (1) there is a clear need for systems that help evaluate disease control measures and help build consensus, (2) that ethics and ethical thinking can have a profound impact on how disease control is executed *and* perceived.

References

Aerts, S. (2006). Practice-oriented models to bridge animal production, ethics and society. Thesis, Leuven, 179 pp.

Conraths, F.J., Gethmann, J.M., Staubach, C., Mettenleiter, T.C., Beer, M. and Hoffmann, B. (2009). Epidemiology of Bluetongue virus serotype 8, Germany. Emerging Infectious Disease 15(3): 433-435.

Federal Ethics Committee on Non-Human Biotechnology (ECNH) (2001). Available at: http://www.ekah.admin.ch/fileadmin/ekah-dateien/dokumentation/publikationen/EKAH_Wuerde_des_Tieres_10.08_e_EV3.pdf. Accessed 10 March 2009.

Zichy, M. (2008).Gut und praktisch. Angewandte Ethik zwischen Richtigkeitsanspruch, Anwendbarkeit und Konfliktbewältigung. In: Grimm, H. and Zichy, M. (eds.) Praxis in der Ethik. Zur Methodenreflexion der anwendungsorientierten Moralphilosophie. Berlin/New York, pp. 87-116.

Hartnack, S., Grimm, H., Doherr, M.G. and Kunzmann, P. (2009). Massentötungen bei Tierseuchenausbrüchen – Tierärzte im Spannungsfeld zwischen Ethik und Seuchenbekämpfung. Deutsche Tierärztliche. Wochenschrift 116, 152-157.

Kunzmann, P. (2007). Die Würde des Tieres zwischen Leerformel und Prinzip. Verlag Karl Alber. Freiburg. 144p.

Mepham, B. (2008). Bioethics, an introduction for the biosciences.Oxford University Press, Oxford, 418 pp.

Mepham, B., Kaiser, M., Thorstensen, E., Tomkins, S. and Millar, K. (2006). Ethical Matrix. Manual. LEI, The Hague, The Netherlands.

Siep, L. (2008). Konkrete Ethik – zwischen Metaethik und Ethik-Kommissionen. In: Grimm, H. and Zichy, M. (eds.) Praxis in der Ethik. Zur Methodenreflexion der anwendungsorientierten Moralphilosophie. Berlin/New York, pp. 47-69.

Van der Zijpp, A.J., Braker, M.J.E., Eilers, C.H.A.M., Kieft, H., Vogelzang, T.A. and Oosting, S.J. (2004). Foot and Mouth Disease. New values, innovative research agenda's and policies. EAAP Technical Series No.5. Wageningen Academic Publishers, Wageningen, the Netherlands.

Can improved ethical labelling boost the consumption of animal welfare-friendly meat products? Experiences from the Danish market for eggs and pork

L. Heerwagen, T. Christensen and P. Sandøe
Institute of Food and Resource Economics, University of Copenhagen, Rolighedsvej 25, DK- 1958 Frederiksberg C, Denmark; lh@life.ku.dk

Abstract

When animal welfare-friendly meat products do not sell, irresponsible consumers are often blamed. The consumers claim to care about the welfare of farm animals, but when welfare-friendly products are actually offered to the consumers, most of them choose the conventional products instead. Even though there is a mismatch between attitudes and actions, and consumers indeed have responsibility for their choice, this still only presents one side of the issue. Another side worth considering is the way in which animal-derived foods are labelled and marketed to the consumer. The objective of this paper is to discuss ethical labelling with regard to the welfare of farm animals as a factor that can make a difference to the consumption of particular products. The labelling and consumption of eggs and pork in Denmark are used as cases. It is argued that the ethical labelling of eggs is better than the labelling of pork on a number of points. Compared with the labelling of pork, the labelling of eggs is found to be simpler, to have a higher level of relevance, and to facilitate the purchasing of animal welfare-friendly products. Consequently, the paper concludes that the ethical labelling of pork can be improved in these ways. Furthermore, the extent to which it may be possible to obtain results similar to those achieved for sales of eggs through an improved system of ethical labelling of pork will be discussed. One clear problem is that pork is a much less uniform product than shelf eggs, and that consumers' expectations of the qualities of pork differ from their expectations with regard to eggs.

Keywords: animal welfare, food ethics, food labelling

Introduction

A recent Eurobarometer survey on animal welfare showed that Danish citizens paid particular attention to the welfare of farmed pigs. Of the Danes surveyed, 63% regarded the welfare of farmed pigs to be 'very' or 'fairly' bad (European Commission, 2005: 18). This picture was also seen in a qualitative study in which Danish citizens characterised farmed pigs as 'unhappy and suffering' (Lassen *et al.*, 2006: 226). These ethically promising Danish citizens turn, however, into disappointing consumers. A number of animal welfare-friendly products, including organic products, are available on the Danish market but their market share is estimated to be less than 5% (Friis, 2005: 12).

Potential explanations for the rather small market share of animal welfare-friendly pork have been discussed, and major topics in that discussion have been the quality and price of the product and the consumers' willingness to pay. Another factor, somewhat ignored however, is labelling. The labelling of pork has mainly been driven by producers and retailers. As a result of competition, many retailers have found it important to distinguish their pork with their own label. This has led to a large selection of labels with shifting emphasis on animal welfare, taste and other qualities.

Labelling provides an opportunity for the consumer to choose an animal welfare-friendly food product instead of products without such consideration. Therefore it seems reasonable to believe that inadequate labelling will hinder the consumer in purchasing animal welfare-friendly products.

Hence, the objective of this paper is to make a contribution to the discussion of whether and how the consumption of animal welfare-friendly pork can be increased by bringing labelling schemes into focus. This discussion will take place in the context of a comparison of the labelling of pork and eggs. The case of egg labelling is interesting because to a large extent it has been developed in order to enable the consumer to purchase eggs according to the degree to which animal welfare is considered during management of the hens (Nordic Council of Ministers, 2004: 53). This 'ethical labelling' has been successful, because eggs that are perceived to be animal welfare-friendly, namely barn eggs, free range eggs and organic eggs, despite demanding a higher price than eggs from hens raised in battery systems, have obtained a market share of about 40%, according to the Danish Meat Association. A comparison of the labelling of pork and eggs could therefore show whether the labelling of pork can be improved to facilitate the purchase of animal welfare-friendly products.

The labelling of eggs and pork

A note on ethical labelling

The nexus of the following comparison is 'ethical labelling', which means that 'a product is labelled with information as to whether the production process respects ethical values' (Nordic Council of Ministers, 2004: 9). For the purposes of the present discussion ethical labelling is restricted to animal welfare considerations only. The definition excludes, first, labelling that informs the consumer about issues such as the weight of the product, date of production, nutrition and health, etc. Second, it excludes labelling that informs the consumer solely about ethical values connected to, for instance, protection of the environment and fair trade.

The labelling of eggs

According to EU regulations all eggs sold as such (not as ingredients in other products) must be labelled according to the system of production, of which there are four. This means that the egg itself and the egg tray are labelled as either organic eggs, free range eggs, barn eggs or battery eggs. This labelling programme was optional from 1975 to 2004 and has been mandatory since then (Nordic Council of Ministers, 2004: 53). Table 1 shows some aspects of the labels used according to the various systems of production.

On the eggshell the system of production is indicated by a printed number ranging from 0 (organic eggs) to 4 (battery eggs). On the egg tray the system of production is written in words. In addition to the number that indicates the system of production the eggs also carry a so-called 'traceability code'. The consumer can look up this code on an official website and obtain information about the farm on which the egg was produced. Instructions on how to use the traceability code are found on the inside

Table 1. Egg labels and animal welfare characteristics (Danish Meat Association, 2009b).

Egg labels	Battery egg	Barn eggs	Free range eggs	Organic eggs
Beak trimming permitted	Yes	Yes	Yes	No
Outdoor access	No	No	Yes	Yes
Max number of hens per m^2 (indoors)	16 (in cages)	9 (in barn)	9 (in barn)	6 (in barn)
Max number of hens per m^2 (outdoors)	Not applicable	Not applicable	4	4

of the lid of the egg tray. The egg trays of organic eggs are labelled additionally with the official Danish 'Ø-label' for organic production methods, and in some cases also with the similar official EU label.

The labelling of pork

The Danish Meat Association distinguishes between two types of pork on the Danish market, namely 'common Danish pork' and 'speciality pork'. The 'common Danish pork' is the conventional product without any special qualities, for example in relation to animal welfare. Speciality pork indicates a product that in some way or another is produced differently from the conventional product in order to obtain special qualities. Over a period of 20 years Danish consumers have been presented with at least 25 labels indicating speciality pork (Jensen, 2008: 12). The table below shows a small selection of the current labels, compared with those for common Danish pork, with regard to selected animal welfare issues:

As Table 2 shows, the selected speciality pork labels all have some sort of animal welfare quality compared with the conventional products. The level of attentiveness to animal welfare varies, though, from the label 'Antonius', which is relatively close to the conventional product, to the label 'Meadow pig', which includes a number of welfare requirements that distinguish it from the standard product. Seven labels for speciality pork, including those selected in the table, can be found on the homepage of the Danish Meat Association accompanied by specific information on animal welfare. Information on other labels is much harder to trace. The label 'Bornholmergrisen' (the Bornholm pig), for instance, is being marketed with the qualities 'more taste and welfare'. However, what this improved welfare consists of is explained in relatively vague sentences such as 'the pig feels better since it has extra living space'.

It should be noted that most of the speciality pork labels are self-certified. Exceptions are for example the labels 'Meadow pig' and 'Organic pig', which are controlled and approved by the Danish Society for the Protection of Animals. The latter is labelled additionally with the official Danish label for organic production.

Can the ethical labelling of pork be improved?

It must first of all be stressed that the labelling of pork, as well as that of eggs, is ethical in the sense that it makes it possible for the consumer to identify and purchase animal welfare-friendly products. However, there also seem to be major differences between the labelling of eggs and the labelling of pork. These differences can be presented under the headings simplicity, relevance and normativity.

Table 2. Pork labels and animal welfare characteristics (Danish Meat Association, 2009a).

Pork labels	Antonius	Organic pig	Meadow pig	Common Danish pig
Tail docking	No	No	No	Yes
Outdoor access	No	Yes	Yes	No
Space	0.85 m^2 (at the weight of 100 kg)	1.2 m^2 (at the weight of 100 kg)	No information	0.65 m^2 (at the weight of 100 kg)
Grinding of teeth in piglets	Yes	Yes	No	Yes
Castration	No later than first week of life	No later than first week of life	None	No later than first week of life

Simplicity – The number of labels on eggs is much fewer than those on pork, and this has been the case for a long period of time. In order to distinguish between the different labels that indicate animal welfare-friendly pork, the consumer currently has to be familiar with eight systems of production, in comparison to the four systems for eggs. Furthermore, because the labels on animal welfare-friendly pork in many cases do not imply that the system is indeed animal welfare-friendly, the consumer is dependent on information that cannot be obtained in front of the cold counter.

Relevance – The labelling of pork includes references to all kinds of variables. Some labels refer to the system of production, as in the case of 'Free range pig'. Other pork labels, however, refer to the place of production, such as the island of Bornholm in the case of the 'Bornholm Pig', and still others to historic characters, as in the case of 'Antonius', named after the saint of swineherds, St Anthony. The labelling of eggs, on the other hand, includes a mandatory reference to the system of production on the eggshell as well as on the egg tray. This means that if consumers have at least some knowledge of what the production system, for example 'free range', denotes, they will have an immediate impression of the origin of the product as soon as they have identified the label. With regard to pork this immediate impression is only obtained in the case of free range and organic pork.

Normativity – The main reason for the introduction of obligatory labelling of eggs was the demand of consumers to be able to differentiate between battery eggs and 'alternative' eggs (Nordic Council of Ministers, 2004: 89). Although the labels on eggs only inform the consumer about the system of production, this prescribes good and bad products *de facto* according to how animal welfare is perceived by many consumers. Although this distinction between good and bad products is quintessential to the notion of ethical labelling, and to ethics in general, it is not present in the case of pork. As described above, the pork products are basically divided into a conventional product and speciality pork. While the former gives no information about animal welfare and how the product was derived, the latter is at least in some cases labelled with regard to animal welfare. The consumer only receives information about the 'good' product – not the bad.

Simplicity, relevance and normativity are important in relation to the willingness of consumers to purchase animal welfare-friendly products. As stated in a recent qualitative study on ethical consumption in Denmark, consumers in general would like more information that enables them to identify 'bad' products; this information should be standardised in a way that makes comparisons possible and that clearly indicates attention to ethical considerations during production (Coff *et al.,* 2005: 14). If the labelling is opaque consumers develop a sense of mistrust and distance themselves from the products. This may make them give up the whole idea of purchasing animal welfare-friendly products (Lassen *et al.,* 2002: 54).

Eggs and pork as products – a perspective

Improved ethical labelling of pork would seem to imply that the labelling, which until now has been driven largely by the wishes of producers and especially those of retailers, should be directed towards the needs of the consumers. This could mean a mandatory labelling system that resembles the one for eggs, with standardised labelling that distinguishes 'caged pigs' from free range and organic pigs.

However, how much are consumers really interested in animal welfare when they are in front of the cold counter looking for pork? Studies show that consumers not only expect attention to have been paid to animal welfare when they purchase organic pork, but they also, and to an even higher degree, expect improved taste, nutrition and health qualities (Grunert *et al.*, 2004). Organic eggs, on the contrary, seem to be purchased mainly because of their animal welfare attributes (Husmer *et al.,* 2003: 61). Does that not imply that pork is a different kind of product from eggs? And does it mean that improved ethical labelling would not succeed in influencing the market for animal welfare-friendly pork?

Pork is without any doubt a different kind of food product from eggs. It appears in a greater variety of forms, it has a different place in meals, and its price is much higher. All this should of course be taken into consideration when assessing the idea of an improved ethical labelling of pork. However, one can hardly be surprised that consumers demand much more than animal welfare, if the labelling in regard to animal welfare is perceived to be poor. Animal welfare can only become a significant theme when consumers can identify pork according to the animal welfare aspects of its production in a transparent and reliable way. In this regard, improved ethical labelling seems to be a key requirement.

References

Coff, C., Walbom, L.C. and Mikkelsen, E. (2005). Forbrugere, etik og sporbarhed – om forbrugernes holdninger og handlinger på fødevareområdet. Copenhagen, Denmark, 35 pp.

Danish Meat Association (2009a). Overview of speciality pig labels. Available at: http://www.danskeslagterier.dk/smmedia/Rettelser_Specialgrise_pdf?mb_GUID=DE96FABC-4541-4E0E-A5A0-AFA6DA7B5AE2.pdf. Accessed 20 March 2009.

Danish Meat Association (2009b). Overview of egg labels. Available at: http://www.danishmeat.dk/Husdyrproduktion/viden_om/Produktion%20af%20aeg.aspx . Accessed 20 March 2009.

European Commission (2005). Attitudes of consumers towards the welfare of farm animals. Special Eurobarometer 229, 137 pp.

Friis, A. (2005). Forvirring ved køledisken. Spir – forbrugermagasinet om mad og økologi, 4 årgang nr. 4: 12.

Grunert, K., Bredahl, L. and Brunsø, K. (2004). Consumer perception of meat quality and implications for product development in the meat sector – a review. Meat Science 66(2): 259-272.

Husmer, L., Jensen, M.L., Poulsen, J. and Hjelmar, U.(2003). Miljø og forbrugeradfærd. Miljøprojekt nr. 280, 184 pp.

Jensen, F.L. (2008). Specialgrise – succes'er og fiaskoer. Svineproducenten 4: 12-15.

Lassen, J., Kloppenborg, E. and Sandøe, P. (2002). Folk og Svin – en interviewundersøgelse om danske borgeres syn på den danske svinesektor og svinekød. Frederiksberg, Denmark: 66 pp.

Lassen, J., Sandøe, P. and Forkman, B. (2006). Happy pigs are dirty! – Conflicting perspectives on animal welfare. Livestock Science 103: 221-230.

Nordic Council of Ministers (2004). Ethical labelling of food. Ekspressen tryk og kopicenter, Copenhagen, Denmark, 135 pp.

Bringing evidence into public deliberations on animal research: the case of research animals used in the development of food for obese people

T.B. Lund, J. Lassen and P. Sandøe
Danish Centre for Bioethics, Department of Human Nutrition, Faculty of Life, University of Copenhagen, Rolighedsvej 25, 1958 Frederiksberg C, Denmark; tblu@life.ku.dk

Abstract

Previous research has tentatively suggested that the use of animals in research into a disease that is related to lifestyle, such as obesity, is not publicly justified to the same extent as research into other life-threatening diseases. The view is that obesity is self-inflicted, and therefore obese people should change their lifestyle instead of sacrificing animals in research that is aimed at the development of new food products and ingredients targeting obesity. Such a view is challenged by strong evidence showing that obesity and obesity-related diseases can develop as a result of factors that have nothing to do with individual choices and habits. It may be hypothesized that the negative stance on the use of animals in research on obesity is based on a lack of awareness of this evidence. It may also be hypothesized that people who are willing to accept animal research into other life-threatening diseases will change their opinion in favour of obesity-related animal research when they have been confronted with this evidence. These hypotheses are explored with the help of data from five focus group meetings held in Denmark in which ordinary citizens debated the ethical justification for the use of animals in medical research. The results of this deliberation process are reported and are juxtaposed with the views of anti-vivisectionists and the medical profession.

Keywords: animal testing, experiment, attitude, focus group, lifestyle disease

Introduction

The use of animals in medical research has always been a controversial topic in which stakeholders have competed for control of the ethical basis of political regulation. Similarly, opponents and advocates of animal research have competed for control of public perceptions of animal research, because significant pressure from the public will result, inevitably, in changes in legislation. A new source of contention seems to have emerged in recent years, which is a result of changes in the disease profile of modern societies. As a consequence of this transition, public health discourse today stresses the role that an unhealthy lifestyle plays in the increased prevalence of non-infectious diseases. Opponents of animal research increasingly point to this fact, and consequently recommend the introduction of initiatives to encourage lifestyle modification instead of animal research.

There has been little investigation into public acceptance of the use of animals in research into lifestyle-related diseases. In the few articles that point to the relevance of the topic to laypersons, the subject was not central to the investigation. However, the question of individual or societal responsibility is certainly a hot topic when debating the use of animals (Lassen and Jamison, 2006). This article will present preliminary results from focus group meetings held in Denmark in which laypersons' perceptions of animal research into lifestyle diseases were examined in greater detail, taking the case of obesity as an example.

After outlining the methods used in the study, in the subsequent section we will examine how representatives of the Danish public balance the issues of personal responsibility, disease severity and animal suffering in their evaluation of animal experimentation used for lifestyle diseases. In the last section of the paper we will discuss the findings in light of the following questions: Will members of the Danish public reject animal research into lifestyle diseases, as anti-vivisectionists recommend? If yes, how is the rejection argued? How will the rejecters react to factual evidence showing that people are not always responsible for 'lifestyle' disease? If the view of the public differs from the views of those involved in the study of lifestyle diseases and those of the opponents of animal research, how are these differences explained?

Data and methods

In order to investigate the answers to the questions listed above, this paper reports some preliminary results from five focus group meetings that were carried out in Denmark in February and March 2009 (N=33). The number of participants in each group varied between four and nine, and they were recruited by a commercial recruitment company. A random purposeful sampling strategy was used, because the purpose of the qualitative data gathering exercise was to obtain typical animal-related political views, argumentative strategies and emotional reactions among Danish people.

The results from the focus groups that are discussed in this article are based on the following procedure. Participants in all groups were introduced to five cards showing examples of particular animal experiments. The five examples varied on four parameters: the medical aim, the animal species used, the degree of pain resulting from the experiment, and a description of the experimental method used on the animal. One of the five research aims dealt with obesity: 'testing medicines aimed at the severely overweight' in which it was explained that the medicine was intended to be integrated in foodstuffs. Obesity was chosen as the example to illustrate a lifestyle-induced disease for two reasons. First, the cause-effect relationship between overeating and obesity is well known to the Danish public. Consequently, this case was chosen to maximize the likelihood that diseases associated with lifestyle and personal responsibility would pop up in the discussion, should they be relevant to the participants. Second, animal research is frequently carried out with this type of aim; therefore obesity is a relevant research scenario.

The participants were asked to rank the experiments (from 1 to 5) in order of acceptability and to make a note of their personal ordering. Subsequently, the participants were asked to clarify the principles they used in assigning ranks and were encouraged to discuss their rankings with the other participants. The results that involved discussion about the ethical justification of animal research into lifestyle diseases, particularly obesity, will be reported primarily here. If the perception that animal research into lifestyle diseases is totally unacceptable remained unchallenged by a focus group, a pre-prepared text was introduced to the participants by the moderator. The text was intended to explain to the participants that there is evidence that lifestyle diseases are not always self-inflicted. This was to ensure that the discussion about animal research into lifestyle diseases would involve factual evidence.

Results

Four attitudinal profiles

Generally speaking, four subgroups of attitudinal patterns were distinguishable among the participants. The first group we call *opposers*. Members of this group were either very critical of animal research in general or used strategies related to animal welfare in their ranking of examples of acceptable research. These *opposers* used completely different rules of ethical evaluation to those of the other subgroups. Importantly, they did not seem to take particular notice of the question of personal responsibility for lifestyle-related diseases, because the sole focus for these participants was the animals. The three remaining groups used identical overall decision rules in that they first and foremost sorted the examples in order of disease severity. Although there were minor differences between participants in the specific evaluation of diseases that were most important to research, the basic cost/benefit decision principles were the same. With minor exceptions the same ordering of the five examples was chosen by all three subgroups (Figure 1). Experiments on cancer and Alzheimer's disease were considered the most important endeavours, despite the fact that the cancer example was the example in which the animals involved were suffering the largest amount of pain.

Migraine was the third most acceptable example. However, migraine was also considered in general to be a highly acceptable research topic, despite the knowledge among participants that it is not a deadly disease. In this respect, the three examples on the right-hand side of Figure 1 can be interpreted as belonging to the same category of severe diseases that, as seen through the eyes of a majority of this group, are legitimate topics for animal-based research.

The testing of deodorant safety, on the other hand, was the least acceptable research example. Participants perceived the use of deodorant, and cosmetics in general, to be a luxury activity in which the use of animals was not warranted. The basic argument against this sort of animal research was that there are tangible and easily carried out alternatives: namely not using deodorants and phasing out synthetic chemicals in cosmetics.

Obesity research: balancing disease severity, alternatives and responsibility

The 'test of medicines aimed at the severely overweight' was also deemed less acceptable than applications of animal research for migraine, cancer and Alzheimer's disease. As with cosmetics, the argument was that there are alternatives – namely, other strategies that do not involve research animals. A better alternative was thought to involve the cessation of unhealthy lifestyles:

> 'And that one with severe overweight, well maybe people should consider changing their diets. That actually makes it possible to lose weight.' (Blamer in Group 2)

Apart from pointing to alternatives, such arguments are also based on the notion that obesity is a self-inflicted condition, and thus the responsibility of the individual. There is direct blaming of the obese for being mentally and physically lazy. However, a second group of participants, which we refer to as *all ins*, countered these arguments. Although they did not dismiss the argument about personal responsibility,

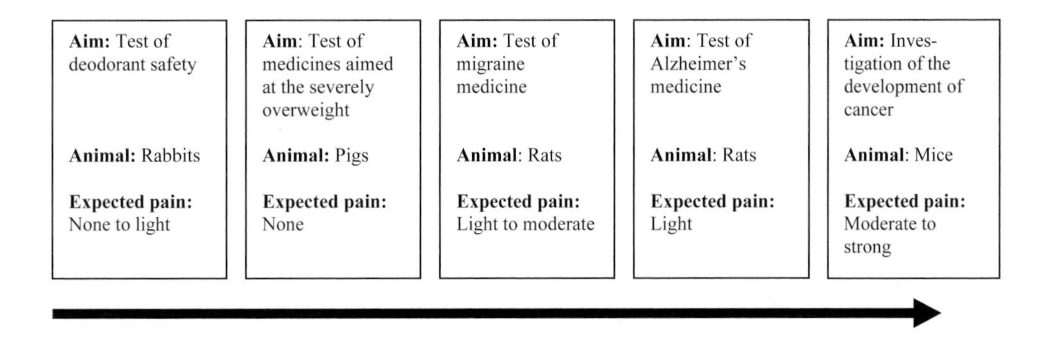

Figure 1. Common ranking of five examples of animal research by the three subgroups All ins, Shifters and Blamers.

they perceived severe overweight as an important disease to control, because obese people do suffer, have a reduced quality of life, and ultimately may die of their disease. In some instances arguments supporting the use of animals in obesity research were backed with reference to economics. The reasoning was that societies spend many resources inappropriately on the treatment of diseases that develop as side effects of obesity, and consequently that animal research is highly relevant.

Evidence for responsibility in the case of obesity

Many of these *all ins* also widened the context of the discussion by noting that some obese people are not solely responsible for their obesity. For them, this was reason enough to accept this sort of research. This line of argument was also acknowledged by some of the critical participants, who comprised a third subgroup that we refer to as *shifters*. In joint consideration of the suffering that obese people experience and the fact that obesity in many cases is not self-inflicted, these previously sceptical participants made a point not to reject the use of animals in obesity research as such. Rather, they ranked this research endeavour lower than cancer, Alzheimer's disease, and migraine research because of the existing alternatives available in the fight against obesity.

> 'I also think that it [the example of medicines aimed at the severely overweight] is completely acceptable, but in weighing it up against the others I just find it less important.' (Shifter in Group 1)

Another specific reason for the acceptance of the particular obesity example that was introduced to the focus groups was that the card stated that the pigs would not experience any pain.

> 'Well, as long as the pigs do not suffer, it's all right. But it is the worst research aim.' (Shifter in Group 1)

It is unclear whether the *shifter* subgroup, who changed their minds in the direction of favouring animal experimentation during their deliberations (and after being presented with evidence of the suffering of obese people and evidence that not all obese people are responsible for their condition), referred to the degree of animal suffering in order to retreat from their earlier, less positive, attitude.

In any case, many *shifters* stated that they were willing to accept the research into severe overweight because there would be no animal suffering. In this respect, it is also unknown how they would have responded to this type of research if there had been pain involved.

Finally, a fourth subgroup, referred to as *blamers*, was also identified. Although they endorsed animal research into severe human diseases, they remained unconvinced of the acceptability of such research into lifestyle diseases, even after the introduction of factual evidence relating to responsibility and the severity of the condition of obesity. Here the perception that obesity is self-inflicted remained at the forefront of their thinking. In this respect the perception of human folly was an overarching interpretative principle among the *blamers*.

Discussion

The results from this investigation in part reproduce previous findings: it is well known that a majority of the public view animal experimentation from a cost/beneficial viewpoint (Herzog *et al.,* 2001). In particular, the more severe the disease in question is perceived to be, the more acceptable animal research will be. However, in general the public remains ambivalent about animal research, and if alternative methods could bring about the same results, or other routes could be taken that would render the use of animals superfluous, a majority would find this preferable. One topical example of successive waves of public and anti-vivisectionist pressure against animal research that was perceived to be unnecessary can be found in the European Union (European Union, 2003). Here a phasing out of the use of animals in the safety testing of cosmetic products was agreed in 2003, following a similar regulation that was introduced in the UK in 1998.

The results reported here demonstrate, furthermore, that consideration of a wider span of current animal research endeavours may uncover thoughts attached to discourses of lifestyle, and in turn may provoke thoughts about alternatives and the unnecessary sacrifice of animals.

The question is how these findings relate to the ongoing controversies about the use of animals in research. We will turn to this in the remaining part of the article (see also Figure 2).

Official arguments for and opponent arguments against research into lifestyle diseases – and its public compatibility

From the viewpoint of the medical profession and their advocates, lifestyle diseases constitute a highly legitimate research area, because these diseases are lethal. The basic reasoning is that the results of successful research may save lives. Anti-vivisectionists, however, appear to have moved the idea of lifestyle diseases to the forefront of their agenda. This can be seen by conducting a simple web search (e.g. 'stop animal testing' AND 'lifestyle'). The fundamental argument against this sort of animal research is that there are alternatives. If only lifestyle modification were carried out effectively, the associated diseases would not be epidemic in character. Given this fact, opponents then introduce an argument of resource efficiency (Marston and Rosser, 2007). They argue that the most efficient way to save lives would be to invest resources in lifestyle modification programmes. Given that this alternative exists, it is argued that lifestyle-related animal research is not justified.

It may be hypothesized that the branch of the anti-vivisectionist movement that uses this argumentative path is also trying to convert the public to their overall cause, namely to abolish animal research altogether, by appealing to well known human traits of reproachfulness and blame. The paradox in

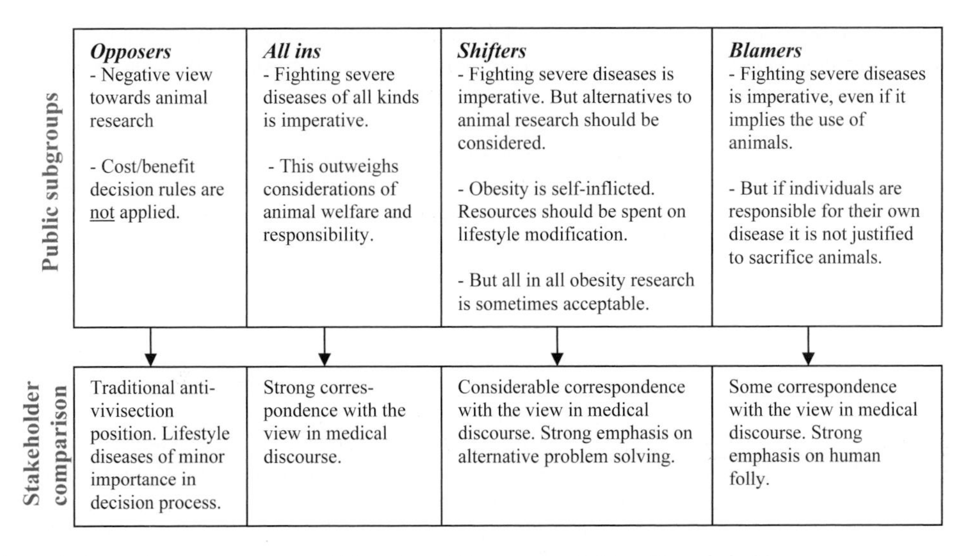

Figure 2. Four subgroups among participants and their compatibility with stakeholders.

this strategy is that their primary allies in the public, the pure *opposers*, with whom they seem to share philosophical views on animal rights, do not attach particular importance to questions of responsibility. On the other hand, *blamers* are attracted to these arguments. However *blamers* are by no means against the use of animals in research that targets vital human diseases, as long as these are not self-inflicted. Consequently, it seems highly unlikely that the anti-vivisectionists' goal of abolishing animal research with the help of public opinion will succeed with the use of this communication strategy.

The *all ins* constitute non-critical allies among the public, as seen from the point of view of the medical profession. They have human progress and the fight against human suffering as their number one priority; they perhaps even surpass the views of biomedical research ethics on this point. The last group, the *shifters*, seem to buy into what may be labelled the official view regarding animal experimentation. From this viewpoint, it is morally acceptable to use animals for research provided that the following two conditions are in place: (1) the research must be relevant to issues of vital importance, such as new ways of preventing, curing or alleviating serious human diseases; (2) there must not be another way to achieve the same result that is less harmful to the animals, and every effort must be made to ensure that the animals involved are not caused more discomfort and suffering than is strictly required by the experiment (Sandøe and Christiansen, 2008). Despite this agreement in the criteria for ethical evaluation the *shifters* attached strong importance to the question of alternatives. As the labelling indicates, they shifted their opinion only when evidential factors were introduced into the equation. Hence, it seems that the discourse of lifestyle disease carries strong emotional and cognitive appeal. The medical profession and its spokespersons should be aware of this challenge in order to ensure the continued endorsement of their use of animals in research into lifestyle-related diseases. They cannot take for granted, as they sometimes seem to do, that all biomedical research really is of vital importance.

Acknowledgements

The study reported in this paper is financed by the Danish Agency for Science, Technology and Innovation under the Ministry of Science, Technology and Innovation.

References

European Union (2003). Directive 2003/15/EC of the European Parliament and of the Council of 27 February 2003 amending Council Directive 76/768/EEC on the approximation of the laws of the Member States relating to cosmetic products. Official Journal of the European Union 11.3.2003: 26-35.

Herzog, H.A., Rowan, A. and Kossow, D. (2001). Social attitudes and animals. In: Salem, D.J. and Rowan, A.N. (eds.) The state of the animals. Humane Society Press, Gaithersburg, MD., pp. 55-69.

Lassen, J. and Jamison, A. (2006). Genetic technologies meet the public. Science, Technology and Human Values 31: 8-28.

Marston, H. and Rosser, H. (2007). Animal Experiments – a Vicious Circle? Paper presented by Helen Marston at Animals and Society II – Considering Animals, 3-6 July 2007, Hobart, Australia, pp. 7.

Sandøe, P. and Christiansen, S.B. (2008). Ethics of Animal Use. Wiley-Blackwell, UK, pp. 192.

Can killing be justified: a dismissal of the replaceability argument

T. Visak
Ethics Institute, Utrecht University, P.O. Box 80103, 3508 TC Utrecht, The Netherlands; T.Visak@uu.nl

Abstract

I evaluate a utilitarian argument that justifies killing animals: the Replaceability Argument. According to utilitarianism, killing an animal, which would otherwise have continued to live a pleasant life, amounts to a welfare loss. The idea of the Replaceability Argument is that bringing another animal into existence, which is at least as happy as the killed animal, can compensate for the welfare loss. The Replaceability argument depends on the view that the possible welfare of possible individuals, such as the possible second animal, counts. This is the Total View. An alternative view is the Prior Existence View, which accepts as moral objects only individuals who already exist or who *will definitely exist*. Those whose existence *depends* on the moral choice in question do not count. So, the Prior Existence View would not take into account the possible welfare of the possible second animal while considering whether killing the first animal is acceptable. This is because the second animal does not exist and its existence depends on the moral choice that is contemplated, i.e. whether to kill the first animal. The Prior Existence View is initially attractive. It does not accept the creation of happy beings as a way of maximizing welfare that counts on a par with making existing beings more happy. The fact that a possible being would be happy is no reason for bringing it into existence. Though initially attractive, a powerful argument has been brought forward against the Prior Existence View. It has been argued that the Prior Existence View cannot accept the expected misery of a possible individual as a reason against bringing this being into existence. This problem for the Prior Existence View has famously been brought forward in the Case of the Wretched Child (Parfit, 1984). I argue, however, that the Prior Existence View can deal in an intuitively plausible way with the Case of the Wretched Child. The Case of the Wretched Child does not dismiss the Prior Existence View. The Prior Existence View has not been dismissed as a plausible view within utilitarianism. The Prior Existence View does not support the Replaceability Argument. Utilitarianism therefore might offer animals a stronger protection against killing than both utilitarians and their critics deemed possible.

Keywords: utilitarianism, prior existence view, total view, character, evaluative focal points

Introduction

Utilitarianism justifies putting animal welfare high on the agenda. First of all, utilitarianism accepts *welfarism*. This is the view that the only value that is ultimately relevant for ethical evaluation is welfare. Secondly, utilitarianism requires the *maximization* of welfare. This means that agents should choose the action, amongst alternatives, that results in the highest amount of welfare. Finally, utilitarianism is both *universalist* and *impartialist* with regard to the issue of whose welfare counts in determining the rightness of actions (Timmons, 2002). This means that the welfare consequences for *everyone that is affected* by an action must be taken into account. If animals are affected in their welfare, this has to be taken into account as well. What is more, every affected being counts *equally*. If we are to reduce suffering and promote welfare, there seems to be much to win in the field of animal husbandry. This holds the more so, if animal suffering counts equally to human suffering.

What about the killing of animals in animal husbandry? As explained, utilitarianism evaluates actions by their consequences in terms of welfare. The consequences of every action, and the consequences alone, determine whether it is right or wrong. There are no categories of actions that are obligatory or forbidden as such and under all circumstances. Therefore, utilitarianism does not rule out killing. Killing can be justified if it promotes overall welfare. Its positive effects can in principle compensate its negative effects. Although killing is not categorically forbidden, killing an animal that would otherwise have had a pleasant life counts as a welfare loss. The welfare that the animal would have experienced is lost. Letting the animal live would be recommended by utilitarianism, because it would maximize welfare.

There are exceptions, in which killing would be allowed or even required. Any welfare loss can be compensated by an equal welfare gain that is also part of the outcome. However, in case of animal husbandry, it has been tentatively argued elsewhere and will be assumed here that the enjoyment that is caused by the killing cannot compensate the utility loss due to the killing. Alternatives are available that score better in terms of utility. Neither, I assume, is the killing necessary in order to avoid greater welfare loss.

There is one other argument that specifies conditions under which the welfare loss due to the killing can be compensated and the killing can be justified. The welfare loss caused by the killing can in principle be compensated by the welfare that is gained by other individuals. In animal husbandry, a new animal that would not otherwise have existed replaces every killed animal. The idea is that this can compensate for the welfare loss due to the killing. This argument is known as the Replaceability Argument.

The Replaceability Argument:

It is permissible to kill an animal, provided that the following conditions are met:

- The life of the animal is pleasant.
- The animal will be replaced, at or after death, by another animal, whose life is at least as pleasant and which would not otherwise exist.
- The killing does not have any unbalanced negative side effects (such as fear or suffering for the animal or others).

Total view versus prior existence view

The Replaceability Argument rests on an assumption that is disputed within utilitarianism. It rests on the assumption that the potential welfare of the potential animal that might replace the first one should count when the decision about killing the first one is made. This second animal does not yet exist, when the moral choice is made. Consider, for instance, the cow that will be caused to exist if and only if the first cow is killed. This cow does not yet exist, when the moral choice in question, i.e. whether or not to kill the first cow, is made. Utilitarianism requires taking into account the consequences of an option in terms of welfare. Killing the cow amounts to a welfare loss. However, that welfare loss might be compensated by the welfare of the second cow that will come into existence and that would not otherwise have existed. It is disputed whether the Replaceability Argument is acceptable, because it is disputed whether the potential welfare of the not yet existing second cow should be allowed to enter into the calculation. After all, the second cow is nothing, when the decision about killing the first cow is made. It is a possible being.

Should the possible welfare of possible beings be allowed to enter into the calculation? In animal ethics, two opposing positions on this issue are discussed: the Total View is opposed to the Prior Existence View. On the Total View,

> '[...] we aim to increase the total amount of pleasure (and reduce the total amount of pain) and are indifferent whether this is done by increasing the pleasure of existing beings, or increasing the number of beings who exist' (Singer, 2005).

On the Prior Existence View, we

'[...] count only beings who already exist, prior to the decision we are taking, or at least will exist independently of that decision' (Singer, 2005).

Singer explains the difference of those views in practice:

'The utilitarian verdict on killing that is painless and causes no loss to others [...] depends on how we chose between the two versions of utilitarianism [...]. If we take what I called the "prior existence view", we shall hold that it is wrong to kill any being whose life is likely to contain, or can be brought to contain, more pleasure than pain. This view implies that it is wrong to kill animals for food, since usually we could bring it about that these animals had a few pleasant months or even years before they died – and the pleasure we get from eating them would not outweigh this. The other version of utilitarianism – the "total" view – can lead to a different outcome that has been used to justify meat eating' (Singer, 2005).

Discussion about the expected misery argument

That a possible child would be happy is no reason for having it, according to the Prior Existence View. But then, is the knowledge that a possible child would have a miserable life a reason *against* having it? This seems intuitively plausible. It has been argued against the Prior Existence View that it cannot accommodate this intuition. It cannot accept the expected misery as a reason against having a child. Otherwise, the view would be inconsistent, because it does not accept the expected happiness of a possible child as a reason for having this child. This major challenge for the Prior Existence View has been brought forward with the Case of the Wretched Child.

The Wretched Child: Some woman knows that, if she has a child, he will be so multiply diseased that his life will be worse than nothing. He will never develop, will live for only a few years, and will suffer pain that cannot be wholly relieved (Parfit, 1984).

Sapontzis defends the Prior Exisence View against the Expected Misery Argument as brought forward in the Case of the Wretched Child. The real evil, according to Sapontzis, is only done when the child is allowed to suffer. As soon as the child can suffer, it is a presently existing sentient being and its welfare counts morally. The Prior Existence View can then fully account for the child's suffering, even if the child is still a foetus. Before that time, when the child is conceived, or when the prospective parents intend to conceive the child, no harm is done. Yet, if the parents' intention is to keep the child, the parents can, according to Sapontzis, be accused of having a bad character. So, according to Sapontzis, the Prior Existence View can deal in a plausible way with the Case of the Wretched Child. Sapontzis explains:

'But if such people did exist, they would have a perverted idea of reproducing and parenting and would show, by keeping the child alive for two miserable years, their willingness to use others merely as means to their own satisfaction. [...] Prior existence utilitarianism can account for this intuition, since this is an evaluation of character, and prior existence utilitarianism no more precludes making character evaluations than does the total population view' (Sapontzis, 1987).

Pluhar argues that Sapontzis' character argument it is flawed. Pluhar claims that a couple that would deliberately conceive the Wretched Child believing that 'God makes no mistakes' could not be accused of bad character. Then, Pluhar claims that even if the couple acted from monstrous intentions, their character could not be dismissed *on utilitarian grounds*. Pluhar (1995) explains: 'Utilitarians can judge character traits to be good or bad solely by reference to their consequences.' Pluhar specifies: 'Specifically, one must ask how much harm (disutility) such traits are apt to cause.' Pluhar goes on as follows:

'For a utilitarian, the bad characters of the parents do not "give the project a strong immoral value": it is the project that makes their characters bad' (Pluhar, 1995).

On utilitarian grounds, according to Pluhar, the project cannot be judged to be bad independently of the character argument. Therefore, according to Pluhar, the character argument fails. So, according to Pluhar, the character argument fails, because that argument cannot be derived from utilitarianism in the way suggested by Sapontzis. Pluhar concludes:

> 'Given the [...] prior existence view, then, there are no grounds for saying that a couple initiating the project of creating a wretched child are deficient in good character *until the wretched child exists.* We are barred from saying that what they have done prior to that point is wrong in any way, even in Sapontzis's indirect way' (Pluhar, 1995).

Defending the prior existence view against the expected misery argument

Now, I will argue that Pluhar is wrong. Pluhar is mistaken in her idea about whether and how utilitarianism can take character judgments into account. In order to make my point, I will make use of the concept of *evaluative focal points*.

As we have seen, utilitarianism's 'theory of the right' requires maximizing welfare. Now, the question rises, how, more specifically, we should go about in maximizing welfare. What should be our focus? Should we always choose the act that maximizes welfare? Should we follow the rules that generally tend to maximize welfare? Should we adopt motives that are likely to maximize welfare? Should we do all of these?

In order to get a clear idea of what utilitarianism requires, we must be clear about the 'evaluative focal points', the focus of moral evaluation (Kagan, 1998).

Direct utilitarianism applies the principle of rightness directly to whatever focal point is to be evaluated. There are many possible forms of direct utilitarianism, depending on what is taken as legitimate evaluative focal points. There are direct utilitarian theories, which accept only one evaluative focal point, for instance acts. There are also theories, which accept several direct evaluative focal points.

Indirect utilitarianism applies the theory of the right (e.g. 'maximize welfare') only to one central evaluative focal point. Other evaluative focal points are evaluated only indirectly in terms of their relation to the central one. Rule utilitarianism is an example of an indirect utilitarian theory. Rules are evaluated directly in terms of the goal of maximizing welfare. ('Choose the rule that maximizes welfare.') Acts are judged right whenever they conform to the right rules.

If character were accepted as a direct evaluative focal point, then utilitarianism would ask people to adopt a kind of character that is most likely to maximize welfare. Now, what kind of character would generally help to maximize welfare? Would this be a character that is insensitive to the needs and suffering of others? Would this be a character that sanctions the use of others as mere means for their own satisfaction? This is ultimately an empirical question, yet I am confident about my hunch that it would not.

Therefore, the plan of the couple to conceive and keep alive the wretched child *can* be condemned on utilitarian grounds as a sign of bad character. This can be done directly, if character is taken as a direct evaluative focal point, be it the only one or one among others.

What goes wrong in Pluhar's discussion of Sapontzis' character argument? Pluhar's most interesting mistake is her belief that:

> 'For a utilitarian, the bad characters of the parents do not "give the project a strong immoral value": it is the project that makes their characters bad'.

That last claim is not necessarily right of utilitarian theories. It would only be right if actions were chosen as a direct focal point, and character as an indirect one. In that case, actions would be directly evaluated in terms of the welfare they produce. We would be required to choose the action that maximizes welfare. Character would be evaluated indirectly, depending on the action. We would need to know whether an action is right or wrong, and could then conclude whether the character that caused it was right or wrong. That, however, would be a strange kind of utilitarianism. After all, what sense would it make if we could afterwards judge that the character must have been bad, because in this specific case it has led to a bad action, even if generally it leads to good actions?

One can reasonably hold that a person's character influences his or her actions. This would be a reason to choose character as a primary focal point. Choosing character as a primary focal point would, besides other advantages, take away the burden of utility calculation before every action. It is reasonable to judge character with regard to its *general* contribution to good or bad actions. A character that would allow for satisfaction from the prospective of having a miserable child is likely not to contribute generally to the maximization of welfare.

So, one can say that planning to have the miserable child is a sign of bad character. Conceiving the child with the intention of letting it suffer is also a sign of bad character. My point is that this kind of character judgments can be made within utilitarianism.

Now, admittedly, choosing character as an evaluative focal point gives rise to many questions. What exactly is character? Can one choose one's character? Does one's character influence one's behaviour? These kind of questions would need to be answered. Maybe, what we are after is not really 'character', but rather 'motive'. Maybe one should rather say that the parents' motive could be judged wrong, according to utilitarianism. The motive is what motivates them to act. They want to have a child no matter what this means for the child. They are prepared to let their child suffer for the satisfaction of their own desire. That motive can be condemned on utilitarian grounds, as it is not a motive that generally helps to maximize welfare.

Conclusion

I have argued, contrary to Pluhar, that Sapontzis' character argument works. I have thereby provided a new defence of the Prior Existence View against the Expected Misery Argument. So, the Prior Existence View has not been dismissed as a plausible view within utilitarianism. The Prior Existence View does not support the Replaceability Argument, which is needed to justify killing animals in animal husbandry. Therefore, utilitarianism might offer animals a stronger protection against killing than both utilitarians and their critics deemed possible.

References

Kagan, S. (1998). Normative Ethics. Westview Press, Boulder, USA.

Parfit, D. (1984). Reasons and Persons. Clarendon Press, Oxford, United Kingdom.

Pluhar, E. (1995). Beyond Prejudice. The Moral Significance of Human and Non-Human Animals. Duke University Press, London, United Kingdom.

Sapontzis, S. (1987). Morals, Reason and Animals. Temple University Press, Philadelphia, USA.

Singer, P. (2005). Practical Ethics (2nd edition). Cambridge University Press, New York, USA.

Timmons, M. (2002). Moral Theory: An Introduction, Rowman & Littlefield Publishers, New York, USA.

In vitro meat – some moral issues

S. Welin
Linköping University, 581 83 Linköping, Sweden; stellan.welin@liu.se

Abstract

In this presentation I will discuss the possibility of producing meat; not from killed animals as has been the case so far, but to tissue engineer meat from animal cells in bioreactors. I will argue that if this is possible, this new technology will solve some major problems. One is related to the environmental load presented by present day intensive animal meat production, another is the ethically sensitive question of animal suffering and killings of animals. However, the new possibilities opened up will also present us with some moral problems which will be discussed at the end.

Keywords: *in vitro* meat, ethics, technology

Technology creates moral problems...

There is no mystery why new technologies will also increase our moral responsibilities. In the old days, when it was impossible to send food around the world to relieve famine in distant areas, there was no moral obligation to come forward with such support. It was simply impossible to do. There was a moral obligation to come forward and help your neighbours if you were in the lucky position of having excess food. Obviously, at that time it was not clear exactly how much you were obliged to do for your neighbours; nor is this clear today. However, when the technology was missing, there was no moral obligation to send food to distant countries to relieve famine. Transport technology has vastly widened our moral responsibility.

Another example is organ transplantation. In former times, it was not possible to move one kidney from a living person to another. This is possible today. Hence, a new moral question has emerged: should I give my kidney to someone in desperate need of one? As most of us have two functioning kidneys, we have the possibility to give one of our two kidneys to someone with end stage kidney disease. This is a moral question created by the development of modern medicine.

It is a question too seldom asked. Most of us tend to shy away from the fact that modern medical technology has expanded what we can do for our fellow beings. It is not just time and property that may be shared with others; even parts of our bodies may be of value for others in distress.

Organ transplantation, especially from living donors, will generate further moral problems which were formerly completely unknown. Should human organs always be given freely with no financial reward, or can there be a scheme for paying volunteers who are willing to part with one kidney? (Omar *et al.*, 2008) Whatever the view one may take on these issues, they are all created by a new technology.

Another example involves human embryos. After the advent of *in vitro* fertilisation and in particular of human embryonic stem cell technologies, human embryos can be used as a resource to produce potentially medically valuable stem cell lines. When it all started in 1998, the moral issue was immediately there. Was it really morally right to destroy human embryos in order to produce human embryonic stem cell lines? Whatever, your moral position on this issue; the very question did not exist some years ago. Technology can create new moral problems.

...but it may also solve moral problems – and create new ones

Sometimes technological development may partly do away with some moral problems. Recently, new development of induced pluripotent stem cells seems to do away with the issue of the use of human embryos. If some more differentiated adult stem cells can be reversed and turned into more primitive stem cells, ultimately the embryo may not be necessary any longer. (Person and Welin, 2008: 137; Yu *et al.*, 2007)

Let us take the question of whether I should volunteer to give my kidney to someone in need. If tissue engineering can be used to produce human organs for transplantation, there is no longer any need for me to volunteer. If organs may be grown outside the human body from donated cells or from stem cells, then the issue of giving my kidney is no longer pressing. There may be better alternatives.

Obviously, such a development does not come without some drawbacks. Here are a few. First of all, this possibility to engineer organs will in the long run be quite expensive. Today, organs are free – except on the black market and a few other places – but the surgery is expensive. The volume of organ transplantation is restricted because of lack of organs. This is particularly true of organs that cannot be obtained from living donors. If organs can be grown in laboratories, the sheer amount of possible transplantation would be astonishing. Why wait until you have a serious heart failure? Better to get a new one in time. Some of this will of course involve saving expensive advanced heart surgery but my simple guess is that all this will in the end be quite expensive. In systems with national health care systems – like in Sweden and most of Europe – some kind of rationing must be done.

Similar problems will appear if regenerative medicine succeeds. It will probably be possible to prolong life that way. It is hard to know how much the average life span will be extended, but it will all come with a cost. The cost is not just of supporting an ageing population, but one may wonder what the changed ratios of old people to young will mean for our societies.

The problems of conventional meat production

There is a wide consensus that human activity is affecting the global climate and our ecosystems. This includes many activities such as burning of fossil fuels but also agricultural activities. It is not just that we may run out of coal and oil in the future but these activities also produce huge amounts of greenhouse gases that tend to increase global temperatures. The system of animal based meat production based on grain and soybeans – especially if they are grown on open newly deforested land – causes a considerable amount of emission of greenhouse gases. According to a recent FAO report the emission is approximately 18% of all emissions of such gases in CO_2 equivalents and 65% of anthropogenic nitrous oxide (Steinfeld, 2006). A slaughtered animal contains many more parts and needs input for moving around. A conservative estimate is that about 80% of the energy input in animal farming is lost.

Conventional intensive meat production also put pressure on land use. The demand for food for animals (but also for direct consumption) has led to an increase in soybean production. The combination of intensive and extensive free range animals also put pressure on land use. Pastures for cattle grazing have been converted to soybean fields and the cattle have been moved deep into the Amazon (Rother, 2003). This is a threat to biodiversity. Another worldwide problem is the growth of cities which often takes place on good and fertile land, which may affect the global food supply (Ananthaswamy, 2002). The increase of the human population and the expansion of agricultural land for growing plants and for pasture will also involve more and more of the land of the planet. There is simply no free space left, at least not for the big animals. One example of the problems is the troubled lives of wolves in the middle of Sweden.

A general consensus seems to be that animal suffering is an evil thing and should be avoided as far as possible (DeGrazia, 1996). In particular, it seems that much of present day practices in the slaughterhouse evoke negative reactions among the public (Eisnitz, 2006). There is also a discussion on a European level to reduce the transportation of animals to be slaughtered. This can be seen as an attempt to avoid (unnecessary) animal suffering.

As to the slaughtering of animals, the views among philosophers and ethicists are more divided. The same holds probably for the public. Painful killing is generally regarded as non-acceptable but the debate is intense on the issue of (painless) killing (merely sentient) animals who do not have clear sense of a future (McMahan, 2002). It may be added that in this way of thinking on inflicting suffering and killing, an important distinction is made between persons (like human beings and probably higher apes) and 'mere' sentient animals (Cavalieri and Singer, 1994). For persons we have a stricter ethics.

Not all ethicists concur with this view. For a dissenting view, see for example Regan (1983) who claims that most animals should be treated as persons and be viewed as having rights in the way humans have. This is also an idea held by many animal rights groups.

In vitro meat

Tissue engineering plays a large role in the endeavour to find remedies for many medical problems. In the future it may be possible to tissue engineer new organs. There have been some important steps taken.

If one will be able to engineer organs for transplantation, it will obviously be possible to engineer an animal tissue or organs for eating. Such an organ need not be functional; it need just be edible. There are laboratories engaged in developing the technique of growing animal cells in bioreactors in order to produce edible meat products. The work is still in an early stage but there is a patent covering such *in vitro* meat production. At present, the nearest realistic task is perhaps to produce minced meat.

One hope of the *in vitro* meat endeavour is to be able to reduce environmental load by replacing intensive, mainly indoor, meat production, where the animals live in unnatural and crowded circumstances, with *in vitro* meat bioreactors. Hence, such a technology will give environmental benefits as well as reducing animal suffering. However, there are still problems to be overcome in relation to the cell culturing but also in relation to public scepticism. (It is still not known what the price of a kilo in vitro meat beef would be compared to conventional meat.)

Benjaminson *et al.* (2002) were first to publish successfully cultured *in vitro* meat by using skeletal muscle explants from goldfish in different media. NASA (USA) supported this and other research in the interest of sustainable food sources for long distance space travel. New Harvest (www.new-harvest.org), a non-profit research organization founded 2004, also supports the development of competitive *in vitro* meat products. Furthermore, PETA (People for the ethical treatment of animals) has announced a prize for the first research group able to produce a commercially available *in vitro* meat product (www.peta.org/feat_in_vitro_contest.asp).

The real goal is to produce all varieties of meat obtainable from animals. The most immediate thing to do is probably to produce minced meat in a bioreactor and later move on to more complex products. It is important to make early assessments about the contribution of *in vitro* meat processes to lessen the environmental load of meat production. In the beginning these estimates will be rather crude, because of the need to improve the bioreactor design, finding the right cell source, the right nutrition and the appropriate way to handle the process.

In vitro meat production may thus solve a moral problem. We can both enjoy meat eating, be good (or at least better) to the environment and reduce animal suffering.

Why not go vegetarian?

There are studies indicating that it is more environmentally friendly to reduce meat consumption and switch to vegetables and plants. Why should one be interested in introducing *in vitro* meat production? The simple answer is that most people like to eat meat and will probably resist switching. It is already a disturbing fact that meat consumption increases with economic development. (Worldwatch Institute, 2006). While global meat production has more than doubled since 1970, the increase in developing countries is increasing at a higher rate. And presently, the big increase in meat production is by increased confined and intensive meat production. (Worldwatch Institute, 2006).

That the public to a large extent prefer to stay meat eaters instead of turning to a vegetarian diet is (as far as it is true) a statement of fact. This does not of course answer the normative questions: should we become vegetarians for the sake of the environment and the animals? A simple answer would be: perhaps we should stop (reduce) eating meat if this was the only way to avoid serious adverse effects to the environment and to the animals. *In vitro* meat is an interesting way out of the problem. Meat produced by *in vitro* meat technology does not involve animal suffering or killing of animals. No animals are involved. (There is however a least one source animal for the cells in the bioreactor, but the cells can in principle be obtained without killing the animal.) There have been arguments that *in vitro* meat may 'save animals and satisfy meat eaters' (Hopkins and Dacey, 2008).

Challenges to the *in vitro* meat regime

Will the consumer accept *in vitro* meat? There is to our knowledge no discussion of consumer preferences with the exception of the Eurobarometer mentioned above (Eurobarometer 64.1 2006). The scepticism of the public may partly be attributed to lack of knowledge – and at the present stage there is very little knowledge on *in vitro* meat as a consumer product. The general thesis of knowledge deficit and the associated claim that the public would be more favourable if they had been better informed does not hold up very well in other technology areas (Persson and Welin, 2008: 193).

The issue may very well be: is this natural or artificial? My obvious answer would be that *in vitro* meat is both natural (produced from real cells) but also artificial (grown and cultured through a tightly controlled technological process). However, exactly the same can be said of much of present day intensive meat production. It is both natural and artificial. The really natural thing seems to be free grazing animals living out in the countryside. Given the above distinction between animal suffering (always bad) and killing of animals (perhaps not always bad) free grazing animals may be an acceptable source of meat. However, slaughterhouse practices seem to be a problem. *In vitro* meat more truly could be advertised as 'meat without suffering' than any other means of conventional meat production.

There is a sad consequence that may happen. There may be fewer animals on the farms after the introduction of *in vitro* meat technologies. However, there is room for some hope. In Sweden, we no longer use horses to do agricultural work or transports. Nevertheless, there have never been so many horses in Sweden as today. People keep them for fun and because they like them.

Other things will not change. Hunting for fun was a bad idea before *in vitro* meat and will be an equally bad idea after *in vitro* meat. *In vitro* meat production will, however, give rise to a new possibility of having wolves in middle Sweden. They can simply be fed with *in vitro* meat and no longer need to feed on sheep – probably still kept for wool – or the dogs and cats of the human inhabitants. This will of course raise the issue again: is this really natural? Who cares?

References

Ananthaswamy, A. (2002). Cities eat away at Earth's best land. New Scientist 176(2374/2375): 9-12.

Benjaminson, M. *et al.* (2002). In vitro edible muscle protein production system (MPPS): stage 1, fish. Acta Astronaut 51(12):879-889.

Cavalieri, P. and Singer, P. (eds.) (1994). The great ape project: equality beyond humanity. St Martin's Press, New York.

DeGrazia, D. (1996). Taking Animals Seriously. Moral Life and Moral Status. Cambridge University Press, Cambridge.

Edelman, P.D. *et al.* (2005). In vitro-cultured meat production. Tissue Engineering 11(5-6): 659-662.

Eisnitz, G.A. (2006). Slaughterhouse: the shocking story of greed, neglect, and inhumane treatment inside the U.S. meat industry. Prometheus Books, Amherst NY.

Hopkins, P.D. and Dacey, A. (2008). Vegetarian Meat: Could Technology Save Animals and Satisfy Meat Eaters? Journal of Agricultural and Environmental Ethics 21:579-596.

McMahan, J. (2002). The Ethics of killing. Oxford University Press, Oxford.

Omar, F., Tufveson, G. and Welin, S. (2009). Compensated Living Kidney Donation: A Plea for Pragmatism. Health Care Analysis (in press).

Persson, A. and Welin, S. (2008). Contested Technologies. Xenotransplantation and human embryonic stem cells. Nordic Academic Press, Lund.

Peta (2009). PETA Offers $1 Million Reward to First to Make *In Vitro* Meat. Available at: http://www.peta.org/feat_in_vitro_contest.asp. Accessed on March 22, 2009.

Regan, T. (1983). The Case for Animal Rights. University of California Press, Berkeley.

Worldwatch Institute (2006). Available at: http://www.worldwatch.org/node/3893. Accessed on November 22, 2006.

Theme 5 – Innovation in animal ethics teaching

A framework to address conflicts in veterinary responsibilities

Natalie B. Cleton[1] and Franck L.B. Meijboom[1,2]
[1]Ethics Institute, Utrecht University, Heidelberglaan 8, NL-3584 CS Utrecht, the Netherlands; N.B.Cleton@students.uu.nl
[2]Department of Animals, Science and Society, Faculty of Veterinary Medicine, Utrecht University P.O. Box 80.166, NL-3508 TD Utrecht, the Netherlands

Abstract

Conflicting responsibilities are often said to be inherent to veterinary medicine in the sense that the responsibilities of veterinarians towards their patients and their patients owners can conflict. To date veterinary ethics has strived to deal with these inherent dilemmas by focusing on individual cases and reflecting on the conflicts present in those cases. This method has provided a sufficiently strong foundation with which to deal with such conflicting responsibilities. However this case-by-case approach has become the subject of discussion as a result of significant changes in society and the veterinary profession. It has been argued that the traditional case-focused approach fails to provide tools to deal with the structural complexity of the societal and ethical problems that veterinarians are faced with today. By analysing the concepts of responsibility, professionalism and the relation of the veterinary professional with society, this presentation aims to identify the conflicts in veterinary responsibilities, how these occur, and on what level and between which parties they could occur. Based on a conceptual analysis of the concepts of responsibility and professionalism, we establish five categories in conflicting veterinary responsibilities: the area, subject, background, concept and context of his/her responsibilities. A framework has been drawn up that incorporates these categories and shows possible solutions with respect to the veterinarian's responsibilities these categories generate. This framework helps veterinarians to address the conflicts in responsibility they are confronted with in a more structured and efficient way.

Keywords: veterinary medicine, responsibility, conflicts, framework

Introduction

Conflicting responsibilities are often said to be inherent to veterinary medicine in the sense that the responsibilities of veterinarians towards their animal patients and the animals' owners often conflict. In the past, veterinary ethics dealt with these inherent dilemmas by focusing on individual cases. However this case-by-case approach has become the subject of discussion as a result of significant changes in society and the veterinary profession. Traditionally, the veterinarian was focused on animal health only. Currently, he/she is also involved in issues of animal welfare, food safety and public health (cf. Rollin, 2006). As a result veterinarians are confronted with new and complex situations that raise new questions of responsibility (Morgan and McDonald 2007; Sandøe and Christiansen 2008; Tannenbaum, 1995). Furthermore, the position of veterinarians in society has changed. Nowadays, for example, veterinarians can be employed by companies. This means that they consequently have double responsibilities, both to their profession and to the company.

Accordingly, it has been argued that the traditional case-focused approach fails to provide tools to deal with the structural complexity of the societal and ethical problems that veterinarians are faced with. By analysing the concepts of responsibility, professionalism and the relation of the veterinary professional with society, this presentation aims to identify the conflicts in veterinary responsibilities, how these occur, and on what level and between which parties they could occur. Identifying and analyzing these conflicts is of crucial importance, because a profession that is unclear about its role in society and that is not in sync with society's expectations easily loses its credibility and integrity to the public, which are the basic elements for functioning as a trustworthy profession (Koehn, 1994; Meijboom, 2008).

Based on a conceptual analysis of the concepts of responsibility and professionalism, we establish five categories in conflicting veterinary responsibilities that help veterinarians to address the conflicts in responsibility they are confronted with.

Responsibility: a matrix

Defining responsibilities is an important, but difficult procedure. Distinguishing between different important facets of that process is important in order to understand what the issue of responsibility entails. In this process the responsibility matrix by Ropohl (1994) can be helpful (Table 1). Ropohl's matrix distinguishes between different types of responsibility and their normative foundations (Werner, 2002; Ropohl, 1994). The first column addresses the different categories, from (A) to (G), into which responsibilities can be divided. Ropohl has formulated these categories by means of a number of different questions that one can ask about responsibility. This includes the questions: '*who* is responsible?' (A), '*what* an agent is responsible for. (B-C), '*what* is the foundation and reason of responsibility?' (D-E), '*when* the responsibilities are established' (F), and the question of '*how* a person, an organisation, or a society can be responsible?' (G).

The other columns, under numbers (1), (2), and (3), are answers to those questions. Ropohl's investigation of the concept of responsibility led him to provide a threefold answer for every category of responsibility. The matrix seems to imply that the different answers in the first column are correlated, just like those under (2) and under (3) would be. This, however, is not the way one should approach this matrix. One should rather accept the answers (1), (2), and (3) as answers to the categories (A) to (G), while the answers under (1), (2), and (3) are not linked in-between rows. They are independent instances of the eight categories, which Ropohl describes.

This grid provides a multifaceted starting-point for the discussion of veterinary responsibilities. However, because Ropohl drafted his matrix for an analysis of the responsibilities in technology assessment not all of the issues in his matrix are of equal importance for the veterinarian context. On top of this, four elements seem to be missing in relation to veterinary practice. The first element is the question about the definition of the concept from which different responsibilities follow. One's definition of animal welfare, for example, influences one's ideas about who is responsible for the welfare of animals. A second, missing aspect is the area in which the responsibilities are situated. Thirdly, being a 'professional' differs from the

Table 1. Categories of responsibility (Original title: 'Morphologische Matrix der Verantwortungstypen' Ropohl, 1994).

Category	(1)	(2)	(3)
(A) Who	Individual	Company	Society
(B) For what	Action	Product	Neglecting to perform an action
(C) For what	Foreseeable consequences	Unforeseeable consequences	Future and belated consequences
(D) Why	Moral rules	Societal responsibly	Legal responsibility
(E) Reasons	Conscience	Judgements of others	Law
(F) When	In advance	Present	Afterwards
(G) How	Actively	Virtual	Passively

concept of 'individual' in a number of ways. Therefore, for the veterinarian context it is important to add the element of 'profession' to the possible answers to the category of 'subject'. The final missing element in Ropohl's matrix is the role of context in the definition of responsibilities. Every action is performed in a certain situation. The context in which an agent performs, or refrains from performing an action should be taken into account, because of its impact on the definition of the agent's responsibility.

This modification of the matrix results in a customized framework that includes five central elements that comprise the central questions from the original matrix: the Area, Subject, Background, Content and Context of responsibilities.

A matrix for conflicting veterinary responsibilities

Based on the above-mentioned analysis, we adapted Ropohl's matrix into a framework designed to identify and structure elements involved in possible conflicts among veterinary responsibilities (Table 2). This matrix distinguishes between the categories or reasons, which are fundamental to the conflict, which makes it a more appropriate method to deal with the complexity that underlies the conflicts in responsibility. This framework can be used by the individual veterinarian, a group of veterinarians or by the profession as a whole depending on the complexity of the problem. The veterinarian, who wishes to examine the conflicting responsibilities that he is confronted with, can analyse the five aspects that can be the origin of a conflict of responsibilities.

A first step is to establish in what area, or in which areas, the veterinarian seems to have responsibilities that might cause conflicts in responsibility. Veterinarians are involved in many different areas within society, that includes many, sometimes unclear, responsibilities that involve many different stakeholders. A second step is to clarify the subjects who have responsibilities in a particular situation and to explicate at which level the conflicts in responsibilities occur, e.g. is the conflict between individuals or between an individual and the professional association? Thirdly, the notion of 'background' refers to the normative reasons of why a subject has or should take responsibility. It deals with social and moral issues that underlie particular responsibilities. Focusing on these issues helps to understand the normative character of the conflicts. The fourth step in the framework, is to establish the content of the responsibilities. This can be understood in terms of a technical question on what parameters determines, for example, health or animal welfare, but it also includes a moral dimension with respect to the value of these themes. The fifth step is a clarification of the context. Are the conflicts in responsibility the result of the specific context, e.g., an unclear idea about the veterinarians' role in society or a situation of coercion that restricts the freedom of the agent?

It would be idealistic to think that simply structuring conflicts in responsibility will makes all conflicts disappear. This framework still needs further development, but it shows the different categories involved, and therefore helps to show in what directions the conflicts are best addressed. It shows that there is

Table 2. Categories in conflicting veterinary responsibilities.

Categories	Possible answers
1. Area	=> Society, professional association, veterinary practice
2. Subject	=> Individual, profession, society
3. Background	=> Moral, professional, legal
4. Content	=> Animal health, food safety, animal welfare,
5. Context	=> Clarity of professional role, problems of coercion

not just a general problem of responsibility, but that one is confronted with a specific problem, which needs a specific answer. Sometimes, it can be a procedural solution, for example, to introduce operating procedures that help to prevent situations of coercion. In other cases, it might turn out that the main problem is the plurality of moral views on what animal welfare means, which asks for a debate on moral status. Therefore, the framework helps to address problems of responsibility in a more structured and efficient way.

References

Koehn, D. (1994). The ground of professional ethics, Routledge, New York.

Meijboom, F.L.B. (2008). Problems of trust: A question of trustworthiness. An ethical inquiry of trust and trustworthiness in the context of the agricultural and food sector, Dissertation, Utrecht University.

Morgan, C.A and McDonald, M. (2007). Ethical Dilemmas in Veterinary Medicine. Vet Clin Small Animals 37: 165-179.

Rollin, B.E. (2006). An introduction to Veterinary Medical Ethics, Theory and cases, (2nd ed.) Blackwell Publishing, Oxford, UK.

Ropohl, G. (1994). Das Risiko im Prinzip Verantwortung. Ethik und Sozialwissenschaften 5: 109-120.

Sandøe, P. and Christiansen, S.B. (2008). Ethics of Animal Use, Blackwell Publishing, Oxford, UK.

Tannenbaum, J. (1995). Veterinary ethics: animal welfare, client relations, competition and collegiality, Mosby, St. Louis.

Werner, M.H. (2002). Verantwortung, In: Düwell, M., Hübenthal, C. and Werner, M.H. (eds.) Handbuch Ethik, Metzler, Stuttgart/Weimar, pp. 521-527.

How to assess what comes out of ethics teaching in veterinary science programmes

T. Dich, T. Hansen and P. Sandøe
University of Copenhagen, Faculty of Life Sciences, Rolighedsvej 25, DK-1958 Frederiksberg C., Denmark; td@life.ku.dk

Abstract

How should we measure whether the learning and teaching environment has brought any change in student cognition and performance with respect to the defined aims of the course? This general question is pertinent in the case of ethics teaching because here the aims are often ambiguous. Is the aim that the students know different theories, is it that they are disposed to behave in a certain way, or is it both? This paper examines different tools to assess the effects of ethics teaching – all have originated in the field of medical ethics (including dentistry and nursing). They are the Problem Identification Test, the Moral Judgement Interview, the Defining Issues Test, the Test for Ethical Sensitivity in Science, the Four Component Model, the Meta-Ethical Questionnaire, and the Ethical Reasoning Tool. Finally, the methods will be evaluated in connection with a specific course in animal ethics for first-year veterinary students, a course which has now been running for eight years. Which method would be most appropriate, bearing the aim of the course in mind? Can they be combined? Will there be a need to develop a new assessment tool?

Keywords: assessment, professional ethics, student learning, veterinary students

Introduction

It is not possible to discuss the assessment of student learning without relating it to the aim of the course. What competencies do we want to give the students? Is it the ability to identify and analyse ethical issues, to reason and make a moral judgement, to place moral values above personal values, to reflect on one's own standpoints, to apply ethical principles to one's own conduct, to act in a morally correct fashion, or something else? These questions are relevant, but lengthy discussion will not be taken up here.

There is no doubt, however, that as a professional one should be able to handle complex situations and take decisions that involve different actors and interests. As such, teaching should supply the students with tools to handle these situations – to become professionals. It also means that when the performance of the students is assessed, which occurs when they are still not professionals, it will only be their potential behaviour that can be measured.

Assessment of whether the learning and teaching environment has brought about changes in student cognition and performance, with respect to the defined aims of the course, requires at least two measurement points: one at the beginning of the course and one at its completion. If measurements are made later, other factors in addition to the course may have an influence on the ability of the students to engage in ethical deliberation. If there is no initial measurement point it will not be possible to find out how much has been achieved relative to the students' entry level of knowledge, experience, and natural aptitude for the subject.

The following methods of assessment all originate in medical, dentistry, nursing and life science ethics. They were found through a review of the literature on the evaluation of education in professional ethics. The methods are described only briefly, but through the references it is possible to obtain more information on each method. There are other methods, but these seem to be the most common methods that have been well described and tested.

Different methods to assess the effect of teaching professional ethics

Problem Identification Test (PIT)

The test is a questionnaire to measure the ethical sensitivity of medical students and professionals. It is based on vignettes (cases), in which the subjects have to identify ethical issues related to each case. Each vignette is scored according to the number of issues identified within the domains of: (1) autonomy, (2) beneficence, and (3) justice. This number is evaluated as an indication of the ability to identify problems (Akabayashi *et al.,* 2004). The PIT was originally based on the approach of Herbert *et al.* (1992).

Moral Judgement Interview (MJI)

This assessment tool, originally developed by Lawrence Kohlberg, uses dilemmas to assess moral judgement (Crain, 1985). Kohlberg defined a theory of cognitive moral development in which we progress through six identifiable stages that can be classified into three levels of reasoning about moral dilemmas: (1) Pre-conventional: reasoning based on self-interest – the individual aims to win rewards and avoid punishment; (2) Conventional: reasoning based on conformity to society's rules, norms and expectations; (3) Post-conventional: reasoning based on universal ethical principles that are founded on justice. Progression through these stages is regarded as a move towards increasing sophistication in moral reasoning (Latif, 2000). Consequently, tests of moral reasoning that are based on Kohlberg's theory essentially measure a student's form of reasoning when solving an ethical problem.

The MJI consists of a semi-structured oral tape-recorded interview in which students are asked to resolve a series of hypothetical moral dilemmas. Each dilemma is followed by a systematic set of open-ended follow-up questions designed to enable the student to reveal the structure of the logic of his/her responses. The interview yields an overall score that reflects the student's stage of moral reasoning.

The approach has been criticised by Gilligan (1982), who argues that women reason through a caring rather than a justice approach. The logical sequence of the stages has also been challenged (Rest 1994).

Defining Issues Test (DIT)

This is an instrument designed to evaluate the relative degree of moral reasoning according to cognitive developmental theories posited by Piaget, Kohlberg, and Rest (Bebeau, 2002). Instead of individual interview responses being analysed by a trained person, as in the Moral Judgement Interview, the DIT is a multiple choice test that can be administered to a group and scored using a computer. It consists of six hypothetical dilemmas. Each dilemma is followed by a series of 12 statements about the dilemma. For each dilemma, students must select and rank those issues that have, in their opinion, the most significant influence on the resolution of the dilemma. The four highest ranked items are included in scoring the DIT. Of these four items, only those that represent principled thinking are included in a P-score (Principled Score) (Rest, 1994).

The test is based on the same controversial theory of moral development as the MJI. A further problem is that the student is asked to put in rank order existing lists of issues. This says little about the student's own thinking about the issues, or whether they would recognise the moral issues involved.

Test for Ethical Sensitivity in Science (TESS)

This test, developed by Clarkeburn (2002), requires students to respond to an unstructured story, and their responses are scored according to the level of recognition of the ethical issues embedded in the story. Unlike the DIT, there are no answers to choose among and put in rank order. The subjects have to produce their own responses. Given that the test measures the spontaneous recognition of moral issues, the subjects are supplied only with minimal guidelines or pre-established thought patterns. The answers can be collected either verbally in an interview or in a written form (Clarkeburn, 2002).

Four Component Model (FCM)

The model was developed by James Rest (Rest *et al.,* 1999) and is a further development of Kohlberg's theory that goes beyond mere cognition. It operates as a dynamic feedback system that facilitates moral development and subsequently promotes moral behaviour. It consists of the following components (Bebeau, 2002; Morton *et al.,* 2006):

1. Moral sensitivity: interpreting the situation as moral. This is tested by placing the students in real-life situations in which they witness an interaction, e.g. on videotape or audiotape.
2. Moral reasoning/judgement: judging which of the available actions are most justified. This can be assessed by the Defining Issues Test (DIT).
3. Moral motivation: prioritising moral values relative to other values. It is often assessed as professionalism or professional identity with regard to a specific code of ethics, e.g. by letting the students write a short essay on how they see their professional role.
4. Moral character: being able to construct and implement actions that service the moral choice. The assessment of essential ethical implementation skills requires performance-based assessment, which has not been developed.

The Meta-Ethical Questionnaire (MEQ)

The Meta-Ethical Questionnaire was developed from Perry's developmental model, which operates with nine ethical development positions and is a description of the students' moral development from a simplistic view with beliefs in absolute answers and authorities to one that is more pluralistic and contextual (Katung *et al.,* 1999).

The MEQ groups the nine positions into three main types: A ('Safety in dualism'), B ('Distress in relativism'), and C ('Comfort in commitment'). The questionnaire consists of 10 statements within each of types A and C, which represent typical statements expected from individuals at these two stages of ethical development. Type B statements are considered to lie between the two extremes. The respondents are presented with opposite statements (A and C types) and asked to indicate on a five-point response scale their preference on the spectrum of the two polar views. Neutral responses are identified as B types (Clarkeburn *et al.,* 2003).

The Ethical Reasoning Tool (ERT)

The ERT was developed to assess the cognitive reasoning of nurses in response to an ethical practice dilemma. It categorises responses to the case study into three professional response levels (McAlpine *et al.*, 1997):

1. Level 1 (traditional): thinking dominated by the use of personal moral values and beliefs and/or conventional moral reasoning.

2. Level 2 (traditional/reflective): practical considerations moderated by some use of reflective reasoning, indicating recognition of some of the relevant ethical issues, and the need for consideration of more than one's own personal beliefs.
3. Level 3 (reflective): critical thinking about ethical issues, with the use of an ethical framework and recognition of the value of other points of view.

In the first class of a course in ethics a pre-test is given to the students. Students are asked to identify significant ethical issues raised by a case study, state what they think should be done, and justify their answers from an ethical perspective. In the last class the students are given the pre-test case study with the same directions. After completion of the test they retrieve their own pre-tests, and are asked to re-read both their efforts and to write down their reflections on their own responses. Only upon completion of the post-test are the students fully informed about the method of assessment, in order not to distort the results by prior knowledge of what is to come.

Discussion: which method to use?

It can be seen clearly from the brief introduction to the seven assessment methods that the methods differ in important ways. When it comes to choosing a method for the assessment of a specific ethics course we think that there are two key issues to consider: (1) Does the method measure variables which, according to the aim of the course, it is relevant to measure? (2) Is the method practical, given the time and other resources available for the assessment?

To illustrate how these questions may be answered in practice we will consider here the evaluation of a specific course, i.e. our own course on veterinary ethics for first-year veterinary students at the University of Copenhagen.

The overall aim of this course is to introduce different ethical positions to the students and to give them the opportunity to identify, work with and reflect on ethical questions related to the veterinary profession, as a part of their training to become professionals. We proceed on the assumption that the goal of teaching ethics is not to make future professionals behave in certain ways, or to tell them what is right and wrong, but to give them the tools with which to handle ethical situations in a complex world.

Figure 1 shows the expected learning outcomes from our course and the corresponding methods of assessment, which were described and discussed in the above sections.

The ability to apply ethical principles to one's own conduct cannot be measured directly during a course, but it is our intention that, by giving the students tools with which to analyse the ethical problems that face them and their profession, and by confronting them with others' viewpoints, we will make them more capable and confident to deal with ethical issues when they become practising veterinarians.

The assessment methods that build on Kohlberg's theory of moral development (MJI, DIT and FCM) are all based on the assumption that there are universal ethical principles and that these are based on justice. This goes against a key assumption that underlies our course, which is that pluralism is a central 2.order value and there are in most cases no universally valid answers on how to act in a morally correct fashion as a professional.

It can be seen in Figure 1 that there are two existing methods of assessment that are intended to measure all the other expected outcomes from the course. In order to evaluate which method would be the most appropriate it is necessary to look at the different designs and the theoretical framework behind the methods.

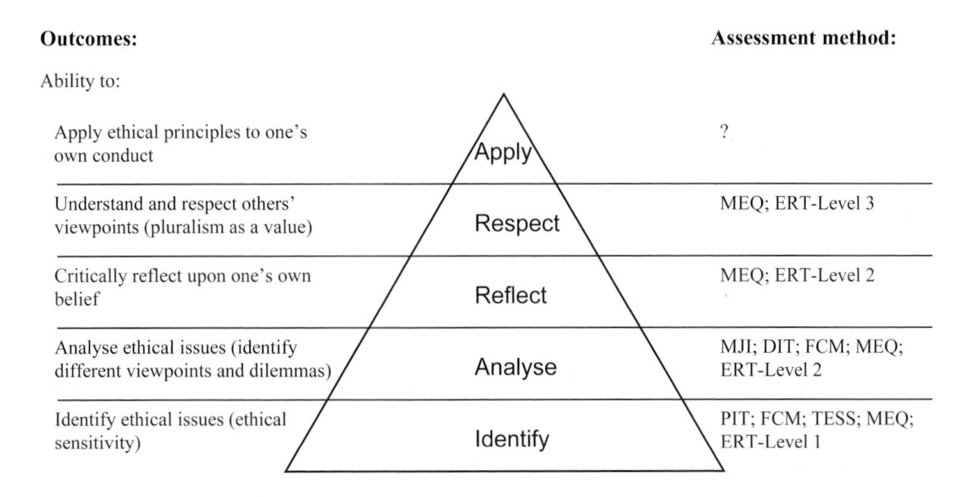

Figure 1. Learning outcomes and matching methods of assessment.

The MEQ is based on fixed statements to which the students have to indicate their responses. The ERT is based on free responses to a case, and reflective responses to the two responses in the pre- and post-test. Each format involves a different set of strengths and weaknesses and extracts different types of information from the respondent. The fixed choice format requires less time to analyse and it is easier to compare between respondents. Open-ended formats, such as the ERT, offer the possibility of discovering unanticipated responses that would be difficult or impossible to identify in the fixed statements. On the other hand, the items in the MEQ format may remind respondents of options that they may otherwise not have considered and reported in the ERT. By reporting their reflections on their two sets of responses in the pre- and post-test, using the ERT, the respondents get an insight into their own ethical development, which adds an extra dimension to the outcome of the assessment method.

The theory behind the MEQ is based on Perry's scheme for cognitive development, which is also reflected in the three professional response levels of the ERT. There are no significant differences in the theoretical framework behind the two methods. They both see ethical development as a cognitive progression from a narrow and self-serving perspective to a potentially reflective, pluralistic and committed perspective.

On the basis of these findings one can conclude that the choice of assessment method for our course in veterinary ethics will depend on the time frame and available resources for the assessment. The ERT is more time consuming, both for the students and the teacher, but both parties could benefit from the learned insight into the development process that is provided by the reflections reported by the students. This could constitute a valid measure of the change in student cognition and the effect of the teaching programme.

References

Akabayashi, A., Slingsby, B.T., Kai, I., Nishimura, T. and Yamgishi, T. (2004). The development of a brief and objective method for evaluating moral sensitivity and reasoning in medical students. BMC Medical Ethics 5: E1 (Published online 29 January 2004).

Bebeau, M.J. (2002). The Defining Issues Test and the Four Component Model: contributions to professional education. Journal of Moral Education 31(3): 271-295.

Clarkeburn, H.M. (2002). A Test for Ethical Sensitivity in Science. Journal of Moral Education 31(4): 439-453.

Clarkeburn, H.M., Downie, J.R., Gray, C. and Matthew, R.G.S. (2003). Measuring Ethical Development in Life Sciences Students: a study using Perry's developmental model. Studies in Higher Education 28(4): 443-456.

Crain, W.C. (1985). Theories of Development. Prentice-Hall, New York, pp.118-136.

Gilligan, C. (1982). In a Different Voice: Psychological Theory and Women's Development. Harvard University Press, Cambridge, Massachusetts, pp.24-39.

Herbert, P.C., Meslin, E.M. and Dunn, E.V. (1992). Measuring the ethical sensitivity of medical students. Journal of Medical Ethics 18: 142-147.

Katung, M., Johnstone, A.H. and Downie, J.R. (1999). Monitoring attitude change in students to teaching and learning in a university setting: a study using Perry's developmental model. Teaching in Higher Education 4: 43-59.

Latif, D.A. (2000). The Relationship between Ethical Dilemma Discussion and Moral Development. American Journal of Pharmaceutical Education 64: 126-133.

McAlpine H., Kristjanson, L. and Poroch, D. (1997). Development and testing of the ethical reasoning tool (ERT): an instrument to measure the ethical reasoning of nurses. Journal of Advanced Nursing 25(6): 1151-1161.

Morton, K.R., Worthley, J.S., Testerman, J.K. and Mahoney, M.L. (2006). Defining features of moral sensitivity and moral motivation: pathways to moral reasoning in medical students. Journal of Moral Education 35(3): 387-406.

Rest, J.R. (1994). Background: Theory and Research. In: Rest, J.R. and Narváez, D. (eds.) Moral Development in the Professions: Psychology and Applied Ethics. Lawrence Erlbaum Associates, Inc., pp.1-26.

Rest, J.R., Narváez, D., Bebeau, M.J., and Thoma, S.J. (1999). New issues, new theory, new findings. In: Rest, J., Narváez, D. and Bebeau, M.J. (eds.) Post-conventional moral thinking: a neo-Kohlbergian approach. Lawrence Erlbaum Associates, New Jersey, pp.99-133.

www.animalethicsdilemma.net: a learning resource for university lecturers in animal sciences and veterinary medicine

A.J. Hanlon[1], A. Algers[2], T. Dich[3], T. Hansen[3], H. Loor[4] and P. Sandøe[3]
[1]School of Agriculture, Food Science & Veterinary Medicine, University College Dublin (UCD), Veterinary Sciences Centre, Belfield, Dublin 4, Ireland; Alison.Hanlon@ucd.ie
[2]Swedish University of Agricultural Sciences (SLU), Sweden
[3]Faculty of Life Sciences, University of Copenhagen, Denmark
[4]imCode Partners AB, Visby, Gotland, Sweden

Abstract

Animal Ethics Dilemma (AED) was launched in 2006, as a free online learning resource. The programme provides a framework for ethical analysis using five ethical perspectives. Based on practical experience, including evaluations from students, it will be argued that through the use of the programme students develop a better understanding for what it means to have a certain ethical view – as distinguished from seeing ethics as a word-game with no real content. The practical experiences presented derive from the use of the programme at three universities in Europe: Dublin (UCD), Copenhagen (UC) and Sweden (SLU). At UCD, the programme is used in three modules: a university wide elective module on animal ethics, the preclinical veterinary programme and an online graduate certificate in canine sports medicine. At the BSc programme in Veterinary Medicine, University of Copenhagen, the programme is used as part of a veterinary course aimed at presenting animal ethics and veterinary roles and tasks. At SLU, the programme is used in three programmes and five different courses: Agricultural Sciences, Agricultural Sciences (elective), veterinary studies: Ethology and Animal Welfare (Level 1 and 3). The use of the programme ranges from providing an introductory lecture to animal ethics and assigning students to complete the case studies in the programme to structuring class discussions and assignments around the case studies in the programme. The time dedicated to teaching animal ethics in each course varies; the programme thus provides a facility to further develop concepts introduced in the classroom.

Keywords: animal ethics, teaching, computer-supported learning tool, role play

Introduction

Bioethics is a core component of the veterinary curriculum in Europe (European directives 78/1026 and 78/1027). However, the availability and content of bioethics amongst European veterinary schools is inconsistent (Edwards, 2002; Gandini and Monaghé, 2002; Von Borrell, 2002). A lack of expertise and staff training in bioethics within the veterinary schools may be contributory factors for this inconsistency. The provision of peer-reviewed teaching resources and frameworks are likely to support the development of new courses on bioethics. Animal Ethics Dilemma (AED) is one example of an established framework that is widely used to support the teaching of bioethics. It is currently available in five languages: English, Danish, Swedish, Dutch and Spanish and has 10,783 registered users (2 March 2009).

AED is a computer-supported learning tool, developed primarily for veterinary undergraduates, but widely applicable to other courses of study such as animal science. It is intended to complement existing courses, and not as a stand-alone such as 'Concepts in Animal Welfare' developed by WSPA and the University of Bristol.

The learning objectives of AED are to promote student understanding of animal ethics, illustrate ethical dilemmas that arise in animal use, broaden the moral imagination and enable students to differentiate between types of ethical arguments.

The programme comprises five case studies:
- the blind hens;
- ANDi the genetically modified monkey;
- euthanasia of a healthy dog;
- animal slaughter;
- rehabilitation of seals.

The case studies are intended to represent ethical dilemmas that can arise in different animal sectors. Each case has been written as a narrative, which has been divided into four levels. Within each level, the student is presented with a statement, an ethical dilemma, followed by four or five responses. Both the statements and responses correspond to different ethical perspectives, in particular: contractarian, utilitarian, animal rights, relational and respect for nature. The narrative or storyline changes depending on the ethical choices selected by the student, and are intended to challenge their perspective. Narrative twists are used as the student progresses through the levels, with the final level giving an outcome to their choices.

On first entering the programme, the student is required to answer a set of 12 multiple-choice questions, based on the ethical perspectives listed above. Once completed, their choices are used to generate a personal profile, to illustrate the proportion of their choices, which are characteristic of a contractarian, utilitarian, animal rights etc. This is represented as a bar chart, which is updated as the students progress through the programme, to reflect changes in their ethical choices.

Following registration, the student can begin to explore the case studies. Terminology used in the case studies appears as highlighted text, enabling the student to learn more, if they so choose. Terminology is also listed in a glossary, which can be viewed separately. In addition, explanations of the theories used are also available for the student to explore.

Pedagogic approach

Special consideration has been given to enhance the pedagogic value of the programme, based on five design characteristics:
- interactive technology;
- student –centred approach;
- case studies;
- role play;
- case study template.

Interactive technology

From the outset, the programme was designed to be interactive, adopting a similar approach as 'gaming' in order to facilitate a higher level of engagement amongst the student community.

Student-centre approach

Students can control their learning by determining the amount and order of information that they receive by selecting how to explore the programme. The programme can be navigated in seven ways:

(1)'My personal profile', (2) Contractarian, (3) Utilitarian, (4) Relational, (5) Animal rights, (6) Respect for nature and (7) 'Show my last choice'. In addition, students can select the 'assist me' option, which explains the basis of the ethical arguments. The navigation choice will determine the ethical dilemmas that are presented such that if a student selects 'Contractarian' they will only be able to progress through the case study by selecting contractarian responses. The 'Show my last choice' option is for students that have previously worked through the case study, and the programme highlights the choices that were made during the previous navigation. This provides the student with an opportunity to reflect on earlier decisions.

Providing on-going feedback to the student further facilitates learner autonomy. For example, the programme is designed to provide feedback to the student through the generation of a personal profile, which assigns their responses to particular ethical perspectives. This facility thus enables the student to reflect on their choices and understand the theoretical framework of their responses i.e. contextualising their learning.

The reality text and a glossary of terminology provide a third aspect of learner autonomy, giving students the opportunity to learn more about the background to each case study and terminology that occurs in the programme.

Case studies

There are a variety of ways of motivating student learning. Reiss (2005) considers case studies to be highly motivating, because students perceive them to be relevant, especially when they are based on real life events. The learning outcomes of using case studies are partly determined by the level of detail provided, such that if too much information is provided, students may find it overwhelming, and too little can appear superficial and demotivate learning (Reiss, 2005). In this context, the programme enables the student to determine the level of information, adapting to the needs of the learner.

Role play

In all of the case studies, the students are required to adopt the role of a particular stakeholder, normally either a junior veterinary surgeon or a postgraduate student. The cases follow the career path of the person, dramatising the potential pitfalls and highlights of their ethical choices. The aim of role-play in the programme is to engage the student, to provide them with a memorable experience (Reiss, 2005).

The case template

A template has been developed to enable students to create their own case study. It is likely that this will facilitate deep learning, providing students with a platform to research, develop and publish ethical arguments on a topic of their choice. In addition, it provides tools to upload photographs and create hyperlinks. The use of the template is currently undergoing evaluation at UCD.

Animal ethics dilemma in the classroom

As a teaching resource, the programme can be used to support the learning of animal ethics in a variety of contexts. Table 1 summarises the application of AED at UCD, UC and SLU. Overall, AED is commonly used to complement didactic teaching such that the five ethical perspectives in AED are presented in lectures and thereafter the students, initially under supervision from the lecturer, are given an opportunity to explore AED. Another learning scenario used at SLU is to schedule a tutorial after students have used AED and the personal profiles generated by AED are the basis for discussion.

Table 1. Summary of contexts in which Animal Ethics Dilemma (AED) is used at three universities in Europe.

University	Course	Programme	Level	Description	Assessment
UCD	Animals in society	Elective	1	The five ethical perspectives used in AED are presented in a lecture; this is followed by a tutorial in a computer suite in which students have to complete one case study in AED.	Students have to prepare a case study using the AED template.
UCD	Animal behaviour and welfare	Veterinary medicine	2	The five ethical perspectives used in AED are presented in a lecture. Studies have to register and complete one case study as part of independent learning.	Questions on animal ethics are typically included in continual assessment and the written exam.
UC	Introductory veterinary course	veterinary medicine	1	The five ethical perspectives used in AED are presented in lectures and the students work with the ethical perspectives in different exercises. One of the exercises is a brief introduction to the programme and then the students have time to use AED.	Students have to write a group based report about an ethical dilemma in the veterinarian field, using at least two of the five ethical perspectives used in AED. Submitting the report is compulsory; the oral examination includes questions on animal ethics.
SLU	Introduction to animal welfare	Agricultural sciences	1	Not used every year, but when included, it is used as a complement to the lecture in animal ethics. No introduction to the programme is given, but students are expected to submit their personal profile.	If included, students have to submit their personal profiles.
SLU	Animal welfare	Agricultural sciences, elective course		To complement lectures in animal ethics. Students work through the programme including at least one case study. They submit their personal profile and the programme is discussed in class.	Students have to submit their personal profiles. Questions on ethics are included in the exam.
SLU	Animal welfare and veterinary legislation	Veterinary medicine	5	No specialised lecture in animal ethics. Course targets students' gaps in general animal welfare knowledge. Students develop their personal profile and work through at least one of the cases at home followed by a discussion.	Compulsory attendance at the course and submission of personal profile.
SLU	Animal protection, welfare and ethics 1	Ethology and animal welfare	1	Short introduction to AED before students spend 2 hours exploring AED; the lecturer is available for questions. Lectures in animal ethics and a whole-day seminar are given later in the course.	The examination for the course includes questions on animal ethics inspired of the tool.
SLU	Animal protection, welfare and ethics 2	Ethology and animal welfare	3	To complement lectures in animal ethics and seminars on animal ethics dilemmas. Students work through the programme.	Students have to give a lecture in animal ethics at a local high school.

Debriefing tutorials create a forum for providing feedback and therefore further support student learning (Reiss, 2005). Further explanation of the ethical framework used in AED is provided by Ethics of Animal Use (Sandøe and Christiansen, 2008).

Evaluation

With any new teaching resource it is important to evaluate the learning benefits. Evaluation of the programme is ongoing. The first stage of evaluation was conducted early during its development, to determine the user-friendliness of the programme. This was achieved by directly observing students whilst using the programme, followed by face-to-face interviews with the students after having used the programme.

Thereafter evaluation has been conducted to assess the learning benefits of the programme. Evaluation using questionnaires has been conducted on the beta version of the programme amongst first year veterinary students enrolled on an animal ethics module at UCD, students at SLU and the UC.

Since its launch, both students and teachers using the programme have been invited to complete an evaluation sheet, which can be accessed from the main menu of the programme. The evaluation results are used to review the programme, both from the mechanistic functions to the pedagogic goals.

Conclusions

Animal Ethics Dilemma is one example of an established framework that is widely used to support the teaching of bioethics. It is adaptable and can be used in a variety of learning contexts, to support student understanding of animal ethics and create a structure for teachers involved in the delivery of bioethics courses.

Acknowledgments

The authors would like to express their gratitude to the organisations that have helped to fund the development of Animal Ethics Dilemma. These include AstraZeneca, The Swedish Animal Welfare Agency, KVL, imCode, Malmö University, SLU and The UCD President's Teaching Award.

References

Edwards, S.A. (2002). A synthesis of animal bioethics teaching in agricultural and veterinary courses in Northern Europe. In: Proceedings Teaching Animal Bioethics in Agricultural and Veterinary Higher Education in Europe. AFANet Workshop, INPL, Nancy France, 23-24 May 2002, pp. 57-61.

Gandini, G. and Monaghé, A. (2002). Teaching animal bioethics in agricultural and veterinary higher education in Europe: a survey in Belgium, France, Italy, Portugal, Slovenia and Spain. In: Proceedings Teaching Animal Bioethics in Agricultural and Veterinary Higher Education in Europe. AFANet Workshop, INPL, Nancy France, 23-24 May 2002, pp. 10-12.

Reiss, M.J. (2005). Teaching animal bioethics: pedagogic objectives. In: Marie, M., Edwards, S., Gandini, G., Reiss, M., Von Borell, E. (eds.) Animal Bioethics: Principles and Teaching Methods. Wageningen Academic Publishers, Wageningen, the Netherlands, pp. 189-201.

Sandøe, P. and Christiansen, S. (2008). Ethics of Animal Use. Blackwell Publishing.

Von Borrell, E. (2002). A synthesis of animal bioethics teaching in agriculture and veterinary courses in West and Central Europe. In: Proceedings Teaching Animal Bioethics in Agricultural and Veterinary Higher Education in Europe. AFANet Workshop, INPL, Nancy France, 23-24 May 2002, pp. 34-39.

WSPA (2008). Concepts in Animal Welfare: An Animal Welfare Syllabus. CD ROM.

A skill-based approach to the teaching of ethics in the life sciences – experiences from Denmark

T. Hansen, T. Dich and P. Sandøe
University of Copenhagen, Faculty of Life Sciences, Rolighedsvej 25, DK-1958 Frederiksberg C., Denmark
tih@life.ku.dk

Abstract

What is the purpose of teaching ethics in the life sciences? It seems that there are two fundamentally different ways to approach this question: (1) character-based teaching, in which the students are taught certain values and how to act in a morally correct fashion; (2) a skill-based teaching of ethics, in which the students are given tools with which to identify and analyse ethical problems. The aim of this paper is to defend skill-based ethics teaching. First, the paper will present experiences from a course in ethics for veterinary students at the Faculty of Life Sciences, University of Copenhagen, Denmark; this course is skill-based. Three important ideas run through the course: (a) the aim of the teaching is to support the ability of the students to recognise and analyse ethical problems and dilemmas that are related to their professional education; (b) ethical theory is taught, together with real life veterinary problems; (c) students work on projects in which they have to view an issue relating to the application of veterinary science from different ethical points of view. It is clear that the course is not 'value free' – one main underlying value is respect for the different ethical viewpoints that have to co-exist in a pluralistic society. Another important underlying value is the idea that conflicts can and should be dealt with through discussion and compromise. However, the course does not promote one set of first order values at the cost of another. It will be argued that such a skill-based approach, in preference to a character-based system, will enable the students to function and flourish in their future professional roles.

Keywords: student learning, veterinary students, animal ethics

Introduction

Imagine that you have dreamt for many years of becoming a student of veterinary medicine and that now your dream has come true. You are looking forward to studying the treatment of animals and to working with animals, which play an important role in your life. Then you realise that the first course in the veterinary study programme is in ethics, and on your first day at the university you are met by a professor of bioethics. That was not what you expected. One may ask: what is the purpose of teaching ethics to students of life sciences? And how shall we teach ethics to life science students?

There is an ongoing discussion about how and why to teach ethics. These discussions are not only connected to the life sciences but cover the teaching of ethics from primary school to university degree programmes (Campell, 2008). Different approaches exist. Each is based on assumptions about best practice, about learners, and about morality itself (Joseph and Efron, 2005). However, at a general level there seem to be two fundamentally different ways in which to view the purpose of teaching ethics: one uses a character-based approach and the other a skill-based approach.

The character-based approach

This approach emphasises that the purpose of ethics teaching is to guide students towards the 'best moral practice' that is available in a given social and cultural context. The approach concentrates on the development of qualities of character, and requires the identification of appropriate values and transmission of these values from teacher to students (Clarkeburn, 2002). The approach rests on the notion that some moral values are essential (Carr, 1996).

The skill-based approach

The purpose of this approach is to support the students in developing tools to recognise, analyse and solve moral problems that can be used in the process of moral decision making. As such the approach concentrates on giving the students moral sensitivity, so that they can recognise morally relevant aspects of a situation and can develop skills in moral reasoning. The approach will often use ethical dilemmas and cases to generate discussion among the students. The skill-based approach is founded on a pluralistic view of ethical values (Clarkeburn, 2002).

The two approaches are described only very briefly here, but the differences in purposes and assumptions are clear. The aim herein is to argue that the skill-based approach to teaching ethics in the life sciences serves the main purpose of such a course from an educational perspective; namely, to prepare the students for the situations and problems they will typically face in their later professional life. To illustrate this, an existing course in veterinary ethics will be examined.

The teaching of ethics to veterinary students at the University of Copenhagen, Denmark

It was decided by the Danish parliament in 2001 that all undergraduate programmes of university education in Denmark should include an introductory philosophy course. The requirement placed on the content of the courses was that they should enable the students to view their subject from a 'general', broader, perspective. Thus a course could view the relevant subject from, for example, a historical, an ethical, an epistemological, or a sociological perspective. At the same time, the courses should have a close connection with research within the relevant field of study (Dich *et al.,* 2005).

At the Faculty of Life Sciences, University of Copenhagen, the first introductory philosophy course as a component of the veterinary study programme was held in autumn 2001. Since then, the course has been a compulsory part of the veterinary study programme. It is timetabled at the beginning of the first year of the veterinary curriculum. Every year 180 students, on average, participate in the course.

The stated aim of the course is to support the students in their ability to recognise and analyse the ethical problems and dilemmas related to their profession. The main focus of the course is on animal ethics. Different ethical perspectives relating to animals are taught, together with other ethical issues that relate to the veterinary profession.

The course is organised as a mixture of lectures, exercises and group-based project work. The lectures are arranged as 'double-lectures', which involve a guest lecturer with a relevant veterinary background and a lecturer with a background in philosophy or ethics. The guest lecturers introduce their research and some current dilemmas in their field of study. The ethics lecturer elaborates on the dilemma from a more explicitly ethical/philosophical angle, and will typically, as part of the lecture, connect the dilemmas to different perspectives of animal ethics: contractarianism, utilitarianism, animal rights, relational views, and respect for nature (Sandøe and Christiansen, 2008). Students then have the opportunity to raise questions and to debate with the two lecturers. During the veterinary course the topics of the lectures include: 'What is an animal?', 'Animal welfare', 'Production animals', 'Transport and slaughtering', 'Food safety and risk assessment', 'Companion animals', 'Exotic and wild animals' and 'Animal experiments'.

The aim of the lectures is twofold: (1) to introduce the first-year students to different fields of activity in the veterinary profession; (2) to illustrate how veterinary professionals encounter ethical dilemmas and problems.

The exercises encourage the students to reflect on constructed ethical dilemmas or to work with 'real world' situations applicable to the veterinary profession. The constructed dilemmas and the 'real world' situations are presented via described cases, a film, or other means. The questions for discussion are then introduced. The exercises are taught in classes of between 30 and 40 students (180 students attend the lecturers). The constructed cases and 'real life' situations will often be discussed in groups of four to six. After the discussions each group summarises their discussion and conclusions for the class. Finally, the tutor will pick up the overall conclusions of the discussion and locate them in a broader ethical perspective, thereby connecting the exercises to the lectures.

An example of a 'real world' case included in the veterinary course is the breeding of Belgian Blue cattle. After watching a film describing the Belgian Blue breed and hearing different veterinarians giving their views on the breed, the students discuss different questions, for example: What could be an ethical argument to justify the breeding of Belgian Blue calves by planned Caesarean? What is the argument against this practice?

The aim of the exercises is to give the students an opportunity to discuss dilemmas in animal ethics with their fellow students and the tutor. The dialogue about ethical questions encourages the students to reflect on their own values and attitudes towards animals.

In addition to the lectures and exercises, the students undertake group-based project work that culminates in a short report (15 pages). The students again work in groups of four to six. Under the guidance of a tutor they select an ethical dilemma or question within veterinary medicine. In their report the students first describe the dilemma or question, and then they are required to analyse it from two different ethical perspectives. Each group has a facilitator who guides them in the project work.

In the group-based project work students have, for example, studied welfare problems of farm animals (pigs, hens and dairy cattle), diseases associated with breeding, and the treatment of dogs with cancer.

As with the exercises, the aim of the project work is to engage the students in discussion and reflection about the use of other species by humans. The project work is differentiated from the exercises by the fact that the groups themselves can choose an ethical dilemma related to a veterinary field in which they have a special interest. The groups have the time to work in depth on their chosen topic, and the goal is to enable the students to analyse an ethical dilemma or question.

Every year the students have to evaluate the course. The general picture is that students view the course very favourably. They find it exciting, inspiring and relevant to them. According to the evaluation, the involvement of the guest lecturers is very important in encouraging a positive attitude towards the course.

Since 2001, four other courses in introductory philosophy have been developed, for study programmes in Natural Resources (formerly Agronomy and Forestry), Biotechnology, Landscape Architecture and Town Planning, and Food Science. The placement of the courses varies from the first to the third year of the different curriculums. The content also differs from that of the veterinary ethics course. In the other courses ethics is taught as just one subject within a broader focus on the philosophy of science. However, the idea of teaching ethics together with real life problems that are relevant to the field of study, the double-lecture system and the exercises are maintained.

Discussion

It should be clear that the course in veterinary ethics uses a skill-based approach, and from an educational perspective there are good reasons for the use of this approach in teaching ethics to life science students. One reason relates to the expectations of the students. Lewis (2005) notes that students of the life sciences expect to study natural science, and a course in ethics can be seen as something peripheral to the core subjects. The students expect 'hard data' and clear, unambiguous answers. It seems that the students typically perceive life science to be related to real world problems and ethics to be more abstract and theoretical. These are all aspects that we have experienced as teachers of courses in ethics at the Faculty of Life Sciences in Denmark. The question is: how can we make ethics relevant for these students?

With regard to the veterinary course, the idea is that by inviting guest lecturers from within the veterinary profession and letting them talk about current dilemmas in their field of research or practice, students will not perceive ethical questions to be peripheral. The aim of describing cases that are similar to real life veterinary problems during the exercises is to illustrate that ethical questions are not necessarily abstract and theoretical.

Ethical questions are indeed a part of the daily work of a veterinarian. Therefore the course is aimed at making the students aware of ethical questions within their field of study. A central aspect of the skill-based approach is, through ethical dilemmas, to enable students to identify ethical issues. The main idea is to integrate ethics into the subject area.

The danger inherent in the introduction of ethical theories through the use of ethical dilemmas and 'real world' cases is that it can produce a rather instrumental way of handling ethical questions. The students see ethical theories merely as a tool to be used to solve problems in their professional life. According to Carr (1996) the instrumental approach may result in a limited conceptualisation of our pre-theoretical notions of moral life. However, using ethical dilemmas and 'real world' cases in ethics teaching in the life sciences is a way to stimulate the students to be interested and engaged in ethical discussions related to their subject. Hopefully this engagement will lead to further reflection on how to handle different values and interests, and thereby it will assist students to develop a professional ethical stance.

When the students later become professionals they are going to act in a pluralistic society with cultural diversity, and they will encounter different values and moral views. If the students, when they become professionals, are to be able to act, take decisions and communicate in such a society it is important to introduce them to different values in their education. The lectures and exercises in the veterinary course introduce different ethical perspectives, and in the project work the students have to analyse an ethical question from two different ethical perspectives. The course, in accordance with the skill-based approach, is based on respect for different ethical perspectives.

From an educational and a societal perspective we have defended a skill-based approach to teaching ethics. Our argument is based on a particular view of the purpose of teaching, which also reveals a particular conception of students. The university is part of a complex and culturally diverse society. This society tolerates and accepts different values and different ethical perspectives. Therefore our role as teachers must be to support our students in developing tools to be used in their future professional life in a society with different values and interests. It is impossible for the course to be value free. Our teaching is based on ethical values such as tolerance, respect and the belief that conflicts have to be solved with discussion and compromise. It also considers the students as autonomous beings whose judgements can be stimulated through ethical discussion.

References

Carr, D. (1996). After Kohlberg: Some implications of an Ethics of Virtue for the Theory of Moral Education and Development. Studies of Philosophy and Education 15: 353-370.

Clarkeburn, H. (2002). The Aims and Practice of Ethics Education in an Undergraduate Curriculum: reasons for choosing a skills approach. Journal of Further and Higher Education 26(4): 307-315.

Campell, E. (2008). The Ethics of Teaching as a Moral Profession. Curriculum Inquiry 38(4): 357-385.

Dich, T., Hansen, T., Christiansen, S.B., Kaltoft, P. and Sandøe, P. (2005). Teaching Ethics to Agricultural and Veterinary Students. Experiences from Denmark. In: Marie, M., Edwards, S., Gandini, G., Reiss, M. and Von Borell, E. Animal Bioethics Principles and Teaching Methods. Wageningen Academic Publishers, Wageningen, the Netherlands, pp. 245-259.

Joseph, P.B. and Efron, S. (2005). Seven Worlds of Moral Education. Phi Delta Kappan 86 (7): 525-533.

Lewis, S. (2005). Philosophizing Incognito: Reflections on Encouraging Students of the Life Sciences to Think Critically. Teaching Philosophy 28(3): 237-247.

Sandøe, P. and Christiansen, S.B. (2008). Ethics of Animal Use. Oxford: Blackwell.

Teaching animal ethics to veterinary students in Europe: examining aims and methods

M. Magalhães-Sant'Ana[1,2], C.S. Baptista[2], I.A.S. Olsson[1], K. Millar[3] and P. Sandøe[4]
[1]Laboratory Animal Science, IBMC – Instituto de Biologia Molecular e Celular, Universidade do Porto, Rua do Campo Alegre 823, 4150-180 Porto, Portugal; mdsantana@gmail.com
[2]Institute of Biomedical Sciences Abel Salazar, Universidade do Porto, Largo Prof. Abel Salazar2, 4099-003 Porto, Portugal
[3]Centre for Applied Bioethics, Division of Animal Science, School of Biosciences, University of Nottingham, LE12 5RD, United Kingdom
[4]Danish Centre for Bioethics and Risk Assessment, Faculty of Life Sciences, University of Copenhagen, Rolighedsvej 25, 1958 Frederiksberg C, Denmark

Abstract

Traditionally, veterinary surgeons have been seen as a collective authority, a homogeneous group of professionals to whom a shared ethical stance was believed to be enough to face the challenges of veterinary practice. However, the profession has changed over recent years from a community dominated by male farm animal practitioners to become increasingly mixed in terms of gender, background and specialization. The development of veterinary specializations reflects recent changes in animal use with more focus on companion animal medicine. This has been accompanied by a greater debate within society about the moral status of animals, where the veterinarian's view is far from uniform and increasingly questioned. This changing environment in combination with an increasing call for greater transparency in decision-making presents a new challenge for the veterinary profession. One of the ways in which this challenge may be effectively met at an institutional level is through veterinary ethics teaching. The aim of this paper is to discuss how teaching veterinary ethics can help future veterinarians deal with an increasingly challenging professional environment. It will be argued that the ability to ethically reflect is of key importance for the profession because it allows veterinarians to understand their different roles and responsibilities as well as appreciating the expectations of different stakeholders in modern society. It is therefore crucial that veterinary training provides students not only with professional role-models, but also with conceptual tools that allow them to see issues from different points of view. Ethics teaching for veterinary students is being developed in many institutions, using different approaches. This paper will present some of the main approaches applied at several veterinary schools in Europe and will explore to what extent they may advance ethical reflection skills in future veterinarians.

Keywords: professional ethics, veterinary training, code of conduct, skills, virtues

Introduction

The veterinary profession has changed immensely during the last few decades. At the start of the last century, veterinary practice was dominated by male veterinarians who focused on the medical aspects of agricultural animal production. Today small animal practices dominate with an increasing gender mix amongst veterinarians. This change has occurred at a time of social debate regarding the moral status of animals and reflects a notable shift in the wider legal and social consideration of animals. Domesticated animals, mainly farm animals, were widely seen as property, with welfare considerations interpreted in terms of productivity. However this has shifted because animal-human interactions in modern urbanised societies are mainly and sometimes exclusively with companion animals. These animals are often seen as family members and are included in the same social connections that people establish with their human conspecifics.

Society in general has evolved to become pluralistic, but alongside this pluralism, citizens are more aware of a professional's responsibilities and at times call for greater transparency in all forms of decision-making. In this context, members of an increasingly heterogeneous veterinary profession are not only seen as animal healers, but also as animal protectors, on whom society relies to defend animal interests.

A key question is therefore, how should the veterinary profession deal with these changing circumstances. Veterinary Schools and Faculties have been at the forefront of scientific knowledge, and as a result responded by introducing new subjects in their curricula such as applied ethology, oncology and neurosurgery. But this new approach to animal health and welfare involves more than just new forms of specialised veterinary training. Another way of helping future veterinarians to cope with this changing environment is through the development of ethical reflection skills that can be effectively supported via the teaching of veterinary ethics. The ability to ethically reflect is of key importance because it allows veterinarians to understand their different roles and responsibilities as well as appreciating the expectations of different stakeholders in modern society.

A number of challenges are faced by those who deliver veterinary ethics courses. There is no such thing as a generally accepted model for veterinary ethics education and the demands of a busy clinical undergraduate timetable often leave little room for the discussion and analysis of the deeper ethical, legal and social aspects of veterinary practice. As a result, it is important to examine the main approaches used in veterinary schools across Europe, explore the aims of these courses and review to what extent they may advance ethical reflection skills in practicing veterinarians.

Case studies: how ethics is taught in three veterinary faculties

Ethics training is becoming an important part of Veterinary Medicine courses across Europe. The European Directive 2005/36/EC states that veterinary surgeons training must include explicit professional ethics teaching. However, this is not a straightforward task since there is no standardised form of professional ethics which is common to every activity within the field of veterinary medicine (Sandoe and Christiansen, 2008). Therefore, a comparison of how this veterinary ethics teaching requirement is being met at a European level is needed. As a starting point to a more extensive analysis of teaching practices across European Veterinary Schools, this paper will briefly examine the activities of three veterinary educational institutions based in Copenhagen (DK), Nottingham (UK) and Porto (PT).

Denmark: Faculty of Life Sciences, University of Copenhagen (Copenhagen)

Ethics is a significant part of the introductory course in veterinary science. The discipline of ethics is placed in the first year of the curricular programme and corresponds to three European Credit Transfer and Accumulation System (ECTS) credits. The person responsible is a trained philosopher who is also responsible for additional lectures during the course. The aim of the first year course is to support the students in their abilities to recognize and analyse ethical problems and dilemmas related to their profession. A main focus in the course is animal ethics. Different animal ethical perspectives are taught together with current 'real life' veterinary issues and problems.

The course is organised as a mixture of lectures, exercises and group based project work (Dich *et al.*, 2005). The lectures are mostly arranged as 'double-lectures' including a guest lecturer with a relevant veterinary background and the aforementioned lecturer with a background in philosophy or ethics. The guest lecturer introduces her research in the field of study and some current dilemmas in her field. The ethics teacher elaborates on the dilemma from a more explicitly ethical/philosophical angle, and will typically, as a part of the lecture, connect the dilemmas to different animal ethical perspectives. During the exercises, students reflect on staged ethical dilemmas or work through 'real life' cases from the veterinarian profession. The constructed dilemmas and the 'real life' situations are presented via described cases, a film and so on. One particular resource used is the internet based learning tool 'Animal Ethics Dilemma' (Hanlon *et al.*, 2006). Besides the lectures and the exercises students deliver a group based project ending with a short report (15 pages) considering an ethical case study. Evaluation includes the written report and a final examination.

United Kingdom: School of Veterinary Medicine and Science, University of Nottingham (Nottingham)

The Nottingham School of Veterinary Medicine and Science is the first new veterinary school in the UK for more than 50 years. The school received its first undergraduate veterinary students in 2006 and as may be expected with a new degree course, the content of a number of modules, including the veterinary ethics component, is continually evolving and under constant review. In terms of veterinary ethics, rather than delivering a single module the ethics, welfare and law teaching is embedded through the five year course. However, some key components are explicitly delivered within a number of prominent modules. This is exemplified by a series of linked lectures and group sessions that are delivered in the Personal and Professional Skills (PPS) module in the second year of the course.

The ethics teaching begins in the first year with a problem solving exercise, based on a series of case studies that are relevant to the students' extra-mural practical experience. The majority of formal teaching is delivered in the second year. Following introductory sessions delivered using a team teaching approach (through the interactions of a bioethicist and veterinarian), students work through a series of clinical relevance cases that are discussed in small groups. It should be noted that these cases are intentionally linked to the parallel body systems teaching that is running at same point during the year, in order to achieve horizontal integration of the curriculum. For example, during the reproduction module, the case relates to the ethics of Belgian Blue breeding.

The early lecture sessions introduce ethical theories and decision-support tools, such as the Ethical Matrix and the 'Animal Ethics Dilemma' website. This is followed by lectures on The Royal College of Veterinary Surgeons (RCVS) Guide to Professional Conduct, current EU and UK animal law, and legal cases. A pluralistic approach to ethics teaching is applied throughout the course, where students are encouraged to reflect on different ethical positions and to develop their ethical reflection skills. Special themes are emphasised within the integrated course such as the ethics of: euthanasia; farm animal and companion animal use; animal experimentation; and professionalism. Within the second year, the integrated ethics component is taught across three modules and is assessed through both exam questions within the Animal Health and Welfare Module and written assessment (ethics clinical relevance essay) in the PPS module. Animal welfare-related assessments also include a short position paper (in the form of letter to a peer-reviewed journal) and an individual presentation. Three individuals, two veterinarians and a bioethicist are responsible for delivering and integrating these elements within the curriculum. In order to enhance learning and expose the student to different views a number of guest lecturers, both veterinarians and animal ethicists are invited to run targeted ethics sessions.

Portugal: Institute of Biomedical Sciences Abel Salazar, University of Porto (Porto)

Ethics topics are addressed in a discipline called 'Veterinary Deontology', placed in the second year of the course, corresponding to one ECTS. These teaching sessions intend to achieve two major goals: (1) to enlighten the students regarding professional rights and duties, according to deontological and ethical rules described in the Portuguese *Veterinary Deontologic Code*; (2) to describe the national and international animal welfare regulations in the context of professional ethics and responsibility of all veterinary surgeons. Both these subjects are approached through a combination of lectures and case studies and promoting discussion among students.

Teaching is done using single weekly theoretical lectures delivered by a junior assistant professor with clinical and research experience. The person in charge of the coordination and planning of all the contents is a full professor, veterinary pathologist and researcher also experienced in laboratory animal management and welfare.

Students are evaluated through group-based project work that tests their knowledge of the *Deontologic Code* and animal welfare issues (animal production, laboratory, wild and companion animals). For each group work, evaluation includes a report (maximum 50 pages) and a 30 minute oral presentation and discussion (seminars are conducted at the end of the course).

Examining underlying learning goals

As the main purpose of ethics training is not the transmission of factual information *per se*, the learning process will be different from that of most other disciplines. The strategy applied to teach ethics to veterinary students ultimately depends on what one considers the task of ethics training to be, within the context of veterinary medicine, and on the didactical objectives. An analysis of the literature suggests a number of partly distinct, partly overlapping approaches, which will be briefly discussed here.

A teaching approach based on a professional code of ethics tends to emphasise rules, where ethics is a normative instrument used to tell the difference between 'right' and 'wrong' actions. Guides to Professional Conduct and animal welfare regulations are used to set the boundaries between good practice and professional misconduct. For this form of teaching approach pedagogical aims not only involve the transmission of rules that should not be 'trespassed' by veterinary students, but also provides the right tools to solve practical problems. Sinclair (2000) presents an example of how a fundamentally rule-based approach is used to analyse ethical dilemmas in veterinary practice in straightforward manner.

Approaches which place the emphasis on values and/or virtues see professional ethics not as an approach which imparts a set of rules, but as a way of improving the understanding of the values and virtues that justify them. The focus is on the understanding of ethical theories that define 'good' or 'right' conduct. Although values/virtues teaching approaches may differ, these are understood as being complementary since it is argued that one cannot transmit values without fostering the corresponding virtues (Steutel, 1997). In the values/virtues approach, ethics is related to the acquisition of clear values concerning moral issues. Teachers often work as role models and their pedagogical concern is that students learn fundamental moral principles. It is intended that eventually students will develop qualities of character (virtues) and appropriate behavioral dispositions that will make them 'morally better people' (Reiss, 2005). It is asserted that a veterinarian possessing these beliefs is in a better position to address ethical dilemmas and confront clients when difficult decisions arise (Morgan and MacDonald, 2007).

Alternatively, the skills-based approach refers to ethics as a means to understand the complexity and ambiguity inherent in human life, rather than internalising codes or values. The didactical objective of ethics teaching is to promote students' moral responsibility by providing conceptual tools that allow them to identify and understand ethical issues from different perspectives. Teachers work not as experts, but as travel guides (Reiss, 2005). The skills-based approach is known to promote students' autonomy and moral development and is claimed by some to be the only acceptable way to teach ethics (Clarkeburn, 2002). By developing an ethical discussion of the different possible decisions that are characteristic of a pluralist society, students become better prepared to cope with professional and social challenges.

Discussion

An initial examination of the three veterinary ethics courses, in light of the approaches outlined above, indicates that each draws on elements of several teaching approaches although there appears to be notable difference in the emphasis. While the course in Porto is constructed around a rules approach (Codes of Conduct and animal welfare legislation), the use of case studies gives an opportunity for students to develop skills. In Nottingham, a skills-based approach is emphasised, however due to the integrated nature of the ethics teaching this is connected to discussions of rules and virtues, while in Copenhagen the aquisition of skills *per se* is a paramount objective. This reflects the fact that the teaching of rules, values, virtues and skills does not need to be exclusive or naturally conflicting. In order to develop skills, students need to understand the values at stake, which are often expressed in normative regulations.

Teaching a discipline is not only a question of methodological approach as discussed above; the details of the course content are just as important as how it is conceptually framed. Training in ethics, it may be argued, should equip students with the necessary knowledge and skills to analyse a difficult situation from different ethical perspectives (such as utilitarian, virtues, contractactarian, deontology, respect for nature) and to take into account the different values involved (such as the autonomy of the client; the welfare of animals; human health related issues). This will help individuals identify and evaluate all available courses of action. Real life scenarios in veterinary practice often involve people with very different, and sometimes opposing, ethical beliefs. Only through an awareness of the diversity of thought and ethical positions can a veterinary surgeon appreciate the expectations of different stakeholders in modern society. Therefore, it may be argued that veterinary ethics teaching should be addressed using a pluralisitic approach.

The nature of the didactic approach applied and the theoretical content of courses are key issues in an analysis of how ethics is taught. Other important issues are the number of hours devoted to the discipline, the integration of ethical considerations into other components of the course (as part of the clinical disciplines in particular), the method of assessment and at which stage of the course the students are taught ethics. This paper has briefly reviewed these key issues through the presentation of three case studies. This review will be included in a more extensive European-wide analysis and discussion of how ethics teaching is integrated into the training of veterinary surgeons. This paper represents a starting point for this wider assessment.

References

Clarkeburn, H. (2002). The Aims and Practice of Ethics Education in an Undergraduate Curriculum: reasons for choosing a skills approach. Journal of Further and Higher Education 26(4): 307-315.

Dich, T., Hansen, T., Christiansen, S.B., Kaltoft, P. and Sandoe, P. (2005). Teaching ethics to agricultural and veterinary students: experiences from Denmark. In: Marie, M., Edwards, S., Gandini, G., Reiss, M. and Von Borell, E. (eds.) Animal Bioethics: Principles and teaching methods. Wageningen Academic Publishers, Wageningen, the Netherlands, pp. 245-258.

Hanlon, A., Dich, T., Hansen, T., Loor, H., Sandøe, P. and Algers, A. (2006). Animal Ethics Dilemma – An Interactive Learning Tool for University and Professional Training. Available at: www.aedilemma.net. Accessed 3 March 2009.

Morgan, C.A. and McDonald, M. (2007). Ethical Dilemmas in Veterinary Medicine. Veterinary Clinics of North America: Small Animal Practice 37(1): 165-179.

Reiss, M. (2005) Teaching animal bioethics: pedagogic objectives. In: Marie, M., Edwards, S., Gandini, G., Reiss, M. and Von Borell, E. (eds.) Animal Bioethics: Principles and teaching methods. Wageningen Academic Publishers, Wageningen, the Netherlands, pp.189-202.

Sandøe, P. and Christiansen, S.B. (2008). Ethics of Animal Use. Blackwell Publishing, Oxford, UK, 178 pp.

Sinclair. D. (2000). Ethical Dilemmas and the RCVS. In: Legood, G. (ed.) Veterinary Ethics – An Introduction. Continuum, London, UK, pp.74-85.

Steutel, J.W. (1997). The Virtue Approach to Moral Education: Some Conceptual Clarifications. Journal of Philosophy of Education 31(3): 395-407.

Theme 6 – Sustainability in food production

Awareness and attitudes of consumers to sustainable food

A. Clonan[1], M. Holdsworth[2], J. Swift[1] and P. Wilson[3]
[1]Division of Nutritional Science, School of Biosciences, University of Nottingham, LE12 5RD, United Kingdom; sbxac4@nottingham.ac.uk
[2]UMR NUTRIPASS, Institute of Research for Development-IRD, WHO Collaborating Centre in Human Nutrition, Montpellier, France
[3]Division of Agricultural and Environmental Sciences, School of Biosciences, University of Nottingham, LE12 5RD, United Kingdom

Abstract

Interest in 'sustainable food' has grown substantially in recent years; however, an official definition for sustainable food in the UK has yet to be agreed upon. Many frameworks rely upon the Brundtland report definition: '*development which meets the needs of the current generation without compromising the ability of future generations to meet their own needs' (World Commission on Environment and Development 1987).* Drawing upon previous qualitative research and using the defining principles of sustainable food provided by 'Sustain: the alliance for better food and farming', this study seeks to assess consumers' understanding of, and attitudes towards, sustainable food. Using a detailed structured questionnaire sent to 2,500 randomly selected Nottinghamshire (UK) residents, a response rate in excess of 33% was achieved. The questionnaire explored shopping habits, attitudes to sustainable food components (organic, fair-trade, local food and animal welfare), dietary intake, stated purchasing behaviour and demographic information. Analysis of the data provides novel findings into which sustainability criteria are prioritised by consumers when purchasing food. Moreover, analysis across demographic categories reveals insightful comparison of consumers' attitudes, purchasing behaviour and dietary intake. Future research will seek to compare and contrast stated and actual preferences by comparing the population survey results to actual purchasing behaviour from supermarket data. Building on the population findings, in-depth case-study interviews will explore in-depth determinants of consumer views towards sustainable food.

Keywords: shopping habits, purchase behaviour, demographic categories, sustainable food

Background

Sustainability is currently a major issue appearing on a wide range of government and industry agendas covering production, manufacturing and service sectors (UK Public Health Association, 2007). In relation to food sustainability, differing definitions exist in the literature; the official definition by Sustain: The alliance for better food and farming (2008) includes components of social, environmental and economic importance. Sustainable food is, therefore, accessible, healthy, nutritious, respects the environment and biodiversity, promotes the use of fair trading practices and respects the rights of workers throughout the food-chain. Currently, different components of sustainability are promoted by different organisations, for example public health nutritionists promote access to healthy, nutritious food, while the Soil Association promote food production without the use of pesticides, and the Fair Trade Foundation promotes fair trading practices. Consumers have numerous value conflicts when making food choices, for example cost, taste preference and convenience (Luomala *et al.,* 2003). Issues of sustainability therefore place further demands upon the consumer. Previous research in this area has investigated some of the individual components of sustainable food, for example consumer attitudes to organic food (Fotopoulos *et al.,* 2003) or ethical food purchases (Marylyn, 2001). However, previous studies have not considered what a consumer definition of sustainable food is, whether it incorporates the issue of healthy eating, or how consumers deal with the demands of sustainability when making food choices. This study therefore aims to address the current paucity of information in this area.

Objective

The main objective covered here was to determine whether socio-demographic differences exist towards consumer definitions of sustainability criteria when purchasing food.

Method

Self administered postal questionnaires were sent to 2,500 Nottinghamshire residents randomly selected from five electoral registers; that encompass both urban and rural areas: Nottingham City, Broxtowe, Rushcliffe, Gedling, and Erewash. 842 completed questionnaires were returned, providing a final adjusted response rate of 35.6%.

Questionnaire validity and reliability

Questions asked were deemed to have content and face validity by a range of experts in sociology, psychology, public health nutrition and environment/agricultural science. The questionnaire was also piloted in a diverse population to ensure that the questions asked were clear and could be answered easily. The final questionnaire consisted of 158 structured questions divided into five sections. This present paper focuses on strands taken from the attitudinal section, dietary intake data and key demographics of the respective participants.

Constructing the attitudinal scale

The constructs included in the attitudinal scale were drawn from 11 qualitative interviews conducted with a broad sample of adults, and were based on the thematic categories that emerged from the qualitative data analysis conducted previously by the same authors. The interview schedule was structured around the consumer guidance for sustainable food provided by Sustain (2008). Following piloting, several statements were either removed or replaced due to poor discriminative nature as demonstrated by frequency analysis. The final attitudinal section consisted of 72 items. In addition some constructs were included from *a priori* assumptions.

Respondents were asked to choose the answer that best suited how far they agreed or disagreed with the statements using a 5-point Likert scale (Likert, 1932), 'strongly agree', 'agree', 'neither agree nor disagree', 'disagree', 'strongly disagree'. They were also given the opportunity to answer 'not applicable'. Attitudinal statements were grouped into constructs that were validated for internal consistency using Cronbach's alpha. The results of one of the constructs (Positive attitudes to sustainable food) are shown in Table 1.

Developing the healthy eating score

A semi quantitative food frequency questionnaire (FFQ) was developed using the five food groups defined in the UK's Eatwell plate (Food Standards Agency, 2008). The FFQ consisted of 25 food items, and respondents were asked to choose how frequently they ate the 25 foods on a scale from never to twice a day or more frequently. Standard food portion sizes were included based on national food portion sizes (Crawley, 1994). Using this information, daily intakes were calculated and individual foods were regrouped into six food groups: fruit, vegetables, protein, dairy, starchy carbohydrates and foods containing fat and or sugar. Using SPSS to establish frequencies, tertiles were set for low medium and high intakes for each food group. These were scored accordingly for each food group e.g.

Table 1. Internal consistency of attitudinal items.

Attitudinal statement	n	Mean[1]	± SD	Cronbach's Alpha (10 items)
I try to buy food that has been produced in the UK	630	1.44	0.60	0.754
I prefer to buy food that has been grown in a way that cares for the environment	630	1.32	0.53	
I always check that the fish I'm buying has come from a sustainable source	630	2.11	0.78	
When buying food, I look for packaging which is compostable	630	1.94	0.78	
Organic food is always the healthiest option	630	2.05	0.76	
I choose organic food wherever possible	630	2.09	0.80	
I try to buy food which is in season in the UK	630	1.36	0.63	
Local food is much better for the environment	630	1.31	0.54	
I always try to buy meat that has been reared in the UK	630	1.50	0.72	
I buy free range meat wherever possible	630	1.56	0.72	

[1]excluding respondents answering not applicable: 1 = strongly agree; 5 = strongly disagree.

'If fruit intake 0-1/day then fruit score = 0'
'If fruit intake 1-3/day then fruit score = 1'
'If fruit intake 3+/day then fruit score = 2'

Consumption of less healthy foods, i.e. those containing fat and/or sugar was scored with the inverse logic, i.e. high consumption was associated with a low score. A healthy eating score was then established by adding together the six scores from all food groups. A higher healthy eating score indicated healthier eating behaviour.

Socio-economic characteristics

Various socio-economic and demographic data were collected at both individual (gender, age, educational level, profession) and household levels (urban/rural, household income). For this paper data, the key demographic variables of age and gender are considered (Table 2).

Table 2. Percentage of respondents by gender and age group.

Socio-demographic characteristics of the sample (n= 842)		
Gender:	Men	40.1
	Women	59.9
Age (y):	18-30	12.2
	31-45	22.4
	46-60	31.8
	61-91	33.6

Data inputting and analysis

Data were entered using EpiData software, v3.1 (Lauritsen *et al.,* 2000). In order to reduce error, a 10% random sample of questionnaires were verified to determine error rate of <1% (Meeuwisse *et al.,* 1999). All data were analysed using SPSS version 16.0 (SPSS inc., 2007). Attitudinal variables were regrouped so that analyses were carried out on three categories: 'agree' (grouped from strongly agree and agree); 'neither agree nor disagree' and 'disagree' (grouped from strongly disagree and disagree). Cross-tabulations were conducted, incorporating the chi-squared test for independence which was used to test the null-hypothesis that there was no difference between gender, age, or respondents with healthier diets. Differences were considered statistically significant at the 5% level.

Results

Gender

Data presented in Table 3 suggests that women have more positive attitudes to particular sustainability criteria when purchasing food.

Age

Strong trends were found between age and positive attitudes towards sustainable food; Figure 1 that shows strength of agreement by age group towards purchasing UK reared meat. This suggests that agreement increases with age, and this trend is apparent in several other positive attitude statements (full data not shown). For all of these attitudinal statements, older adults were more likely to state that they actively thought about different aspects of sustainability when purchasing food.

Healthy eaters

Healthier eaters in the sample were more likely to have strongly agreed or agreed with eight out of the 10 attitudinal statements, which suggests that healthier eaters are more likely to think about aspects of sustainability when purchasing food.

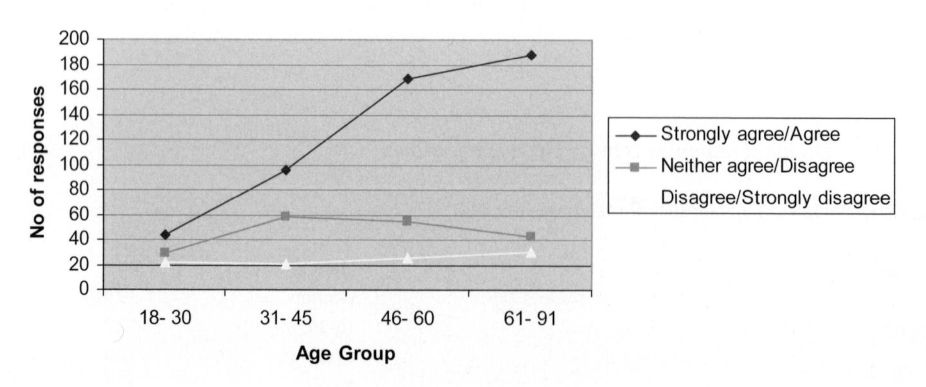

Figure 1. Attitude towards purchasing UK reared meat by age group. $\chi^2 = 34.2;_{df=2}$

Table 3. Gender comparison of positive attitudes.

Attitude statement	Strongly agree/agree				Neither agree/disagree				Disagree/strongly disagree				χ^2
	Men		Women		Men		Women		Men		Women		df=2
	n	%	n	%	n	%	n	%	n	%	n	%	
I try to buy food that has been produced in the UK	197	61.7	305	63.5	101	31.6	155	32.3	21	6.6	20	4.2	2.3
I prefer to buy food that has been grown in a way that cares for the environment	210	65.2	356	74.3	95	29.5	111	23.2	17	5.3	12	2.5	9.4
I always check that the fish I'm buying has come from a sustainable source	65	23.2	126	29.2	103	36.8	154	35.7	112	40.0	151	35.0	3.5
When buying food, I look for packaging which is compostable	84	26.8	191	40.6	132	42.2	168	35.7	97	31.0	111	23.6	16.1
Organic food is always the healthiest option	86	27.6	128	27.3	124	39.7	200	42.6	102	32.7	141	30.1	0.8
I choose organic food wherever possible	73	24.9	139	30.5	102	34.8	157	34.5	118	40.3	159	34.9	3.4
I try to buy food which is in season in the UK	217	66.8	380	77.2	74	22.8	77	15.7	34	10.5	35	7.1	10.9
Local food is much better for the environment	223	69.0	356	73.7	78	24.1	110	22.8	22	6.8	17	3.5	5.1
I always try to buy meat that has been reared in the UK	187	59.7	312	66.1	75	24.0	110	23.3	51	16.3	50	10.6	6.0
I buy free range meat wherever possible	174	57.2	268	58.0	92	30.3	130	28.1	38	12.5	64	13.9	0.5

Discussion

This investigation has presented evidence to suggest that there are gender differences in attitudes towards sustainable food. Table 3 shows that women in this study are significantly more likely than men to consider environmental issues – particularly packaging, in addition to matters such as UK seasonality and UK reared meat. Previous research as shown that gender has a huge influence on sustainable behaviours with surveys highlighting that women are more likely to make ethical purchases or buy organic food (Organisation for Economic Co-operation and Development, 2008).

The age trends apparent in the data illustrate an increasing concern with issues of sustainability around food as respondents get older, perhaps due to the additional time allowances that older people arguably have to do their food shopping. The oldest age group demonstrated the strongest agreement with sustainability statements, and this is perhaps also a reflection of their former interactions within a food system that was not yet globalised, and therefore did not rely on transporting food long distances. Older respondents are likely to have developed the habit of using local, seasonal produce and this would potentially have been mirrored by an increased frequency of food shopping at local stores rather than weekly supermarket trips. There was also a strong trend for healthier eaters to hold more positive attitudes to issues of sustainability when purchasing food, however further analysis is required to establish how much this data is influenced by gender.

Future work will include further analysis of the data set, in particular seeking to compare reported purchasing behaviour with actual purchasing data for key sustainable food purchases.

References

Crawley, H. (1994). Food Portion Sizes. London, Stationery Office Books.

Food Standards Agency (2008). The eatwell plate. Available at www.eatwell.gov.uk/healthydiet/eatwellplate/. Accessed 07/06/2008.

Fotopoulos, C., Krystallis, A. and Ness, M. (2003). Wine Produced by organic grapes in Greece: using means-end chain analysis to reveal organic buyers purchasing motives in comparison to non- buyers. Food Quality and Preference 14: 549-566.

Lauritsen, J.M., Bruus, I. and Myatt, M. (2000). EpiData Denmark Brixton Health: a tool for validated data entry and documentation of data.

Likert, R. (1932). A Technique for the Measurement of Attitudes. New York, Archives of Psychology.

Luomala, H.T., Laaksonen, P. and Leipamaat, H. (2003). How Do Consumers Solve Value Conflicts in Food Choices? An Empirical Description and Points for Theory-building Advances in Consumer Research. 31 (564).

Marylyn, C. and Ahmad, A. (2001). The myth of the ethical consumer – do ethics matter in purchase behaviour. The Journal of Consumer Marketing 18(7): 560.

Meeuwisse, W.H., Hagel, B.E. and Fick, G.H (1999). The impact of single versus dual data entry on accuracy of relational database information. Journal of Informatics in Primary Care June: 2-8.

Organisation for Economic Co-operation and Development (2008). Promoting Sustainable Consumption: Good Practices in OECD Countries. OECD.

Sustain: the alliance for better food and farming. (2008). Consumer Guide to Sustainable Food. Available at www.sustainweb.org/pdf/SFG Consumers 1pp.pdf.

UK Public Health Association (2007). Climates and Change. London.

World Commission on Environment and Development (1987). Our Common Future. General Assembly Reports. United Nations.

Agricultural entrepreneurship and sustainability – is it a good or bad fit?

C.C. de Lauwere
Agricultural Economics Research Institute of Wageningen UR, P.O. Box 35, 6700 AA Wageningen, the Netherlands; carolien.delauwere@wur.nl

Abstract

In today's Dutch agriculture emphasis is put on entrepreneurship, social responsibility and sustainability. But do these fit together? In economic theories entrepreneurs are seen as movers of the markets, seekers of profit opportunities and innovators. Not all farmers however meet these conditions and if they do, there is no guarantee that this goes with socially responsible entrepreneurship and sustainability. In a sociological explorative study a multiform group of 20 pig and 21 dairy farmers – both male and female – were asked about their views on animal welfare and other features of sustainable farming. The group consisted of conventional, organic and free range farmers with different farming styles. Their farms varied in levels of scale, intensity, degree of specialization and participation in quality assurance schemes. In the in-depth interviews, it became clear that the farmers focus on different aspects of sustainability and that multi-dimensional sustainability is not a self-evident aim for all farmers. An economically viable farm is important for all farmers, although farmers with idealist motives stress this aspect less than other farmers. Social sustainability at the level of the farm (work load and schedule, division of tasks, balance work/ family life/ social life) is accentuated by conventional farmers on large scaled specialized farms. At a higher level of social sustainability (fair trade, fair prices, poverty reduction), in particular organic and biodynamic farmers stress that farmers have to take the responsibility to contribute to social equity. The latter group puts also emphasis on their responsibility towards the ecosystem. They, for instance, focus on sustainable cattle, mineral management and nature and landscape conservation. The interviewed large scale conventional farmers on the other hand, see energy production as a potentially profitable option to contribute to ecological sustainability. This means that agricultural entrepreneurs do not 'automatically' take all aspects of sustainability – people, planet and profit – into account. Policy makers who think they can stimulate sustainable agriculture by promoting agricultural entrepreneurship should be aware of this.

Keywords: farmers, diversity, profitability, internal and external social sustainability, social responsibility

Introduction

Dutch agriculture of today is confronted with rapidly changing circumstances. As consumer demands and governmental legislation are becoming stricter, agricultural entrepreneurs are being required to commit increasingly more resources to animal welfare, environmental measures and maintenance of the landscape. In 2000, the Dutch Ministry of Agriculture, Nature and Food Quality (LNV) stated that agricultural entrepreneurs should operate as financially independent units, deliver high quality products produced in a socially sound way, and respect social values. In turn, they should receive societal appreciation (LNV, 2000). In 2008, the Ministry has formulated more far-reaching ambitions with regard to sustainability: '*...Within 15 years, livestock farming in the Netherlands should be sustainable in all respects and have a broad public support. This means that livestock farmers produce with respect for human beings, animals and nature throughout the world...*' (LNV, 2008). This illustrates the major importance the Ministry attributes to agricultural entrepreneurs in the transition process towards sustainable agriculture. In economic theories, entrepreneurs are seen as movers of the markets, seekers of profit opportunities and innovators (Van Praag, 1999) In these theories they are held responsible for economic development through innovations of products, processes, markets as well as organizational innovations (Van Praag,

1999; Shane, 2003). Important questions are whether Dutch agricultural entrepreneurs are able and willing to take the lead towards more sustainable farming systems (do they consider themselves mainly responsible?) and how do they understand 'sustainable agriculture'? The latter question is an obvious question because sustainability is a confusing and contested concept which can be interpreted and conceptualized in many different ways (Boogaard *et al.*, 2008; Van Calker *et al.*, 2005; McGlone, 2001). McGlone (2001) defines it as follows: *'If our systems of production are in harmony with the environment, the animals, the workers and the community and if they are efficient and economically competitive then the system may be said to be sustainable'.* This refers to the multi-dimensional character of the sustainability concept (people, planet, profit). Besides this, sustainability is a multi level as well as a multi actor concept because it can be enacted on farm level, regional level or global level and the involvement of many actors and institutions is needed (Van Calker *et al.*, 2005). This illustrates that sustainability can be conceptualized in many different ways. The study discussed here examines how Dutch pig farmers and dairy farmers interpret the sustainability concept. The question: 'agricultural entrepreneurship and sustainability – is it a good or bad fit?' shall be explored[6].

Research methods

A sociological study was carried out based on in-depth interviews with a multiform group of pig farmers (n = 20) and dairy farmers (n = 21), including conventional pig farmers (n = 11) and conventional dairy farmers (n = 14), organic pig farmers (n = 3) and organic dairy farmers (n = 5), free range or other 'alternative' pig farmers (n =4), biodynamic dairy farmers (n = 2) and pig farmers with more locations, combining conventional farming with organic farming or free range farming (n = 2). Their farms varied in levels of scale, intensity, degree of specialization and participation in quality assurance schemes. Both male (n = 28) and female farmers (n = 13) were interviewed. A main criterion for selection was to maximize diversity. Therefore the sample is not representative for the pig and dairy sectors as a whole.

The farmers were asked about their views on animal welfare and other aspects of sustainable farming, such as nature, landscape, environmental issues and relationships between farmers, society, market and technology. Their views on animal welfare have been previously described (De Lauwere *et al.*, 2007). These will be summarized in this paper. The views of farmers with regard to sustainability were analysed according to an analytical framework based on Van Calker *et al.* (2005), who in cooperation with stakeholders compiled a list of sustainability attributes with respect to economic, internal and external social and ecological sustainability (Table 1).

On the basis of the interviews it was estimated to which extent the farmers 'fit' into the profile of 'real' agricultural entrepreneurship. The 'entrepreneurial features' described by Van Praag (1999) were taken as criteria (being a 'seeker of profit opportunities', a 'mover' of the market' and/ or an 'innovator'). Besides this, it was assessed whether the selected farmers were driven by 'heart' or by economics (Schoon and Grotenhuis, 1999) and whether they were willing to take responsibility for their way of farming completely, partly or not at all.

Sustainable farming according to farmers

The interview data showed that farmers accentuate different aspects of sustainability. The following value orientations with regard to sustainability emerged:

[6] The study described is a part of a larger project called 'A new ethics for livestock farming: towards value based autonomy in livestock farming?', which is funded by the Dutch Organization of Scientific Research (NOW) and the Dutch Ministry of Agriculture, Nature and Food Safety (De Greef *et al.* 2006).

Table 1. Analytical framework to analyse the farmers' views with regard to sustainability based on Van Calker et al. *(2005).*

Overall sustainability			
Economic sustainability	Internal social sustainability	External social sustainability	Ecological sustainability
profitability	working conditions	food safety animal welfare animal health landscape quality use of undisputed products social equity[1]	closing nutrient cycles[2]

[1] This attribute is not mentioned by Van Calker *et al.* (2005).
[2] These attributes for ecological sustainability were too specific and were therefore summarized as 'closing nutrient cycles'.

1. A first value orientation puts a special focus on economic and internal social sustainability. Economic and commercial values are central. The farmers concerned consider animals above all as means of production that are serving human interests. They keep the animals according to minimum legislation and good production and health are the major indicators for animal welfare. Internal social sustainability (work load and schedule, division of tasks, balance work/ family life/ social life) is important for some of these farmers, especially the ones with personnel; some others who cannot afford this, complain about it. The farmers do not feel responsible to contribute to social equity (fair trade, fair prices, poverty reduction). Energy production is mentioned by some of them as a possibility for a profitable contribution to ecological sustainability. This value orientation is wide spread among the conventional pig breeders and some (large) dairy farmers.
2. A second value orientation is partly identical to the first because emphasis is on economic sustainability. Internal social sustainability however is not such an important issue for these farmers. They work hard but this does not seem to bother them because they see it as their moral duty to take good care of the animals they are responsible for. The relation between farmers and animal is central in their farm management. The main difference with the former value orientation is that the farmers are more ambiguous about their way of farming. They keep their animals for example according to minimum legislation, but they would prefer to treat the animals according to higher standards. This however does not 'fit' with the production system for economic reasons. Another difference is that these farmers, in adition to productivity and health, use physical and behavioural features as indicators for animal welfare (the animal looks bright, the animal shines, the animal is lively). Some of them contribute to external social sustainability through nature and landscape conservation, participation in environmental programs and classification schemes for improved food quality. This orientation is to be found among both conventional pig and dairy farmers with family farms or family farms with one co-worker. Dairy farmers in this value orientation often offer summer grazing.
3. In a third value orientation, emphasis is on economic sustainability through contributing to external social sustainability and/or ecological sustainability. This strategy sustains a premium price for the products. The main motive to produce according the requirements of environmental programmes (for example the Dutch ecobrand 'Milieukeur') is economical. Animal welfare is a starting point for farm management and emphasis is put on creating space for animals so that they can express

natural behaviour and on the animal's identity. These values are however based on the demands of the production system and consumers rather than on ethics. The problems of social equity are not denied, but the farmers do not feel responsible. Organic pig farmers and free range pig farmers especially 'fit' in this value orientation.
4. In the fourth value orientation, emphasis is on external social and ecological sustainability. Economic sustainability and internal social sustainability are important as well but it is accepted that profitability or working conditions sometimes are conflicting with for instance animal health and welfare. The farmers concerned join programmes for nature and landscape management; closing the nutrient cycle is another important aspect of their farming style. Social equity is a main concern for them. Like in the third value orientation, animal welfare is a starting point for farm management, and emphasis is put on creating space for animals to express natural behaviour and on the animal's identity. The farmers concerned are driven by heart rather than by economics. This value orientation especially is found among extensive dairy farmers, organic farmers and bio-dynamic farmers.
5. The fifth value orientation is more or less identical to the fourth in that emphasis is on external social and ecological sustainability. In this value orientation, economic values are of less importance because livestock farming is not the main activity of the farmers concerned. Provision of maximal animal welfare is the starting point and the production process is fully adapted to the needs of the animals. Apart from a focus on naturalness, the expressing of natural behaviour and respectful treatment of animals self-realization or self-development of the animals is also considered important. Bio-dynamic farmers or farmers with 'alternative' ways of farming can be found in this value orientation

Agricultural entrepreneurship

Among the interviewed farmers 'real' entrepreneurs could be distinguished according to the definition of Shane (2003) or the 'entrepreneurial features' mentioned by Van Praag (1999). The data however showed that there is no relationship between 'real' entrepreneurship (in the sense of movers of the markets, seekers of profit opportunities or innovators) and value orientation with regard to sustainability. There are entrepreneurs among the farmers for whom the emphasis is on economic and internal social sustainability and others for whom emphasis is on economic and external social sustainability. Besides this, it appeared that being a 'real' entrepreneur is not a guarantee that a farmer also is willing to take full social responsibility. Some of them do, but others tend to shift their responsibility upon consumers, retailers and or the government. It is however obvious that 'real' entrepreneurs emphasize being an entrepreneur (making a profit) rather than on being a stockman (taking good care of the animals) if they are asked to define a 'good farmer' themselves. Farmers who do not meet the entrepreneurial features described by Van Praag, can be described as 'real stockmen' for whom taking good care of the animals is important (although they can fill in 'taking good care of animals' according to different value orientations). Like the 'real' entrepreneurs, some of them do take their social responsibility and others do not. The data showed that stockmen who do take their social responsibility are more often driven by heart rather than economically driven. Besides this, it appeared that those farmers who are ready to take full responsibility might contribute to overall sustainability more than for instance some 'real' entrepreneurs (Table 2 and Table 3).

Agricultural entrepreneurship and sustainability: do they fit together?

In Dutch agriculture an important role with regard to sustainable agriculture is attributed to agricultural entrepreneurs (LNV, 2000 and 2008). The data presented in this study however indicates that 'real' entrepreneurship, according to the definitions of Shane (2003) or Van Praag (1999), is not a good starting point to guarantee overall sustainability. Some entrepreneurs take social responsibility, but there are also 'real' entrepreneurs who tend to shift the social responsibility to consumers, retailers and/ or the government. This is also found by for example Te Velde *et al.* (2002) and Van Huik and

Table 2. Relationship between social responsibility and some 'general' features of the farmers interviewed.

Social responsibility	n	Farming method	Value orientation	Entrepreneurship
Shifted to consumers retailers, government	10	Esp. conventional farming	Emphasis on economic and internal social sustainability	Economically driven
Is accepted but partly shifted	21	All farming methods	All value orientations	Economically driven and part of them also by heart
Is fully accepted	10	Esp. organic or free range farming	Emphasis on external social and ecological sustainability	Driven by heart

Table 3. Relationship between social responsibility and other aspects of sustainability.

Social responsibility	Animal welfare	Animal health	Social equity	Use of GMOs in animal feed	Environment
Shifted to consumers retailers, government	Minimum legislation	Important	Is not worried about it	Is not so worried about it	Not so important
Is accepted but partly shifted	Varies	Important	Is worried, but does not feel responsible	Is worried, but does not reject it	Important; part of them strive to close nutrient cycles
Is fully accepted	Natural behaviour	Not so important	Is worried and does feel responsible	Rejects this	Striving to close nutrient cycles

Bock (2007). On the other hand, there are other farmers who might not completely 'fit' the definition of 'real' entrepreneurs, because for their way of farming more emphasis is put on stockmanship, but they are fully prepared to take social responsibility. So, policy makers who exert much effort to realize sustainable agriculture, should not have implicit faith in 'real' agricultural entrepreneurs. The willingness of agricultural entrepreneurs and other farmers to take social responsibility should be a more important criterion. It might also be worth redefining the concept of (agricultural) entrepreneurship, placing more emphasis on overall sustainability.

References

Boogaard, B.K., Oosting, S.J. and Bock, B.B. (2008). Defining sustainability as a socio-cultural concept: citizen panels visiting dairy farms in the Netherlands. Livestock Science 117: 24-33.

De Greef, K.H., De Lauwere, C.C., Stafleu, F.R., Meijboom, F.L.B., De Rooij, S., Brom, F.W.A. and Van der Ploeg, J.D. (2006). Towards value based autonomy in livestock farming? In: Kaiser, M. and Lien, M. (eds.) Ethics and the politics of food. Wageningen Academic Publishers, Wageningen, the Netherlands, p.61-65.

De Lauwere, C.C., De Rooij, S. and Van der Ploeg, J.D. (2007). Understanding farmers' values. In: Zollitsch, W., Winckler, C., Waiblinger, S. and Haslberger, A. (eds.) Sustainable food production and ethics. Wageningen Academic Publishers, Wageningen, the Netherlands, p. 198-203.

LNV (2000). Voedsel en groen – het Nederlandse agro-food complex in perspective (in Dutch). Dutch ministry of agriculture, nature and food quality, Den Haag.
LNV (2008). Future vision on livestock farming of the Dutch minister of agriculture, nature and food safety (in Dutch), January, 16th., 2008. Den Haag.
McGlone, J.J. (2001). Farm animal welfare in the context of other society issues: toward sustainable systems. Livestock Production Science 72: 75-81.
Schoon, B. and Grotenhuis, R. (2000). Values of farmers, sustainability and Agricultural policy. Journal of Agricultural and Environmental Ethcis 12: 17-27.
Shane, S., (2003). A general theory of entrepreneurship. Edward Elgar, Cheltenham, 327 pp.
Te Velde, H., Aarts, N. and Van Woerkum C. (2002). Dealing with ambivalence: farmers'and consumers' perception of animal welfare in livestockbreeding, Journal of Agricultural and Environmental Ethics 15: 203-219.
Van Calker, K.J., Berentsen, P.B.M., Giesen, G.W.J. and Huirne, R.B.M. (2005). Identifying and ranking attributes that determine sustainability in Dutch dairy farming. Agricultural and Human Values 22: 53-63.
Van Huik, M.M. and Bock, B.B. (2007). Attitudes of Dutch pig farmers towards animal welfare. British Food Journal 109(11): 879-890.
Van Praag, M. (1999). Some classic views on entrepreneurship. The economist 147(3): 311-335.

Food production and sustainability

K. Klint Jensen
Danish Centre for Bioethics and Risk Assessment, University Of Copenhagen, Rolighedsvej 25, DK-1958 Frederiksberg C, Denmark; kkje@life.ku.dk

Abstract

Assessment of the sustainability of food production is a complex matter. I shall identify two major approaches to sustainability: One line of research (*Bottom-Up*) takes sustaining a system over time as its starting point and then infers prescriptions from this requirement. Another line (*Top-Down*) takes the Brundtland Commission's suggestion that the present generation's need-satisfaction should not compromise the need-satisfaction of future generations as its starting point. It then measures sustainability at the level of society as some form of just distribution of need satisfaction or welfare between, as well as within, generations and infers prescriptions from this requirement. These two approaches may conflict and in this conflict, I shall demonstrate, the Top-Down Approach ethically speaking has the upper hand. However, the notion of a just distribution of welfare is rather vague. Moreover, the long term consequences of agricultural technologies are uncertain, and this uncertainty clearly affects the prescriptions that can be derived from sustainability. I shall then introduce food security for all at all times as a minimal requirement for calling a distribution of welfare just. This makes it possible to connect the two approaches, because the focus will be on the management of natural resources, in the widest sense, on which the poor will be dependent for food. The productivity of these resources should be safeguarded and possibly be increased when they are handed over to future generations also dependent on them for food security. This in turn has implications on the second approach. In the concern for future generations, the future food insecure people should get priority. It means that for environmental protection for the sake of the future poor, worldwide equitable sharing of local costs for the poor, evidently, becomes very important. I draw out some implications of sustainability on this more sharpened notion of justice.

Keywords: distributive justice, food insecurity, future generations

Introduction

'Sustainability' has become a notorious buzz-word, referring to something undeniably good. However, the almost universal acceptance of the claim that development should be sustainable comes at a cost, namely that the concept has become almost empty. But this means that it is very unclear what exactly the practical implications of sustainability are, e.g. for food production.

Hence, there is a need to clarify the concept. However, any clarification will also make the concept more controversial. And of course, more than one clarification is possible. The differences between conflicting interpretations of 'sustainability' will uncover disagreements about value judgments as well as about empirical issues. This paper attempts to map the logic of some of the important disagreements.

The first step is to recognize that sustainability is a normative concept. Sustainability is a valuable state – a goal we should strive to realize. This means, firstly, that, that the only way to make the concept more precise is through reflection on what is valuable about sustainability and how this value relates to other important values. And secondly, analysis of the practical implications of sustainability presupposes that the goal has been clarified. To determine the necessary means to achieve the goal of sustainability, once it has been specified, is basically an empirical question. But since the question of implications is concerned with the farther future, there will be uncertainty about the exact consequences of any strategy.

The goal of sustainability

Consideration of the sustainability of activities dates back to the late 19th century, where the question rose within forestry how to determine the maximum cut that could be sustained in the long term; later, similar considerations were developed within fishery, and eventually they spread to many other areas. The underlying principle was the constraint that the resource stock in question should be kept non-decreasing over time, and the simple prescription was to keep the harvest rate per year within some area smaller than, or equal to, the natural regeneration rate per year for the resource within the area.

In a similar vein, one could attempt to determine the conditions under which a farm can sustain its production over time. However, a farm, defined as a system occupying a certain area, is an open system in exchange with its surroundings. The goal of sustainability could then be defined as making the farm self-sufficient in terms of energy, nutrients and other flows of matter; or at least minimizing the input from the surroundings as well as the loss to the surroundings of energy and matter.

One line of research takes its departure from the observation that if a farm is to sustain its production over time, minimizing the exchange of matter and energy with the surroundings is not likely to be sufficient. Roughly, the farm also needs to be sustainable in a social sense. Hence, this research takes the form of uncovering the expectations of all stakeholders who have an influence on the farm's survival over time (e.g. Kristensen and Halberg, 1997; Gibon, 1997). Enhancing the concept of sustainability along these lines corresponds roughly to what Paul Thompson has dubbed 'sustainability as functional integrity' (Thompson, 1997). I shall call the systems approach a *Bottom-Up Approach*, because it starts out by inferring prescriptions for a single farm (area, activity), and then reaches prescriptions for society at large by putting together the prescriptions for all relevant activities.

Another line of research originates in economics. From an economical point of view, the prescription to keep the stock of renewable resources non-decreasing over time leaves many questions unanswered. For one thing, many resources are non-renewable. If they are to be used at all, it is by definition impossible to keep the stock non-decreasing over time. Economists suggest answers along two lines (e.g. Solow, 1974; Hartwick, 1977). Firstly, a reduced stock of exhaustible resources can be compensated for by an increase in renewable resources. This strategy assumes that substitution between different types of capital is possible. Secondly, it may become possible through technological development to achieve increasing efficiency in the use of resources.

The important point is, however, that if we are going to take substitution and technological development into account in order to make more general prescriptions, it is necessary to look at the *value* of resources, i.e. their potential to generate welfare or satisfy needs, rather than simply the stock.

Hence, economic analyses move from the simple prescription of keeping a single renewable natural resource stock non-decreasing over time to the more general prescription of keeping the level of welfare non-decreasing over time. This is inspired by the Brundtland Report (World Commission, 1987), which defined a sustainable development as a development that 'meets the needs of the present without compromising the ability of future generations to meet their own needs'. (Economists prefer to analyze this in terms of welfare rather than need satisfaction, because it is notoriously difficult to draw a line between needs and other goods). It is implicit in this move that it no longer makes sense to apply 'sustainability' on a single activity in isolation, such as a single farm or even farming; rather, we shall judge whether the globe at large and its development is sustainable. Hence, I find it reasonable to dub this approach, a *Top-Down Approach*, because it starts by determining sustainability as a goal at a global level, and then compares different activities' contribution to this goal.

The general idea of a sustainable development underlying the Brundtland Report and the economic discussion is a development that fulfills justice within and between generations. The economists interpret the Brundtland Report's notion of justice between generations as the constraint that the level of welfare should be non-decreasing over time. There is room for discussion of whether this requirement of equality between generations is too rigid. Would it be acceptable if some generations become slightly worse off than their predecessors, if this could lead to a greater overall total of welfare across generations (Broome, 1992)?

Conflicts between the approaches

If the Bottom-Up Approach is based on the implicit premise that the farm system under consideration should sustain its production perpetually, or at least as long as the time horizon reaches, as it often is, it may be in conflict with the Top-Down Approach. The latter has among other things the implication that a form of production, which at some time can be part of a sustainable development, does not necessarily need to be part of it at a later time. For instance, technological development may have made it superfluous or just too costly to maintain.

From an ethical point of view, it is clearly the Top-Down Approach that has the upper hand; whereas the notion of justice within and between generations has great ethical weight, the claim that some isolated form of production, e.g. a farming practice, should be sustained over time at all costs has not. It appears arbitrary to choose some form and level of production at some time as baseline for all further development. Also, it appears arbitrary to consider the area presently connected to a farm as the unit of concern. It seems much more reasonable to be concerned about increasing the overall productivity of resources, leaving room for technological development and the potential for substitution between resources and between products and judging productivity from the perspective of international division of labor.

The Top-Down Approach implies that the sustainability of any activity should be measured by its contribution to the goal of justice within and between generations. However, this very general goal is rather indeterminate. And the problems of population ethics add to this indeterminacy. The economic analyses referred to above assume a constant population over time. But if the population grows over time, as it is likely to do, what does the concept of justice between generations imply for comparing generations of different sizes? Given some fixed stock of resources, is it better that it sustains few people or many? Should we care about the total welfare or the average level of welfare? No-one has so far been able to state a reasonable concept of justice between generations with variable population sizes (Arrhenius, 2000 gives an overview of the discussion). This leaves the concept of sustainability seriously indeterminate.

But even given some precise notion of justice between generations, prescriptions for sustainability will depend on the rate of population growth, the rate of increased efficiency in use of resources and the possible substitutions between types of resources. These are empirical questions, but presently we do not know the answers for certain.

The economic discussion of the possibility of substitution has uncovered disagreements about the likelihoods of different scenarios. Should we believe in the optimistic view that all natural resources sooner or later can be substituted with man-made capital? Or should we believe in the pessimistic view that no natural resources can be substituted? Or should we believe in the assumption that some natural resources can be substituted, but that certain critical natural resources must be preserved in order for ecosystems to function?

It should be acknowledged that, in deciding about the strategy we believe to be sustainable, we are faced with a decision shrouded in uncertainty. Even with agreement about the likelihoods, there might be disagreement about the right strategy in dealing with this kind of uncertainty about the future. A strongly precautionary approach would choose the act with the best worst outcome. Organic farming seems to be based on a precautionary approach like this – it prescribes, roughly, that we should only use technologies which we now know for certain will not endanger our stock of natural resources in the future. A weaker type of precaution would consider the likelihood and the seriousness of the consequences of the different strategies and then seek to balance the degree of precaution against its costs. If we believe that substitutions are likely to become possible to a reasonable degree, the organic strategy may seem too costly and another strategy may seem more reasonable, even though it is more risky.

A sharpening of the goal of sustainability

Given the indeterminacy of the objective of the Top-Down Approach and the uncertainty of its implications, it is very far from being operational. Clearly, a sharper specification is needed in order to make the goal more determinate. I shall suggest the following minimal requirement of justice: *a development is only sustainable if it involves food security for all at all times.*

Even though the Brundtland Report puts weight on justice both *between* and *within* generations, the latter perspective tends to get lost in discussions about sustainability. But equality between generations in terms of an *average* level of welfare, and even an increase of the *average* level of welfare over time, are compatible with the continued existence of food insecurity for some people.

Instead, a sharper specification implies focus on inequality *within* generations and how *this* inequality develops over time. Clearly, other specifications are possible. Moreover, this sharpening will have controversial implications. It implies that sustainability's concern for future food insecure people ought to be given much more weight than the concern for other future people. Hence, it would favor food security for all at all times, even if the implication would involve rich people now and in the future becaming considerably less rich than they otherwise would have been. In terms of the concern for environmental protection for the sake of the future poor, equitable global sharing of local costs and benefits, evidently, must be very important.

What is relevant for the sharpened goal of sustainability is particularly the management of natural resources in the widest sense, on which the poor will be dependent for food. The productivity of these resources should be safeguarded and possibly be increased when they are handed over to future generations also dependent on them for food security. This perspective is clearly compatible with the Bottom-Up Approach and it could benefit from the systems perspective regarding local resource bases. However, equitable sharing of costs is very important. Hence, if future use of natural resources in some area is threatened by some praxis, the local community should be paid compensation in order to protect the endangered resources. Moreover, certain practices might irretrievably compromise water or other resources if they are left unchecked because too many people are dependent on them. In this case, migration and establishing of other livelihoods for the involved people might be the best solution. Again, equitable sharing of costs is clearly very important.

The implication for the rich part of the world is, therefore, a greater concern for poor people. Presently, there is a discussion of the contribution of animal production to global warming. Thus, there might be reason for the rich of the world to reduce emission of greenhouse gases from livestock. This would make food production more sustainable by contributing to a reduction in global warming. However, the effects are likely to be seen only in the long term. Temperatures are likely to continue to grow for the rest of the century. This will primarily endanger the poor in the world and it is likely to make more people food insecure. According to the suggested sharper specification of sustainability, sustainable food production *also* involves transfer of resources to the poor in order to increase their capacity to deal with climate changes and their impact on local resources on which they are dependent for food.

Conclusion

Two approaches to sustainability of food production have been outlined: Bottom-Up and Top-Down. The latter has, ethically speaking, the upper hand. However, it is also rather indeterminate, a sharpening of the Top-Down Approach is suggested where the concern for future generations should concentrate on the poor.

References

Arrhenius, G. (2000). Future Generations: A Challenge for Moral Theory. Uppsala University.

Broome, J. (1992). Counting the Cost of Global Warming. Cambridge: White Horse Press.

Gibon, A. (1997). Addressing livestock farming systems ecological sustainability at the regional level: an example from the Central Pyrénées. In: Sørensen, J.T. (ed.) Livestock farming systems: More than food production. EAAP Publication No. 89. Wageningen Academic Publishers, Wageningen, the Netherlands, pp. 30-41.

Hartwick, J.M. (1977). Intergenerational equity and the investing of rents from exhaustible resources, American Economic Review 66: 972-974.

Kristensen, E.S. and Hallberg, N. (1997). A systems approach for assessing sustainability in livestock farms. In: Sørensen, J.T. (ed.) Livestock farming systems: More than food production. EAAP Publication No. 89. Wageningen Academic Publishers, Wageningen, the Netherlands, pp. 30-41.

Solow, R.M. (1974). Intergenerational equity and exhaustible resources. Review of Economic Studies, Symposium, pp. 29-45.

Thompson, P. (1997) The varieties of sustainability in livestock farming. In: Sørensen, J.T. (ed.) Livestock farming systems: More than food production. EAAP Publication No. 89. Wageningen Academic Publishers, Wageningen, the Netherlands, pp. 5-15.

World Commission on Environment and Development (1987). Our Common Future. Oxford University Press.

'ALMO': a bottom-up approach in agricultural innovation

S. Karner
IFZ – Interuniversity Research Centre for Technology, Work and Culture, Schloegelgasse 2, 8010 Graz, Austria; karner@ifz.tugraz.at

Abstract

'ALMO' is a farmers' bottom up initiative, which established a farmer-business cooperation that aims to overcome the unfavourable situation of alpine oxen farming in a mountainous area in the South-East of Austria. This paper describes the historical development of this alternative agro-food initiative, and analyses contextual factors according to their impact on the emergence and performance of the ALMO initiative.

Keywords: alternative agro-food network, sustainable bottom up innovation, farmer-business cooperation, alpine oxen beef, rural development, alpine areas

Introduction

Within the scope of a current EC Framework Programme 7 project (FAAN-Facilitating Alternative Agro-Food Networks) we investigate a variety of 'counter-movements' dedicated to an 'economy of quality' in contrary to the rationalistic agro-industrial model. This paper presents the first set of results from an Austrian case study on alternative agro-food initiatives (AAFNs). 'ALMO' ('Alm Ochsen') serves as an example of a bottom-up agricultural innovation, which is regarded to be a successful and sustainable strategy for the production and marketing of beef from alpine oxen as a high-quality product.

Which contextual factors influence the initiatives' emergence and development?

This paper sets out the contextual factors limiting and hindering the initiatives' emergence and development. The presented data was gained through semi-structured interviews (n=12) with actors involved in the initiative's network and participatory observation at a network meeting (November 2008; approximately 90 participants). Furthermore, written material (meeting protocols, personal correspondences of ALMO members, press releases, marketing folders) was also analysed.

History of the ALMO' initiative and contextual factors

ALMO is a farmer-business cooperation founded 20 years ago based on the idea of four farmers and the official veterinary surgeon of the region. The association started with 45 members, most of these were local farmers, but the association also included small regional butchers, the official veterinary surgeon, and a former member of the agricultural chamber. To guarantee steady sales for farmers, the initiative approached a larger business partner, who joined the initiative in 1993. ALMO has continued to grow and this has occurred alongside a period of increasing professionalism within the association – especially with regard to marketing strategies. Unlike other farmer-business co-operations, the farmers still retain power and control over their initiative and as a result they gain most of the added value. Their contract with the business partner guarantees a continuous purchase at a stable price, which is above the average price for beef, it is even higher than if the farmers would market the meat via direct selling. The registered brand is product-based and owned by the members of the initiative. Today this vertical network includes about 550 farmers, two small butchers, a large producer and a distributor of meat delicacies. Of particular interest is the collaboration with animal welfare and environmental non-governmental organisations (NGOs) with respect to ensure the production systems meet very high animal welfare criteria and are GM-free.

The initiative is territorially embedded in the 'Almenland Teichalm-Sommeralm' region, which is the largest alpine pasture area in Central Europe, where traditional alpine pastoral agriculture prevails. ALMO was established to overcome the unfavourable economic situation experienced by those involved in pastoral oxen farming in Austria, and we identified five contextual factors, which have been influencing the development of the initiative.

a. Cultural, historical and social factors

Based on tradition many of the ALMO farms have been maintaining the farm business model used by their ancestors, which considerably contributes to their identity as alpine farmers. In this region oxen have been used traditionally as working animals, but the mechanisation of agriculture has continued to displace the oxen as working and draught-animals. In autumn after the harvest and preparation of the fields for winter grain has been completed, the animals were traditionally sold at the so called 'Herbstviehmarkt' ('autumn cattle market').

Although contact with consumers has transferred to intermediaries, the relationship can be characterised as proximate: this initiative has carefully chosen its business partners based on the view that they share the same interests, and that through specialist trade personal contact with consumers is cultivated by emphasizing the quality and origin of the product.

Even though ALMO is a big network, it is still characterised through close social and personal relationships, which are based on transparency and go beyond simple market relations. The business partners know many of the farmers personally and at the beginning of the initiative, social cohesion was an especially important factor. As the members call it 'ALMO is a business, which relies on the handshake-quality of the partners', relationships are based on 'honesty and trust'.

b. Economic and market-related factors

ALMO was motivated by the continuous decrease in marketing opportunities for meat from alpine oxen. Beef production was increasingly becoming intensified and as a result moved from the alpine region into the agricultural plains, which put the typical alpine pasture landscape at risk. According to the emerging trends in cattle farming, extensive ox farming was considered obsolete by many livestock experts and stakeholders. At this time there was also a trend – mainly originating from Italy – towards light and young calf meat. The idea of selling oxen as high quality beef was contradicting mainstream livestock production trends. Previously, after their use as working animals, oxen were sold either for fattening to another region in Austria or Italy, or they were exported as living animals to North Africa, which was heavily subsidised at this time.

Farmers perceived the export to North Africa as an unfavourable development, since this was linked to strenuous and painful transportation conditions for the animals and, despite a high level of public subsidy, the farmers received a very low price. Selling the animals for fattening to other regions was also not an attractive marketing strategy, since the meat of the relatively old animals was considered to be of low quality. If the animals were sold to Italy, the meat was further processed, but only the Italian processors benefitted from the added value.

Food scandals, which are linked to the conventional food production system, can sometimes offer a market opportunity for alternative systems. This was the case for ALMO, in 2001 despite the general crisis in the marketing of beef due to BSE, nearly twice as much ALMO oxen was sold as in the previous year. Conventional meat was increasingly being imported from outside of the region and being identified as a standardized anonymous product. An early strategy of the ALMO farmers was to increase awareness of regional and high quality products amongst consumers, and to gain consumers' trust through greater transparency with regard to the origin of food. Even before the accession of Austria to the European Union in 1995, pictures of the farmers and their addresses were put in the display case where the meat was stored for sale. This strategy was not only intended to point out an alternative choice and to increase trust in local products, but also to encourage a sense of solidarity for the local farmers.

c. Political factors

Agricultural experts at this time supported the development of intensive bull breeding and fattening in lower regions of Austria, which was linked also to subsidies. Instead of giving support to extensive pasture farming, the original planning proposal was that the alpine areas should be re-seed as forest. The idea of keeping alp oxen and preparing a market for this product was not seriously considered, and it was neither approved by the livestock experts nor by decision-makers within the agricultural chamber. Thus the ALMO initiative did not get financial or organisational support from these groups.

Despite the lack of support from official experts, some individuals from the Federal Ministry for Agriculture and the agricultural chamber were in favour of the idea of alpine ox farming and provided financial support by 'initiating' a special funding for the procedure of meat maturing. This so called 'Reifungsprämie' ('maturing bonus') has been integrated in the Austrian Agri-environmental Programme, which supports sustainable production methods. After Austria's accession to the European Union, LEADER offered a good opportunity for territorially focused developments. The LEADER II and LEADER+ programmes gave useful support for the further development of the initiative. e.g. the local abattoir was bought and rebuilt according to organic farming standards. Furthermore a good practice code for rearing quality cattle was developed and subsequently qualification and training for farmers was also offered. LEADER+ supported the establishment of professional marketing for these types of products, as a result a marketing company was set up to promote ALMO and the Almenland region.

d. Location-related and geographical factors

Due to the geographical conditions of the region the range of farming options is limited, pasture farming is best suited to mountain areas, as these areas are not favourable for any other kind of cultivation.

Another crucial factor is the 'embeddedness' of this practice within a region. ALMO is strongly integrated in the regional development of the 'Almenland' region, which subsequently became the 'ALMO Genussregion' ('ALMO region of culinary delight'). Tourism plays an important role in the alpine regions and this also provides the possibility to market regional and high-quality speciality food. The cooperation between farmers, gastronomic specialists and other providers of tourism facilities aims to foster a regional strategy through the cross-linking of different sectors and ALMO has become the lead initiative within this strategy. This is of special relevance since the ultimate aim for the region and its future development is to becoming a 'Slow (Food) Region'.

e. Knowledge related factors

Innovations within ALMO have not been explicitly planned, but they have resulted from searching for a strategy to overcome obstacles. The drivers for innovations have always been members of the initiative or the network that ALMO is embedded in: through cooperation different actors have brought in different sets of knowledge and different expertise. Thus innovations have been nurtured by different types of knowledge. An illustrative example is related to the quality of the meat. Through sterilisation the oxen grow much slower, thus the animals are much older at the time of slaughtering than intact bulls or calves. Due to the slower growth rates of the animals, the meat contains more fat, has a more intense flavour and is more fine fibred than bull-meat, but it has to be matured to reach this quality. One of the local butchers was able to apply a specific traditional maturing process and contributed this knowledge to the development of the initiative.

There is not a great deal of scientific expertise or research relating to oxen farming, thus the farming knowledge base still very much lays in the hands of the local farmers. Many innovations, especially the implementations of these innovations, come from social relations, and knowledge is shared more often informally via social contacts, than more formally, e.g. through training courses.

Conclusion

In conclusion it can be stated that a diverse set of contextual factors have a different impact on the emergence and performance of the Network's initiatives' by influencing them in an either more positive or negative way.

According to the ALMO members' perceptions of these factors, social factors have an unexpectedly high impact on the development of the ALMO initiative. In particular, personal relationships – both within the initiative as well as within the environment it is embedded in – have stimulated the development and the continuous success of the initiative. The main driving and supporting structure of ALMO is ascribed to single individuals, their biographies, their social capital and their skills. The advantage of setting up a network is not only linked to increasing market power and the co-operative use of facilities, but also to the added value of connecting different skills that are able to handle contextual factors. The cultural and historical background, and the high local/regional identity enabled the initiatives establishment and the setting of a specific direction for the Network's development. Due to the geographical location the initiative had only a few options as alternative strategies, but these matched traditional practices.

Economic and market-related factors have been the main drivers for change. ALMO was initiated as an alternative strategy to overcome a challenging market situation. Under the given circumstances this response to the changing market conditions might not have been as effectively handled by following a mainstream productionist paradigm.

Political factors have been hindered and also supported this initiative. The mainstream political context posed several obstacles for the initiative, but with the Austrian accession to the EU the second pillar of the agricultural policy became increasingly important for providing financial support for the further development of ALMO. Moreover it fostered within the LEADER programme its embeddedness into the region. Finally the resulting linkage to other regional activities led to a stronger vertical arrangement of the network, which turned out to be a stabilizing factor.

Sustaining ethical trade in farmed aquatic products between Asia and the EU

D.C. Little[1], F.J. Murray[1], T.C. Telfer[1], J.A. Young[1], L.G. Ross[1], B. Hill[2], A. Dalsgard[3], P. Van den Brink[4], J. Guinée[5], R. Kleijn[5], R. Mungkung[5], Y. Yi[7], J. Min[7], L. Liping[7], L. Huanan[7], L. Yuan[7], Y. Derun[7], N.T. Phuong[8], T.N. Hai[8], P.T. Liem[8], V.N Ut[8], V.T. Tung[8], T.V. Viet[8], K. Satapornvanit[9], T. Pongthanapanich[9], M.A. Wahab[10], A.K.M. Nowsad Alam[10], M.M. Haque[10], M.A. Salam[10], F. Corsin[11], D. Pemsl[12], E. Allison[12], M.C.M. Beveridge[12], I. Karunasagar[13], R. Subasinghe[13], M. Kaiser[14], A. Sveinson Haugen[14] and S. Ponte[15]

[1]University of Stirling, United Kingdom; dcl1@stir.ac.uk
[2]CEFAS Weymouth Laboratory, United Kingdom
[3]University of Copenhagen, Denmark
[4]Wageningen University, The Netherlands
[5]Leiden University The Netherlands
[7]Shanghai Ocean University, China
[8]Can Tho University, Vietnam
[9]Kasetsart University, Thailand
[10]Bangladesh Agricultural University, Bangladesh
[11]World Wildlife Fund, Denmark
[12]World Fish Center, Malaysia
[13]Food and Agriculture Organisation, Italy
[14]University of Bergen, Norway
[15]Danish Institute for International Studies, Denmark

Abstract

Trade in aquatic products is the largest global food sector, by value, with Asia representing the main external source of aquatic products into the EU. Current EU policy supporting international trade between Asia and Europe concentrates on issues of food safety as measures of quality, whilst market-forces drive development of standards and labels that identify social and environmental parameters. The SEAT (*Sustaining Ethical Aquatic Trade*) project proposes to establish an evidence-based framework to support current and future stakeholder dialogues organised by third party certifiers. This will contribute to harmonising standards, helping consumers to make fully informed choices with regards to the sustainability and safety of their seafood. The 'Ethical Aquatic Food Index' (EAFI), a qualitative holistic measure of overall sustainability intended to support consumers' purchasing decisions, will be based on detailed research centred on a Life Cycle Assessment (LCA) of current processes. Systems thinking will be applied to analyse livelihood impacts along the global value chains of four farmed aquatic products, tilapia, shrimp, freshwater prawns and *Pangasius* catfish in four major producing countries China, Thailand, Vietnam and Bangladesh. Initial assessments of environmental impacts by and on aquatic production and processing systems, and impacts on product safety and social equity will lead towards prioritisation of critical issues and supportive action research. Micro, small and medium enterprises (MSMEs) based in the EU and Asia will participate in this process, enhancing their relative competitiveness. By strengthening the knowledge base surrounding EU-Asia seafood trade the project will provide the evidence required to support further expansion whilst ensuring a fair deal for producers who are meeting appropriate social and environmental goals and offering a safe and sustainable product for consumers.

Keywords: aquaculture, ethical aquatic food index, ethics, life cycle assessment

Farmed aquatic products and sustainability

Trade in farmed species is growing rapidly and becoming a significant component of global levels of seafood supplies. Growth in production and export of fin and shellfish from Asia to European markets has accelerated over the last decade and this contributes a major share to what is now the most important internationally traded food commodity sector (FAO, 2006). In value terms, fishery and aquaculture products are the most traded food in the world. In 2004 some 140 million MT – 38% of total production – was traded internationally (FAO 2006). Developing countries accounted for over half the value of all exports, while developed countries accounted for 81% of all imports (US$ 75 billion), the EU, Japan and the United States together accounting for 73%. The EU is now the world's largest seafood market.

However, there are major issues regarding the sustainability of this trade from ecological, public health and broader ethical perspectives. Existing concepts and approaches to sustainability are limited in scope and interpretative power. Bio-economic interpretations of sustainable aquaculture use profitability based on sound environmental management as their key objective (Boyd and Schmittou, 1999); in practice tradeoffs are likely between environmental impacts and profitability. The trade-off between policy and practice reflects weak and strong sustainability and has become a constant tension. The key issue has moved from sustainable production to sustainable consumption. Acknowledging the importance of market drivers and consumer expectations, these trends have occurred in parallel with a wider rise in environmental consciousness and 'ethical' consumption. Nevertheless, tensions remain. Ruttan (1994) commented that 'sustainability has emerged as an umbrella under which a large number of movements with widely disparate reform agendas have been able to march while avoiding confrontation over their often mutually inconsistent agendas.'

In the last few decades aquaculture has changed from a small-scale practice to large commercial/industrial enterprises in certain locations in Asia. Typically responding first to increased local urban demand, production has often increased dramatically to supply international markets. It is acknowledged that aquaculture can enhance both food security and employment in developing countries, increasing supplies of affordable fish (Ahmed and Lorica 2002; Bailey 1997). However increased levels of international trade in farmed fin and shellfish potentially affects very large numbers of people far removed from the sites of production, with wider implications for sustainability. Further sustainable development of trade in aquaculture between Asia and Europe has potential to help the region make significant progress towards the strategic Millennium Development Goals of the international community particularly those related to poverty alleviation, gender equality, environmental sustainability and global partnerships for development.

Reducing risks of global trade

Consolidation of food commodity chains by fewer larger vertically integrated organisations is a growing trend, although micro, small and medium enterprises (MSMEs) still constitute the major proportion of the business. Increasing sophistication and regulation of these supply chains raises questions regarding distribution of benefits from increased growth and broader governance and policy issues relating to trade rules. On the ground this trend is being matched by geographical clustering of production and processing capacity. This phenomenon raises particular challenges to environmental sustainability and food safety but there are also potentially beneficial impacts of such 'hot spots', including impacts on household livelihoods and broader socioeconomic implications. Environmental and spatial modelling techniques have emerged as a predictive management tools to both assess risk and inform management to reduce negative environmental impacts. These include both modelling of the risks of impacts of aquaculture on the environment (e.g. Perez *et al.,* 2002) and the impacts of external contaminants on culture fish and shellfish (Van Den Brink and Kater, 2006).

While production of farmed fish and shellfish has climbed rapidly in Asia, there has been a concurrent increase in fish consumption in Europe. This increase has been driven by a combination of more favourable attitudes towards fish, often related to perceptions of health benefits, but also more adverse dispositions towards many traditional alternatives which have been the subject of food scares. Public trust in food products, particularly those transported in processed form over large distances is under increasing threat however. Most economic analyses of food safety still assume that there are 'objective' notions of safety, risk, and hazard. In reality, notions vary dramatically between individuals, and across time, countries, and cultures (Freidberg, 2004) and are embodied in regulations.

The trend to standards and food certification is increasingly confusing and, the expectations and definitions of 'quality' by retailers and consumers in export markets have dramatically broadened and become more complex. In addition to expectations of food safety, taste and freshness, so-called 'credence characteristics' have become important (Grunert *et al.,* 1996). These embrace perceptions of what constitutes safe and healthy food but also extend to environmental, social, ethical or religious characteristics among consumers who want their consumption choices to reflect their commitments to principles in these areas. Traceability of the product through a value chain in which compliance for the qualities in demand can be demonstrated is becoming essential.

Wider initiatives to stimulate more sustainable food production include the standards setting and certification efforts of various Government, Inter-Government and Non-government entities. Many of these efforts appear driven primarily by commercial incentives to gain market advantage, most are clearly based on an insubstantial evidence base and there are concerns that they will further marginalise smaller scale producers (McClanahan *et al.,* 2009). Standards are important for developing country farms and firms because they determine the mechanisms of participation in specific global value chains and shape market access to specific countries (Daviron and Ponte, 2005). On the one hand, standards set entry barriers to a value chain for new participants and raise new challenges to existing developing country suppliers. On the other hand, the challenge of rising standards may provide the opportunity for selected suppliers to add value, assimilate new functions, improve products, and even spur new or enhanced forms of cooperation among actors in a specific industry or country (Jaffee and Henson, 2004; Jaffee and Masakure, 2005).

The project approach

Until now a range of different sustainability measures has been used to assess aquaculture production systems. Increasingly LCA (Life Cycle Analysis) is used for industrial and agricultural production (Rebitzer and Ekvall, 2004). While the diffusion of LCA has been driven by ISO standardisation, its application alone is considered limited in terms of practical decision making. Inclusions of other tools that are complementary and incorporate empirical, spatial and temporal factors are required; acknowledgement of social and economic aspects of producers, intermediaries and consumers are crucial. Furthermore in addition to acknowledging 'strong' and 'weak' approaches to sustainability assessment, a systems analysis that can (1) set local and international system boundaries and (2) incorporate Asian producer perspectives in the context of emerging global standards and trends in consumer behaviour in Europe is required. A more holistic assessment of sustainability developed with consideration of broader fair-trade, ethical and possibly other concerns is the objective of an ethical aquatic food index (EAFI), that would be a major outcome of the SEAT project. Such an index would attempt to synthesise and weight the values related to food production, processing and marketing as an outcome of a transparent negotiation between the stakeholders involved.

Improving the overall sustainability of aquaculture production in Asia will require addressing the emerging market realities in Europe and the different values and capacities among producers in Asia.

Most of the actors involved in the production and trade are MSMEs and it is with them that this research will focus as the point for developing technological and management improvements in current practice. Even where larger scale organisations have emerged, MSMEs often still remain significant components of the chain. Enhancing their capacity to increase sustainability through improved information of current and future consumers' perceptions is vital for their continuance. Consideration is also required of the relative competitiveness of European-based production and opportunities for partnership and mutual benefit also demands analysis. These would include the potential for further development in the trade of European farmed fish products within luxury markets in Asia. In addition there is potential for MSMEs involved in relevant services to understand market needs and opportunities in the expanding Asian-based production technologies.

The SEAT project aims to focus on four major cultured aquatic commodities and four producer countries in Asia (Table 1) to provide a context for assessing sustainability of these aquatic food chains. It was a rapid increase in farmed shrimp production (originally based initially on Black Tiger shrimp, *Penaeus monodon*, but increasingly the white shrimp *Penaeus vannamei)* in South and Southeast Asia that attracted attention to environmental and social problems associated with aquaculture development (Goss *et al.*, 2000). In contrast, freshwater prawn and tilapia production in Asia has increased with little controversy. Freshwater prawn (*Macrobrachium rosenbergii*) is indigenous in three countries and culture production systems have evolved in response to local opportunities and resource constraints. The introduction of Nile tilapia (*Oreochromis niloticus*) to Asia in the 1970s has resulted in their widespread production in China, Thailand and Vietnam (Belton *et al.*, 2007, Belton and Little, 2008). Tilapia exports from China dominate global trade, other producer countries still focusing on meeting domestic demand but with ambitions to expand into a rapidly growing export market. Export trade in *Macrobrachium*, particularly from Bangladesh, continues to grow rapidly while maintaining a relatively high value niche market. Catfish (*Pangasius* spp.) production from the Mekong Delta has also expanded rapidly as market focus has moved to Europe, which now consumes more than 50% of production. Indeed, both tilapia and catfish have now become more mainstream white fish commodities in Europe.

A shared strength of all four target species is their inherent culture characteristics: consistent growth and food conversion and tolerance of a range of environmental and management variables. The species are omnivorous, in contrast to most species cultured in Europe, North America and Japan. A high proportion of production costs in intensive aquaculture is for feed and formulated diets for omnivores, however herbivores diets are much lower in animal protein and less dependent upon captured fishmeal and oil, with important consequences for the environmental costs of production and the risks of contamination (Naylor *et al.,* 2000).

Table 1. Project scope.

Species	Tilapia	*Pangasius* catfish	*Penaeid* shrimp	*Macrobrachium* prawns
China	√√	√	√	?
Vietnam	√	√√	√√	(√)
Thailand	√	√	√√	√
Bangladesh	(√)	√	√√	√√
Europe	(√)	(o)	(o)	(o)

Notes: √√ - Major export industry; √ - Significant domestic industry; (√) – Limited local production/consumption; (o) - consumption only.

The selected mix of countries and species (Table 1) gives opportunities for assessment of a range of key sustainability issues in a four stage process. (1) An initial systems analysis (Fitzhugh and Byington, 1978) will enable identification of appropriate case studies from which (2) specific longer term studies regarding Life Cycle Analysis, environmental modelling, social and economic factors and food safety and public health will be investigated. This will assess all parts of the food chain from input procurement to marketing and consumption in Europe. (3) The findings will be exposed to a process of rigorous debate with stakeholders, especially with regard to local and international perceptions of 'values' and broader ethical principles base around an ethical matrix approach (Kaiser and Forsberg, 2000). The key prioritised weaknesses identified (4) will then be subjected to action research (Friere, 1990) with appropriate actors within the MSMEs that constitute the production and marketing networks. Outcomes from this process are expected to be taken up by the MSMEs involved and communicated to a broader group of the same stakeholders through trade associations and other channels. The outcomes will also contribute to policy and standards development with appropriate stakeholders.

Improving the transparency of sustainability attributes for these rapidly expanding aquaculture systems is a key objective. Providing an improved scientific evidence base for on-going certification efforts and development of standards is a critical output; so far, standards developed for aquaculture products have tended towards narrow, single interests, and as a result, niche markets. Moreover their stringent requirements, typically imposed from Europe, have made compliance impossible for the majority of producers in Asia. A related and important constraint is the emergence of asymmetries in access to trade related information available to many actors involved in Asia-Europe seafood trade despite the development of considerable resources, much of it web-based. Information asymmetries affect the value of credence products, such as organic foods in the market (Giannakas, 2002) that in turn affect interest in value addition by actors removed from timely and accurate information. Unequal access to information also undermines collaboration between stakeholders (Chi and McGuire, 1996). Consequently opportunities to raise quality standards for most of the produce imported to Europe have not been fully realised to the potential cost and detriment of MSMEs and consumers.

The overarching objective of the SEAT project is to increase levels of trust among European consumers in farmed seafood from Asia as an ethical and sustainable choice, requires engaging stakeholders, including Asian producers and processors, in open dialogue and improving access to relevant information. Rational consumer choice requires an understanding of the values that underlie any type of certification as well as what is omitted and an important outcome of the research will be more accessible and transparent standards. The research will also explore ways in which communications to aid consumers' understanding might be developed. In this respect the recent proliferation of standards may be counterproductive; although clearly part of a wider trend towards 'political consumerism', consumer confusion is a major risk.

References

Ahmed, M. and Lorica, M.H. (2002). Improving developing country food security through aquaculture development – lessons from Asia. Food Policy 27: 125-141.

Bailey, C. (1997). Aquaculture and basic human needs. World Aquaculture 28(3): 28-31.

Belton, B., Little, D.C. and Grady, K. (2007). Reassessing Sustainable Aquaculture through the Lens of Tilapia Production in Central Thailand, Society and Natural Resources (in press).

Belton, B., and Little, D. (2008). The Development of Aquaculture in Central Thailand: Domestic Demand Versus Export Led Production. Journal of Agrarian Change 8(1).

Boyd, C.E. and Schmittou, H.R. (1999). Achievement of sustainable aquaculture through environmental management. Aquaculture Economic Management 3: 59-69.

Chi, T and McGuire, D.J. (1996). Collaborative ventures and value of learning: Integrating the transaction cost and strategic option perspectives of the choice of market entry modes. J. International Business Studies, 27.

Daviron, B. and Ponte, S. (2005). The Coffee Paradox: Global Markets, Commodity Trade and the Elusive Promise of Development. Zed Books: London, UK and New York, USA.

FAO (2007). The State of World Fisheries and Aquaculture (2006). Food and Agriculture Organisation of the United Nations, Fisheries and Aquaculture Department.

Fitzhugh, H.A. and Byington, E.K. (1978). Systems approach to animal agriculture. World Animal Review 27: 2-6.

Freire, P. (1990). A critical understanding of social work (M. Moch, Trans.). Journal of Progressive Human Services 1(1): 3-9.

Freidberg, S. (2004). French Beans and Food Scares: Culture and Commerce in an Anxious Age. New York: Oxford University Press, Oxford, UK.

Giannakas, K. (2002). Information Asymmetries and Consumption Decisions in Organic Food Product Markets 50(1): 35-5.

Goss, J., Burch, D and Rickson, R.E. (2000). Agri-food Restructuring and Third World Transnationals: Thailand, the CP Group and the Global Shrimp Industry. World Development 28(3): 513-530.

Grunert, K.G., Larsen, H.H., Madsen, T.K. and Baadsgaard, A. (1999). Market Orientation in Food and Agriculture Springer: Berlin, Germany, 300pp.

Jaffee, S. and Masakure, O. (2005). Strategic use of private standards to enhance international competitiveness: Vegetable exports from Kenya and elsewhere. Food Policy 30(3): 316-333.

Jaffee, S. and Henson, S. (2004). Standards and agro-food exports from developing countries: Rebalancing the debate. World Bank: Washington, DC.

Kaiser, M. and Forsberg, E.M. (2000). Assessing fisheries – using an ethical matrix in a participatory process. Journal of Agricultural and Environmental Ethics 14: 91-200.

McClanahan, TR., Castilla, J.C., White, A.T. *et al.* (2009). Healing small-scale fisheries by facilitating complex socio-ecological systems Reviews in Fish Biology and Fisheries 19(1): 33-47.

Naylor, R., Goldburg, R., Primavera, J., Kautsky, N., Beveridge, M., Clay, J., Folke, C., Lubchenco, J., Mooney, H. and Troell, M. (2000). Effect of aquaculture on world fish supplies. Nature 405(6790): 1017-1024.

Pérez, O.M., Telfer,T.C., Ross, L.G. and Beveridge, M.C.M. (2002). Geographical information systems (GIS) as a simple tool to aid modelling of particulate waste distribution at marine fish cages. Estuarine Coastal and Shelf Science 54: 761-768.

Rebitzer, G. and Ekvall, T. (eds.) (2004). Scenarios in Life-Cycle Assessment (LCAS). ISBN978-1-880611-57-9, 68 pp.

Ruttan, V.W. (1994). Constraints on the design of sustainable systems of agricultural production. Ecological Economics 10: 209-219.

Van den Brink, P.J. and Kater, B.J. (2006). Chemical and biological evaluation of sediments from the Wadden Sea, The Netherlands. Ecotoxicology 15: 451-460.

Use of traditional techniques in food production: market competitiveness and sustainability in an Italian case study

C. Malagoli and B. Scaltriti
University of Gastronomic Sciences, piazza Vittorio Emanuele Fraz. Pollenzo, 12060 Bra (CN), Italy; c.malagoli@unisg.it; b.scaltriti@unisg.it

Abstract

The aim of this paper is to explore if the use of traditional techniques can result in the production of sustainable food that is also economically competitive. The concept of sustainable food production is often understood as ecological sustainability but it should also be considered from a wider point of view, taking into account the local socio-economic role of food production. The case study discussed here is the production of a so called Slow Food Presidium, the 'Castagne essiccate nei tecci' in Italy. Slow Food Presidia are defined as a set of endangered speciality foods, that should be preserved due to their influence on regional biodiversity, their role in the maintenance of local systems of economic exchange, and as part of the preservation of cultural heritage, etc.. 'Castagne essiccate nei tecci' are dried chestnuts produced in five municipalities of Liguria (Italy) using traditional techniques and local inputs, renewable sources of energy, old restored buildings, etc. The first part of this analysis concerns economic competitiveness. The costs of production and the market have been analysed in order to understand what are the best ways of assuring the economic viability of such small production systems. The second part of the analysis deals with sustainability of the production system and this is evaluated, through a multicriteria analysis of the traditional techniques compared with the conventional techniques of industrialized production. The challenge for this type of production systems is to combine sustainable production (in some cases it results in higher costs) with market competitiveness. The conclusion of this paper is that the use of traditional techniques can increase the environmental sustainability of food production. The critical point is that it is difficult to cover the higher costs of a traditional production system without special conditions. This is represented in this case by the 'Slow Food' network, which is able assure the presence of 'Castagne nei tecci' in several market at an adequate price level.

Keywords: chestnuts, local economy, slow food

Introduction

The recent economic crisis highlights the importance of finding a new food paradigm for the future. The inclusion of a house garden at the White House is the latest sign (and one of the most famous ones) of changing attitudes relating to the food system. Van Der Ploeg (2008) describes an agriculture of the empires *vs* the agriculture of the peasants, and other scholars write about agricultural changes (Scaltriti, 2008; Lang *et al.*, 2004; Sachs, 2007) using words such as war, conflict, revolution, etc. The forecasts about the future of food production are not clear. However many scientists agreed that sustainability is a key concept for the food system of the future, although sustainability itself is not a clear concept. Sustainability is sometimes connected with organic agriculture but also with local food production. There are different methodologies applied to measure sustainability, such as the food miles or the ecological footprint (Blanke and Burdick, 2005). In this paper we would like to propose the definition of sustainable agriculture (or sustainable food system) given by the economist John Ikerd (1993), who defines it as '*capable of maintaining its productivity and usefulness to society over the long run. ... it must be environmentally-sound, resource-conserving, economically viable and socially supportive, commercially competitive, and environmentally sound*'. In this definition we would like to stress the contemporary presence of environmental issues together with economic ones. There are a number of traditional production systems which are environmentally friend that use artisanal techniques with a high labour input, and low energy consumption, however many of these have suffered from commercial problems, where sale price does cover the production costs, as in the case of Parmigiano Reggiano cheese (De Roest, 2000). The aim of the paper is to discuss a case study where an economically competitive product can be produced sustainably.

Fruit chestnuts production in Italy: local and tradition

Italy is claimed to be one of the prominent regions in Europe in terms of the levels of production of fruit chestnuts and the quality. The biodiversity and the culture of territorialisation of production has resulted in the presence in Italy of several chestnuts products that have been processed traditionally using old techniques that require a high labour input (and low energy consumption). They are three main groups of products (Table 1):

- slow food praesidia;
- PDO and PGI products;
- traditional products protected by Italian Agriculture Department.

Before discussing the production of fruit chestnuts it is important to briefly describe the nature of a Slow Food Presidium. The objectives of the Slow Food Movement is to safeguard local food, local breeds and varieties, and traditional methods of production. It is intended that these objectives will protect food production methods that are threaten by the modernity of productionist agriculture and protect biodiversity and the wider environment. The products that are produced under this system are called Slow Food Presidia and can have an economic role in keeping some marginal rural areas alive (Corigliano *et al.*, 2002). Currently there are 200 Slow Food Presidia products global, For example red and white cows in Val Padana in Italy were under threat due to changes in dairy farm management where this breed was been increasing replaced by Friesian cows in Parmigiano Reggiano cheese production. However, due to the speciality market create Slow Food Presidia the breed has survived. Similar cases have been seen with plant varieties and artisan food.

In terms of chestnut production, all three types of chestnuts (Table 1) are partially processed using traditional techniques with the use of renewable sources of energy, in restored buildings, maintaining cultural tradition and safeguarding biodiversity. The supply chain is very short at least until the processing step of the production chain.

Table 1. Traditional Italian chestnuts (www.politicheagricole.it).

Slow Food Presidia	Castagne essiccate nei tecci
PDO and PGI chestnuts	Castagne del Monte Amiata (PGI)
	Castagne di Montella (PGI)
	Castagne della montagna cuneese (DOP
Traditional products protected by the Italian Department of Agriculture	Castagna d'antona, carpinese
	Castagna del prete
	Castagna fresca e secca di granaglione
	Castagna mondigiana del Pratomagno
	Castagna pistolesa, bianchina
	Castagna della Val di Bisenzio

In terms of the case study, Castagne essiccate nei tecci are produced in five municipalities in Liguria, a region in the Northern Italy, more famous for its coastline and olive oil rather than its forests and mountains. The original cultivar is Gabbiana or Gabbina, a local variety present in Liguria from many centuries. The chestnuts are picked in November and then sold as fresh product or dried. The traditional drying process takes place in old stone made buildings called tecci, where the chestnuts are dried at a temperature range of 25-35 °C for three weeks. The source of heat derives from chestnut timber. Starting from these dried chestnuts, a whole set of products is produced: chestnut flour, beer, cakes, sweets, conserves and ice creams.

Production costs

Since one of the main criticism of speciality methods of production, e.g. Slow Food, organic, etc, is the profitability of the system, the paper will examine in more detail the production costs. Therefore in this part a calculation of the production costs of dried chestnuts 'Castagne essiccate nei tecci' is conducted. In order to determine producer costs, the cost calculation was carried out using data obtained from a questionnaire completed by chestnut growers. The survey was conducted as a personal interview survey with 12 participants. These data were supplemented by addition information from depth interviews with the local technicians. Interests have been calculated considering a rate of 2% (the current interest rate use in the area).

Costs have been aggregated as three main groups according to Marchesini *et al.* (2005):
- total explicit costs;
- total calculated costs;
- total production costs.

The operations that involve labour costs are the critical point of the cost calculation. If we consider a full salary (based on Italian Trade Unions indications) the productions costs are quite higher. This is the fairest option as all forms of labour should be acknowledged and valued, but it is not the common one. Free labour is often involved in these production systems, such as the common involvement of relatives in picking, for example, and at other times friends of the farmer can be involved. In the Table 2 labour costs relating to the picking process are considered to be zero, while in Table 3 it is possible to see what the impact on the production cost would be if labour was charged, according to the level of the salary.

Commercialisation

The economic operators that can sell 'castagne essiccate nei tecci' are all members of the Cooperativa 'Il Teccio'. They are the only economic operators who can sell these kind of chestnuts. The members of the cooperative are farmers and other operators of the chestnut supply chain. The cooperative decides the price and imposes it on all the members, in order to have only one price in the market. With a continuous reviewing of the production costs, this mechanism helps all the producers cover their production costs. In 2007, taking in to account the previous production costs, the price was equal to 4.00 €/kg. The cooperative (and also the Presidium) was founded in 2002 with 26 members, it has subsequently grown to reach 36 members.

The role of the cooperative in the price formation ends with the production of the dried chestnuts. At this point, the pricing of all of the derivatives (beer, flour etc.) is set within free market conditions, however the price of the raw materials and the conditions of the 'disciplinare' make the prices very similar. The presence of the cooperative, but especially the role of Slow Food and the reputation of all the Slow Food Presidia give to the 'Castagne essiccate nei tecci' the network and labelling to reach different market, some of which are unusual for small local productions. In fact commercialisation plays out at three different stages:

Table 2. Farmer-self reported production costs for 'Castagne essiccate nei Tecci' on an average farm (350 kg).

	Partial costs			Total costs		Unit costs
	Raw materials	Machinery equipment	Labour	Euro	%	Euro/kg
1 Land operations				24.00	5.97%	0.07
Hedge trimmering		24.00				
Raking						
2 Pruning			176.00	176.00	43.78%	0.50
3 Picking and transport			***			***
4 Drying				119.00	29.60%	0.34
Teccio		84.00				
Wood	25.00					
Tractor		10.00				
5 Peeling						
Machinery equipment		35.00		35.00	8.71%	0.10
Manual sorting						***
A Explicit costs	25.00	153.00	176.00	354.00	88.06%	1.01
6 Land capital maintenance insurance				22.00	5.47%	0.06
7 Management				22.00	5.47%	0.06
8 Taxes and other dues				0.00		
9 Interest on imprest capital				3.98	0.99%	0.01
10 Rent of the teccio				0.00		
B Total calculated costs				47.98	11.94%	0.14
A+B Total costs				401.98	100.00%	1.15

Table 3. Production cost and salary for picking.

Type of salary for picking	Production cost (euro/kg)
Not considered	1.15
Informal salary	3.12
Full salary	4.77

- in the production area (producer, small shops, agri- tourism farms and restaurants);
- in the region Liguria (Coop Liguria) and in the surroundings (Italy);
- international market (Coop Suisse).

The last market opportunity is somewhat unusual for small local production, like that seen with Castagne nei tecci, and this is believed to be due to the Slow Food Presidia brand.

Sustainability of 'Castagne nei tecci'

The second part of the analysis deals with the sustainability of food production. There are several methodologies that can be applied to describe and to measure the sustainability of farming systems. In this case we chose a methodology to evaluate sustainability on the base of a multicriteria analysis. Different specialists (technicians, university professors and economic operators of the supply chain; n=8) were asked to assess through an email questionnaire the dried chestnuts production, considering some basic factors:

- primary energy requirements;
- use of chemicals;
- origin of raw materials.

The comparison between production systems such as 'Castagne nei tecci' with other traditional and artisanal techniques and industrial ones, show that according to the specific three factors mentioned above, 'Castagne nei tecci' were assessed more sustainable than other techniques by all experts. However, it should be noted that these are quite a narrow set of 'factors', a more comprehensive sustainability assessment would need to be conducted before prominent claims about the overall sustainability of these systems could be made.

Conclusions

'Castagne nei tecci' is a good example to examine whether sustainable food production at a local level is possible. The challenge is to ensure that sustainability objectives are linked with market competitiveness. For this case, the presence of a cooperative that can increase the market power of agricultural (and small artisan) operators and the added value of a network such as Slow Food, helps to assure competitiveness by opening up market opportunities that include also the foreign international market of Switzerland, which is quite unusual for small local productions. Networks such as Slow Food may be able to build a form of good practice that can be followed also by other groups, like public institutions, associations of producers or consumers' movements.

References

Antonioli Corigliano, M. and Vignaò, G. (2002) I presidi Slow Food: da iniziativa culturale ad attività imprenditoriale. Available at: www.fondazioneslowfood.it/eng/presidi/economici.lasso accessed 25th March 2008.

Blanke, M.M. and Burdick, B. (2005). Food (miles) for thought: energy balance for locally-grown versus imported apple fruit. Environmental Science and Pollution Research 12(3): 125-127.

De Roest, K. (2000) The production of Parmigiano-Reggiano Cheese, Van Gorcum, Uitgeverij.

Ikerd, J. (1993). Two Related but Distinctly Different Concepts: Organic Farming and Sustainable Agriculture. Small Farm Today 10(1): 30-31.

Lang, T. and Heasman, M. (2004). Food wars. The global battle for mouths, minds and markets. Earthscan, London.

Marchesini, S., Hasimu, H. and Canavari, M. (2005). Production costs of pears and apples in Xinjiang (China). DEIAgra Working paper 003. Vol. 1.

Rigby, D. and Bown, S. (2003). Organic Food and Global Trade: is the market delivering agrcultural sustainability? Discussion Paper Series. School of Economic Studies University of Manchester. No 0326 August.

Sachs, W. (2007). Fair future Zed Books Lmt, London.

Scaltriti, B. (2008). Gastronomic Sciences and degrowth. Slow Food Revolution vs Gene Revolution. Proceedings of the I international Conference on Degrowth. Paris.

Van der Ploeg, J.D. (2008). The new Peasantries: Struggles fro Autonomy and Sustainability in an Era of Empire and Globalization. Earthscan, London.

Care and responsibility as key concepts of agricultural ethics

Franck L.B. Meijboom
Ethics Institute, Utrecht University, Heidelberglaan 8, NL-3584 CS Utrecht, the Netherlands; F.L.B.Meijboom@uu.nl

Abstract

Traditionally farmers had specific values and norms. With respect to animals, soil, and the environment, they had action guiding moral beliefs and principles. Nonetheless, farmers recently have been accused by the public of having no morality, but instead relay on an economic evaluation framework to steer their behaviour. To understand this tension between the traditional values and the accusations, this paper introduces the distinction between farmers' ethics and agricultural and food ethics. One can argue that farmers traditionally have norms and values that enable them to act morally in individual cases, while an increasing number of ethical issues in agriculture are beyond the individual level, such as the definition and interpretation of animal welfare, of sustainable food production or of nature and water management. Consequently, an individual case based ethics does no longer suffice for professionals in agriculture. Based on previous research into existing moral beliefs and values of farmers, this paper presents the notions of care and responsibility as key concepts for an agricultural ethics, which enables professionals in agriculture to deal with the ethical challenges entailed by sustainable food production.

Keywords: agriculture, care, responsibility, sustainability

Introduction

Traditionally farmers had specific values and norms. With respect to animals, soil, and the environment, they had action-guiding moral beliefs and principles. Nonetheless, farmers recently have been accused by environmental and animal welfare organisations of having no morality, but an economic evaluation framework that steer their behaviour. To understand this tension between the traditional values and the accusations, this paper introduces the distinction between farmers' ethics and agricultural ethics. The distinction has similarities with the one between medical ethics and health ethics. If we define medical ethics as mainly focussed on ethical issues within the clinical relationship between the physician and her patient, health ethics contains a wider realm and includes questions of prevention and just distribution of scarce resources. Similar, one can argue that farmers traditionally have norms and values that enable them to act morally in individual cases, while an increasing number of ethical issues in agriculture are beyond the individual level, such as the definition and interpretation of animal welfare, of sustainable food production or of nature and water management. Consequently, an individual case based ethics does no longer suffice for professionals in agriculture. Based on previous research to existing moral beliefs and values of farmers, this paper presents the notions of care and responsibility as key concepts for an agricultural ethics, which enables professionals in agriculture to deal with the ethical challenges entailed by sustainable food product

Farmer's ethics

Being a farmer traditionally included a whole range of moral beliefs and ideals about treating animals and nature, about good food and the value of farming. Farming often has been related to notions as stewardship, care for and connection with the country. Farming was not just a job, it was '.... a vocation or profession that prides itself on the unquestionable value, even nobility, of its work.' (Dundon, 2003: 427). Based on a recent exploratory investigation on Dutch farmers' values and norms regarding the care for and treatment of farm animals, their perceptions of 'good farming' in a more general sense and how they evaluate this, it became clear that most farmers still have clear moral beliefs and ideals. Of course there is an obvious plurality of moral views, in the sense that one cannot easily speak about one ethics for all farmers. Nonetheless, it is equally obvious that the current farmer in the Netherlands has moral ideas about how to deal with ethical questions with respect to his animals, with nature or social justice (De Rooij and De Lauwere, in press; see also Grimm, 2006). Therefore, it is striking that farmers only play a modest role in the public debate on the future of food production. Furthermore, they regularly are accused of being 'motivated purely by profit, not by any compassion for animals or traditional ethic of animal care.' (Fraser, 2001: 636) This is remarkable given the results of the in-depth interviews with a large number of farmers. In these interviews farmers present themselves as entrepreneurs, who work within the economic market, but also as persons who are genuinely concerned with the environment and their animals, and who often acknowledges the importance of sustainable food production.

From farming to agriculture

To understand the tension between the role farmers currently play in the public debate and the input they could have, given their moral beliefs, it is relevant to sketch some developments in farming and agriculture.

In this context the situation at the end of the Second World War is crucial. At that time, there was a direct need to increase food production. Food security was paramount on the policy agenda and required a central policy. Given the problems, food production could no longer depend on the ideas or expertise of individual farmers only. It required a rationalised and industrialised way of agriculture, steered by national and European policy. As a result, the individual farmer became part of an agriculture that had as its central main aim to supply sufficient safe food at the lowest possible cost. This goal had many consequences, for instances, it required 'farmers to continuously adopt new technologies that enhance production.' (Chrispeels and Mandoli, 2003: 5) Consequently, 'agriculture has grown from a primary basic sector of the economy to a secondary technical sector with complex joint interactions with the rest of the economy.' (Kunkel, 1984: 21) This change also has consequences for the moral competence a farmer needs to function in agriculture. One the one hand, there is more and more complexity. For instance, farmers are no longer producing for their own region, but their products are exported to all parts of the world. As a result, one's own moral ideas about animal welfare are confronted with those that (implicitly) underlie trade policies and guidelines. On the other hand, there is a reduction of issues that are considered as morally relevant in agriculture. With food security as the main goal, economic considerations received a central position and other aspects of farming and food production were often left out of consideration or even deliberately externalized. For instance fiscal policies and subsidies were often used to 'externalize most of the environmental costs of food production.' (Chrispeels and Mandoli, 2003: 5) In this context, the (implicit) morality of the market became central and farmers got used to speaking and counting in market terms only. They still had moral beliefs beyond the economic considerations, but these were mainly held as private opinions.

Agriculture after food security

Even though it would be naïve to state that food security is no longer an important issue with an estimated world population of 8.3 billon in 2030 (cf. Beddington, 2009), in Europe the exclusive focus on quantity has changed to one that takes food quality and sustainability into account too (e.g., Brom *et al.*, 2007). The central task of agriculture is still food production, but a 'transition to agricultural sustainability' is considered as essential for the future of agriculture (Ruttan, 1999: 5960). Consequently, the public expectations towards agriculture change. 'At one time, the role of agriculture was to produce food, but now many people expect agriculture to be carried out in an environmentally friendly way that maintains the rural economy.' (Dundon, 2003: 427) On top of this, the scope of the tasks of agriculture is broadened and now can include nature management and water management, the so-called green and blue value of agriculture (SER, 2008).

In this context farmers, as professionals of agriculture are asked to reflect on a broad range of themes including the use of technologies and the issues that were externalized up until recently, such as animal welfare and climate impact of food production. This confronts a farmer with two general problems. On the one hand, he is asked to reflect on issues that were absent in the past, such as the climate impact of food production. On the other hand, the number of involved parties is increased enormously. The farmer now operates in a full network of other agents. Consequently, his personal moral beliefs have to be discussed with others and need to meet the standards of validity and legitimacy on a public level. Kunkel accurately describes the situation accurately when he writes:

> 'Farming and ranching have historically enjoyed autonomy in the United States. Personal ethical responsibilities have been intrinsic to farming and rural life. But, now, new issues of fairness, conflicts of interests and their resolution, considerations of values, and integrity come into play and the future will likely force broader perception of ethics related to agriculture in aspects of national policy.' (Kunkel, 1984: 20)

Twenty-five years later issues of fairness, the resolution of conflicts of interests, considerations of values, and integrity are central themes in the public debate on the future of food production. However, in practice farmers are hesitant to introduce their moral beliefs as input for this debate. This illustrates the need to enable farmers to deal with broad and complex moral issues of agriculture. There is no escape, since neither agriculture nor farming can be value free or consist of activities that have a morally neutral character. Therefore, my claim is that the current situation asks for an ethics of agriculture as an independent addition to a farmer's ethics, which is focussed on individual cases. The norms and values that enabled farmers traditionally to act morally in individual cases have to be redefined in order to be able to deal with issues of animal welfare, of sustainable food production and of nature and water management in a public level. An ethics of agricultures should enable farmers to handle '...the multiple and often conflicting demands that sectors of the public press on agriculture.' (Dundon, 2003: 427)

Care and responsibility as key concepts

Ethics is often about choice and action. Combined with the economic context of food production, ethics with respect to agriculture have regularly been interpreted in terms of goal rationality and a utility calculus. Consequently, Kunkel notices that in agriculture we often decide what to do 'in a semi-utilitarian manner.' (Kunkel, 1984: 20)

However, it is remarkable that empirical evidence from our research project shows that farmers in daily practice do much more than making utilitarian evaluations and do not define their actions only in terms of goals rationality. It is beyond the scope of this paper to evaluate a utilitarian approach to the ethical questions of agriculture, but I do not see it as a weakness of farmers that they also include other dimensions that do not fit in a strict utilitarian approach. Furthermore, it is interesting to see that they often do not frame their moral beliefs in terms of abstract principles, such as those often used in bio-medical ethics. This does not imply that their actions are arbitrary; they often have clear and deep beliefs about how to treat animals or nature. These beliefs, however, have a more dynamic character and cannot easily be defined in terms of general principles only.

In the ethics of farmers the concepts of care, commitment, and responsibility play a central role. Even though a farmer considers his animals as part of his economic activities, the animal is considered as part of a relationship and therefore needs due care. They often formulate what they owe to nature and animals in terms of commitment rather than in terms of what they ought to do according to abstract rules or principles. For instance, animals are cared for, because they are considered to be important. Partly of course, because they are of economic value, but at the same time because of they are part of a relationship and because they are constitutive for farming and agriculture. The problems farmers had with killing healthy animals as part of the control of animal diseases highlight this point. Even though they were compensated in financial terms, they did not feel at ease with the situation. The animals were killed for a reason that did not fit in the normal relation of the farmer and his animal, into their 'way of life' (Stafleu *et al.,* 2004). In the same manner, they consider having a responsibility in a strict sense: good care in response to what the animals and nature contribute to their work. This does not imply that principles, such as fairness or concepts as sentience do not play any role. They play a role, but they are continuously defined in terms of specific relationships.

Up until now these views are mainly focused on the practice of the farm, but I argue that they could be helpful for the broader issues of agriculture too. I see at least three advantages of this approach. First, the focus on relationships can take the interests of current stakeholders seriously, but need not to deny those of others, including animals and future generations. Second, the focus on relationships can inform the discussion on responsibility. Defining responsibilities is no longer an abstract discussion, but depends on the kind of relationship one has towards the other. In situations of clear dependence a definition of responsibility in terms of due care is more appropriate, while in another relationship merely preventing losses would be a sufficient interpretations of one's responsibility. Third, the approach has a dynamic character and can be responsive to change (cf. Van der Burg, 2003), which is relevant, because of the ongoing moral debate on sustainability agriculture. As long there is a clear moral pluralism with respect to the future of agriculture, a dynamic approach that is open to reflection and debate is necessary.

References

Beddington, J. (2009). Speech during the Sustainable Development UK 09 conference, London.

Brom, F.W.A., Visak, T. and Meijboom, F. (2007). Food, citizens and market: the quest for responsible consuming. In: Frewer, L. and Van Trijp, H. (eds.) Understanding consumers of food products. Woodhead Publishing, Cambridge, UK, pp. 610-623.

Chrispeels, M.J. and Mandoli, D.F. (2003). Agricultural Ethics, Plant Physiology 132: 4–9.

Dundon, S.J. (2003). Agricultural Ethics and Multifunctionality Are Unavoidable. Plant Physiology 133: 427-437.

Fraser, D. (2001). The New Perception of animal agriculture: Legless cows, featherless chickens, and a need for genuine analysis, Journal of Animal Science 79: 634-641.

Grimm H. (2006). Animal Welfare in Animal Husbandry – How to Put Moral Responsibility for Livestock into Practice. In: Kaiser, M. and Lien, M. (eds.) Ethics and the Politics of Food. Wageningen Aacademic Publishers, Wageningen, the Netherlands.

Kunkel, H.O. (1984). Agriculture Ethics -- The Setting, Agriculture and Human Values, 1/1, pp. 20-23.

Ruttan, V.W. (1999). The transition to agricultural sustainability, Proceedings of the National Academy of Sciences of the United States of America, 96/11, pp. 5960-5967.

SER, Sociaal Economische Raad. (2008), Waarden van de Landbouw, Advies 05/08, Den Haag, SER.

Stafleu, F.R., De Lauwere, C.C. and De Greef, K.H. (2004). Respect for functional determinism. A farmers interpretation of respect for animals. In: De Tavernier, J. and Aerts, S. (eds.) Science, Ethics and Society, Leuven: CABME.

Van der Burg, W. (2003). Dynamic ethics. The Journal of Value Inquiry 37: 13-34.

Governing scarce water: what should retailers do?

S. Ripoll, P. Steedman and T.C. MacMillan
Food Ethics Council, 39-41 Surrey Street, Brighton, BN1 3PB, United Kingdom; santiago@foodethicscouncil.org

Abstract

Water scarcity will be one of the greatest challenges facing the world in coming decades and the food sector will feel this keenly. By 2025 an estimated 1.8 billion people will be living without enough water to survive. Already many rivers and aquifers are too dry to support ecosystems of livelihoods that previously depended on them. Because irrigation uses 70 percent of abstracted freshwater, food is on the front line of this problem. As concern over water escalates, food production will be affected and blamed in equal measure. Some major food companies are seeking to address this: water shortages already strain production in some regions, so they are concerned about material scarcity as well as about their reputations for corporate responsibility. However, it is not clear how food businesses should act on their responsibilities for water stewardship, given that place-specific social and environmental impacts of water scarcity frustrate initiatives based on regulating the 'embedded water' in food products. In this paper we report on a current project to pilot a dialogue-based approach to governing water use through international food supply chains. The project brings together a major UK retailer, a supplier in Spain and civil society organisations in both countries and operating across Europe. The project is funded under an EU Framework Programme 7 initiative to support innovative research co-operation involving civil society organisations.

Keywords: governance, uncertainty, virtual water, stakeholder, dialogue

Introduction

The Food Ethics Council (UK), the Fundación Nueva Cultura del Agua and Pablo de Olavide University (Seville, Spain) are collaborating on research that seeks to help address the contribution that food production makes to water scarcity. The work is part of CREPE, a project funded under the EU Framework Programme 7 that promotes research by civil society organisations (CSOs).

The research aims to clarify what it would entail for consumers, businesses and policy makers to take full responsibility for the water used to produce food. The work takes a case study approach, focusing on the export to the UK of fresh produce from Andalucia in southern Spain, and working with a major UK retailer and one of their suppliers in that region.

This paper summarises our work and findings to date, which support our premise that the technical and moral complexities of water scarcity defy management approaches based simply on indicators of water use. Our work will pilot an approach to governance that feeds diverse technical analyses of water use into a dialogue between stakeholders about how responsibilities for water stewardship should be shared.

Water scarcity, 'virtual water' and food

In many regions, including parts of Europe, water scarcity will be one of the most immediate environmental, social and economic challenges of the coming decades. International dimensions of water scarcity arise especially in the food sector. Globally, agriculture uses 69% of all freshwater extracted from rivers, lakes and aquifers, while in some southern European countries it accounts for up to 90% of all water use. Trade in agricultural products has been described as a trade in the 'virtual water' taken to produce them (Allan, 1999). Academic researchers have proposed that the import of virtual water to water scarce areas will alleviate pressure on the environment in those areas (Turton, 1999; Allan, 1997).

Environmental CSOs in the EU have appropriated the concept of virtual water to support their work to mitigate the causes and negative impacts of water scarcity. They have challenged academic arguments that virtual water trade, from regions abundant in water to those which are scarce, will alleviate environmental pressures, pointing to evidence that water users may deplete already scarce water for export production despite also importing water-intensive products (Yang *et al.*, 2006). Instead, research undertaken and commissioned by CSOs has developed the concept of virtual water as a tool for public awareness campaigning, policy advocacy and negotiation with supply chain stakeholders.

The 'virtual water' concept is extensively used by CSOs in the UK. The mass media have prominently reported the impacts of UK food consumption on water scarcity in countries of origin of horticultural and agricultural imports. Several CSOs – including WWF-UK, WWF-Spain and WaterWise – have undertaken or commissioned substantial research in this area to support campaigning and advocacy (e.g. WWF-Spain, 2006; Zygmunt, 2007). However, concern over water scarcity has not yet prompted a major shift in consumer or supply-chain behaviour.

Case study: UK imports from Spain

Southern Spain is one of the most water-stressed regions of Europe, exporting large volumes of water-intensive produce to other EU countries, including the UK. Over 20% of the Spanish tomato growing area is concentrated in Murcia and Almería, for example, while over 95% of Spanish strawberry production comes from Doñana. CSOs in Spain are attempting to reduce the over-exploitation of water resources for export. However, supply-side water governance has proved difficult, as seen by the numerous illegal wells in Almeria/Murcia compared with legal ones.

Therefore demand-side mitigation strategies, of the kind being advanced by CSOs in the UK, are considered crucial to mitigating water scarcity in southern Spain. But significant knowledge gaps constrain their successful development. This project seeks to address such gaps.

The research will tackle four main questions:
- How does UK food consumption contribute to water scarcity in Andalucia?
- What are the impacts of water scarcity?
- What should food businesses do to manage the problem responsibly?
- How could EU policy contribute to better water governance?

Our initial work to address these questions has focused on:
- A review of approaches to water governance.
- A review of the literature on the environmental and social impacts of water scarcity.
- A review of the concepts of 'virtual water' and 'water footprint' – central to the discourses of water scarcity and water resource management – to underpin an analysis of the trade in 'embedded water' through food between Andalucia and southern Spain.

While the case is important in its own right, the study also aims to draw out wider lessons for demand-side strategies for mitigating water scarcity problems exacerbated by trade between EU member states.

Water governance

As the severity of water scarcity becomes more widely recognised, governments, international agencies, businesses and CSOs are investing heavily in addressing it. One of our first tasks in this project has been to review existing initiatives and other projects in the pipeline in order to ensure our work builds on them rather than 'reinventing the wheel'. Organisations involved in initiatives to improve water governance in UK food supply chains include:

- The World Business Council for Sustainable Development, which has developed the concept of Fairwater Stewardship.
- The Water Stewardship Initiative, based in Australia.
- WWF, which had played a prominent role in promoting research and industry action on water scarcity.
- The Water Footprint Network, which works closely with WWF.
- Water Witness, a spinoff from research work at the University of East Anglia.
- The Alliance for Water Stewardship, which includes several of these other initiatives and major CSOs.
- WaterWise, a UK CSO supported by water companies.
- Major food businesses including M&S, SAB Miller, Unilever and PepsiCo.

The focus of most efforts to date has been on measuring water use and improving water resource management.

Emerging priorities

Despite the impressive effort now being devoted to addressing water scarcity, it remains unclear what consumers, businesses and policy makers should do to improve water governance along supply chains. It is evident that each group of stakeholders shares responsibility, but in what proportion they do so and how the responsibilities should be met is uncertain.

Organisations working to improve water governance, in the public sector, private sector and civil society – including many of those listed above – are becoming increasingly aware that simply seeking to reduce water use or water scarcity may have unintended consequences. Scarcity of a resource is not intrinsically bad – it becomes a problem when the resource is essential for meeting human needs and sustaining healthy ecosystems, as water clearly is, and when there is no guarantee that those needs will be met.

Indeed, there is an emerging understanding that, in addition to the primary problem of not having enough water for human and environmental needs, water scarcity can cause secondary problems including knock on-effects of mitigation efforts such as:

- Economic – e.g. industrial flight driven by regulation or abstraction costs.
- Social – e.g. deprivation due to unemployment.
- Environmental – e.g. high energy consumption to desalinate water.

The priorities for work to govern better the complex impacts of water use and to clarify the responsibilities of different stakeholders include to:

- Focus on the impacts of water use rather than its volume.
- Recognise that these impacts are place-specific, shaped by hydrological, ecological, social and economic conditions, and cultural values.
- Take into account secondary effects of water scarcity.
- Recognise the opportunity costs of water use and mitigation efforts.
- Understand how the capacity to govern water use is distributed among stakeholders within the regions of use and along supply chains.

The complexity of primary and secondary impacts of water use means that labelling embedded water is unlikely to offer a solution.

A dialogue-based approach

Our work will seek to address those priorities by bringing together growers, retailers, regulators and CSOs from Spain, the UK and the European Commission in a collaborative process of analysis and problem-solving. We will pilot a dialogue-based approach to water governance, working with one major UK retailer and one of its Andalucian suppliers.

The dialogue process will centre on a stakeholder group that also includes sectors in the region who compete with horticulture for water, relevant regulatory bodies and CSOs.

The dialogue will discuss how the stakeholders involved share responsibility for governing the water use associated with horticultural exports to the UK. It will be informed by research provided by the project team and it will provide a space for stakeholders to reach a common understanding of their shared responsibilities and form new partnerships to fulfil them.

The project will seed this stakeholder dialogue and help those involved to continue dialogue after the project ends if they have found it useful.

A small number of other organisations are also currently exploring approaches to water governance in food supply chains that similarly employ stakeholder dialogue to navigate a way through moral and technical uncertainty. The approach we are developing in this work package will be unusual because it establishes an iterative relationship between the research team and the stakeholder group. Specifically, the research team will support the stakeholder dialogue by:

- Analysing the virtual water trade between Andalucia and the UK and the water footprint of British consumers, highlighting how different methods allocate responsibility to consumers or producers.
- Reviewing existing water governance approaches so the stakeholder group can learn from past experience.
- Analysing power and governance capacity across relevant stakeholder groups, highlighting the opportunities and constraints that each is working with.
- In particular identifying opportunities for EU policies to contribute to better water governance, highlighting any major problems they currently present.
- Reviewing draft analyses on the above in light of issues emerging from the first stakeholder meeting.

We will facilitate three stakeholder workshops starting in October 2009. The work is due to be completed in May 2010.

References

Allan, J.A. (1997). Virtual Water: A long term solution for water short Middle Eastern economies? Occasional Paper 3, School of Oriental and African Studies (SOAS), University of London.

Allan, J.A. (1999). A convenient solution. The UNESCO Courier 52(2): 29-31.

Turton A.R. (1999). Precipitation, People, Pipelines and Power: towards a virtual water based political ecology discourse. MEWREW Occasional paper 11, Water Issues Study group, School of Oriental and African Studies (SOAS) University of London.

WWF-Spain (2006). Illegal water use in Spain: causes, effects, solutions. Madrid: WWF/Adena.

Yang, H., Wang, L., Abbaspour, K.C. and Zehnder, A.J.B. (2006). Virtual water highway: water use efficiency in global food trade. Hydrology and Earth System Science Discussions 3: 1-26.

Zygmunt, J. (2007). Hidden Waters. London, Waterwise, www.waterwise.org.uk.

Food distribution: exploring ethical futures

P. Steedman, S. Ripoll and T.C. MacMillan
Food Ethics Council, 39-41 Surrey Street, Brighton, BN1 3PB, United Kingdom; tom@foodethicscouncil.org

Abstract

That food reaches plates in the UK is a logistical feat in a country where fewer than 2% of population grow it and many ingredients travel thousands of miles. Yet the way food travels cannot be sustained. More important than the direct environmental, social and economic costs of transporting it is how the UK's food distribution infrastructure cements in place production, consumption and trading practices that destroy the environment, harm animals and are deeply unjust. The focus of this paper is to describe the 'scenario' method that has been used by the Food Ethics Council to explore challenges and opportunities with stakeholders in food distribution. The scenario method was pioneered in the business community and is now widely used in policy, including in UK policy development relating to food and farming. We reflect on its value as a tool for ethical deliberation.

Keywords: scenarios, food system, strategy, deliberation, stakeholders

Introduction

We cannot predict the future at the best of times. At present, with instability in the economy, geopolitics and the biosphere all looming large, this uncertainty seems particularly threatening. Yet the need to plan for a sustainable, ethical food system has never been greater. Improving food security, tackling hunger, obesity and injustice, supporting animal welfare, and preserving the environment, all demand urgent action in pursuit of long-term aims.

How can people striving for a better food system – civil society groups, public policymakers and businesses – begin to plan campaigns, policies or products for the future if it looks so uncertain? And how can we avoid the opposite trap, of assuming that the future will be like today, only more so? This is where scenarios can help our thinking. Businesses such as Shell (multinational energy and petrochemical company), and the UK government through its Foresight programme, use scenarios to help them think creatively about future threats and opportunities and to avoid the danger of assuming that the future will be much like the present.

The Food Ethics Council (FEC) developed a set of four scenarios to help the organisation think about how the UK's food system could develop over the next couple of decades. What might we be eating in 2022, and why? Where could it have come from? Where might we be eating it – and how might it get there? These scenarios formed part of the evidence for the Council's project on food distribution. The project examined the impact of food distribution networks on our environment, economy, culture and communities, and their contribution to climate change (FEC, 2008). The work resulted in the development of a sustainable vision for the future of food distribution, and provides a roadmap for government, business and civil society.

A toolkit

It is valuable to share these scenarios as they were found to be useful and FEC feel that other groups may wish to apply them in other settings. In addition, it can take considerable effort and time to develop scenarios such as these. FEC believe that a free toolkit such as this can make many of the benefits of scenarios accessible to other organisations at a minimal cost to themselves and to wider stakeholders. This paper therefore presents details of the four scenarios.

FEC's scenarios are not predictions or projections. Instead they represent four plausible stories about how the future could turn out, based on current trends and factors likely to drive change, as well as on significant uncertainties about how they are likely to play out. None of the scenarios is likely to 'come true' *per se*, but elements in each may well do. FEC and other organisation can use these scenarios to ask 'what if…?' questions, to help FEC and others 'future-proof' strategies: 'if the world turns out like this, will our policy still work or be a priority?' Scenarios can also help organisations think about the desirability of certain futures: 'if that happened, would our food system be better or worse for people, animals and the environment? And what do we need to do to encourage or avoid it from happening?'

The scenarios

FEC has developed four scenarios for the future of UK food in 2022. They can be summarised as follows: (1) *Pass the VatBeef*™ *QuikNoodle; (2) Carry on consuming; (3) Cash poor, time rich, experience hungry; and (4) A lot of allotments.*

Scenario 1: pass the VatBeef™ QuikNoodle

All the latest technology, including biotech, in-vitro meat and milk and hyper-efficient closed loop recycling systems are brought to bear on growing problems of nutrition and hunger in the UK. Cost and convenience are key consumer priorities, with general disdain shown towards any pretension around food – as long as it's safe and filling, consumers are not concerned where it comes from, unless it carried a heavy toll in carbon credits.

Scenario 2: carry on consuming

Personalisation and segmentation are key within this scenario. An explosion of brands sees ever tighter targeting of products at smaller and smaller groups, including the widespread marketing of nutraceuticals and functional foods. Supply chains are dominated by a handful of companies, but global competition and the demands of Corporate Social Responsibility (CSR) policies on carbon and food safety have led to greater reliance on production and processing within Europe.

Scenario 3: cash poor, time rich, experience hungry

Enter a world of seemingly limitless choice, gourmet bragging and web-based recommendations and retail. Only those with the luxury of time to spare can personally track down the most exclusive eco-friendly purchases, or chat with the artisanal producers now gaining increasing power in the food system. Still, the vast majority of people benefit from automatically ordered, cleverly managed doorstep deliveries. Concern about 'quality' food – low-input, traceable, fairly produced – is generally high, even if knowledge about whether the system is really that low-carbon is limited.

Scenario 4: a lot of allotments

Food growing has penetrated and surrounded the urban environment. With multi-storey farms and a widespread commitment to growing your own food (wherever there is a space), people's understanding of where their food comes is higher than it has been for generations. Food is a key part of the social and cultural – as well as physical – fabric of towns and cities. High street retail is very fashionable and localism is a dominant theme, with the biggest retail and foodservice companies turning their attention to larger markets and margins elsewhere in the world.

Reflections

This scenario process highlighted possible future trends that have received relatively limited attention in policy and business initiatives on food distribution, despite their potentially significant implications. These scenarios cause reflection on future possibilities, including:

- The prospect that scarcity of natural resources will become an overriding driver of change in the food sector. Steep rises in energy and food prices over the past year have built awareness of resource scarcity, but the possibility that very tight constraints on fuel, water and other factors of production could force a radical reconfiguration of the food system is rarely considered. Similarly, the challenge posed by resource scarcity to the presumption of continual economic growth is routinely ignored.
- People in the UK valuing the cultural and social dimensions of food much more strongly – a society where most people are 'foodies', and 'fuelies' (individuals who see eating as a necessary inconvenience) are few and far between. 'Foodie' citizens may be more alert to the climate change, animal welfare, health and labour rights implications of what they eat, and more willing to pay for them, making it easier to meet a host of government objectives.
- The hugely significant changes in global trade and geopolitics that could arise as China, India and other fast-growing economies become increasingly influential in the international arena. The part that this played in recent food price rises is often overstated but how profoundly it could alter Europe's future place in the world, and our influence (good and bad) on domestic and international food security, is scarcely considered.
- The rapid pace at ways of shopping change and the need for planning policy to think ahead about such changes in order to ensure they support sustainable development.
- Technologies that could overturn assumptions about what is and is not sustainable. An example might be in-vitro 'meat', already being cultured experimentally in laboratories. The ethical implications of this technology are yet to be explored in depth, but it would be relevant to consider how far in-vitro meat production could challenge assumptions about the inefficiency of converting plants into animal protein.

However the most striking lesson learnt from exploring these scenarios was that the future can hit us sooner than we think. Over the course of this food distribution project – from the scenario workshops in May 2007 to the time of writing in 2009 – events that experts had told us were possible in more than a decade's time, but pushing plausibility, have already happened. One of these was a major upswing in food prices. Straining to see a long way ahead can shed light on what's just around the corner.

References

Food Ethics Council (2008). Food distribution: an ethical agenda. Food Ethics Council, Brighton.

Theme 7 – Concepts of risk, trust and safety in the food system

Food industry: risk factory in the risk society?

Z. Lakner and G. Kasza
Corvinus University of Budapest, Department of Food Economics, Villányi út 29-43, 1119 Budapest, Hungary; gyula.kasza@uni-corvinus.hu

Abstract

Food safety has become a popular issue in the media since the early 1990s. It has gained enormous importance in the European Union due to the significance of BSE that resulted in the extensive reorganization of the systems applied to control risk. Among many new disciplines, the risk analysis approach has emerged to provide a methodical paradigm for authorities. Risk analysis has been used in catastrophic loss management for a long time and is rooted in financial decision-making. Food safety, however required new control methods. The White Paper on Food Safety defined risk assessment, risk management and risk communication as distinctive but coherent and cooperative areas. The role of risk assessment and risk management is relatively clear. On the other hand, risk communication is often neglected in everyday policy processes. Its role can be limited to a reaction to press and crisis communication in many cases, although it is well known that 8 out of 10 patients treated with food borne illness are infected by food at home that has not been handled correctly. Fortification of consumer consciousness is also an important issue, which can be fostered by well designed communication campaigns and routine bilateral communication. This paper covers many of the important food safety cases (such as BSE, avian flu, red pepper intoxication, guar gum scandal, food frauds) that have occurred in Hungary from 2000. The analysis presents primary research data derived from different measurement methods utilized in previous years to indicate public resonance of the cases and the communication of different institutions.

Keywords: food safety, risk communication, risk perception, BSE, avian flu

Introduction

The food safety issue has become a focal point of media communication and scientific research in the last twenty years all over the world. In former centrally planned economies (like Hungary) this process occurred in parallel with drastic changes in politics and social life. Put another way: risk-minimising in food consumption gained considerable importance for consumers of developed states, while in former socialist countries this occurred in parallel with the loss of the illusion of paternalistic, risk-minimising state, highlighting the importance of individual decisions.

The food safety issue is relatively new for Hungarian consumers, because-as a consequence- of (1) an autocratic food import policy in socialist times; (2) strong state control of the press by the state, prohibiting the publication of any potentially panic-creating news; (3) strict, rigid and centralised food safety inspection system. Such scandals, like the glycol fortification of wine in neighbouring Austria (Mayer-List, 1985) have been cited in Hungarian press as dirty tricks by western profit-seekers (Heltai, 1985), suggesting that such things would not occur in Hungary. That is why news on food safety problems has created considerable resonance in the Hungarian society. The aim of this paper is to offer a summary of results of secondary and primary research carried out by the authors concerning food safety related to Hungarian consumer behaviour with specific attention to lessons and experiences.

Methodology

Four food safety scandals are analysed from the point of view of consumer science. Beside the analysis of relevant documents, we have carried out four consumer surveys. The survey population was not representative in terms of education levels (people with higher level of education were over-represented), but the collected data still conveyed relevant information on consumers' ways of thinking and attitudes.

The survey method we applied was a questionnaire based face-to-face consumer survey. Selection of the sample was random, the places the surveys were conducted include traffic nodes, non-food related events and homes. The response rate was relatively high, because experienced interviewers helped our work (Table 1). We also offered some incentives (such as ball pens or candy bars) for the respondents in return. Some of the surveys were funded by national sources (BSE; 2001 and partly Guar gum; 2006), while others were financed partly or totally by companies (such as Avian flu; 2005). In some cases we did not have funding (Paprika; 2004). The driving force behind organizing these surveys was to gather enough data to understand the reaction of Hungarian consumers in regard of different food safety cases. These experiences were transmitted to policy makers, translated to media and utilized in food engineer training (especially at training of food safety experts).

The BSE scandal

Although food additives, preservatives and chemical residues have a certain history in news and tabloid communications, the first food safety issue that caused a global impact was BSE (*bovine spongiform encephalopathy*), also more commonly referred to as *mad cow disease*. It has attracted wide attention because this issue contained several elements that raised public concerns or interest.

This issue was notable as it resulted in 'a touch of moral dismay' as the disease originated in the animal husbandry practice of industrialized farms, where cows had been fed animal proteins (including processed cattle carcasses) in order to increase the cost effectiveness of production. Although similar methods have been in use for decades, this practice was described in the media as 'unnatural', or by the phrase of 'indirect cannibalism' or 'enforced cannibalism'. Overall, the media strongly suggested that economic drivers have undermined the fundamental ethical values of traditional agriculture: *'The short-sighted, greed-driven, truth-denying farmers, meat renderers and slaughter-house owners who were responsible for producing the killer filth.'* (Sunday Business Post, 04.11.2000).

The BSE scare had its consequences: serious economic losses, impacts on consumer confidence in governmental and academic institutions, major changes in the organizational structures and policies both at EU and national levels. There have, however, also been less obvious consequences: the recognition of food scares as valuable news, which can easily raise public interest. BSE proved its value in providing front page stories for years even in distant and BSE-free countries. Furthermore, many journalists acquired

Table 1. Sample sizes and dates of surveys.

Acronym of survey	Date	Number of respondents
BSE	2001	865
Paprika	2004	420
Avian Flu	2005	640
Guar Gum	2007	1,577

a basic expertise and a contact list in the field of food safety during the years of the BSE scare, which could then be used when dealing with any new food crisis.

There were no BSE cases in Hungary. In spite of this, the population was still very much aware of the scare. Approximately 99% of the population (n=865; 2001) were well informed about the BSE crisis, as primary data from our 2001 research set shows. Slightly less than half of respondents (48.72%) reported that their beef consumption had changed as a consequence of the BSE disease. These people were mainly women and older people.

The consumption-change has been different according to self-estimation of respondents (Table 2) (the statistical system of Hungary is unable to convey reliable information of change of consumption for different reasons, therefore it has not been possible to conduct a time-series analysis of beef meat consumption and prices).

As the results show, almost half (exactly 48.2%) of the respondents decreased beef consumption to different extents. These people previously used to consume beef (we analysed only this segment of the sample at this point). Almost 14% of the questioned people stated that they ceased beef consumption due to the BSE crisis. Note that about 0.5% ate more beef than before. During the interviews they addressed discounts (possibly but not necessarily in connection with the crisis) as the main reason for the increase in consumption. Later, during the analysis of cases such as the avian flu, the guar gum case or a country-wide food fraud (called MegaTrade scandal), we found also that a segment reporting consumption growth seems to be always present, consisting of respondents with higher than average and often technical oriented education, besides another, price-sensitive consumer group, coming from mainly relatively low income, urban families.

As a summary it can be stated, that the BSE crisis considerably influenced consumer behaviour in Hungary, in spite of the fact, that the Hungarian livestock was BSE-free: (although BSE-positive livestock was imported) Hungarian agri-marketing communication had not been able to utilize this fact to increase level of confidence nor loyalty among Hungarian consumers for domestic products.

The paprika scandal

One of the best-known 'Hungaricums' is the seasoning/ground paprika, which is widely utilized in the preparation of industrial meat products and in the Hungarian cuisine. The official ban on the retail trade of paprika on 28 October 2004, as a consequence of identifying that in some paprika powders the level of aflatoxin B1 was higher than allowed, caused consumer alarm and became a headline in the press all over the world.

Analysing the electronic and printed press it is clear that the Hungarian government had been reluctant to communicate the fact that the Hungarian paprika producers imported a large quantity of paprika from South America. During the long distance transport the relative humidity of products increased,

Table 2. Changes of beef consumption in regard of the BSE crisis.

Eats more beef since the outbreak of the crisis	0.5%
Eats somewhat less	16.4%
Eats much less since the outbreak	17.9%
Ceased the consumption	14.0%

and this created a favourable climate for toxin-production. For our analysis, the most important lessons learned from the questioning of Hungarian consumer knowledge of the paprika crisis were as follows:

- The majority of consumers (98%) had some information on the food safety problems regarding Hungarian paprika. Most of the respondents (80%) were rather unsatisfied with the authorities' activities, because they were not clear.
- An overwhelming majority of consumers had knowledge of the paprika problem after 72 hours of the official withdrawal of the products.
- The exact cause of the food safety problem was known by more than half of the respondents. The rate of correct answers to the question about the cause of the problem was higher in the better-qualified and younger respondent group.
- There were considerable differences between the different socio-economic groups of respondents regarding the information sources on the paprika problem. For the younger generation the most important information source was the internet, the middle-age respondents reported television as their primary source, and the eldest respondents reported radio broadcast as their first reference point.

The paprika problem appeared to be a frequent topic of discussion among consumers at that time, both in families (80%) and at public and work places (76% of the respondents stated so). It was most frequently cited in the case of small town and village dwellers, which can be explained by the fact that the interpersonal relations between non-family members in small towns and villages are much more intensive compared to those in larger towns and in the capital. The most important lessons are: (1) The governmental information system on import was too complex and not up-to-date (for various reasons); (2) The official quality inspection organisations were not sensitive enough to 'weak signals' (e.g. forums on the internet) of the coming problems; (3) ISO and HACCP systems of companies did not offer sufficient security for the consumer that the product in the packaging matched the quality declaration; (4) Globalisation of food trade and the increasing complexity of food chains involve the extension of product paths, which highlights the importance of continuous product monitoring; (5) Quality standards of the imported products should be systematically updated; (6) A well-defined risk and crisis communication strategy was not available on governmental and company level. The debates of the different institutions regarding quality inspection and food safety considerably decreased public trust; (7) Hungarian consumers attribute great importance to the problem of food safety. In spreading information widely, the modern methods of communication gain increasing importance. These forms of media should be more effectively utilised in risk communication, too; (8) There are differences in risk acceptance and risk awareness between different consumer segments. These differences should be taken into consideration when planning crisis communication; (9) It has been proven that on the basis of consumers' risk acceptance and risk awareness, the intended food consumer behaviour can be forecasted (Siegrist *et al.*, 2006). This can be an important tool in the development of a risk communication strategy on governmental as well as entrepreneurial level.

The avian flu problem

In 2005 a new food safety problem has been created out of a basically non-food issue (a basic heat treatment terminates the virus, making even the contaminated food safe, assuming it has not made any contact with raw eaten ingredients, such as vegetables or fruits), partly as a result of some governmental responses and the influence of media producers, which gained considerable consumer attention in Hungary: avian flu This lead to some people becoming afraid to consume poultry products. As identified in our direct question measurements the level of poultry consumption significantly decreased (Table 3), but the consumer responses were divided. Note that our results match to the Eurobarometer 257 (Special issue on avian influenza, 2006) very well, with the slight difference that Eurobarometer used a 3-way scale (more/less/just as much), while we differentiated two levels of 'less'.

Table 3. Changes of poultry consumption after the avian flu outbreak.

Our poultry consumption from the outbrake of the disease		Level of qualification			Total
		Elementary	Maturity	BSC, MSC	
Considerably diminished	% within opinion	13.9%	52.8%	33.3%	100.0%
	% within qualification	11.1%	6.8%	7.8%	7.5%
Diminished	% within opinion	6.3%	62.0%	31.6%	100.0%
	% within qualification	11.1%	17.5%	16.2%	16.5%
Has not changed	% within opinion	9.8%	58.7%	31.6%	100.0%
	% within qualification	77.8%	75.0%	73.4%	74.7%
Increased	% within opinion		33.3%	66.7%	100.0%
	% within qualification	0.7%	2.6%	1.3%	
Total	% within opinion	9.4%	58.5%	32.2%	100.0%
	% within qualification	100.0%	100.0%	100.0%	100.0%

Based on level of acceptance of different statements, formulated on a Likert-scale, the combination of categorical principal component analysis and the cluster analysis of three clusters have been identified: (1) the conscientious consumers, who read the product labels more carefully than before, and try to minimise their poultry consumption. Their share in the population has been approximately 70%. (2) the take-it-easy buyers, who do not attach considerable importance to avian-flu problem; and continue their consumption (~20%) and the price-sensitive buyers, who consider the avian-flu problem as possibility to buy cheaper products (10%). Interestingly, the majority of respondents considered poultry from family farms to be safer, despite the fact that they are more exposed to the virus, if one hypothesises that the virus is transmitted via wild birds.

Relationship between consumer attitudes, norms and perceived control has been analysed by structural equation model, based on Fishbein-Ajzen theorem (1974). The structural model describes two types of relationships: the relationships between observed variables and latent variables, and that among latent variables. The directly observed variables are indicated by ellipses. The continuous latent variables (attitude, norms, perceived control) are indicated by rounded rectangles. The behaviour itself (marked by rectangle) has been measured by four indicators.

These indicators were marked by pentagons. The graph shows the unstandardized coefficients. Each unstandardized estimate represents the amount of change in the outcome variable as a function of a single unit change in the variable causing it. By definition, the first estimate in each group of variables is set as 1. Based on the data above, structural equation modelling was elaborated to determine the influence of different factors on the efficiency of a farm. The chi-square test showed that fitness of the model was not significant, indicating that the null hypothesis, that the model fits the data, cannot be rejected. This finding was corroborated by the Root Mean Square Error of Approximation (RMSEA) statistics. According to Muthén (2000, 2004) the recommended cut-off value is 0.06. The RMSEA estimation was 0.1, therefore the model does not fit perfectly, but in our view the findings are still important for practical lessons, because of the trends they indicate. Values on the arrow indicate the coefficients of equations.

Analysing Figure 1 demonstrates that consumer behaviour is a highly complex issue influenced by numerous factors. We have not been able to determine a significant relationship between the perceived control variables and the intended behaviour. However, the results highlight the importance of subjective norms.

Summary

Food production and consumption are activities with inherent risk which is increasingly recognized and emphasized – in some cases, over-emphasized – in public discourse. Some food producers apply high food safety standards and act in a responsible way when problems occur. However the activities of some food industry actors can represent a source of risk, which needs continuous monitoring in order to keep the food chain safe. Control of these risks are only partly dealt with by governmental measures, whereas another part of the preventative work is done by the consumers (and competitors).

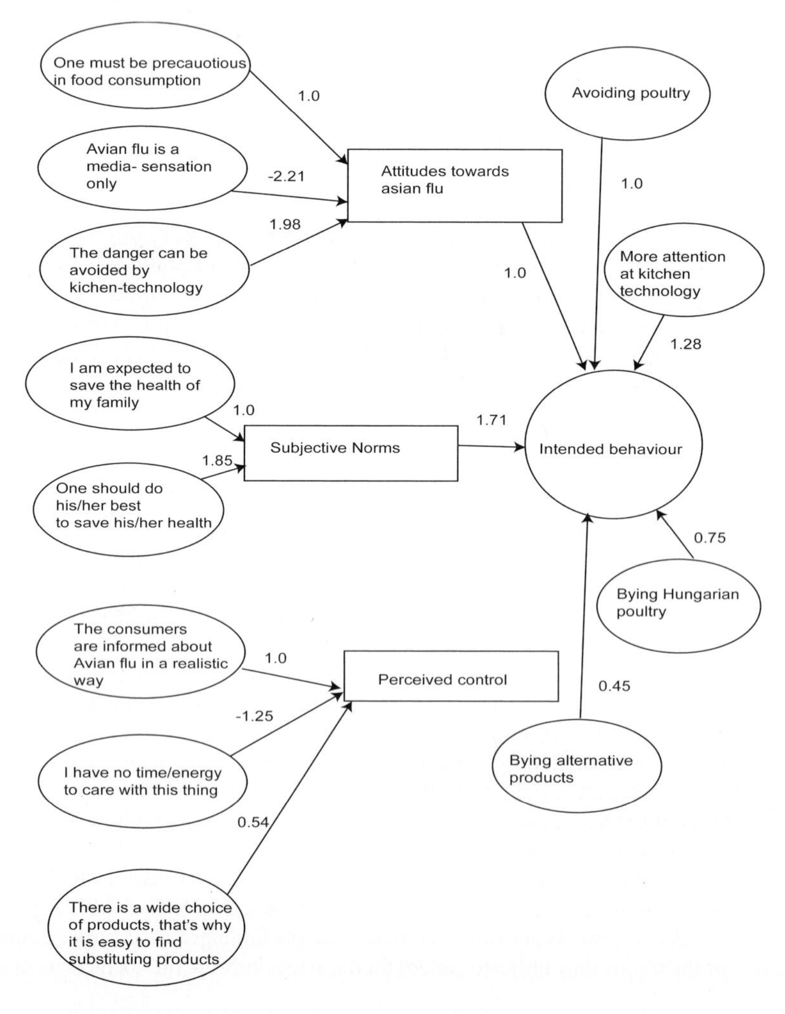

Figure 1. Structural equation model for describing consumer behaviour in avian-flu crisis.

However, in most cases it is easier for citizens to shift responsibility to the state than to take the necessary pro-active and preventive measures. Some of our research data show that consumers consumed even the suspicious products like adulterated goods without reporting these products to the authorities. On the other hand, governments should prepare a more coherent strategy for food safety risk communication. The experience of food crises highlights the overlaps in the governmental communication system. This fact can be explained by the complex nature of food safety (agricultural, veterinary, human, trade regulatory, etc.), but multi-channel communication in a critical situation could be extremely dangerous. In Hungary, it can be considered as a positive fact, however, that in the last years the governmental communication on food safety issues were concentrated at the Ministry of Agriculture and Rural Development. The Hungarian act on food chain and authority control (year 2008 number XLVI) defines the food control authority as the only responsible authority for food and feed control, from soil to the consumer shopping basket (earlier it was shared among numerous authorities and unnecessary overlapping of scope of duties burdened the daily work of enterprises and governmental funds). The Health and Medical Officer Service still plays an important role in surveying and analysing human food borne illness cases in the catering trade (from the 'door of the store room'). The Hungarian Authority for Consumer Protection also remained to inspect labelling information and economic losses caused to consumers (size of portions, etc.). The changes resulted in a concentrated food control system, which is responsible for keeping up multilateral risk communication and plays a cardinal role in crisis communication.

References

BBC (2002). Living and dying with CJD. http://news.bbc.co.uk/1/hi/health/background_briefings/bse/82975.stm Last accessed: 21.04.2009.

European Commission (2006). Eurobarometer 257 – Special issue on avian influenza.

Fishbein, M. and Ajzen, I. (1974). Attitudes towards objects as predictors of single and multiple behavioural criteria. Psychological Review 81(1): 29-74.

Heltai, A. (1985). Ausztria borbotránya (Wine scandal of Austria). Magyarország, 28(4): 4.

Mayer-List (1985). Die Tricks der Weinmischer (Tricks of wine producers) http://www.zeit.de/1985/34/Die-Tricks-der-Weinmischer Last accessed: 22.03.2009.

Muthén, B. (2002). Beyond SEM:General latent variable modelling. Behaviometika 29: 81-117.

Muthén, L.K. and Muthén, B. (2004). MPlus Statistical Analysis with latent variables, User's Guide, Muthén and Muthén, Lons Angeles, pp.1-505.

Siegrist, M., Keller, C. and Kiers, H.A.L. (2006). Lay people's perception of food hazards: Comparing aggregated data and individual data. Appetite 47: 324-332.

Five steps to trustworthiness

F.L.B. Meijboom
Ethics Institute, Utrecht University, Heidelberglaan 8, NL-3584 CS Utrecht, the Netherlands; f.l.b.meijboom@uu.nl

Abstract

Problems of trust are evident in the agricultural and food sector as well as in society in general. To deal with this situation, I claim that they have to be addressed as problems of trustworthiness. This paper presents five central steps that follow from a study on trust and trustworthiness (Meijboom, 2008) and that enable individuals to deal with problems of trust both from a practical and a theoretical point of view. These five steps do not solve all problems of trust, but provide clarity with respect to trustworthiness, which is a necessary condition to build and maintain trust.

Keywords: food, risk, trust, trustworthiness

Problems of trust

Problems of trust are evident in the agricultural and food sector as well as in society in general. All over Europe and in the United States trust in public and market institutions is in the spotlight. This does not imply that we are confronted with a crisis of trust. The problem is that people have to rely on others, but often do not know whom to trust. I call this the 'problem of trust'. To deal with this problem of trust, it has to be addressed as a question of trustworthiness. This poster presents five central steps that follow from a study on trust and trustworthiness (Meijboom, 2008) and that enable individuals to deal with problems of trust both from a practical and a theoretical point of view.

The step from trust to trustworthiness

I propose a shift to conceptualising the problem of trust as a problem of trustworthiness. For this step there are conceptual and moral reasons. On the one hand, it has a conceptual background. An individual cannot decide to trust. One may want to trust, but one cannot trust at will. On the other hand, the autonomy of the individual provides a strong moral reason for this shift. A lack of or hesitance to trust should be acknowledged as a legitimate point of view, rather than as a failure only. This does not imply that the truster cannot be wrong, but shows that the burden of the proof also lies at the level of trustworthiness. Even though a trusting relationship is by definition asymmetric and includes differences in knowledge and power, the truster should be treated as a person who is capable of autonomous agency. Consequently, the main question is not how the individual can be changed so that he will trust, but what conditions the trustee has to fulfil in order to be worthy of trust. This first step has direct consequences for how the 'problem of trust' has to be addressed.

The step from risk to trust

Trust is relevant in situations of uncertainty. As a consequence, enabling consumers to deal with risks is regularly seen as an effective answer to questions of trust. The idea is that if individuals are able to address a danger as a risk, one has the opportunity to decide how to deal with the situation rather than the restricted choice to take or leave the danger (cf. Luhmann, 1988). This approach seems promising. If a situation of danger turns into one of risk, it becomes an object of action, because one has the option of making an assessment of the danger and of addressing it as a risk that one can analyse and manage. Consequently, one has the opportunity to decide on the best way to deal with the uncertain situation.

Therefore, providing information on risks and enhancement of transparency is often proposed as the most efficient (regulatory) approach to this problem. Despite the importance of both aspects, there are two problems. First, transparency and risk communication already presume some levels of trust. Only if one already considers the provider of information reliable, the information becomes useful. Information about the risk at stake will only help if someone already considers the provider of the information reliable. Therefore, an overriding emphasis on risk communication only begs the question.

Second and even more important, trust is fundamentally different from taking risks, even though they can be relevant in the same situation. Trust is not the outcome of an assessment of the risks and benefits of trusting in the light of the aims and goals one pursues. In contrast to someone who takes risks, a truster is not counting, but coping with complexity. He or she is not calculating risks, but dealing with the uncertainty he or she is faced with. Therefore, better risk assessment and more risk information do not necessarily lead to more trust. To help this person it can be useful to translate the problem of known uncertainty into one of risk. Consequently, he or she can make his personal assessment and decide whether it is worth to take the risk involved in relying on another agent or not. However, this risk-benefit analysis does not provide a direct reason to trust (Lagerspetz, 1998). It can only show that, given his preferences, the risk is worth taking. Trust has a different focus. It starts where a risk focus ends. It arises in situations that remain uncertain despite the attempts to turn the uncertain aspects into risk factors. The aim is not to try to make a risk-benefit analysis, but to personally assess the competence and motivation of the trusted agent. A truster runs a risk, but does not take risks. Therefore, being trustworthy is different from enabling individuals to take risk or to reduce the risks at stake.

The step from reliability to trustworthiness

Trust is a way of dealing with the uncertainty that comes with the freedom of agency, rather than with uncertainty as such. If a trustee would invariably act according to a predictable pattern determined by his or her nature or its organisational structure, it would be like relying on a machine. Machines perform, they do not deliberate; machines are programmed, not motivated. Consequently, they can be reliable, but not trustworthy. To be trustworthy, the trusted agent should not merely act in a predictable way, but should be motivated to respond to what is entrusted. Thus, if one says that some (collective) agent is trustworthy, it means that he or she is not just predictable, but worthy of trust even though they have the power to do otherwise. If we address the problem of trust as a question of trustworthiness, the problem cannot be reduced to a lack of predictability or a need for structures on which a truster can anticipate.

The step from competence and motivation to the inclusion of respect for autonomy

Trustworthiness begins from the duty to show due respect for the truster as a person who is capable of autonomous agency. Despite the vulnerable status of the truster, trustworthiness is predicated on recognition of the truster as a moral agent and as a moral equal. In general, it implies that a truster ought not to be considered just as a vulnerable person, but that he should be treated as person who has the capacity to choose his own goals and values. This respect results in some clear constraints on what counts as trustworthy behaviour. If trust were about dealing with uncertainty as such, then power, coercion, or controlling behaviour would be relevant methods, which would help to establish trust. However, if we take freedom, agency and a participant attitude as constitutive for trust, these options are incompatible. Genuine trust starts from a participant attitude and from the recognition of the other as an autonomous agent. Consequently, any form of exploitation or manipulation of the truster by the trustee rules out trustworthiness. Furthermore, this conclusion implies the demand to take the moral dimension of what is entrusted seriously. Trustworthiness has to include the ability and willingness to deal with the moral dimension of trust implied in the recognition of the truster as moral agent.

The step from arbitrary compromises to acting in a trustworthy manner

The final step is to cope with the problem of conflicting moral expectations that results from the respect of the truster as a moral agent and a moral equal. This requires a balance between accommodation and integrity. If one starts out from respect for the truster, her views have to be taken seriously even if they are in conflict with those of the trustee. This requires accommodation, which implies that the trustee should be open to the other's moral view, should be prepared to change his view and should be willing to actively search for new ways to deal with the conflict. To deal with such a conflict making compromises is often inevitable. The trustworthiness of an agent who makes compromises in every situation can be questioned. Compromises easily get a character of arbitrariness if the truster does not have legitimate reasons for the decision whether or not to act according to the expectation of the truster. Moreover, on an institutional level, it is sometimes unclear to whose view one should accommodate given the many trusters and many expectations. Nonetheless, it is possible to remain trustworthy and make the compromises that are sometimes necessary from the perspective of accommodation. Having integrity is essential with respect to this. Integrity does not just complicate the demand of accommodation. It can provide reflected constraints on the demand of accommodation. Integrity understood as a sincere commitment of the trustee with 'those projects and principles which are constitutive of one's core identity' or with the tasks and aims that are constitutive for an institution, lead to constraints on the demand of accommodation that are not arbitrary. They are not beyond debate, but they are reflected and the trustee can give legitimate reasons for the decision whether or not to act in the expected way. Consequently, not everyone will trust this agent, but the agent is trustworthy despite the confrontation with the moral conflict. In practice, it can be frustrating to lose trust although one has legitimate reasons not to act in the expected way. Despite this frustration, this is better than trying to look for ways to get people to trust an agent although this agent is not trustworthy, either because he or she is not competent or not adequately motivated. If one is trusted, but not trustworthy, the problem of trust will return in the end.

These five steps do not solve all problems of trust, but helps a truster to assess who can be trusted (a) in those situations that remain uncertain even after the uncertain aspects have turned into risk factors as much as possible, (b) when predictability fails, (c) when conflicting moral values are at stake, and (d) when the hesitance to trust starts in the belief that one is disrespected as an autonomous person. In these cases the trustee's reflection and communication on its competence and motivation can make the difference. The communication provides the truster with crucial information for the assessment of the trustee. Moreover, this communication creates expectations about the future behaviour of the trustee, which result in principles that require the trustee to protect the truster from manipulation and serious losses. This does not solve all problems of trust, but provides clarity with respect to trustworthiness, which is a necessary condition to build and maintain trust.

References

Lagerspetz, O. (1998). Trust: The Tacit Demand, Dordrecht: Kluwer Academic Publishers.

Luhmann, N. (1988). Familiarity, confidence, trust: problems and alternatives. In: Gambetta, D. (ed.) Trust: Making and breaking cooperative relations, Oxford: Basil Blackwell, pp. 94-107.

Meijboom, F.L.B. (2008). Problems of trust: A question of trustworthiness. An ethical inquiry of trust and trustworthiness in the context of the agricultural and food sector, Dissertation Utrecht University.

Cisgenic crops: more natural, more acceptable

H. Mielby and J. Lassen
Danish Centre for Bioethics and Risk Assessment, Department of Human Nutrition, Faculty of Life Sciences, University of Copenhagen, Rolighedsvej 30, 1958 Frederiksberg C, Denmark; hmb@life.ku.dk, jlas@life.ku.dk

Abstract

Previous research indicates that the public does not reject genetic technologies as such; but rather base their assessment of a given application on judgements regarding usefulness, naturalness, possible alternatives as well as a number of other moral and risk related issues. In this context the development of cisgenic crops (where the plant is transformed only with its own genetic material) has been suggested to provide a preferable alternative to traditional transgenic applications. Since cisgenics does not transgress borders of species perhaps they will be considered 'more natural', and therefore more acceptable. Based on five focus group interviews (n=35) drawn from the general public of Denmark we examine whether this is the case.

Keywords: cisgenic crops, GM crops, naturalness, public attitudes, focus group

Introduction

It is already well documented that a discrepancy between lay-people and experts exists with regard to the perceived acceptability of genetically modified crops. In spite of experts' assurances, the public remains sceptical about the usefulness, naturalness and safety of said crops. At the policy level this has led to the adoption of numerous new strategies. One is that of involving lay-people or the general public in the decision making process – or perhaps even externalizing risky decisions entirely by turning them into questions of consumer preferences (as in some cases of labeling). Another strategy involves proceduralising risks trough various control systems and explicitly risk-based approaches – applied to wider and wider definitions of risks (Power, 2004). Finally, the proposed move towards 'more natural' technologies could be interpreted as such a strategy as well: If more acceptable products were developed, then maybe the lay-expert discrepancy would resolve itself.

The latter strategy presupposes that there is indeed a positive correlation between 'more natural' and public propensity to accept. However, to talk about something as 'more natural' requires that naturalness is understood as a continuum rather than a dichotomy, an understanding not necessarily shared by all interpreters. This already affirms that the concept of naturalness is difficult to discern. Naturalness may for example refer to different forms such as history-based, property-based or relation-based naturalness, and to different entities: Ecosystems, plants, traits, events, products, production methods and so on can be seen as natural or unnatural as the case may be (Siipi, 2008). With regard to the concerns about gene technology being unnatural there seems, however, to be similarities as well. The basic concern seems to be that gene technology is in conflict with some fundamental values or principles of order (Lassen, 2006).

In this paper the case of cisgenic crops is considered. Cisgenic crops are crops genetically modified with genes from within the same species or from sexually compatible donor plants. There are other defining technicalities as well, however, the central idea is that they do not transgress borders of species, whereas transgenic crops do (Schouten, 2008). Supposedly this brings them in closer alignment with the 'principle of order' narrowly interpreted as 'species'. The aim of the paper, then, is to examine first whether that makes them 'more natural' in the eyes of the public, and secondly whether 'more natural' in this context equals 'more acceptable' than traditional transgenic crops.

Data and methods

Empirically we draw on five focus groups of 6-8 persons (total n=35) conducted in June 2008. Each focus group was defined by educational background, and mixed with respect to age and gender. The first three sessions were conducted in Copenhagen and the remaining two in Aalborg. All participants were recruited through an analysis company in a topic blinded way. The focus group discussions each lasted approximately 150 minutes and were organised around three exercises.

First participants were asked to write down in keywords their immediate associations to the word 'genetic engineering' and then to elaborate on the significance of each association.

In the second exercise participants were presented with a deck of cards describing various examples of genetically modified plants, and requested to make prioritised lists going from what card they 'liked the most' to what they 'disliked the most'.

In the final exercise the participants were exposed to four extracts from recent newspaper articles and were are asked to comment on each article in turn. The articles were deliberately selected so as to present both positive and negative views on genetic engineering. Furthermore, the articles were selected so as to make it very difficult to relate to them without relating to the trustworthiness of different actors.

The focus group interviews were recorded and transcribed *verbatim*. Inspired by Stephen Toulmins' (Toulmin, 2003) approach to argumentation analysis we are in the process of identifying the core arguments that people use when talking about genetic technology in general. In the following, however, we will draw specifically on those arguments related to the perceived naturalness or unnaturalness of cisgenic crops.

Results

Whether cisgenic crops are believed in fact to be more acceptable may be answered with reference to the second exercise mentioned above. Here the participants were presented with eight different examples of modified crops. These examples were selected so as to differ with respect to objective, the origin of inserted genes and the method of transformation. The example representing cisgenesis was a cisgenic wheat with improved nitrogen uptake intended to decrease the amount of applied fertiliser required.

While for the purpose of this paper we do not wish to go into a much detail about objective-related differences, clearly we need at least to consider these differences, and how they were used to distinguish between the different examples. Typically the participants differentiated between three kinds of benefits: Benefits to public health (both at home and in the developing world), convenience related benefits (like longer shelf-life) and benefits to the environment. The participants differed as to whether public health or environmental benefits were to be preferred but agreed that convenience related benefits were not that valuable. Thus there is no doubt that variations in perceived benefits are affecting the perceived levels of acceptability. If we want to assess the effect of naturalness on acceptability we need therefore to isolate the effect of benefits.

The easiest way to do this is to compare the cisgenic wheat solely to those examples associated with similar benefits. That is to say crops which are also aimed at providing environmental benefits. More specifically a transgenic maize with resistance to insects and a transgenic herbicide resistant rapeseed was used. On almost all accounts these two crops were associated to the same category as the cisgenic wheat, as illustrated by this participant:

'There are some of them where the purpose is to reduce the use of pesticides and fertilisers. Here rapeseed and maize and [cisgenic] wheat align. It is the same purpose that is intended in these cases. You avoid destroying nature.'

To determine whether the cisgenic wheat is more acceptable, we have then only to look at how they were ranked against each other: When asked to rank the examples going from what they 'liked the most' to what they 'disliked the most', more than half of the participants preferred the cisgenic wheat to both the maize and the rapeseed. This provides at least some indication that the cisgenic wheat is indeed seen as more acceptable than the corresponding transgenic applications, which is also exemplified in statements like the following:

A: '[The purpose is roughly comparable, but] it could be argued that the one with maize, it is from a bacteria, whereas the one with nitrogen uptake is not. I would say that I do prefer the latter one to the former.'

B: ' As do I. It seems to be just a slightly more advanced form of crossbreeding. [...]'

Hence, the question is whether we can be certain also that the favourable ranking can be accredited to cisgenic wheat being more natural. To answer this question affirmatively we must first turn our attention to whether the participants even recognized the difference in origins and transformation methods, and to whether these differences were discussed in terms of some applications being more natural than others, as is evident in this dialogue:

A: '[Didn't you suggest earlier] that ... well that it made a difference whether the genes came from the plant itself, or from related plants, or if you mix... what to call it... totally different genes... plant and animal kingdom together, right?'

B: 'Precisely, I do not think that they have anything to do with each other. If you begin mixing [them] that just seems too unnatural.'

While the relative naturalness of the different methods was debated in all focus groups, some participants did not attach decisive value to these differences. Instead they believed GM crops to be GM crops regardless of the method of transformation; the difference therefore being regarded as unimportant. Others again did not know what to think, feeling trapped between reason and intuition. This is reflected in the following quotation:

'Well, I [picture] a fish swimming around, and then actually comes a tomato. That can't be right. I don't know if that is exactly 'natural'. [... Then again] it is totally unimportant where those genes are derived from in principle, as it is just a part of a long DNA strand. In principle it doesn't matter if it is derived form a fish, a bacterium, a dog or whatever. It is a trait you are transferring. But when it says derived from a fish – then a certain picture does come to mind.'

Having thus established that the participants did at least consider these distinctions on the grounds of some transformation methods being perhaps more natural than others, the next question is whether cisgenic crops (when considered in this light) were generally preferred to their transgenic counterparts. Again this was quite obviously the case, as concluded by this participant:

'[...] if we are looking for what is the more natural, then I clearly think that it is [the cisgenic one, where the card reads:] 'Through genetic modification a gene is inserted into the plant, which improves nitrogen uptake form the soil. The new genes are derived from the plant itself or from closely related species'. Well, the way I see it, it does not become much more natural than that.'

Concluding discussion

In the focus group interviews differences in transformation methods were both recognized and discussed in terms of some being more natural than others. In general, the adding of 'something else' was believed to render an application unnatural, and even more so when inserted genes were derived from non-plant organisms. While not all participants attached decisive value to these differences, a majority of the participants did prefer the example of cisgenic wheat to the equivalent examples of transgenic maize and rapeseed. These findings indicate that cisgenic crops are indeed seen as more acceptable because they are considered more natural.

Although public acceptability apparently correlates with the perceived naturalness of different transformation methods, it is not a result only of said naturalness. Usefulness, necessity, risks and so forth need to be taken in to consideration as well. Furthermore, acceptability may depend on factors not related to characteristics of the crops themselves, but rather to the believed reasons for developing them. If the development of cisgenic crops is taken to indicate a shared concern about the unnaturalness of transgenic crops, then the outcome is inclined to be more positive than if it is interpreted as merely a semantic excuse for introducing more of the same. It is therefore difficult to asses at this point whether the strategy of developing more natural crops will be a successful one.

Acknowledgements

This work is funded by The Danish Food Industry Agency

References

Lassen, J. and Jamison, A. (2006). Genetic Technologies Meet the Public: The Discourses of Concern. Science, Technology and Human Values 31: 1-8.

Power, M. (2004). The Risk Management of Everything. Demos. Available at: http://www.theiia.org/download.cfm?file=9876. Accessed 22 March 2009.

Schouten, H.J. and Jacobsen, E. (2008). Cisgenesis and intragenesis, sisters in innovative plant breeding. Trends in Plant Science 13(6): 260-261.

Siipi, H. (2008). Dimensions of naturalness. Ethics and the Environment 13: 1-71.

Toulmin, S.E. (2003). The Uses of Argument. Updated edition, Camebridge University Press, New York, USA.

Ambiguous food products: consumers' views on enhanced meat

N.K. Nissen[1], P. Sandøe[2] and L. Holm[1]
[1]Sociology of Food, Institute of Human Nutrition, Faculty of Life Sciences, University of Copenhagen, Rolighedsvej 30, DK-1958 Frederiksberg C, Denmark; Nikn@life.ku.dk
[2]Danish Centre for Bioethics and Risk Assessment, Faculty of Life Sciences, University of Copenhagen, Rolighedsvej 25, DK-1958 Frederiksberg C, Denmark

Abstract

In Denmark it is becoming increasingly common for cuts of meat to be enhanced by injection or by tumbling with a solution of water, salt, sugar, and other ingredients such as phosphates, antioxidants and acids. This is primarily done to make the meat more moist and tender. The introduction and spread of this technology has generally gone unnoticed. However, some criticism has been raised about this method of processing meat and the way it is marketed. On the other hand, a sensorial test has shown that consumers prefer enhanced meat to non-enhanced meat. This study investigated consumers' conceptualisations of enhanced meat. Four focus group interviews were conducted involving 27 consumers from various socio-demographic backgrounds. The focus group interviews concerned the consumers' knowledge of enhanced meat and their ways of conceptualising and assessing this type of meat. Unprompted, the consumers had very little knowledge of enhanced meat. When introduced to enhanced meat they expressed ambivalence. On the one hand, the consumers liked the taste, tenderness and moisture of the enhanced meat. They also liked the fact that enhanced meat remains tender and moist when cooked because it is insensitive to harsh cooking. On the other hand, enhanced meat was considered by the consumers to be unnatural and unauthentic, and was assumed to be associated with long-term health risks. Furthermore, the interviewed consumers associated the enhanced meat with foul play because it is sold under the misleading label 'neutrally marinated meat'. When new food products enter the market they will after a while be positioned in relation to existing categories of foods (as high quality or discount products, as everyday food or festive food, etc.), but the final position of enhanced meat in the categorisation of foods in Denmark is not yet settled. Consumers request that enhanced meat is labelled and marketed in an honest way. However, even if that occurs, and even though this type of product is in some contexts seen as useful, it may in most contexts still be viewed as potentially problematic.

Keywords: marinated meat, consumers, acceptability, ambivalence

Introduction

In Denmark it is becoming increasingly common for cuts of meat to be enhanced by injection or tumbling with a solution of water, salt, sugar, and other ingredients such as phosphates, antioxidants and acids. Pork, beef and especially poultry are being enhanced, and in this process the weight of the meat is typically increased by 5-15%. The idea is to make the meat more moist and tender, and some of the ingredients prevent rancidness and the growth of bacteria. This method of processing meat is in Danish called 'neutralmarinering' which can literally be translated as 'neutral marination' (Danish Meat Association, 2006; Dörffer, 2003; Rosenvold, 2006). In this paper we use the terms 'enhancement of meat' and 'enhanced meat', because these are the most common terms used in English.

The introduction and spread of enhanced meat in Denmark has generally gone unnoticed. However, some criticism has been raised in the media and in the Danish Parliament about this way of processing meat and the way it is marketed (Folketingets Lovsekretariat, 2006, 2007; Hvilsom, 2006; Jydske Vestkysten, 2005; Politiken, 2006). On the other hand, a blinded sensorial test showed that consumers prefer enhanced meat to meat that has not been enhanced (Søndergaard, 2007).

While many studies have investigated consumers' sensorial assessments of enhanced meat through blinded tests (Carr *et al.*, 2004; Robbins *et al.*, 2003; Søndergaard, 2007), our literature search showed that no studies appear to exist that concern consumers' conceptualisations and acceptance of enhanced meat.

This study investigates the views of consumers about enhanced meat from a sociological perspective. The aim is to explore how enhanced meat is conceptualised, and what these conceptualisations mean to the way in which consumers value enhanced meat.

Methods

The views of consumers about enhanced meat were investigated by qualitative and semi-structured focus group interviews. Four focus group interviews were conducted, which involved 27 consumers. To ensure representation of different views all the focus groups included consumers from various social backgrounds; two of the interviews took place in the region of the capital city of Denmark, while the other two took place in a rural area. The participants were recruited through a market research firm. At recruitment the participants were informed that the overall topic of the interviews was food, but the more specific topic of enhanced meat was not revealed.

The focus group interviews lasted approximately 250 minutes each and they were organised around exercises. After some initial talk about food and meat in general, the participants were asked to write down using keywords their spontaneous associations with the word 'neutral marination' ('neutralmarinering'). After this the participants were given a short introduction to the nature of enhanced meat. They were then guided through two exercises that requested them to prioritise different types of enhanced and non-enhanced meat and to assess if they would like to serve or be served enhanced meat in different social contexts. Finally, the participants were asked to discuss the consequences of enhanced meat for various stakeholders, for example the meat industry, supermarkets, restaurants, and lay people with, respectively, little money, poor health, poor cooking skills and other characteristics. In addition to these exercises, samples of enhanced and non-enhanced pork were served during the focus group meetings, and non-blinded tasting of this meat formed the basis for further discussion. All interviews were recorded and transcribed, and the transcribed text was coded into themes as a basis for the analysis.

Findings

Knowledge, and demands for information and clarity

The focus group interviews demonstrated that most of the interviewed consumers knew nothing, or very little, about enhanced meat. Many were not familiar with the term 'neutral marination' ('neutralmarinering') and they had no idea that this way of processing meat exists and that it is rather common. Among those who recognised the word 'neutralmarineret' from the labels in supermarkets or from the media, many did not know what it meant.

A few of the interviewees had personal experience of eating and cooking enhanced meat. However, this had always happened by chance, because the enhanced meat was cheaper than other meat, and/or because the interviewees were not aware of what the words on the label of the meat packaging meant. None of the interviewed consumers had chosen enhanced meat deliberately.

The interviewees both with and without prior knowledge about enhanced meat found the word 'neutral marinated' ('neutralmarineret') self-contradictory and misleading, and their associations with the word were negative. They felt cheated and deluded, because they had not been informed clearly about what enhanced meat is or told that a lot of the meat sold in Denmark is enhanced.

Eating and cooking enhanced meat

All the interviewed consumers assumed spontaneously that enhanced meat tastes of nothing, or tastes worse than meat that has not been enhanced, and they did not have particular expectations with regard to tenderness and moisture. However, when they tasted the samples of enhanced pork and non-enhanced pork during the focus group interviews most of the interviewees stated that they liked the taste, tenderness and moisture of this enhanced meat very much. Thus, in general, the interviewees preferred the enhanced meat to the non-enhanced, although a few thought that the tenderness and moisture were excessive, or that the taste or texture was wrong.

With regard to the preparation and cooking of meat, those interviewees who had personal experience of cooking enhanced meat were from the outset very critical. For example, they spoke about water seeping from enhanced meat when fried. However, because the participants were informed during the focus group interviews that enhanced meat remains tender and moist, even if it is fried for too long or warmed-up several times, many began to concentrate on the advantages of cooking enhanced meat. They emphasised that people with poor cooking skills may benefit from enhanced meat because it is insensitive to harsh cooking. Furthermore, they welcomed the fact that enhanced meat makes it easier for anyone – including those who can cook – to cook meat properly; that is to say to produce meat that is moist and tender and with a good flavour. They liked the thought of not having to be so careful when cooking, and of being more certain of a successful outcome, especially when serving food for guests.

However, some of the interviewed consumers still had reservations about the advantages of cooking enhanced meat. These interviewees expected most people to be able to cook good meals, even using meat that has not been enhanced, and thought that those who are not able to do so should learn the skill. Moreover, some interviewees reported that they much preferred to prepare their food from scratch themselves, even when this is not the most convenient method. They preferred meat that has undergone a minimum of industrial processing, and they found it easier to control the freshness and quality of meat that has not been produced industrially.

It also became clear from the focus group interviews that consumers view enhanced meat differently according to the social context that the meat is part of. With regard to special occasions, when eating out or when serving food for guests, the interviewees were reluctant to accept enhanced meat. Although some of them favoured convenient cooking, enhanced meat did not sit well with their idea of good cooking. When eating at good, expensive restaurants or serving food for guests the food is expected to be composed of wholesome products and to be made from scratch, and enhanced meat does not fit into this picture.

On the other hand, the interviewees in general tolerated the idea of enhanced meat better when considering everyday meals. This was the case for everyday meals cooked at home, as well as for meals eaten in cheap restaurants and similar places. Accordingly, most of the interviewees accepted the use of enhanced meat in large kitchens that cater for many people, for instance in hospitals, nursing homes and work canteens. Enhanced meat was seen as appropriate in the everyday context and especially in large kitchens, because shortage of time and money was seen as a basic constraint on everyday meals.

Health risks, unnaturalness and lack of authenticity

The health risks associated with eating enhanced meat were given much attention by the participants in the focus group interviews, and this topic caused speculation and anxiety. Many of the consumers interviewed saw some kind of connection between the consumption of enhanced meat and risks of specific diseases such as cancer, diabetes and allergies. Furthermore, many expected enhanced meat to be connected with other long-term risks that have not yet been uncovered. By contrast, the interviewees rarely expressed concerns about the more immediate and direct risks of diseases caused by bacteria.

The interviewees were especially concerned about the health risks that susceptible people may be exposed to when eating enhanced meat. They were worried about children and about people who already suffer from health problems, especially cancer patients, kidney patients, people with allergies, and elderly people, who in general have poor health. Some interviewees also worried that those who eat enhanced meat might run the risk of protein deficiency.

Nevertheless, some of the interviewees also pointed out potential health advantages of eating enhanced meat. They suggested that consuming enhanced meat could have a slimming effect because the intake of meat is decreased and the intake of water is increased. They also suggested that some elderly people might avoid loss of weight by eating enhanced meat, because it is easier for them to chew this type of meat and therefore they will eat more.

The conceptualisation of enhanced meat as something associated with health risks is linked to an understanding of enhanced meat as unnatural and unauthentic. When the interviewed consumers spoke about enhanced meat they used words such as false, synthetic and alarming, whereas meat that has not been enhanced was referred to as the real, pure, authentic and natural meat. Meat that has not been industrially processed is viewed as safe because it is as basic and untouched as possible, and the interviewees emphasised that they preferred to know what has happened to the meat they are eating and who has done it.

The negative conceptualisation of enhanced meat as unnatural and unauthentic was not only to do with the potential health risks, but also seemed to be an expression of the interviewees' feelings of the loss of the quality and cultural values that are linked to food that has been through only a minimum of processing and that has a known history.

Resources – economic and environmental

Another topic discussed in the focus group interviews was the consequence of the consumption of enhanced meat with regard to different kinds of resources. The interviewed consumers were in general greatly concerned about the pricing of enhanced meat compared with meat that has not been enhanced. However, the interviewees did not have a clear picture of the price level themselves, and they did not get information about this from the interviewer. This frustrated the interviewees because they saw the level of pricing as very important to their general attitude to enhanced meat.

According to most of the consumers interviewed, enhanced meat should be cheaper than non-enhanced meat. They assumed that the cost of production is lower for enhanced meat, because water is added and sometimes poor cuts of meat are used. Some interviewees stated that they would prefer enhanced meat to other meat products if it were cheaper, because price is an important consideration when choosing meat.

Nevertheless, many interviewees believed that enhanced meat might actually be more expensive than non-enhanced meat. They suspect the producers and retailers of enhanced meat to impose surcharges and to increase their profit on enhanced meat, despite the consequences for consumers with regard to cost, health, etc. The interviewees were upset about this and felt that they were being forced to pay for more than they got; they felt as though they were being defrauded.

The interviewed consumers were not very concerned about the questions concerning resources other than economics. However, a few interviewees considered that production and consumption of enhanced meat might influence environmental resources. Some suggested that the environment could benefit from enhancement of meat, because addition of water might decrease the amount of meat consumed. On the other hand, some interviewees reflected on whether enhanced meat will do harm to the environment, because this method of processing meat might demand more resources.

Discussion

Our focus group interviews with consumers provide an insight into a wide range of views on enhanced meat. Despite the fact that the interviewed consumers in general had none or very little prior knowledge of enhanced meat, they turned out to have strong opinions about this type of meat after being introduced to it during the focus group interviews.

In these opinions a clear ambivalence is displayed. On the one hand, the overall impression is that many of the consumers liked the taste, tenderness and moistness of the enhanced meat. They also liked the fact that it is easy to cook successfully, and that it remains tender and moist when cooked. However, there was a clear tendency among the consumers to be anxious about the health risks of enhanced meat, and they were sceptical because they considered it to be unauthentic and unnatural. Along with this the consumers associated this type of meat with foul play, because of the misleading and inadequate pricing, labelling and information provided.

Contradictory views on enhanced meat were expressed because the interviewed consumers had different opinions and weighed arguments differently. Additionally each of the consumers, as mentioned, individually possessed contradictory views on enhanced meat, which led to feelings of ambivalence. This ambivalence, for example, became obvious when the consumers talked about their positive experiences when tasting enhanced meat, and at the same time expressed anxiety about the long-term health risks of eating it.

The different views and the feelings of ambivalence indicate that consumers' views about this type of food product cannot be assessed solely from their experiences of taste and texture. Consumers have a much broader perspective on quality, and they assess food from the viewpoint of a broad range of topics. Naturalness and authenticity are generally very important aspects of the understanding of meat quality, because meat that has been processed less and has a known history is much more highly valued than enhanced meat that has been taken through obscure industrial processes. Authenticity and naturalness are seen as important values in their own right, but are also related to widespread anxiety about the potential long-term health risks associated with eating enhanced meat.

The consumers' feelings of ambivalence were intensified by their experiences of being defrauded and not being given sufficient information. In particular, the misleading and non-informative labelling attracted criticism. The interviewees suspected that the meat industry and retailers produce and sell enhanced meat only for profit and without any thought for the consequences for consumers.

These different views of ambivalence also have consequences with regard to the social categorisation of enhanced meat. Food products are usually positioned in a range of food categories, for example as high quality or discount products, as everyday food or festive food, etc. These categorisations are based on, but also provide a basis for, social life in general. For example the distinctions between everyday and festive occasions are marked by different food products being associated with, or having different meanings in, the different social contexts. The final position of enhanced meat in the categorisation of foods in Denmark is not yet settled. Given the ambivalence of consumers, enhanced meat does not easily fit into the existing categories, and this was demonstrated particularly by the uncertainty and contradictions in the attempts of the consumers to assess enhanced meat differently in various social contexts.

Conclusion

Unprompted, consumers have very little knowledge about enhanced meat, and when introduced to it consumers view enhanced meat as an ambiguous food product. The place of enhanced meat in the food categorisation system in Denmark has not yet been settled, and as such it is not associated clearly either with everyday and discount food or with festive occasions and gourmet food. The consumers liked the enhanced meat because of its taste and texture and because of its advantages with regard to preparing and cooking the meat. However, these positive views were in the main overruled by the consumers' views of enhanced meat as unnatural and unauthentic, and by its potential association with long-term health risks. Moreover, the consumers were sceptical, because absent and misleading information made them feel defrauded.

Consumers request that enhanced meat is labelled and marketed in an honest way. But even if this occurs, and even though this type of product is in some contexts seen as useful, it will probably in most contexts still be viewed as potentially problematic.

Acknowledgements

This study is part of the larger study 'Culinary meat improvement by marination technologies: quality and molecular mechanisms', which is financed by The Danish Food Industry Agency, a part of The Danish Ministry of Food, Agriculture and Fisheries.

References

Carr, M.A., Crockett, K.L., Ramsey, C.B. and Miller, M.F. (2004). Consumer acceptance of calcium-chloride-marinated top loin steaks. Journal of Animal Science 82(5): 1471-1474.

Danish Meat Association (2006). [Enhanced beef demands new frying habits] Marineret oksekød kræver nye stegevaner. URL:http://www.danishmeat.dk/Forside/publikationer/Ny_viden_om_s/2006/Nr__2/Marineret_oks.aspx (read 5 March 2009).

Dörffer, M.T. (2003). [Enhanced meat – a challenge to fresh meat?] Marineret kød – en udfordring til det ferske? URL:http://www.danishmeat.dk/Slagteri_og_Foraedling/Produktkvalit/Spisekvalitet/Publikationer/Marineret_koe.aspx (read 5 March 2009).

Folketingets Lovsekretariat (2006). [Question S 4151] Spørgsmål S 4152. Folketinget, 2 May 2006.

Folketingets Lovsekretariat (2007). [Question 32] Spørgsmål 32. Folketinget, 5 December 2007.

FoodSam (2007). [Marination of meat] Marinering af kød. URL:www.foodsam.dk/data/files/fakta_om/fakta_om_kod/18_Marinering_af_kod_august_2007.pdf (read 21 February 2009).

Hvilsom, F. (2006). [Meat is being pumped with water] Kød bliver pumpet med vand. Politiken 19 April 2006.

Jydske Vestkysten (2005). [Water chicken on the menu] Vandkylling på menuen. Jydske Vestkysten 23 May 2005.

Politiken (2006). [Country – oh water chicken] Land- øh... Vandkylling. Politiken 23 April 2006.

Robbins, K., Jensen, J., Ryan, K.J., Homco-Ryan, C., McKeith, F.K. and Brewer, M.S. (2003): Consumer attitudes towards beef and acceptability of enhanced beef. Meat Science 65(2): 721-729.

Rosenvold, K. (2006). [Tender meat by mechanical tendering and marination] Mørt kød med mekanisk mørning og marinering. Ny viden om oksekød, Slagteriernes Forskningsinstitut 3(1): 1-2.

Søndergaard, M. (2007). [Thema: Chicken] Tema: Kyllinger. Tænk 80: 26-33.

How can consumer trust in organic products be enhanced?

F. Schneider[1], M. Stolze[1], A. Kriege-Steffen[2], J. Lohscheidt[2] and H. Boland[2]
[1]Research Institute of Organic Agriculture (FiBL), Ackerstrasse, CH-5070 Frick, Switzerland
[2]Justus Liebig University Giessen, Institute of Rural Sociology and Extension, Senckenbergstrasse 3, D-35390 Giessen, Germany; flurina.schneider@fibl.org

Abstract

The study presented explores consumer trust in organic food and the effectiveness of enhancing consumer trust by communication strategies on traceability. The research is based on the general finding that trust is one of the most crucial aspects when consumers decide whether to buy or not to buy organic products. However, there are few empirical studies which analyse in detail consumer trust in organic food and the ways it can be enhanced. First, based on a quantitative inquiry involving 600 people in Germany the study presented investigates consumer trust in the different actors involved in the organic supply chain (farmers, processors, traders, labels), through distinct attributed qualities such as benefits for health, ecology and animal welfare as well as the customers' criteria for assessing trustworthiness of the organic products. Empirical data is analysed by multivariate statistics such as cluster analysis to identify distinct consumer groups with respect to their trust characteristics. In a second step qualitative research methods using interviews combined with a visualizing technique will be used. The aim of this method is to better understand the consumers' attitudes towards the supply chain of organic food and the complex construct of trust. The results will help to develop communication strategies for enhancing consumer trust in organic food.

Keywords: consumer trust, organic food, qualitative and quantitative approaches

Introduction

When consumption of organic products should be widened, it will be important to motivate new groups of consumers to buy organic food. A number of previous studies on the consumption of organic food identified trust as one of the most crucial aspects when consumers decide whether to buy or not to buy organic products (e.g. Zanoli, 2004). The importance of trust in buying organic foods can be explained by the fact that consumers generally cannot distinguish organic products from conventional ones by their appearance or taste. Neither before nor after purchase are consumers able to directly discern typical product attributes of organic food such as advantages for the environment, animal welfare or health. When buying organic food, they have to trust in the significance of the organic labels or in the words of the salespersons. They have to trust farmers, processors and retailers not to cheat, and certification bodies to do their job. To speak in terms of Darby and Karni (1973) organic products have a high degree of 'credence attributes' which are not directly observable by consumers. This is in contrast to 'search attributes' which can be discerned by consumers before purchase (e.g. price, colour and size), and 'experience attributes' which can be verified after purchase (e.g. taste and shelf life).

There are relatively few studies which explore trust in organic food. Bech-Larsen and Grunert (2001) investigated credence attributes using the example of organic products. They conclude that in case of credence attributes the credibility of information and information carriers (neutral guarantees) plays a crucial role. Karstens and Belz (2006) and Kaas and Busch (1996) argue for transforming credence attributes in 'quasi-search' attributes such as labels, self-declarations, product brands, corporate brands, personality and internet presence. There are other studies which analyse credibility of assurance schemes such as organic certification (Van Amstel *et al.,* 2008), however, there are very few studies which analyse the credibility as perceived by the consumers. Nilsson *et al.,* (2004) state that the majority of consumers have faith in organic labels, but they are uncertain about what the concept organic is about, and consequently, they require more information. Other authors, however, conclude that information is not enough in establishing trust (Meijboom *et al.*, 2006). Eden *et al.,* (2008) even found that assurance schemes and information may increase scepticism rather than to strengthen trust, as people start to reconsider food production and regulation processes.

While there is general agreement on the importance of trust in decisions on organic food, the investigation of trust in organic food system is still at the beginning. In particular, there is little knowledge on why consumers trust/distrust organic food, what consumers put their trust in, how consumer trust in organic food can be gained and what is the relationship between consumer trust and buying behaviour. Furthermore, there are few attempts to conceptually grasp the construct of 'trust in organic food' by linking it to the broad literature on trust. The aim of this study is therefore to close this gap by an explorative study on consumer trust in organic food. As a first step, we intend to build a comprehensive framework for conceptualising 'trust in organic food'. This framework will be explored, tested and refined in two empirical studies in Germany. On the one hand, consumer trust in organic food is investigated by means of a quantitative inquiry. On the other hand, there will be a qualitative study using interviews combined with a visualizing technique. Special emphasis will be placed on how consumers evaluate trustworthiness of organic products and what the role of traceability is.

This paper begins with an overview of the conceptual reflections, followed by a discussion of the research design and concluding with final remarks.

Conceptualisation of trust in organic food

By looking for a suitable conceptual foundation of 'trust in organic food' it is striking to notice the number of different ways to approach the phenomenon of trust. Trust is investigated in many different disciplines ranging from philosophy, sociology to psychology and marketing research. There is neither complete agreement about its definition, nor about the conditions that determine its development as well as its measurement. Against this background, we explored the literature on trust in organic food to identify basic dimensions, processes and structures of trusting relationships with regard to organic food. Subsequently, we attempt to link these insights to existing trust concepts.

We found that if consumers trust in organic food, different forms of trust may be relevant. Consumers may base their trust on the normative foundation of the organic movement, namely the shared values and convictions related to ecological and healthy food. Furthermore, consumer trust may also depend on personal trust in the competence and integrity of known representatives of the organic movement such as neighbouring farmers, political advocates, doctors, and other consumers. Moreover, a more abstract and institutional form of trust will probably also be involved. Consumers may base their trust on the confidence that all actors of the organic supply chain will act in the desired way, because there are clear guidelines and sufficient control measures. Finally, consumers may also trust in what we term the 'brand personality of organic food'.

With regard to personal trust between the consumers and actors of the organic movement we can refer to a broad set of research. There is a prominent view stating that personal trust can be seen as multidimensional construct consisting of two distinct, but inter-related dimensions: the trusting intention and the trusting beliefs of the trustor (McKnight *et al.*, 2002). The trusting intention is seen as the cognitive, emotional or habitual willingness of the trustor to depend on somebody else in a risky situation. The trusting belief is the associated belief that the object of trust is trustworthy. It is the trustor's perception that the trustee possesses characteristics that would benefit the trustor. In the literature, trustworthiness is often described as consisting of three dimensions: competence (ability of the trustee to act as expected by the trustor), integrity (honesty, promise keeping and acting according to its stated values) and benevolence (caring and acting in the interest of the trustor).

The importance of personal trust in organic food is reported by several research projects. For example, consumers currently state that they have more confidence in small shops where they experience the competence, integrity and benevolence of the sales staff. Or, consumers prefer to buy organic products directly from the farmer to be sure that they really are buying organic food (Zanoli, 2004).

Furthermore, trust in organic food has similarities in what Lahno (2001) describes as trust in an 'organisation' or 'institution'. According to Lahno, institutional trust is predominantly directed to the efficacy of the rules and principles in guiding the behaviour of people. In this regard, consumers trust the actors of the organic supply chain not only for their personal trustworthy characteristics, but also for the existence of efficient guidance and control systems. Some consumers state, for example, that they trust the organic labels because they know that farms are regularly checked. Others say that that they doubt that organic labels guarantee organic production as farmers may try to cheat to earn more money (e.g. Zanoli, 2004).

However, institutional trust is not only about the reliability of some mechanisms ('technical dimension'). Lahno proposes to speak of institutional trust only if this technical dimension is associated with the normative dimension that the rules are actually valid and justified demands and obligations. 'In this way, institutional trust is characterized by the perception of being connected to the people whose behaviour is being determined by the institution, in sharing their respect for the normative foundation of the institution' (Lahno, 2001). Consumers currently mention aspects which refer to this normative dimension of institutional trust: On the one hand, they mistrust the stated benefits of organic food; this is if organic production is really better for human health, animal welfare, taste or ecology. On the other hand, consumers do not consider these advantages as meaningful for their life. Thus, they do not share the normative foundation of organic rules.

So far, we have looked at forms of trust which focus on direct or indirect relationships between the consumers and other actors of the organic supply chain. However, trust in organic food may also emerge from the customers' interaction with the 'brand personality of organic food'. In marketing research there is a growing stream of research which argues that consumer trust in brands may not only come from institutional trust in the brand organisations, but from trust in the 'brand personality' (Wünschmann and Müller, 2006). A 'brand personality' can be defined as a set of human characteristics associated with a brand (Aaker, 1997). The concept is based on the premise that brands can have personalities in the same way as humans. As a result, consumers can perceive a brand as trustworthy, because they regard it as competent, with integrity and benevolent. In other words, consumers can have very similar relationships with brands as with humans and they can trust in brands as much as in humans.

Against the background of the previous considerations, we conceptualise trust in organic food as a multidimensional concept embracing the consumers' willingness to depend on other actors be they farmers, retailers, certification bodies or labels ('trusting intention'), on the one hand, and consumer beliefs in the trustworthiness of these actors ('trusting beliefs') on the other hand. (Figure 1). Consumer trusting intentions and trusting beliefs towards organic food are seen as a result of their general disposition to trust, personal trust in specific actors of the organic movement, and institutional trust in the effectiveness and justification of the organic rules and the associated feeling of being connected to the actors of the organic movement. These different forms of trust are not regarded as independent from each other, but present an interplay between them. Institutional trust is formed in personal encounters with salient representatives, but the existence of institutional trust may also strongly influence whether you trust a person or not.

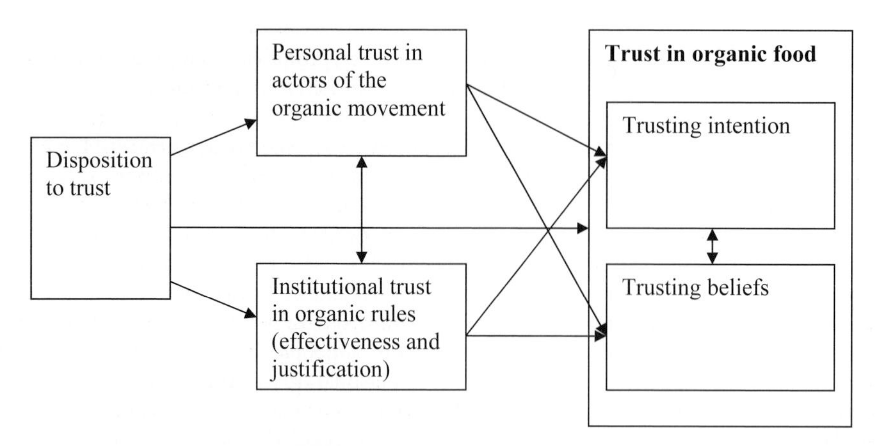

Figure 1. Model on trust in organic food (following McKnight and Chervany, 2001).

Research design

In the following we will give a short overview of the research design of the empirical part of the study. The research project is subdivided in two main tasks:

Task 1, quantitative inquiry

Within task 1 consumer trust in organic food is investigated based on a quantitative inquiry of 600 persons. Data collection will be conducted in April 2009. The inquiry aims to explore different dimensions of trust in organic food. Special emphasis will be placed on the following research questions: To what extend do consumers trust in organic food? In which organic quality attributes (benefits for health, ecology and animal welfare etc.) do consumers trust? In which actors of the organic movement (e.g. farmers, processors, traders, labels) do they trust? How do consumers evaluate trustworthiness of organic products? What is the role of label information, personal contacts and traceability? How can trust in organic food be enhanced?

Empirical data will be analysed by multivariate statistics such as cluster analysis and regression analysis. Using regression analysis we will explore causal relationships between consumer trust in organic food and different trust building variables. By means of cluster analysis we will identify distinct consumer segments with respect to their trust characteristics.

Task 2, qualitative interviews

Based on the results of the quantitative questionnaire the interviewed consumers will be classified according to their extent of trust into three groups: consumers with high, middle and low trust in organic food. Approximately 25 to 30 of them, 6 up to 10 out of each group, will be chosen for the second part of the study. The aim is to get deeper and more detailed information about the influencing factors according to the three trust levels for a better understanding of this complex construct. Furthermore it is expected to get information about trust-building factors to develop trust-building strategies for the actors involved in the supply chain of organic food.

The applied method is the structure-formation-technique. It is a qualitative research method characterized by a two step process, first the investigation of the knowledge of the interviewee about a special theme and second a structure-laying-process (Scheele and Groeben, 1988). The person questioned describes his or her activities and there is a dialogue between the interviewer and the person questioned in which the interviewer can check if he has understood everything in the right way. As a result diagrams are put up which show the contents of the concepts of the person questioned and the formal relations between them. The concepts are linked with the formal relations (Dann, 1992). So the presented study starts with an interview with open questions. The consumers will be asked about their attitudes towards organic food and the evoked feelings, motivations etc. Furthermore they will be asked to discuss their trust in the different actors involved in the organic supply chain and the system of traceability.

In the second step the attitudes and answers identified in the interview will be structured. For this procedure they will be written down on cards during the interview. The possible relations between them (e.g. from very important to unimportant; positive or negative) are also prepared on cards by the interviewer. The result of this process is a structure of the answers and attitudes and the relations between them.

One expected result of this qualitative method is to find out if trust refers to the organic product or to the supply chain of organic foods. Another expected result is to identify the crucial determining factors and characteristics influencing the decision making process to trust and the expectation in the given trust. In particular it may resolve, what are the main expectations of consumers regarding systems of traceability and do these have an influence on trust or not?

Final remarks

The research project will provide deeper insights into the complex construct of trust in organic food from both qualitative and quantitative perspectives. Synthesizing the results of the two streams of inquiry will aid the development and refinement of communication strategies for enhancing consumer trust in organic food.

References

Aaker, J.L. (1997). Dimension of Brand Personality. Journal of Marketing Research 34: 347-356.

Bech-Larsen, T. and Grunert, K.G. (2001). Konsumentscheidungen bei Vertrauenseigenschaften: Eine Untersuchung am Beispiel des Kaufes von ökologischen Lebensmitteln in Deutschland und Dänemark. Marketing 3: 188-197.

Dann, H.-D. (1992). Variation von Lege-Strukturen zur Wissensrepräsentation. In: Scheele, B.: Struktur-Lege-Verfahren als Dialog-Konsens-Methodik. Münster: Aschendorffsche Verlagsbuchhandlung GmbH & Co., S.2-41.

Darby, M.R. and Karni, E. (1973). Free Competition and the Optimal Amount of Fraud. Journal of Law and Economics 16: 67-88.

Eden, S., Bear, C. and Walker, G. (2008). Understanding and (dis)trusting food assurance schemes: Consumer confidence and the 'knowledge fix'. Journal of Rural Studies 24: 1-14.

Kaas, K.-P. and Busch, A. (1996). Inspektions-, Erfahrungs- und Vertrauenseigenschaften von Produkten. Theoretische Konzeption und empirische Validierung. Markeitng ZFP 18: 243-252.

Karstens, B. and Belz, F.-M. (2006). Information asymmetries, labels and trust in the German food market: A critical analysis based on the economics of information. International Journal of Advertising 25: 189.

McKnight, D.H. and Chervany, N.L. (2001). Conceptualizing Trust: A Typology and E-Commerce Customer Relationships Model. Proceedings of the 34th Hawaii International Conference on System Sciences.

McKnight, D.H., Choudhury, V. and Kacmar, C. (2002). The impact of initial consumer trust on intentions to transact with a web site: a trust building model. Journal of Strategic Information Systems 11: 297-323.

Meijboom, F.L.B., Visak, T. and Brom, F.W.A. (2006). From trust to trustworthiness: Why information is not enough in the food sector. Journal of Agricultural and Environmental Ethics 19: 427-442.

Nilsson, H., Tunçer, B. and Thidell, Å. (2004). The use of eco-labelling like initiatives on food products to promote quality assurance--is there enough credibility? Journal of Cleaner Production 12: 517.

Scheele, B. and Groeben, N. (1988). Dialog-Konsens-Methoden zur Rekonstruktion Subjektiver Theorien. Tübingen: A. Francke Verlag GmbH.

Van Amstel, M., Driessen, P. and Glasbergen, P. (2008). Eco-labelling and information asymmetry: a comparison of five eco-labels in the Netherlands. Journal of Cleaner Production 16: 263-276.

Wünschmann, S. and Müller, S. (2006). Markenvertrauen:Ein Erfolgsfaktor des Markenmanagements. In: Bauer, H.H., Neumann, M. and Schüle, A. (eds.) Konsumentenvertrauen. Konzepte und Anwendungen für ein nachhaltiges Kundenbindungsmanagement. Verlag Vahlen, München, pp. 117-133.

Zanoli, R. (2004). The European Consumer and Organic Food. School of Management and Business, Wales.

Theme 8 – Global food security

Global food security: should agricultural research change its perspective?

M. de Lattre-Gasquet[1], P.-H. Duée[2] and S. Duboc[3]
[1]Gasquet, Cirad, 42 rue Scheffer, 75116 Paris, France; marie.de_lattre-gasquet@cirad.fr
[2]Inra, Versailles, France
[3]Common Advisory Committee for Ethics in Agricultural Research, Paris, France

Abstract

Agricultural production and transformation in Europe and North America made extraordinary progress after the Second World War until the 1980s due to consensus among the stakeholders (farmers, unions, agro-industrialists, government, consumers, researchers) about the aims to be reached, and due to agricultural policies, scientific and technical progress, leading to high pressure on the ecosystems. Since the 1980s, emerging diseases and sanitary crises, the growing coexistence of national, regional and international policies, environmental concerns, evolution in proprietary regimes, changes in the academic world, etc. have broken down the consensus on priorities. Recent food riots have revived the question of the capacity of the world's natural resources to feed future generations. Today, according to FAO, more than 920 million people suffer from hunger in the world. But food security does not have only a quantitative aspect, it also has also a qualitative aspect. Obesity and malnutrition are found all over the world which leads to high mortality rates. Increased consumption of animal proteins is leading to high pressure on natural resources. Climatic change is putting extra pressure on food production thus leading to migration and rural exodus. Research has a very important role to play in answering the quantitative and qualitative challenges linked to food security, but in order to do that, priorities need to change. Many of the challenges facing agriculture currently and in the future will require more innovative and integrated applications of present knowledge, science and technology (formal, traditional and community-based), as well as new approaches for agricultural and natural resource management. Innovative institutional arrangements are essential for the successful design and adoption of ecologically and socially sustainable agricultural systems. The paper describes the work done by the Joint Ethics Committee of Inra (Institut national de la recherche agronomique) and Cirad (Centre de coopération internationale en recherche agronomique pour le développement) on the question of food security.

Keywords: food security, ethics, agricultural research

Introduction

Food insecurity, which is one of the many facets of poverty, returned to the spotlight on the global stage in the early years of the millennium. Unfortunately, this merely served to remind those working to improve living conditions of poor populations in the least advanced countries – and also in the developed countries – of the task facing them. Inra (Institut national de la recherche agronomique) and Cirad (Centre de coopération internationale de la recherche agronomique pour le développement), two French agricultural research institutions, are among them.

The first section of this paper discusses how the concept of food security has evolved over time and geographically. The second section presents the question that the Common Advisory Committee for Ethics has been asked to look at. The third section gives a historical perspective and looks at the way Inra and Cirad have considered food security. Lastly, some of the recommendations proposed by the Committee are presented.

Food security: a multi-dimensional concept and a worldwide problem

Food security is a flexible concept as reflected by the many attempts to define it, particularly as related to research and policy practice. Food security as a concept originated in 1974, in the discussions of international food problems at a time of global food crisis. The initial focus of attention was primarily on food supply problems, i.e. to assure the availability and to some degree the price stability of basic foodstuffs at the international and national level. In 1983, FAO expanded its concept to include securing access by vulnerable people to available supplies, implying that attention should be balanced between the demand and supply side of the food security equation. In 1986, the World Bank introduced the distinction between chronic food insecurity, associated with problems of continuing or structural poverty and low incomes, and transitory food insecurity, which involved periods of intensified pressure caused by natural disasters, economic collapse or conflict. The concept of food security was then elaborated in terms of 'access of all people at all times to enough food for an active, healthy life'. By the mid-1990s food security was recognized as a significant concern, spanning a spectrum of issues from the individual to the global level. The definition was broadened to incorporate food safety and also nutritional balance, reflecting concerns about food composition and minor nutrient requirements for an active and healthy life. Food preferences, socially or culturally determined, became a consideration (FAO, 2003). Four main dimensions of food security can be identified: physical availability, access to food, food utilization, and stability of the other three dimensions over time. For food security objectives to be realized, all four dimensions must be fulfilled simultaneously (FAO, 2008).

Today, despite substantial improvement in terms of agricultural productivity, the situation is very worrying. In 2009, nearly one billion persons suffer from hunger in the world, and many more suffer from malnutrition leading to problems such as obesity and diseases. Hunger and obesity are two expressions of the same problem facing poor people: either they do not have enough money to buy food, or they can only buy calories and unbalanced diet at a low price (Chourot, 2008). In the South, recent food riots have brought back the question of our capacity to feed the world in the medium and long-term, but even in developed countries, eating is becoming a financial issue for a growing part of the population and poor people have to take advantage of free meals given by a variety of institutions when they exist. On top of that, a succession of sanitary crises, emerging diseases and zoonoses (mammalian neurodegenerative diseases, avian influenza, etc.), of fraudulent practices (dioxin, melamine, adulterated sunflower oil, etc.) and controversies related to GMOs have led to increasing suspicion towards the actors of the food chain despite increasing European and international rules and regulations.

It should be added that food preferences are part of cultural models. Food habits evolve over time and spatially due to different factors like migration, social status, work organisation, health policies, advertisements and communication, training, etc. (Poulain, 2002).

The Common Advisory Committee for Ethics in Agricultural Research

The Common Advisory Committee for Ethics in Agricultural Research was created in 2007 and reports to the Chairmen of the Board of Cirad and Inra. It comprises fourteen members exterior to these research agencies and is chaired by Mr. Louis Schweitzer. This joint committee follows on from the Inra-IFREMER Ethics and Safety Committee on Agricultural Research Applications (COMEPRA) founded in 1998 and which ended its mission in 2007 on the one hand, and from the Ethics Committee of Cirad founded in 2001 and which ended its mission in 2005, on the other hand. Its missions are to reflect, advise, raise awareness, and – if need be – alert. The committee studies ethical issues that may arise from research activities in and outside France in the areas of agriculture, food and nutrition, the environment, and sustainable development, especially in aspects touching on the link between science and society. It takes into account the specific missions and activities of Cirad and Inra, particularly concerning research for the development of Southern countries. The committee formulates opinions and recommendations. The committee handles only ethical issues, as questions of professional conduct fall under the remit of Inra and Cirad management.

The ethics of research and the search for ethics is 'a necessary and endless worry' (Sicard, 2006). Ethics tend to establish criteria to judge if an action, in its core, its motives, its means and its consequences, is good or bad. Ricoeur's (1990) definition of ethics is 'The goal of a good life with and for others within just institutions'. Ethics concerns everyone when choices have to be explicit and responsibilities taken. Encouraging and engaging in ethical reflection is very important for research organisations.

In 2008, the chairmen of Inra and Cirad decided to ask the Common Committee for Ethics to look at the question of food safety since the changes in the evolution of the global, the ecological, the scientific and the institutional contexts enhance the responsibilities of research institutions. Research can no longer be expected to be only a knowledge producer and a technical solution provider to socio-economical issues. The main issues of concern leading them to raise ethical questions are the following:

- The evolution of the global context:
  - Climatic variations are not new but their effect is potentially increased by population growth. What can research do about the resulting rural exodus, migratory movements and food insecurity affecting millions of people?
  - There is increased interdependence of problems, as shown by the on-going crisis, and the number of groups which are outside the economic sphere is growing. What should be done for groups like very small farmers and nomads? Is it possible to insure greater equity?
  - Uses of pesticides in farm production, uses of new technologies and produces, genetically modified foods, and spillovers from animal diseases to humans have made consumers apprehensive of the impacts of food on human health. Surroundings of production processes, such as animal welfare and environmental concerns, have become more relevant to consumers (Von Braun, 2005). Diseases are transferred from animal to man and vice versa, and have considerable economic consequences. Research is questioned by the growing lack of trust in food safety.
- The evolution of the ecological context:
  - Resources, such as water or soil, are becoming rare, thus limiting agricultural production.
  - Agricultural research in the context of sustainable development is evolving quickly. For many years, productivity was the principal objective. Sustainable development is now a widely recognized objective. That means that agriculture must be at the same time competitive, provide attractive jobs, respect environment, and allow for safe and sound nutritious food. How to make sure that the objectives of sustainable development do not put aside the urgency of food security?
- The evolution of the scientific context:
  - Since agricultural research is being conducted increasingly by scientists affiliated with private corporations, consumer groups are suspicious of research outcomes (Waast, 2002), even if they come from the public sector.
  - Growing scientific complexity and increased interactions between disciplines have consequences on scientific methodological approaches and practices. There is a need for epistemological reflectivity on behalf of researchers.
- The institutional context.
  - The first goal of the Millenium Development Goals (MDG) that the members of the United Nations unanimously adopted in 2000 is 'Halve the proportion of the people in extreme poverty and suffering from hunger between 1990 and 2015.' Agriculture is central to achieving the MDGs and the recent International Assessment for Agricultural Science and Technology for Development (IAASTD) shows that we are far away from reaching them. Which obstacles prevent the elimination of hunger and malnutrition? Is it more efficient to import cheap food? Should developed countries diminish subsidies and trade barriers to allow more imports from

developing countries? How to continue supporting farmers in the North (i.e., the Inra's main partners), and not necessarily the largest farmers, while at the same time ensuring market access for poor farmers of third countries (i.e., the Cirad's main partners)?

The Inra and the Cirad's approach to food security: a historical perspective

The origins and histories of the two institutions are very different, but a historical perspective allows us to see some common concerns and similarities. Inra was created after WWII to develop and modernize agricultural production, whereas Cirad is the culmination of a process of evolution of French agricultural research for the tropical countries.

Inra was created in 1946 when 33% of the active population in France was not self-sufficient. The mission that was then given to Inra was to develop and modernize agricultural production, especially through genetic improvement of plants and animals. In twenty years, France became the first agricultural producer in Europe and the first exporter. Beyond the efforts made by Inra, there was also consensus among the stakeholders (farmers, unions, agro-industrialists, government, consumers, researchers) about aims and adapted agricultural policies. However, improvements also meant high level of pressure was placed on ecosystems.

From the 1960s, the ethos of Inra was similar to the initial objectives of the Common Agricultural Policy (CAP): improvement of agricultural productivity, equitable lifestyle for all the rural population, market stabilisation, security of supplies, reasonable prices for all consumers. This gave an ethical caution to agricultural productivism. At the end of the 1970s, the environmental problems caused by the intensification of production systems and high differences in agricultural revenues led to the loss of the moral caution of agricultural productivism. In the 1990s, the CAP changed orientation from agricultural commodity support to broad rural development objectives. Today, Inra's research is guided by development in scientific fields and focuses on worldwide challenges related to food and nutrition, the environment and land use facing the agriculture and agronomics, such as climate change, competition between food and non-food crops, the exhaustion of fossil resources, etc.. Global agricultural research must address the sustainable food supply.

French tropical agricultural research was born over a hundred years ago (policies of 'enhancement' of the French colonies), and really took off once the importance for the national economy of expanding and promoting tropical and intertropical agriculture and agricultural products was acknowledged. Some forty years ago, French agricultural research was primarily the reserve of nine institutes. All of them had a dual vocation: specialized research and scientific and technical cooperation with producing countries. Most of them dealt with export crops; only one of them dealt with food crops. Their ethos was based on the development of export crops and enhancement of local territories.

Colonial science left a considerable legacy in terms of knowledge, organizational models and strategic choices (agriculture and health favoured). This legacy was inherited and enhanced after the countries' independence (Waast, 2002). During that period, in the 1960s and 70s, the attention of agricultural policies and research was geared towards quantity as shown during the 1974 World Food Conference.

The founding of Cirad, in 1984, was concomitant to the appearance of a new generation of researchers. They sustained a new process of scientific production: 'national science', whose main characteristics were as follows: science is for the public good; the State funds most of the budget; the direction of that science is determined by the country's most pressing needs; research scientists are civil servants and have the right to pursue careers; they are imbued with national values as well as scientific ones; besides the peer community, the recipients of the product are principally the public authorities (CIRAD, 2001).

The direct users of the product were hardly involved. The Cirad's mission took these characteristics into account; there was a switch from sector and biotechnical approaches to a more global approach of agriculture and rural development. The Cirad's researchers became an instrument of 'public aid for development' and to switch from an attitude of 'substitution' (doing the work for) to 'cooperation' (doing the work with).

This system lasted a few years, until signs of a profound change began to bubble through. The free market ethos meant that States reduced their intervention. The expected source of progress became innovation in private companies and no longer the discoveries of science. Agriculture and food security were no longer regarded as a priority nor was public aid for development. At the beginning of the 1990s, Cirad set increasing priorities on the preservation and improvement of the natural environment. Generally speaking, a rift has opened up between the researchers with the 'national' ethos attached to their old established values and researchers open to the 'market', who are paid for their service and who are connected to the worldwide sectors working on leading speciality subjects. At Cirad, some researchers continued to be guided by the desire to help the 'third world', others were guided by the desire to publish, and others started to operate as if they were in a bureaucratic framework. This led to debates about rationality oriented by finalities and rationality oriented by values.

In 2008, agriculture and food security return as research priorities when the World Bank underlined the importance of the contribution of agriculture for development and when food riots erupted across the globe. Accordingly, in the 2008-2012 strategic vision of Cirad (Cirad, 2008), three out of six strategic lines dealt with food security and food consumption: helping to invent ecologically intensive agriculture to feed the world, innovating to make food accessible, varied and safe; supporting public policies aimed at reducing structural inequality and poverty.

As part of a global sustainable development perspective, Inra and Cirad have recently founded the 'French Initiative for agricultural research' to promote, in connection with partners from Northern and Southern countries, a joint overview of the changes taking place in agriculture, food and nutrition, environment and the rural world, and to propose scientific projects.

Recommendations made by the Common Ethics Committee

Before making recommendations, the Committee defined its values and positions. These are:

- Our universe is reasonably limited: our responsibility is to pass it to future generations without major and irreversible degradations. This principle implies that we consider at any one time both long and the very long-term perspectives.
- The world constitutes a system. An action on one element of the system influences the other parts of it. For this reason, the problems have to be preferably addressed at a worldwide level.
- The universal value of human dignity has to be the first consideration and has to find concrete application in the Committee's recommendations. Solidarity and reciprocity within human relationships have to be favoured.
- The idea of progress socially, economically and technically speaking has to be deeply analyzed, explicitly chosen and equally shared.

In view of the global situation and the history of Cirad and Inra, the Committee is currently making recommendations relative to the following elements:

Carrying out food security-oriented research

The first recommendation of the Committee is that Inra and Cirad should take full consideration of the fact that they carry out mission-oriented research. They should orient more of their research towards local and global food security while promoting ecologically-sustainable practices. This could promote interest and facilitate consensus amongst other stakeholders. Improving food security in a sustainable fashion is a public good.

Mission-oriented research implies research to improve the knowledge relative to an object or a process, whether physical, chemical, biological, economic or social; the lack of knowledge is considered the limiting factor to harmonious economic, social and environmental development. It shows explicitly the linkage between knowledge production and the problems of societies. Research questions evolve with time and space. Gearing research towards food security requires carrying out both fundamental and applied research. This approach means that Inra and Cirad has full responsibility over all stages of their work, and even after their work is finished. More precisely, this recommendation means that:

- During the elaboration of research questions, food security challenges should be estimated. The diversity of food patterns should be increasingly recognized and promoted by research as this is realistic from an economical point of view and also culturally beneficial.
- During the research, working with local partners and reinforcing the implication of rural and urban actors and consumers in research should be improved. In 2004, the Ethics Committee of Cirad made a recommendation on the conditions of intervention in rural societies of developing countries. As food security problems are found in rural and urban conditions, in the North and the South, the Committee recommended implication of local groups in food security oriented research.
- At the end of research projects, impacts should be evaluated, decision makers alerted, and light shed on civil society discussions.

Using foresight, especially the Agrimonde Platform, to detect signals relative to Food Security

Inra and Cirad have launched a foresight platform on farming and food systems worldwide in 2035 (Agrimonde). The study serves to foresee the role of French and European agriculture in different global change scenarios and to pinpoint the fundamental issues with which agricultural research will be faced. It is based on the results of the Millennium Ecosystem Assessment and fits in with the work of the International Agricultural Assessment of Science and Technologies for Development (IAASTD). The Committee recommends the use of Agrimonde to maintain a high level of attention on food security issues, and to develop a communication and alert system around it.

Training researchers for collective ethical reflection

Researchers, from Inra and Cirad, but also from partner institutions especially in the South, have to become more familiar with the concepts of ethics and to think about the ethical consequences of their acts. The Committee recommends the organization of training on these topics.

Conclusion

The Common Committee for Ethics does think that research in general, and Cirad and Inra in particular, should change their perspective on food security. It will present its report to the Boards of Trustees. Within each institution, in-house debates will be organised around the recommendations and therefore this should lead to increase awareness of the questions and recommendations.

References

Cirad (1991). Le projet d'entreprise du Cirad. Renouveler notre coopération dans un monde qui change.

Cirad (2008). Strategic vision 2008-2012. Available at http://www.Cirad.fr/en/le_Cirad/pdf/Cirad_Strategie_GB_web.pdf.

FAO (2003). Trade Reforms and Food Security. Conceptualizing the Linkages. Chapter 2. Available at http://www.fao.org/docrep/005/y4671e/y4671e00.htm#Contents.

FAO (2008). Basic Concepts of Food Security. Available at http://www.foodsec.org/docs/concepts_guide.pdf.

Chourot, J.M. (2008). La fracture alimentaire. Projet No. 307, novembre 2008.

Poulain, J.P. (2002). Manger aujourd'hui. Attitudes, normes et pratiques. Paris: Ed. Privat.

Van Braun, J. (2005). Agricultural Research – On ethics and Responsibility of Science for Poverty Reduction and Food Nutrition Security. Paper presented at Deutscher Tropentag, October 11-13, 2005, Hohenheim.

Waast, R. (2002). The State of Science in Africa. Available at http://www.ird.fr/fr/science/dss/sciences_afrique/pdf/synthese/english/an_overview_english.pdf.

The food professional's dilemma: a healthy diet or a healthy profit?

A.E.J. McGill
The International Union of Food Science and Technology (IUFoST): Future for Food, 89 Melvins Road, Riddells Creek, Victoria 3431, Australia; albert.mcgill@futureforfood.com

Abstract

The varying roles of food professionals are described and their involvement in the provision of food dietary guidelines and in the development of new food products is explained. The conflict that may be set up in these differing tasks can lead to a professional dilemma. An explanation of the dichotomy between sectors of the food professions indicates how the dilemma may occur and also how it might be resolved. A more integrated approach to the challenge of development of foods and diets for healthy living is suggested.

Keywords: dietary guidelines, new product development, conflict resolution

Introduction

Most professional food scientists are employed by the food industry and have their professional status ratified by their national institution. In the United Kingdom that is the Institute of Food Science and Technology (IFST) and in the United States it is the Institute of Food Technologists (IFT). All of these national institutions have codes of conduct and guidelines for ethical behaviour and most of these institutions are members of the International Union of Food Science and Technology (IUFoST). These member institutions or Adhering Bodies (ABs), as they are termed, share their technical expertise and compare professional standards. Most countries, recognizing the relationship between food and health, have developed national dietary guidelines aimed at improving their nation's health and preventing premature death. Food professionals are involved in the development of these guidelines. Despite every effort so far, most developed and many developing countries appear to be unable to prevent the growth of obesity and the spread of type 2 diabetes among their populations.

The global food industry has developed rapidly over the last century and mergers and acquisitions have produced large monopolies and, more recently, very large retail enterprises with great purchasing power and market influence. Their success has been fuelled by increasing sales and the consistent high profitability of processed foods, many high in fat, sugar and carbohydrates, inconsistent with or directly contrary to dietary guidelines. The development of many of the fast food and instant commodities presently available for purchase has been in the hands of food professionals.

The dilemma facing many food professionals in the industry is how to apply their skills and abilities in the development and marketing of products that may negate the very dietary guidelines they know can bring health and wellbeing to consumers, including themselves.

Training food professionals

In any consideration of this process there must be some agreement as to the definition of both 'profession' and 'food'. The New Shorter Oxford English Dictionary (1993) defines 'profession' as 'A vocation, a calling, especially one requiring advanced knowledge or training in some branch of learning or science, specifically law, theology or medicine; generally, any occupation as a means of earning a living' and a 'professional' as 'Engaged in a profession, especially one requiring advanced knowledge or training'. 'Food', from the same source, is defined as, 'Substance(s) (to be) taken into the body to maintain life and growth, nourishment; provisions, victuals.' These definitions are broad and it follows that the compass of food professionals will be great. However, in most instances a basis of scientific study at a university of three to five years duration, to degree standard or above is a minimum requirement for membership of most food related professions. Full professional status requires proof of experience in the area of professional competence, usually through approved internships, supervised employment and, in some cases, further examination of the candidates' professional, rather than academic, knowledge. Whilst this process is common for those entering the medical, dietetic, nutritional, food science and technological sectors, some areas, particularly in the hospitality related areas, may rely much less on academic rigor and much more on practical and artistic skills. Nonetheless, each related professional association will have its own rules of membership and related codes of professional conduct, whose failure to observe may result in expulsion from the profession and loss of employment. Many of these professions operate both nationally and globally, where for example the Institute of Food Science and Technology (IFST) from the United Kingdom, the Institute of Food Technologists (IFT) from the USA and the Australian Institute of Food Science and Technology (AIFST) register professional members from their respective countries but are, themselves, members of the International Union of Food Science and Technology (IUFoST), a global organisation. Recognition of professional status in one country may well give similar recognition in many others, where common standards of training and conduct apply.

Development of dietary guidelines

All governments have concerns for the health of their populations, not only as a matter of general wellbeing, but also as a means of improving productivity and reducing the costs of medical care and welfare payments. As part of the strategy for ensuring good health many governments develop health policies. Such policies often emanate from surveys of consumers and their use of food. In the UK the National Diet and Nutrition Survey (NDNS) was established as a joint venture between the Ministry of Agriculture, Fisheries and Food (MAFF) and the Department of Health (DH) in 1992, the role of MAFF devolving to the Food Standards Agency (FSA) in 2000. The surveys are used 'to develop nutrition policy at both national and local levels and to contribute to the evidence base for Government advice on healthy eating.' In addition the surveys '. . . are designed to provide detailed information on the current dietary behaviour, nutritional status and oral health of adults in Great Britain.' From the results of the surveys an Advisory Committee develops the national dietary guidelines. A similar approach is used by the United States government, placing the task in the hands of the Department of Health and Human Services (HHS) and the Department of Agriculture (USDA). It is important to note the requirements for membership of the US Dietary Guidelines Advisory Committee (Federal Register, 2008). Committee members should be '. . . knowledgeable of current scientific research in human nutrition and be respected and published experts in their fields. Familiar with the purpose, communication and application of Dietary Guidelines and have demonstrated interest in the public's health and wellbeing through their research and/or educational endeavours. Expertise in, but not limited to: Cardiovascular disease, cancer, paediatrics, gerontology, epidemiology, general medicine, overweight and obesity, physical activity, public health, nutrition biochemistry and physiology, nutrient bioavailability, nutrition education, and food safety and technology.' The paucity of representation of food science and technology or other elements of food services should be noted against the dominant presence of medical and nutritional specialists. In examining the profiles of the members of the Dietary Guidelines Working Committee, Australia (National Health and Medical Research Council, 2008), no member had any qualifications in food science, technology or service and no experience in any part of the food industry. A similar pattern emerges for the United Kingdom committee. These food professionals concentrate on the whole diet outcomes as reflected in the populations' health.

Development of 'new' food products

A food retailing organisation is required to attempt to increase profits continually. To achieve this there has to be a gain in market share and/or and increase in margins. The first scenario requires the development of 'new' products and the second requires increases in efficiency or lower commodity costs to the manufacturer. Such pressure can result in the demise of farming enterprises and the failure of processing companies.

In developing new products it must be realised that for the food market >95% of all new products fail. However, for those that succeed there is great market share to be gained, so the risk is considered worthwhile. New products can emerge from different sources, some from the discovery of novel ingredients or processes that match a perceived need. The majority come from identification of opportunities by the marketing departments of either the retail organisations or their suppliers. Frequently the food technologist is requested to formulate a product to a market design niche either on intrinsic quality or on cost, sometimes both. When there may be a related health claim for dietary advantage, particularly of a weight-loss nature, nutritionists may be involved. In most cases there is not much attention, if any, paid to the product's role in a healthy diet. This is not true for weight-loss meals, but this represents a small part of the industry's product range. Whilst 'low fat', 'reduced sugar', 'low salt', 'no artificial colours', 'low cholesterol' are all labels that attract the eye of the consumer, it requires much attention to the small print on the label to ascertain how this particular product will affect the diet as a whole. A label of 'low fat' on apples is always correct but of little value unless a diet is very rich in apples at the expense of high fat components.

The development of 'new' food products lies in the hands of food scientists, technologists, chefs and food service professionals, influenced greatly by the marketing exercises of the employing organisation. There is some evidence of the use of the healthy eating concept to promote some whole meal ideas (Condrasky *et al.*, 2008) but generally the focus is on a product, not a diet.

Professional divergence

In considering the roles of food professionals in the development of dietary guidelines and of new food products it can be seen that there is a divergence of both the roles and the perspectives of two separated groups. In the dietary guidelines task those professionals involved are drawn from the medical, dietetic, nutrition and health sector of the continuum. In spite of all their work with dietary advice, the increase in illnesses such as obesity, cardiovascular disease, late onset diabetes, stroke and hypertension continue and even seem to accelerate in parallel with increasing affluence. Thus guidelines, however well designed, without implementation of appropriate strategies, yield no favourable results.

By comparison, the tasks of product development, which is the development end of food processing and service, the professionals are drawn from the food science, technology, engineering and hospitality sector of the same continuum. As such any dilemma may be more apparent than real as few professionals from one sector cross over to work in the other. The effects of individual food items or even meals is chronic rather than acute (unless allergies are at work) and all are guilty of over-indulgence at some time. The event only becomes a health problem when it becomes a habit. This issue was reported some time ago in relation to fast-food and the diet (McGill, 1989).

The ethical stance

Lest it be considered that the explanation of divergence amongst food professionals salves any conscience, there must be some consideration of the effect of Codes of Professional Conduct of the various professional bodies and any impacts they might have on the behaviour of consenting professionals. In the case of one example, the IFST's Code of Professional Conduct (IFST, 2005), which in total occupies 36 pages of text, a significant requirement is for the professional member 'to take legitimate steps through proper channels to ensure (or assist in ensuring) the wholesomeness of any food with which he or she is concerned'. If the definition of 'wholesome' is accepted as 'conducive to general wellbeing; beneficial, safe; promoting physical health, and formerly also, curative and medicinal', then the requirement on the professional is significant with regard to health. Further examination shows that the guidelines of the Code deal specifically with Nutrition Value. 'The total diet of any consumer should be such as to provide and adequate balance and amount of known nutrition requirements. Each consumer's choice of a number of foods, and of quantities of each, however, represents one of the virtually infinite number of permutations and combinations. Such choice by the consumer is entirely outside the control of the food scientist or technologist concerned with a particular food or group of foods. With certain exceptions, discussed below, therefore, it is not normally possible to specify that an individual food must have particular nutrition characteristics. The exceptions are as follows:

- Foods for which legislation specifies minimum nutrition standards.
- Foods for which nutrition claims are made.
- Foods which are generally recognised as being valuable sources of specific nutrients.
- Novel food products which may significantly replace foods of nutrition significance.
- Food products intended for particular nutrition purposes.

How then is the food professional engaged in product development or even manufacture, able to continue to work and keep within the Code of Conduct? Another element of the Code states that the Professional should behave so as 'to respect any confidence gained in his or her professional capacity.' Some comfort must be taken from within the detail and the reminder that 'no one food is a diet' and 'no diet is all of health'.

A solution by integration

An experience from the IUFoST conference held in Seoul, South Korea in 2001, may give some direction to a solution of the apparent dilemma. One of the keynote speakers at a plenary session was the then President of the Chinese Peoples' Republic Medical Association. His inclusion seemed to be more for political than scientific reasons as this was a food conference. Although he had to speak through an interpreter, he recognised with some amusement, our irritation at his apparently inappropriate inclusion in the programme. He observed that in our history, no medical practitioner had been included as a speaker, yet in his country, no medical doctor would consider the treatment of a patient before first examining their diet and then would begin any treatment required through dietary manipulation in the first instance.

A way must be found and strategies deployed to bring a closer integration of the work of food professionals in the development of effective dietary help to improve the health of all people. There are signs of a more co-ordinated approach to improving health through good diet and the right foods in the recent discussion paper from the UK government (The Strategy Unit, 2008).

References

The Strategy Unit, U.K. Government Cabinet Office (2008). Food: an analysis of the issues. Revision A 29 January 2008. http://www.cabinetoffice.gov.uk/upload/assets/www.cabinetoffice.gov.uk/strategy/food/foodanalysis.pdf. Accessed 6 March 2008.

Condrasky, M., Warmin, D., Wall-Bassett, B. and Hegler, M. (2008). Building the Case for Healthy Menus. Food Technology 62: 47-59.

Federal Register (2008). US Dietary Guidelines Advisory Committee 68: 26280.
Institute of Food Science and Technology (IFST) (2005). Code of Professional Conduct. http://www.ifst.org 36pp. Accessed on 31 March 2009.
McGill, A.E.J. (1989). Nutritional Value of Fast Food. In: Bush, P.B., Clarke, I.R., Kort, M.J. and Smith, M.F. Proceedings of the National Food and Nutrition Conference. Technikon Natal Printers, Durban South Africa. ISBN 0-958-3062-1-4, 175pp.
National Health and Medical Research Council (NHMRC) (2008). Membership list, Working Committee, 2008-2010. http://www.nhmrc.gov.au/your_health/healthy/nutrition/committee.htm#mem. Accessed 31 March 2009.
Brown, L. (ed.) (1993). New Shorter Oxford English Dictionary. Oxford University Press, Oxford, United Kingdom. ISBN 0-19-861134-X. 3801 pp.

Food insecurity, human rights, and gender inequalities

V. Sodano
University of Naples Federico II, Department of Agricultural Economics, via Università 96 80055 Portici, Napoli, Italy; vsodano@unina.it

Abstract

In order to cope with the recent food crisis, FAO, the World Bank, US and EU governments have suggested some prescriptions, but these may only open new opportunities for corporate power to continue the commodification of food initiated by the globalization process and the neoliberal agenda. The paper reviews the different interpretations of the 2008 food crisis and shows how the 'official' interpretations and suggested interventions fail to capture the real nature of the food insecurity problem that is political and ethical before being economic. Stemming from the Nussbaum's capability approach the paper asserts that the problem of food insecurity cannot be successfully addressed without considering food as a human right, and, accordingly, tailoring food and agricultural policies based on the principle of food sovereignty. Particular attention is given to the role of gender inequalities as cause of food insecurity in many poor countries.

Keywords: food crisis, women, neo-liberalism, capability approach

The 2008 food crisis: mainstream and heterodox interpretations

Starting from March 2007 until April 2008, grain prices sharply rose. In one year wheat rose 77% and rice 16%, with peaks registered at the beginning of 2008, when for example rice prices soared by 141% and some types of wheat by 150%. Simultaneously in 2008 world grain stocks reached their lowest level for twenty years. As a first dramatic effect the poorest people living in countries dependent on food imports faced shortages and hunger, as a result of not being able to afford higher food costs. Food riots erupted in countries all over the world, with numerous victims registered in Haiti, Cameroon, Egypt, Philippines, Cote d'Ivoire. Also in richer countries, the poorer members of the population who devote a higher share of their income to basic goods suffered an extraordinary high increase in their food bills. Soon after these events media, governments, international bodies and civil society all over the world have addressed the food crisis in order to give explanations, forecasts, and policy advice. The FAO conference held in Rome in June 2008 was the supreme show of 'international community' interest in facing the food crisis.

Explanations of the current food crisis given by participants at the FAO Conference (FAO, 2008), attributed different relative importance, but did agree on at least two causes: (1) the structural changes in demand associated with the high economic growth rate of the emergent capitalistic countries (China in particular); (2) the strong pressure on the energy market, this latter aspect inducing both rising costs of the very fuel dependent food system and a strong competition between food/feed and biofuel crop cultivation. With regards to a third cause, the role of the financial market crisis and its effects on the grain futures market, there was instead a strong disagreement. Together with the price increases in 2007 and at the beginning of 2008 there have been some distortions in the futures markets. What has been observed is that futures prices have risen faster than spot prices, the basis becoming larger and more volatile. Furthermore, on many occasions futures prices have been far higher than spot prices near the expiration day. Notwithstanding this evidence and the very strange coincidence between the food and financial crisis, the latter has not been bound in such a close relationship, especially in official documents by international bodies and mainstream academics.

In contrast to the 'official' interpretation of the crisis, heterodox analysis, as reported by ECTgroup and PANAP (ECT group, 2008; Guzman, 2008) identifies 3 important underlying points missing in official documents of FAO, national governments and the World Bank (WB) that should be considered when trying to understand and cope with the food crisis.

The first point is that the food emergency did not emerge overnight, and did not begin with record-high price. It has been affecting poor countries for 20 years already. In the early 1960s developing countries had an overall agricultural trade surplus approaching $7 billion per year (FAO, 2004). By the end of 1980s the surplus has disappeared and many countries became net importers of food. This shift has been the consequence of US and European policies that have favored corporate agribusiness by keeping commodity prices low, dismantling trade barriers and marginalizing millions of small scale farmers.

The second point is that there exists a strong food-financial crisis nexus. The reason for food 'shortages' has been speculation in commodity futures, following the collapse of the financial derivatives markets. Desperate for quick returns, dealers have been taking trillions of dollars out of equities and mortgage bonds and have ploughed them into food and raw materials. The amount of speculative money in commodity futures ballooned from US$5 billion in 2000 to US$175 billion in 2007. It is called the 'commodities super-cycle' on Wall Street and its latest illustration is the 2008 'land grab' by rich governments and corporations (GRAIN, 2008).

The third point, finally, is that whereas shortage of supply has been pointed at as a main cause of the recent price surge, this may not be the case. Looking at actual data and forecasts on production and consumption, production seems to outpace consumption, on average on a two years basis, for all types of food. Price rises have been driven by the international food trade, notwithstanding the fact that global food trade has been estimated to be only around 10% of global food production. Because global food trade is controlled by a few TNCs that have gained exceptional profits from the last price peaks (as reported by Lean, 2008, Cargill's net earnings soared by 86 per cent from $553 million to $1.030 billion over the same three months, and Archer Daniels Midland increased its net earnings by 42 per cent in the first three months of the same year) it is likely that high prices have been the consequence, besides the speculation on financial markets, of the exercise of a strong market and buying power by these leading companies.

In other words the heterodox interpretation contends that global food crisis is political-economic in nature and not the mere consequence of unbalanced supply-demand movements. According to this view, the food inflation that has pushed millions of people into poverty and worsened the life of the 2.5 billion people already living on less than $2 a day, has been the consequence of: (1) excess of market/ buying power exercised by the big corporations of the agribusiness; (2) process of financiarization of the world economy, that has made food commodities markets vulnerable to financial crisis; (3) twenty-five years of lasting neoliberal policies that have worsened inequalities and created food import dependence in less developed countries.

According to the official interpretation of the crisis, FAO, WB and US and EU governments suggested the following prescriptions to cope with the food crisis: further trade liberalization; enhancing agriculture productivity by shifting from smallholders farms to labor-intensive commercial farming; relying on the private sector as provider of agricultural services; promotion of innovation through science and technology; developing high-value markets (i.e. food sold through supermarkets) for domestic consumption; facilitating input markets in order to assure better access to improved seed and fertilizers; improving the land market to facilitate agriculture consolidation processes; enhancing the performance of producer organization to achieve competitiveness of smallholders; linking local economies to broader markets and shift from self-consumption and self-employment to production for the market and to wage employment; investing in safety nets for poorest people, preferring targeted cash transfers and in-kind food distribution.

Most of these suggested interventions are likely to continue commodification of food that has been initiated by the globalization process and the neoliberal agenda accordingly with the Washington Consensus 'credo': privatization, liberalization, deregulation, decreasing public social expenditure. As far as they reinforce the true causes of the food crisis, – i.e. corporate power, neoliberal ideology and financiarization- they are unlikely to prevent further future food crisis and promote food security. When the true causes of the food crisis are taken into account other forms of interventions should be considered, based on a different perspective that would recognize the limits of the neoliberal project and place human rights and food sovereignty at a premium.

Food insecurity: the human right and capability approaches

In article 25 of the Universal Declaration of human rights the human right to adequate food is explicitly recognized as part of the broader human right to an adequate standard of living, with this latter included among economic, social and cultural rights in addition to political and civil rights.

A Human Right Approach (HRA) is a very important guide in designing policies to tackle hunger and malnutrition. First it helps to underline that fighting hunger is about upholding human dignity and not merely about meeting biological needs. Second it focuses on the universality, and therefore somehow the international solidarity and commitment of food security policies. Third, because human rights are indivisible and interdependent it recognizes the importance of achieving food security goals without violating any other human right. Fourth, as long as in the international human right law the primary burden for upholding human rights is assigned to nation states or governments, that means that food security policies should be carried out by public bodies within a general framework of welfarism.

The Human Development Capability Approach (HDCA) reinforces and completes the HRA to food security at least into two ways. Primarily, The HDCA complements the international human right framework by providing normative support for positive obligation and duties (Vizard, 2006); in other words it supports the idea that even in the western liberal tradition and consistently with its ideals of liberty and individualism there is room for positive (besides negative) obligations. In the HDCA this is made possible stemming from its particular definition of liberty that entails a concept of freedom as the range of valuable things that a person can do and be. This definition, relying upon ethical principles consistent with Kant's categorical imperatives, goes far beyond the definition of liberty given by the classical utilitarianism of liberal theories. Nussbaum has made this point very clear in addressing the problem of adaptive preferences (Nussbaum, 2000). Secondly the HDCA, being more focused on material and real aspects of human dignity and freedom, indicates more practical ways of how to incorporate moral concerns into actual public policy and individual actions. The HDCA takes account of the both the intrinsic and instrumental value of human rights (Vizard, 2006), as for example highlighting the instrumental role of civil and political rights in the context of famine reduction interventions, and therefore helps to integrate the ethical and political discourse in the political economic one.

Notwithstanding the normative and practical contributions that the HRA and the HDCA can give to food security policy design and implementation, they have not been widely taken into account by national and international policy makers so far (Rae, 2008). In the US during the last 30 years various groups of scientists and exponents of civil society have proposed to make the human right to food the moral and legal cornerstone of US domestic and international initiative in the area of food security, without any success. The U.S. government has consistently opposed formal right-to-food legislation as overly burdensome and inconsistent with constitutional law (Messer and Cohen, 2007). More in general,

The U.S. government has repeatedly asserted that economic social and cultural rights are not part of American legal and political culture, whose liberal ideals would conflict with the agenda requested for the upholding of positive rights.

Such a narrow interpretation of liberal ideals has not always been accepted. Before the advent of the neo-liberal era, during the period of the embedded Liberalism of the 50s, 60s and 70s (Harvey, 2005) Liberalism has evolved to become a more inclusive political theory and advanced industrial democratic countries became welfare states. Welfare states have been quite successful in realizing several social and economic rights at least within the boundaries of their nation-states, but major successes have been achieved in those democratic states more prone to an ideal of socialist (instead of liberal) economic organization, such as the social-democracies of Northern Europe. Anyway this progress registered during the era of embedded Liberalism began to fade during the successive Neo-liberal era, whereby a strong attack against state economic and social intervention and the concept itself of the welfare state has been launched.

As contended in the previous section, the food crisis of 2008 has been the result of more than 25 years of harsh Neo-liberalism that have, left food markets at the mercy of corporate power and financial crisis, reduced the food self-sufficiency of many poor countries, and deprived people of social security and public goods. A human right and capability approach to food security helps to better understand the nexus between food insecurity and the dissolution of public realm (Clarke, 2004) and the commodification of food supported by neo-liberalism. More important, it helps to clarify once again that hunger is a political problem that requires political solutions. Political solutions that were not been suggested at FAO summit or at the level of other international bodies such as the World Bank and the International Monetary Fund. As already noticed in official documents addressing the food crisis, the proposed interventions are aimed at reinforcing the neo-liberal model, and food is considered as an ordinary private goods instead of a human right (that to say food as commodity instead that food for community).

Food crisis, gender and human rights

The acknowledgement of the political roots of the food crisis and of the pivotal role of human rights and capability approaches in food security policy is particularly fruitful for understanding and coping with the very negative effects of the food crisis on women.

It is a matter of fact that as long as women represent the poorest swathe of the world population (worldwide, of the 1.2 billion people living on less than $1 a day, 70% are women), the negative effect of the crisis are borne mainly by them. Moreover women are hit more by the food crisis because of social and cultural restraints: (1) as the ones in charge of food provision and meal preparation, women are confronted directly with the problem of rising food prices, trying different strategies to secure food for their family (from roaming about in search of 'better prices' to engagement in informal jobs and even prostitution); (2) when food is scarce men and boys have precedence over women and girls in satisfying nutritional needs; (3) every economic crisis can exacerbate men's violence against women and result in a great prominence of traditional religious norms leading to women subjection and inferiority; 4) when basic needs are not fulfilled, problems concerning women rights in the poorest countries are given even less importance. Moreover, because women work in agriculture (more than half of the world's food is grown by women and in many poor countries 80% of food need is met by women's agricultural work) but with much less land ownership and access to agricultural services and extension with respect to men (World Bank, 2008b), as farmers they are likely to suffer the most from agricultural price volatility and pressure on land markets induced by the food crisis.

The recent food crisis represents only a small amount of the total price paid by women because of neo-liberal globalization (Attac France, 2005; Bakker, 1994; Bakker, Gill, 2003; Beneria, 2003; Aguilar, Lacsamana, 2004; Kingfisher, 2002; Sassen, 2007). More specifically, notwithstanding some positive effects, the economic empowerment of women associated with their growing participation in the labor market has had many negative effects such as: the loss of social security due to the policies of structural adjustment and the consequent burden on women of unpaid work for social reproduction; the renaissance of fundamentalism and patriarchal norms as a means to secure the capitalist class against social instability; the deterioration of terms and conditions of women's work together with an increase in work segregation reinforcing gender stereotypes.

Despite the fact that the effects of the food crisis are not gender-neutral and despite the important role of women in producing and providing food, the gender effects of food crisis are not sufficiently taken into account by official documents. As an example, the two recent documents of the World Bank on the food crisis and agriculture either do not ever quote women (World Bank, 2008a) or do quote women and their important role as farmers but with very contradictory suggestions on policies targeted at women (World Bank, 2008c). Moreover they use the 'neutral' term 'household' when referring to the target subjects of policies or to the people affected by the crisis. Women are also not quoted as central actors for improving food security and agriculture sustainability.

Despite the fact that 70 percent of economically active women in low-income, food-deficit countries are employed in the agricultural sector and play a pivotal role in growing, processing, and preparing food, the international response to the food price crisis has been gender-blind. It has failed to recognize that women are agricultural producers who face specific challenges and constraints, including lack of access to land, agricultural inputs and complementary assets. The institutional response to the food crisis should reflect instead a higher gender mainstreaming within development organizations, and should mobilize adequate financial resources for gender equality and women's empowerment within the agricultural sector. Agricultural policies should be specifically targeted to women, in particular through: defending (de jure e de facto) women's right to the land (ActionAid, 2008); facilitating women's access to credit; promoting innovation and technologies that fit women's work peculiarities; offering public agricultural extensions and services directly to women. Consistently with the HDCA, food security cannot be achieved without meeting people's basic needs and rights, and, particularly, without ending the discrimination against women all over the world.

Conclusion

The recent food crisis is one of the negative outcomes of 25 years of tough Neo-liberalism. The first way to combat it and prevent future crisis is to change the nature of the shift, produced by Neo-liberalism, from public to private, and from socio-political to an economic agenda. That is to say that the social and the public sphere should be again placed first with respect to the economic and the private sphere, at a discursive as well as a political level.

In contrast to the World Bank's policy advice on how to cope with the food crisis and to eradicate hunger, three simple objectives/strategies should be pursued: food as a human right, food sovereignty, and gender equality and women's empowerment.

These three objectives/strategies rely on one another and cannot be achieved separately. They sharply contrast with those suggested worldwide at national and international level, which are aimed at a further commodification of food (through trade liberalization and privatization of public goods), and do not take into account women as central actors in all issues related to food security.

References

ActionAid (2008). Securing women's rights to land and livelihoods. Action Aid international briefing paper, June. Available at: http://www.actionaid.org/.
Aguilar, D. and Lacsamana, A. (eds.) (2004). Women and globalization. Humanity Books, New York, 427 pp.
Attac France, AAVV (2005). Quand les femmes se heurtent à la mondialisation. Editions mille et une nuits, Paris, 189 pp.
Bakker, I. (1994). The strategic silence. Zed Books, The North-South Institute, 170 pp.
Bakker, I. and Gill, S. (2003). Power, production and social reproduction. Palgrave MacMillan.
Beneria, L. (2003). Gender, development and globalization. Routledge, New York, London, 212 pp.
Clarke, J. (2004). Dissolving the public realm? The logic and limits of Neo-liberalism. Journal of Social Policy 33(1): 27-48.
EtcGroup (2008). Who owns nature? Corporate power and the final frontier in the commodification of life. Available at: www.etcgroup.org.
FAO (2008). Soaring food prices: facts, perspectives, impacts and action required. FAO Conference on world food security, Rome, 3-5 June.
GRAIN (2008). Seized! The 2008 Land Grab for Food and Financial Security. GRAIN briefing, October.
Guzman, R.B. (2008). The global food crisis: hype and reality. PANAP, July 200, issue no. 7.
Harvey, D. (2005). A brief history of Neo-liberalism. Oxford University Press.
Kingfisher, C. (ed.) (2002). Western Welfare in decline: globalization and women's poverty. Philadelphia, University of Pennsylvania Press.
Lean, G (2008). Multinationals make billions in profits out of glowing global food crisis. The Independent, Sunday, 4 may.
Messer, E. and Cohen, M.J. (2007). The human right to food as a U.S. nutrition concern 1976-2006. IFPRI Discussion Paper 00731.
Nussbaum, M. (2000). Women and human development, Cambridge University Press.
Rae, I. (2008). Women and the right to food, FAO, http://www.fao.org/righttofood.
Sassen, S. (2007). A sociology of globalization, Norton & Company inc.
Vizard, P. (2006). The HDC Approach and human rights. Human and Development and Capability Association, briefing note. Available at: http://www.capabilityapproach.com.
World Bank (2008a). Agriculture for development, world development report.
World Bank (2008b). Gender in agriculture, sourcebook.
World Bank (2008c). Rising food and fuel prices: addressing the risks to future generations. October 12.

Food security and genetically modified crops: oversimplified answers to complex riddles?

A. Tsioumanis
Democritus University of Thrace, Department of Agricultural Development, Lahana 31, 54638, Thessaloniki, Greece; steriost@auth.gr

Abstract

Although food security is not a new concern for a great part of the developing world, global food shortages start posing new questions, even for relatively economically stable nations. For the first time since the early seventies, market stability for a number of commodities is seriously questioned. Under the circumstances, the debate on the potential for adopting techniques of modern biotechnology to combat food insecurity is, once more, heating up. Agricultural productivity is essential in order to combat food insecurity. However, a series of observations reveals that increasing agricultural productivity may be necessary but does not constitute, by any means, a panacea. Characterising genetically modified crops as 'the only solution', due to alleged productivity gains, seems to ignore a series of questions that proponents of genetic modification are unable to answer. Furthermore, serious market distortions in the form of heavy subsidies in the past decades have rendered agricultural investment less profitable for many countries. The effect of protectionist policies has not ceased as it is obvious in the case of biofuel policies. This paper tackles the fluctuation of market prices of basic agricultural products that have surged to record levels recently and tries to address the complexity of factors behind this. Approaching the expectations arising from the applications of modern biotechnology in the field of agriculture, in the form of companies' banners such as 'we will feed the world'; the paper concludes that they will not be able to fulfil their promises.

Keywords: food security, modern biotechnology, food prices

Food security status quo

Food security has been a priority in a chain of international meetings during the last few decades. Gradual eradication of poverty and hunger, including specifically the reduction of people suffering from hunger by 50% between 1990 and 2015 has been set as one of the UN Millennium Goals, following the awareness that the right to food in a quantity and quality sufficient to satisfy the dietary needs of an individual constitutes a fundamental human right.

As the year 2015 draws much closer, it is evident that the UN Millennium Goal will not be achieved. During the period 1990-92, which is used as the base period, there were approximately 840 million people facing food insecurity, meaning that the target was set at 420 million undernourished people. FAO estimates show that the number of undernourished people in the developing world actually augmented, reaching 923 million in 2007 and an estimated 963 million in 2008 (FAO, 2008).

This increase is undoubtedly influenced by population dynamics; it is a fact that the share of undernourished people in the developing world has declined by a few percentage points during the first years of the 21st century. However, even this progress was not sustained and challenges that lie ahead have remained qualitatively unaltered. Ahmed *et al.* calculate the number of people living in ultra poverty – less than US 0.5$ a day – as 160 millions. Even worse, in certain regions including Sub-Saharan Africa as well as Latin America, the number of ultra poor people has increased, while in aggregate and proportionately, their share has decreased more slowly than the share of poor people, who are defined as those living on less than US 1$ per day (Ahmed *et al.*, 2007).

Studying food security during the last decades has led to the basic assumption that the concept does not only include food availability but also economic access to food. This is perfectly portrayed in the definition adopted at the World Food Summit in 1996. *Food security exists when all people, at all times, have physical and economic access to sufficient, safe and nutritious food to meet their dietary needs and food preferences for an active and healthy life* (FAO, 1996). Nowadays it is rather evident that food security is heavily dependent on availability, accessibility and adequacy as researchers have underlined many decades ago.

While some countries can not meet the minimum daily requirement of 2,350 calories per person via production and imports, there are cases where, even though a surplus of food products exists, the access to a satisfying amount of food is hindered by the conditions of absolute poverty under which a large part of the population lives. *Poverty is the main cause of food insecurity, and hunger is also a significant cause of poverty* (WHO, 2005). Together they are leading to a downward spiral as food insecure or malnourished people are not able to fulfil their potential and develop their skills, lagging thus in productivity. All these render them unable to earn a livelihood.

Food prices: to distort or not to distort?

Food prices started to decine during the second semester of 2008 but it is evident that this recent decline may not camouflage the bigger picture. The index of nominal food prices, provided by FAO, has doubled between 2002 and 2008, while the real food price index started rising in 2002, following thirty years of stable prices and mainly declining trends. Price increases became dramatic during 2006 and 2007, leading to record highs for a number of crops. By mid-2008, real food prices were 64% higher than their 2002 levels (FAO, 2008). Since 2000, prices of wheat, butter or milk have tripled, prices of corn, rice or poultry have doubled.

The effect of rising food prices on food security is therefore crucially important. At the microeconomic level, the outcome of higher prices depends on whether the household is a net buyer or a net seller of food. FAO empirical analysis from developing countries indicates that about 75% of rural households and 97% of urban households are net food buyers. Thus, although generalisations should be carefully made due to regional differences, it is rather safe to assert that the vast majority of poor households are the main victims of price fluctuation, with the landless being the most vulnerable (FAO, 2008). The impact of increasing food prices on food-insecure and poor households is already dramatic and will continue in the same direction in the near future.

Regmi *et al.* (2001) calculated that for every 1% increase in the price of food, food consumption expenditure in developing countries decreases by 0.75%, an estimate that may be further aggravated by recent price pressures. They also compared consumption spending response to price changes among developed and developing countries and their findings show that food consumption spending is much more elastic in low-income countries (Regmi *et al.*, 2001). Poor people, once more, bear the burden.

High food prices have social repercussions that are often less visible in affluent societies. While demonstrations over the price of tortillas in Mexico are relatively well known, protests and riots over food prices were not geographically confined Yemen, Pakistan, Mauritania, Senegal, Cameroon, Argentina, India, and Burkina Faso constitute an indicative list of countries that have also witnessed protests over the last two years alone. Facing this turmoil and risking their popularity and thus electoral results, local governments often opt to protectionist measures in order to help domestic consumers and control food price inflation. Jordan, Bangladesh and Morocco, are increasing subsidies and reconsidering their tariff regimes, India and Egypt are restricting exports, China centrally controls domestic food prices, while Russia has implemented price controls on basic foodstuff. Export tariffs on wheat and other basic food products become once more popular; countries import more than they actually need in order to build their stocks. The ban on US soybeans exports imposed by Richard Nixon back in 1973 does not look as distant or as alien as it did a decade ago. Neither do the bread intifadas in Egypt and Morocco in 1977 and 1984 respectively.

While market distortions may provide a relief for domestic consumers in the short run, they are economically unsustainable and fail to prioritise the greater good. Price controls and export tariffs render production less profitable, damaging further the supply side. Subsidised consumer prices delay social unrest but simultaneously create an artificial economic environment, stimulating demand and creating even more problems in the long run. In February 2008, Kazakhstan's government announced its plan to restrict wheat exports in order to ensure domestic supply. As a result, global wheat prices augmented by 25% in a single day (Time, 2008).

It is true that food prices rise for a variety of reasons. Lowering stock levels that induce price volatility, production shortfalls due to adverse conditions that are often correlated to climate change, oil prices, changing diet patterns in regions that have become more affluent in recent decades, especially in Asia, trade policies that often contradict one another, and financial speculation in food markets all play a distinct role creating today's reality. However, it is of crucial importance to agree on the goals before discussing the means to achieve them. Policies concerning biofuels perfectly portray this urgent need.

Feedstock represents the principal share of total biofuel production costs. For ethanol and biodiesel, feedstock accounts for 50–70% and 70–80% of overall costs, respectively (IEA 2004). Thus, prices of agricultural commodities used in biofuel production are increasingly connected to energy prices, leading to even more fluctuations. As increasingly more land is devoted to biofuel production, competition between food and fuel is transparent. Future competitiveness of the biofuel sector will determine the extent to which land, water and capital will be diverted away from food production. Forecasts related to biofuel production is not an easy task; IFPRI has developed IMPACT, which stands for International Model for Policy Analysis of Agricultural Commodities and Trade and has generated two future scenarios, one based on the current rate of biofuel increase, the other assuming a drastic raise. Under the planned biofuel expansion scenario, international prices increase by 26% for maize and by 18% for oilseeds. Under the more drastic biofuel expansion scenario, maize prices rise by 72% and oilseeds by 44%. Under both scenarios, the increase in crop prices resulting from expanded biofuel production is also accompanied by a net decrease in the availability of and access to food, with calorie consumption estimated to decrease across all regions compared to baseline levels (Von Braun, 2007).

Competition between food and fuel is heavily influenced by protectionist policies regarding biofuels. In general, subsidies for biofuels that use agricultural production resources are extremely anti-poor because they implicitly act as a tax on basic food, which represents a large share of poor people's consumption expenditures and becomes even more costly as prices increase (Von Braun 2007). Analysing in depth externalities that may render a certain degree of protectionism on biofuels social benefits is out of the scope of this paper. However, the intensity and multidimensionality of the subsidies is not compatible with the challenge of feeding a growing population. *The bewildering array of incentives that have been created for biofuels in response to multiple (and sometimes contradictory) policy objectives bear all the hallmarks of a popular bandwagon aided and abetted by sectional vested interests. Understanding the consequences of these changes before any further damage is inflicted is the only responsible way forward* (IISD, 2007).

'We will feed the world' and other stories

Facing a new food crisis, calls for production increases are predictable. Ban Ki-moon, the United Nations secretary-general, called for a 50% increase in global food output by 2030. Since an increase of this magnitude is a huge task, once more proponents of genetically modified crops have thrived. 'We are a piece of the solution. Seeds are the starting point,' pronounces Hugh Grant, chief executive of Monsanto.

On one hand, it has to be said that judging from the above quotation, some progress has been made as far as arrogance is concerned. Not long ago, President Bush was urging *European governments to end their opposition to biotechnology for the sake of a continent threatened by famine* (Bush, 2003) and the pro-biotech sector was arguing that those opposing biotechnology are to bear the moral responsibility for the physical extinction of millions of people in the developing world. On the other hand however, the potential to combat food insecurity via the biotechnological sector as it stands today, has remained unchanged.

Focusing solely in increasing production is a narrow path that may not solve existing problems regarding food security because it does not alleviate any of the structural weaknesses related to distribution of economic power and wealth. The 'green revolution', which constitutes a popular term used to describe the introduction of improved seed strains, fertilisers, and irrigation as a means of producing higher yields in crops such as rice, wheat, and corn offers some valuable lessons towards this direction.

Productivity gains that followed the introduction of new varieties and in particular its effect on the cultivation of wheat and rice, was the main factor that contributed to the rapid increase of production in the period 1960-1990. During this period the global production of cereals was doubled, the available amount of food per capita increased by 37% and the actual values of these products decreased on average by 50% (McCalla *et al.,* 2000). Swaminathan uses the example of India pointing out that since land cultivation commenced over 4000 years ago until the end of colonial rule in the late 50s, Indian farmers developed the capacity to produce 7 million tonnes of wheat per year. The introduction of semi dwarf strains in irrigated areas from 1964 to 1968 drove wheat production up to 17 million tonnes (Swaminathan, 2000). Problems, mainly associated to local disparities did not cease to exist. Despite the increased productivity and the amelioration of the conditions of life in many developing countries (mainly in Asia), in Africa the available amount of food per capita during the same time interval decreased, as a result both of the reduced rate of agricultural production and the rapid population growth of the continent. Notwithstanding regional differences, one should not forget that it is in Asia, precisely where Green Revolution seeds have contributed to the greatest production success, that roughly two-thirds of the undernourished in the entire world live.

Although in the case of the Green Revolution, productivity gains were hindered by the negative effects of unequal distribution rendering thus hunger elimination goals unattainable, they became very popular *per se*, so that defenders of the use of modern biotechnology in the field of agriculture often draw parallels to the earlier agricultural revolution. The new technological leap looks more ominous than the previous one due to fundamental differences between the green and the gene revolutions.

Research that led to the green revolution was conducted by the public sector using public financing and improved seeds were available for multiplication and distribution for free. On the other hand, the research related to the applications of modern biotechnology on agricultural production is directed mainly by the private sector and the genetically modified improved seeds are promoted for commercial use under patent granting. Research during the green revolution was mainly adapted to the conditions prevailing in the developing countries and was aiming to be important for these countries, cultivations such as, rice, corn, tropical fruits and vegetables. In contrast, the target-group of agricultural biotechnology is the farmers of the developed world and also, the production is mainly pushed to the markets of the developed countries (Pinstrup-Anderesen *et al.,* 2000). Agricultural biotechnology has not yet produced a crop variety that has any direct connection to hunger or nutritional needs and the popular Golden rice is no exception. The most prevalent genetically modified crops remain corn, soy, cotton and canola and the dominant traits are herbicide tolerance and insect resistance. Concluding, the green revolution has taught us that as long as the agricultural research is not focusing on the needs and the actual conditions of the developing countries, the adoption of new technologies not only cannot guarantee the solution of the existing problems but it may also deepen the inequalities in given geographical areas.

Concluding remarks

The fact that one in six human beings remain malnourished, although humanity has reached an unprecedented state of technological development is not a phenomenon that can be approached via the power and applications of a single discipline. Not only does it involve a multiplicity of factors concerning technical, economical and social considerations, but it also surrounds our deepest notions on ethical behaviour and morality. Focusing on a single aspect of the problem, as for example increasing production, is simply not adequate, not even in simpler economic terms.

It could be argued that the increasingly modest approach of the pro-biotech sector as being just part of the solution as opposed to the solution *per se*, is nothing but a shift in strategy. While that should not be surprising, the pressing nature of the current status quo remains unchanged.

Technological advances provide a picture of what is technically attainable, not socially desirable. Sabbato (1951) argues on the immoralist character in the modus operandi of scientific knowledge pointing out that *science on its own guarantees nothing as when it comes to the realization of a major achievement, its moral concerns are absent.* Beyond a discussion of the essence of science, easily visible problems also exist. The urgent need to protect invested capital, the existence of vested interests and pressure groups and the uncomfortable reality that our unanimous goal is to maximise short or medium term fiscal profit render a genuine social dialogue unattainable.

Even when productivity gains via the use of genetic modification can be argued beyond doubt, a series of changes regarding research, priorities, goals, testing, implementation and policy-making would also be needed. Such a drastic shift is not foreseeable and thus, even if agricultural productivity exceeds the most optimistic expectations, without other interventions people will continue to starve.

References

Ahmed, A., Hill, R., Smith, L., Wiesmann, D. and Frankenburger, T. (2007). The world's most deprived: Characteristics and causes of extreme poverty and hunger. 2020 Discussion Paper 43.Washington, D.C.: International Food Policy Research Institute.

Bush, G. (2003). quoted in Journal of Agricultural and Environmental Ethics 16: 525-529, 2003, Editorial

FAO (2008). The State of Food Insecurity in the World, Rome Italy.

International Energy Agency (IEA) 2004. Biofuels for transport: An international perspective, Paris, France.

IISD (2007). Biofuels- At What Cost?, Global Subsidies Initiative, International Institute for Sustainable Development, Geneva, Switzerland.

McCalla, A.F. and Brown, L.R. (2000). Feeding the Developing World in the Next Millennium: A Question of Science?, Agr. Biotechnology and the Poor, pp. 32-36.

Pinstrup-Andersen P. and Cohen, M.J. (2000). Modern Biotechnology for Food and Agriculture: Risks and Opportunities for the Poor, in G.J. Presley and M.M. Lantin, eds., Agricultural Biotechnology and the Poor.

Regmi, A., Deepak, M.S., Seale Jr., J.L. and Bernstein, J. (2001). Cross-country analysis of food consumption patterns in Changing structure of global food consumption and trade, ed. A. Regmi.Washington, D.C.: United States Department of Agriculture Economic Research Service.

Sabbato, E. (1951). Hombres y Engranajes. Reflexiones sobre el dinero, la razón y el derrumbe de nuestro tiempo. Buenos Aires, Sur, 1951. Edición definitiva: Barcelona, Seix Barral, 1991.

Swaminathan, M.S. (2000). Genetic Engineering and Food Security: Ecological and Livelihood Issues. In: Presley, G.J. and Lantin, M.M. (eds.) Agricultural Biotechnology and the Poor, pp. 37-42.

Von Braun (2007). The World Food Situation, New Driving Forces and Required Actions, International Food Policy Research Institute, Washington D.C.

WHO (2005). Modern Food Biotechnology, Human Health and Development: an evidence based study, World Health Organisation, Geneva, Switzerland.

Theme 9 – Participatory methods and ethical analysis in biotechnology appraisal

Evaluating participatory methods

B. Bovenkerk
Ethics Institute, Utrecht University, Heidelberglaan 8, 3515 CA Utrecht, the Netherlands; b.bovenkerk@uu.nl

Abstract

Several reasons to call for public participation in decision-making about emerging biotechnologies can be noted: (1) biotechnology is a field with many uncertainties, and not only risks (known unknowns) but also unknown unknowns, and in this context the primacy of expert knowledge has been challenged; (2) involving the public has proven to be indispensable in receiving public trust; (3) biotechnology could be characterized as an unstructured policy problem because disagreements not only exist on the level of possible solutions, but on all levels, about facts and values, from problem demarcation to interpretation of research results. If these circumstances are not taken seriously policy problems may become intractable. Different methods of public participation have been proposed, such as citizens' juries, consensus conferences and deliberative polls. My central question is how should we evaluate these methods? I have developed four broad conditions for the evaluation of public debates that lead to more specific criteria: (1) the debate should be open ended (the debate can in principle be re-opened), (2) it should be inclusive (it should not exclude groups or arguments), (3) there should be an absence of power imbalances (pernicious group dynamics should be avoided), (4) the results should be of a high quality (the debate should lead to preference transformation). In this paper, I will defend these criteria on the basis of the theory of deliberative democracy under conditions of pluralism.

Keywords: biotechnology, public debate, deliberative democracy

The call for public participation

Public participation in decision making about emerging biotechnologies has become more and more popular. Several reasons for the call for public participation can be noted. First of all, members of the public are becoming better informed and more highly educated, creating a wish to become involved in political and social issues. At the same time, expert knowledge has become increasingly questioned in the wake of what many see as failures of the technological paradigm, resulting in environmental degradation on a scale never experienced before, and an increased perception of the risks we are exposed to. As most risks are largely invisible and it is often hard to establish the cause of environmental harms, the public relies on expert knowledge in order to discover risks. At the same time, the public is confronted with conflicting accounts of risk assessment by different experts and is aware that some predictions by experts have turned out to be false. Conflicting expert accounts have drawn attention to the fact that science is not value-free and objective (Slovic, 1999). In risk assessment, value choices are made in the problem definition, the research methods, and the interpretation of outcomes. In this context, it is acknowledged more and more that expert knowledge is not purely objective, that it is not infallible, and that lay persons can contribute certain experiential knowledge that is often overlooked by scientific experts. When we are dealing with GMO's we are not only dealing with risks in the sense of known or quantifiable unknowns, but also with unknown unknowns. In other words, it is such a new technology that we do not even know what possible risks to look for. When doing field trials to establish environmental effects of GM crops, unpredictable consequences are out of necessity left out. Science can only deal with *anticipated* uncertainties and out of necessity discounts unanticipated ones (Wynne, 2001). However, scientific risk analysis cannot incorporate such variables and therefore tends to ignore them, while for the public their possibility are an important reason to call for caution. For example, it has been lay people who have drawn attention to the possibility of unanticipated consequences of growing genetically modified crops.

As Brian Wynne (Wynne, 2003) argues, the value of lay experiential input lies not just in the ability to supplement expert knowledge with factual knowledge, but also in the fact that it tends to be critical of institutional structures and to reflect on broader issues than merely technological ones; in effect, lay expertise can play a role in challenging the validity of present expert knowledge.

Secondly, public acceptance of gene technology has proven to be influenced, amongst other things, by the amount of trust that citizens have in the institutions that govern the use of biotechnology (Paula and Birrer, 2006). Simply providing the public with information, for example about the risks (or lack thereof) of consuming genetically modified food, is not enough to build trust in the institutions regulating biotechnology. After all, how one interprets this information is already dependent on the level of trust one has in these institutions in the first place. Franck Meijboom *et al.,* (2006) convincingly argue that trust is not simply a characteristic of individual citizens or consumers, but that trust is established in a relationship, and that government institutions, therefore, need to show they are trustworthy. This calls for what is referred to as 'two-way transformative learning'; rather than a cognitive deficit model where public participation is simply a different word for top-down 'education' of the public, two-way transformative learning assumes that something can be learned in both directions; opinion transformation can take place on all sides of a dialogue between government, stakeholders, and the public (Wickson, 2006).

Thirdly, biotechnology could be characterised as an unstructured policy problem. According to Matthijs Hisschemoller and Rob Hoppe (1995), in a well structured problem there is broad agreement about the facts and values involved and solving the problem essentially comes down to applying standard techniques and procedures to the problem; policy decisions can then largely be left to bureaucrats and experts. In unstructured problems no such agreement about facts and values is present, and even the question as to how the policy problem should be defined in the first place is disputed; disagreement exists on all levels, from problem demarcation to the interpretation of research results. Hisschemoller and Hoppe argue that unstructured problems call for a so-called 'learning strategy', which involves broad public participation at an early stage, for example in the form of public debate, in which experts have a similar status as lay people, and which eventually feeds into political decision making. The benefit of this approach is that misunderstandings about different parties' motives and beliefs are cleared up and this means that even if consensus is often unlikely, at least the policy problem is defined more realistically and will be perceived as more legitimate by those involved. If unstructured policy problems are erroneously treated as structured ones and decisions are left up to bureaucrats and experts without public input, the public will not feel that their views have been taken seriously. According to Hisschemoller and Hoppe (1995: 48) 'the crucial matter is whether a given problem construction takes into account all the differences of opinion about the problem and its possible solutions'. If this is not the case, this eventually backfires and this may lead to an intractable policy problem with the risk of escalation into protest, sabotage, or possibly even violence.

Public participation can come in many forms, such as consensus conferences, citizen's juries, deliberative polls, and planning cells. It should be possible to compare the success of these different 'deliberative microcosms' in avoiding that policy problems become intractable. Their relative effectiveness, therefore, needs to be evaluated. Different checklists for evaluation have been proposed (for example Rowe and Frewer, 2000), but no coherent argumentation has been given for why particularly these criteria were deemed important. Here I want to develop such an account; I will propose four broad conditions that will each lead to further specific criteria and I will defend these on the basis of my view of deliberative democracy – which is, after all, the theoretical underpinning of the call for public participation – under conditions of pluralism.

Deliberative democracy

As a general definition, we could say that deliberative democrats hold that a decision has been reached legitimately if it was the result of a procedure in which all those potentially affected by the decision have had a free, equal, and fair chance to influence the outcome through rational deliberation. An important aspect of democracy for deliberative democrats is, then, that all persons should be treated as free and equal and for them this amounts to an equal opportunity to participate in the drafting of decisions that will influence their lives. Like traditional liberal democrats, deliberative democrats are well aware of the pluralism of opinion that characterises modern democracies, but many of them think that they can actually deal with this pluralism better than liberals can. Amy Gutmann and Dennis Thompson (1990), for example, argue that the liberal view of what properly belongs to the domain of politics precludes too many issues about which moral disagreement exists and that this does not take reasonable pluralism sufficiently seriously. Political liberals, in short, argue that in the face of pluralism governments should remain neutral between different conceptions of the good life and the way to do so is by constraining dialogue about these conceptions. They make a distinction between those issues that are part of people's comprehensive notions of the good, which should remain in the private realm, and those issues that should be universally shared and that should be discussed in the public realm (Ackerman, 1989).

However, this distinction between the public and the private can be criticised. There is no neutral a priori way of deciding what should belong to the public and what to the private spheres. Furthermore, we cannot expect citizens to leave their comprehensive notions of the good life behind when entering the public sphere, because their moral viewpoints rely on these. The requirement of conversational restraint demands of people to hold back exactly on those issues that they deem very important. Moreover, where public goods, such as natural or environmental goods, are involved, the state cannot simply refrain from interference, because decisions on them will have to be made. However, many conflicts about how to deal with public goods are based on the kind of disagreements about values and worldviews which liberals would like to relegate to the private sphere. If governments decide not to take a stance on such issues they are in effect giving priority to some people's values over those of others. For example, if government decides to leave the determination of CO_2 emission levels up to individuals they are favouring companies who profit from emitting CO_2 over environmentalists who want to see these levels cut. These problems with the public/private distinction reveal the failure of liberalism to remain completely neutral; it has an inherent bias towards certain values, such as individual autonomy. In the opinion of deliberative democrats, conversational restraint is neither desirable nor necessary in order for governments to remain a neutral position, nor for citizens to show each other respect as free and equal persons. The liberal condition of exclusion of comprehensive views is based on a narrow view of moral argumentation, which presupposes that citizens have a more or less stable and unified point of view (Grotefeld, 2000). For if two citizens each have a stable, unchangeable point of view, there seems to be no point for them to provide each other with insight into their views. In that case, they have no hope of ever convincing each other of the correctness of their views and they might as well leave their comprehensive views out of public debate. If, on the other hand, participants in a debate are allowed to explain the broader picture to demonstrate where their viewpoints are coming from, there is more chance that participants will reach mutual understanding. After all, how can we achieve mutual understanding if we do not get to properly explain our views?

A problem with the liberals' static view of opinions or preferences is that it precludes the possibility of preference or opinion transformation. People's 'real' preferences may change after they have been confronted with opposing viewpoints, and decisions preceded by an exchange of arguments will, therefore, tend to more accurately reflect these preferences. Also, through deliberation the range of alternatives to choose from may change or alternative solutions may be found and this opportunity is lost under conversational restraint. Deliberative democracy, on the other hand, proposes public deliberation especially because this is thought to lead to preference or opinion transformation. Moreover, public deliberation offers a way to find shared values or at least clarify the nature of a moral conflict. The liberal static view of citizens' opinions is reflected in actually existing liberal democracies' most common mechanisms of social choice, voting and bargaining, that are criticised by deliberative democrats. The aggregation of preferences through voting is inherently problematic, because when there are more than two options to choose from, there is no clear and non-arbitrary method for aggregating preferences in a way that does justice to the voters' positions, because of problems such as voting cycles (Dryzek, 2000). Bargaining is regarded problematic because it favours powerful elites who have more bargaining power and because it 'treats even matters of principle as though they were conflicts of interest' (Schlosberg, 1999: 21). This would require one to treat one's fundamental convictions as something negotiable.

Criteria for evaluation

From my elaboration of deliberative democracy four conditions can be formulated that could form the basis of a set of criteria to evaluate how well participatory methods can deal with disagreement resulting from pluralism. Firstly, the condition *of non-exclusion* states that no reasonable views should be excluded from public deliberation. I think more views should be included in the range of 'reasonable' views than traditional liberals allow when they are using this term. My arguments against conversational restraint support this criterion. Moreover, this criterion is based on the respect that governments owe their citizens and this includes respect for their opinions and their intellectual capacities. Respect for persons is a normative premise that I take as a given, because as has been argued elsewhere, each political theory in a democracy should adopt it (Kymlicka 2002).The confrontation with others and their diverging points of view is necessary to stimulate these intellectual capacities. Finally, given the fact that in the real world we are confronted by certain limitations – it is the *condition humaine* that we have limited knowledge and time, that facts can be multi-interpretable, and that we are dealing with a complex world – it is always possible that certain entrenched beliefs or views turn out to be mistaken. In such a world it is of utmost importance that there are different lines of thought that can be tested against one another. Even if engaging with views opposed to our own does not change our views, at least it helps us clarify and strengthen them. As a corollary to the non-exclusion of views we should also wish not to exclude any persons or groups. As was argued convincingly by Anne Phillips (1996), in the past it was believed to be sufficient to represent the views of different, often marginalised, groups in politics; as long as the whole spectrum of different beliefs and opinions were taken into account in a decision equality had been achieved. She terms this view the 'politics of ideas' and contrasts this with a 'politics of presence', in which it is held that the physical – perhaps even proportional – presence of members of all the different groups is necessary in order to stimulate real equality. After all, we cannot easily divorce ideas from the persons who hold these ideas; ideas are shaped by experiences and differently situated groups in society will have different life experiences. Through past experience it has become clear that having an elite of highly educated white middle-class men represent the ideas of all groups in society leads to too much homogeneity and an exclusion of difference. The condition of non-exclusion supports evaluation criteria such as 'inclusion of all relevant views', and 'opportunity of marginalized groups to voice their concerns'.

Secondly, the condition of *absence of power differences* states that in public deliberation decisions should not be forced by powerful groups upon people with less power. This criterion is based on the critique of bargaining. Also, if, as I have argued, less emphasis should be placed on the social choice mechanism of aggregation and more on deliberation, we should ensure that this does not counteract the most important benefit of voting, namely the fact that everyone has an *equal* vote. It is important, therefore, that everyone gets an equal opportunity to take part in the deliberations of the decision-making process, but this is only possible when power balances are avoided as much as possible. Moreover, this condition follows from the view that decision making should not be solely left up to bureaucrats and experts and that lay persons should be able to challenge expert knowledge. This condition leads to criteria such as 'independence of organisers and facilitators', 'absence of distorting group dynamics', 'presence of open expert contestation', and 'influence of outcomes on policy decisions'.

Thirdly, the condition of *open-endedness* states that the outcomes of a public deliberation should in principle be revisable. This criterion follows from the view that opinion transformation should be possible before decisions are taken. Again, in the real world we are faced with limitations – on human knowledge, interpretation, and the complexity of problems – and this entails that we are likely to make wrong decisions at times. Decisions should, therefore, be revisable when new information or arguments come to light. Moreover, if we want to show respect to all persons and their opinions we should keep the possibility open that any group's views may at some point be favoured. In order to give those whose preference was not realised in a specific decision – those who lost the conflict – the opportunity to revise decisions in future these should in principle be open-ended. As James Bohman (1996: 96) argues, deliberation is fair if it makes 'possible continued participation of all groups in a common deliberative framework'. This condition supports criteria such as 'transparency' and 'review rights'. Finally, the condition of *quality of debate* states that the participatory method was successful if certain outcomes were reached, most notably if preference or opinion transformation in fact occurred, if mutual understanding was reached, if the source of disagreements has become clearer, or if possible middle ground or alternative solutions were formulated that all or most participants could accept.

References

Ackerman, B. (1989). Why Dialogue? The Journal of Philosophy 86(1): 5-22.

Bohman, J. (1996). Public Deliberation: Pluralism, complexity and democracy. Cambridge, Mass.: MIT Press.

Dryzek, J.S. (2000). Deliberative Democracy and Beyond: liberals, critics, contestations. New York: Oxford University Press.

Grotefeld, S. (2000). Self-restraint and the Principle of Consent: Some Considerations on the Liberal Conception of Political Legitimacy. Ethical Theory and Moral Practice 3: 77-92.

Gutmann, A. and Thompson, D. (1990). Moral Conflict and Political Consensus. Ethics 101: 64-88.

Hisschemoller, M. and Hoppe, R. (1995). Coping with Intractable Controversies: The case for problem structuring in policy design and analysis. The International Journal of Knowledge Transfer and Utilization 8(4):40-60.

Kymlicka, W. (2002). Contemporary Political Philosophy. Oxford, Oxford University Press.

Meijboom, F., Visak, T. and Brom, F.W.A. (2006). From Trust to Trustworthiness: Why information is not enough in the food sector. Journal of Agricultural and Environmental Ethics 19: 427-442.

Paula, L. and Birrer, F. (2006). Including Public Perspectives in Industrial Biotechnology and the Biobased Economy. Journal of Agricultural and Environmental Ethics 19: 253-267.

Phillips, A. (1996) Dealing with Difference: A Politics of Ideas, or a Politics of Presence? In: Benhabib, S. (ed.) Democracy and Difference. Contesting the Boundaries of the Political. Princeton, Princeton University Press.

Rowe, G. and Frewer, L.J. (2000). Public Participation Methods: A Framework for Evaluation. Science, Technology, & Human Values 25(1): 3-29.

Schlosberg, D. (1999). Environmental Justice and the New Pluralism: The Challenge of Difference for Environmentalism. Oxford, N.Y., Oxford University Press.

Slovic, P. (1999). Trust, Emotion, Sex, Politics, and Science: Surveying the risk-assessment battlefield. Risk Analysis 19(4): 689-701.

Wickson, F. (2006). From Risk to Uncertainty: Australia's environmental regulation of genetically modified crops. Wollongong, University of Wollongong.

Wynne, B. (2001) Public lack of confidence in science? Have we understood its causes correctly? In: Pasquali, M. (ed.) EurSafe 2001: Food Safety, Food Quality and Food Ethics. Florence, Italy, A & Q Milan.

Wynne, B. (2003). Seasick on the Third Wave? Subverting the Hegemony of Propositionalism: Response to Collins & Evans (2002). Social Studies of Science 33: 401-417.

New epistemologic perspectives in biotechnology ethics: the case of GM fish

L. Coutellec
Institut National des Sciences Appliquées de Lyon, Département des Humanités, 69621 Villeurbanne Cedex, France; leo.coutellec@insa-lyon.fr

Abstract

To anticipate the possible presence of genetically modified fish (GM fish) in France, a network of scientists – specialised in the transgenesis of fish, the socio-economics of the fish industries, the sociology of consumption, the detection of GMOs, legal regulation, the philosophy of contemporary science and ethics – proposes to examine the various aspects of the technology, its impacts and the strategies to implement, before any controversy arises. The question of the commercial use of GM fish is complex in that there are a number of social and food-related issues involved. It is difficult to anticipate all possible trajectories, where: multiple disciplines are involved; the relations between cause and effect are not linear; the consequences cannot be ascertained in the short term; and there is a storm of uncertainties at so many levels. In this paper, I propose to report some of the reflections regarding the epistemology and ethics of this project. The question that concerns us for this work is the following: What types of scientific approaches and knowledge can facilitate close interaction with ethics and a more effective social intervention? In this sense, the basic premise is that technological ethics must be accompanied by renewed epistemological reflection. Indeed, it is possible to renew our representation of the concept of science and its relationship to technology by adopting the position of epistemological pluralism. In the face of a complex problem, this conception of epistemological pluralism will enable us to consider ways of producing knowledge that can tolerate uncertainty, and constructive research approaches like interdisciplinarity. I submit the hypothesis that these new perspectives in epistemology create the possibility for new relationships between ethics and epistemology, and open the way for technological ethics taking into account multiple parameters, multiple strategies or scientific and technological trajectories. I present a part of this approach discussing the possible case of accidentally imported GM fish in France. Our research is a theoretical and practical contribution for ethical analysis in a context of biotechnological emergence.

Keywords: epistemological pluralism, uncertainty, interdisciplinarity, GMOs

Introduction

Technological ethics is forced to take into account reflections on the concept of science and cannot escape scientific research approaches. *The basic premise of this paper is that technological ethics must be accompanied by renewed epistemological reflection,* the preferred route for those who want to avoid technological ethics being merged with scepticism or prevailing relativism. I define the representation of science by conventional epistemology according to four features (not exhaustive): (1) being mainly focused on theory or on a combination of theory/experience; (2) producing certainty or absolute probabilistic ambitions; (3) root-type relations of identity, between disciplines or with society; (4) a conception of science that is 'value-free' and a clear dichotomy between facts and values. To go beyond this conception, I adopt the position of epistemological pluralism which, faced with a complex problem such as GM fish, will enable us to consider ways of producing knowledge that can tolerate uncertainty and constructive research approaches like interdisciplinarity.

Epistemological pluralism

Because scientific approaches, methods and criteria of knowledge are increasingly complex, because science and technology – not to be confused – are increasingly intertwined, epistemology must measure these multiple means of producing knowledge; there is no 'one' scientific method or 'one' representation of science. I conceive epistemological pluralism as an alternative to positivist epistemology (theoretical, absolutist monism) and relativistic epistemology (sociologically imbued), which aims to take into account a more comprehensive conception of science and its methods, and as a means of considering positive interactions between ethics and epistemology. Modelling and experimentation (Dubois *et al.*, 2007), technique (Feenberg, 2006), research methods such as action-research (Coutellec, 2009) and interdisciplinarity, but also rich ingredients such as hypotheses or uncertainties, are important examples that call for renewed epistemology. The challenge is to understand these new means of producing knowledge without lapsing into relativism. Faced with a complex problem, the aim is to articulate a plurality of knowledge, ingredients of scientific process, disciplines and strategies of research. From a more general viewpoint, two trains of thought also feed the idea of epistemological pluralism. These are the questioning of the representation according to which science is exclusively value-free (Lacey, 2001), and the desire to go beyond the overly strict dichotomy between fact and value (Putnam, 1999). According to H. Lacey, the three components of the idea that science is value-free are *objectivity*, *neutrality* and *autonomy*. For example, *autonomy* is a value according to which scientific practices, institutions, the characteristics of scientific methodology, the adoption of strategies in research, and the priorities and direction of basic research are (or should be) free from external interference and the disproportionate influence of any particular value (ethical, political, economic, etc.) outlook (Lacey, 2001). This may be an ideal but remains an illusion in the current context, and, what is more, would prevent us from considering interactions between epistemology and ethics. However, science and technology also convey and presuppose values, and there are interactions between epistemic values and ethical and social values (Douglas, 2000). In this sense, it is pertinent to note that if mainstream modern science has been built almost exclusively on one strategy, which H. Lacey calls the 'Baconian ideal' – which conventional epistemology still draws widely upon as we have seen – it is important to stress that knowledge can be acquired through other strategies, which are related to other values, such as social welfare and environmental sustainability, and whose 'forms of knowledge' (Hacking, 1999:183). For example, while probabilistic reasoning as a form of reasoning, and the disciplinary approach as a research strategy, have vastly dominated reflection on contemporary science, we can try to go beyond them in the context of epistemological pluralism. I now illustrate this by detailing a research approach such as interdisciplinarity and presenting an opening it offers through the possibility of tolerating uncertainty.

A vision of the interdisciplinary process

I am working on the concept of interdisciplinarity in the light of our willingness to consider epistemology and ethics together. Faced with a complex problem, the *epistemological bias* would be to make a discipline the referent to understand the problem as a whole. As I have pointed out, there is a prevailing traditional conception of science that focuses exclusively on the disciplines, according to a *root-type relation of identity*. The latter is thought to guarantee the objectivity and neutrality of science, versus ideologies. I show that from an epistemological pluralism perspective, it is possible to consider truly interdisciplinary research strategies that are also objective and impartial without lapsing into ideology, but with somewhat different values. I hold that a discipline brings both attributes of independence and proximity with other disciplines involved in the same interdisciplinary process. I call this the *identity-relation* of a discipline, similar to the idea of *identity-rhizome* of French philosopher Deleuze and contradicting the idea of *identity-root*. This implies that a discipline can change by exchanging with the other, without being lost or distorted. The phases of disciplinarity and interdisciplinarity do not oppose one another but are mutually enriching. I can now justifiably use the term *interdisciplinarity* as such. Often, the words *inter*-disciplinarity, *pluri*-disciplinarity and *trans*-disciplinarity are considered interchangeable. However, beyond the subtlety of semantics, there are major epistemological differences between these three concepts. I do not choose interdisciplinarity at random but by choice, conviction: this is a certain *form of knowledge* and a research strategy bearing values. *Inter*-disciplinarity, which I hope will subsist in this project, is an *interaction logic* that allows for each discipline and its relationship, a process of interaction between *disciplinary identities-relation*. To express this approach in time, I talk about *iterative interaction process*. During this process, each field of knowledge is called upon as points of view which are re-interpreted by other disciplines. From these boundary movements, the elements of the problem are reorganised and the various disciplinary fields are restructured by way of successive arbitration by the various critical or harmonic positions that accompany them. In the case of GM fish, this vision of a truly interdisciplinary research approach is somewhat unsettling and brings uncertainties to light.

Tolerating uncertainty

While uncertainty has always been more or less the starting point of scientific undertakings, what is new in contemporary situation is that it is also constitutive of multiples interactions between science, technology and society, with situations where predictability is inherently limited. In the latter case, we must adopt responsible ways of managing uncertainty, and ethics has a role to play in this. This is what I call tolerating uncertainty. It is not a matter of resigning ourselves to it (scepticism), or ignoring it (positivism). The consequences of taking into account uncertainty are epistemological and ethical. Epistemological in that it plays a determining role in the knowledge production process, and ethical in that a certain kind of uncertainty raises an ethical question as to its management. According to a typology established by W.E. Walker, there are uncertainties that are epistemic – 'uncertainty due to the imperfection of our knowledge, which may be reduced by more research and empirical efforts' – and ontological or variability uncertainties – 'uncertainty due to inherent variability, which is especially applicable in human and natural systems and concerning social, economic, and technological developments' (Walker *et al.,* 2003:13). The classic distinction between risk on the one hand – which is quantifiable, measurable by means of probabilities, insurable, and which implies that we are aware of all possible situations – and uncertainty on the other hand – which is defined by the negation of these conditions – becomes a major theoretical challenge. The tendency to dilute uncertainty in the calculation of risk shuns the need, in the face of 'new risks', to build intellectual strategies tolerating uncertainty. This tendency implies that science and expert techniques, about theory and facts, merely identify, objectify and quantify the risks and that ethics, about values and practice, merely assesses and recommends risk management (which is down to policy-makers). The collective, contradictory and conflictual possibility of objectifying risks and uncertainties – bringing into play science and ethics upstream, as soon as hazards are identified, and no longer simply at the level of their management or the 'negotiation' of their acceptability – is excluded. The *science of risk* also involves a selection of parameters, of hypotheses and causes, a restriction on methodologies, that demarcate its territory; it acts as a divider and limiter. It is, however, possible to avert this 'ideology of risk' or *political overdetermination of risk*. The apprehension of risk and uncertainty always stems from social construction (Levasseur, 2006: 182-183), for example seen in economic and probabilistic reasoning which is dominant today. The research strategy that aims to collectively objectify uncertainties through an interdisciplinary process is an alternative. It is akin to risk analysis that is weighted ethically with clarified epistemic values (Shrader-Frechette, 1991). This strategy does not consider risk as a key concept today replacing ethics to define the relationship between man and the world. The challenge is rather to consider technological ethics going beyond the boundaries of the vision of conventional sciences and that *tolerates uncertainty*. The aim is to restore the epistemological status of the uncertainty in a context of interdisciplinary research in technological ethics, in accordance with the positive definition of the precautionary principle proposed by Matthias Kaiser and Jeroen P. van der Sluijs, whereby 'when human activities may lead to morally unacceptable harm that is scientifically plausible but uncertain, actions shall be taken to avoid or diminish that harm' (Kaiser and Van der Sluijs, 2008: 222).

Perspectives for GM fish

I build on the assumption that these different points can be structured according to the following logic: the premise of epistemological pluralism enables us to consider and to sustain a truly interdisciplinary process. From this emerges a series of questions, frictions between disciplines and the moving of boundaries. The latter then results in uncertainties and these uncertainties require further interdisciplinarity research. This can be illustrated through the issue of GM fish, with the idea that the way research is conducted is just as important as the results.

The interdisciplinary process concerning GM fish

Examining the plausibility, characteristics and multiples impacts of an accidental presence of GM fish in France – each discipline of the DOGMATIS network[7] first develops its response to the issue: geneticists seek to ascertain the technical reality of transgenesis; detection chemistry and biology assess the transferability of traceability techniques from GM crops to GM fish; socio-economists try to understand the possible flow of imports of GM fish onto the French market; sociologists study the public perception of risks associated with this import; legal advisers assess the state of the law regarding the issue and the relevance of the solutions. At each stage of the process, the relationship between the discipline and the initial issue is reworked, and the problems and issues are then reorganised. Below are examples of the results of these interdisciplinary movements resulting in a series of questions: How stable is the genome after integration of the transgene? Are the detection techniques used for GM crops appropriate for GM fish? Can we measure and characterise the risk of accidentally imported GM fish in France? On what representations does the public perception of risk (consumers, NGOs, professionals, etc.) depend? How relevant are the solutions regarding GM fish from a legal viewpoint and can we re-examine legal concepts and categories? How can the uncertainties and pluralities be reported to enlighten the ethical debate? All of these questions are interdisciplinary. In order to illustrate this point, let us consider the possibility and characteristics of understanding the risks of accidentally imported GM fish in France. We must first assess the risk of GM fish passing from the laboratory to the farm. This requires high grade knowledge of the global context of GM fish research; the contribution of geneticists is thus essential. We must then identify the potential trade vectors of GM fish flows, which require a firm understanding of international development and the structure of aquaculture. To measure risk, we must also assess the process of GM fish identification and traceability. This implies in-depth legal considerations regarding system failure and responsibilities in the event of accidental presence or release. We must consider how the risk will be perceived by the public concerned. And finally, the philosophers must ponder the social and environmental consequences of such a technological trajectory.

From these different levels of questioning and by taking into account the multiple dimensions of a problem, it becomes possible to highlight the *inductive risks* (Douglas, 2000) and the uncertainties. This brings us into the realm of non-epistemic values (social and ethical values), such as the *tolerance of uncertainty* in the face of a complex problem.

[7] This network is called DOGMATIS (Défi des OGM AquaTiques Impacts et Stratégies). It is a multidisciplinary project, financed by the ANR (Agence Nationale de la Recherche) and supported by INRA (Institut National de la Recherche agronomique), for 2007-2010.

Tolerating uncertainties of GM fish issues

The technological trajectory of GM fish involves both epistemic and ontological uncertainties. For the former, these may be uncertainties in the process of transgenesis itself. According to an FAO report, 'introduction of a transgene into an animal is not a precisely controlled process, and can result in a variety of outcomes regarding integration, expression and stability of the transgene in the host. [...] The expression of the transgene can have pleiotropic effects, that is, effects upon multiple traits of the host' (FAO, 2003: 5). So, the epistemic uncertainties of the transgenesis 'arise from the lack of control over the number of sequences and sites of insertion, the rate of expression of the transgene, the complexity of interactions between the gene networks, the multiplicity of gene functions, epigenetics and the interactions with environment'. At present, the control of the transgene is unpredictable (Le Curieux-Belfond, 2009: 5). These observations do not facilitate the objectifying and quantifying of environmental risks of potential transgenic releases. In the same way, in order to facilitate the identification of the new organism and enhance traceability, it is necessary to know the number of copies and insertion point of the transgene (Bertheau, 2009). In light of previous findings, the assessment can get complicated when the origin of any GM fish is not necessarily known. If research is to continue on this issue, to improve the identification and traceability of GM fish and effectively predict the environmental impacts, these epistemic uncertainties can have a significant ontological dimension. For example, predicting structured population dynamics in changing environmental conditions, which is inherently variable, is very difficult (Hayes *et al.,* 2007: 192). In addition, the accidental presence of GM fish may be linked to unknown or non-predictable vectors. Indeed, beyond the *evident vectors* (aquaculture, fish trade flows, research laboratories), there may be *non-evident vectors* (unauthorised laboratories [*biohacking*], aquarists, amateur fishing of escaped GM fish, etc). Once again, we are confronted with uncertainties that are due to variability. These can also be unexpected medium and long-term side-effects of this technology. They are also linked to the acceptance of technological pluralism regarding options available for the development of aquaculture. Indeed, at a time when the quantities of aquatic products stemming from fishing and aquaculture are virtually the same, at a time when we are seeing what is now commonly known as the *blue revolution* or the *age of aquaculture*, the consequences of introducing a new technology into the entire aquaculture system is difficult to predictable. What will the consequences be on small aquaculture structures in the northern and southern hemisphere? What impacts will this have on food levels production over the century? It is fair to say that 'the use of these technologies in aquaculture raises a number of unique questions, including issues relating to environmental impacts, fish welfare, food safety and distributive justice' (Millar, 2006: 442). To this is added a societal variability, due to the 'chaotic and unpredictable nature of societal processes' (Walker *et al.,* 2003), individual behaviour (consumers and citizens), uncertainties surrounding legal solutions (increasingly complex responsibilities)... a series of problems that require tolerance of inevitable and inexhaustible uncertainty.

The interdisciplinary process underway will enable us to move forward on these issues, to characterise more effectively the uncertainties whether they are epistemic or ontological and the risks whether real or perceived. The quality of the knowledge generated – dependent epistemic and non-epistemic values – becomes a relevant means of enlightening the democratic debate and conditions the possibilities of effective citizen participation, which will begin at the earliest possible stage of the process research.

Conclusion

Technological ethics cannot be summed up as a mere evaluation or management of the quantifiable risks conducted in hindsight. I propose that it should be enriched by epistemological pluralism, another conception of science, particularly through the application of interdisciplinarity. Only the latter allows uncertainties to emerge and be collectively objectified. I believe that this epistemic position which takes into account identifies and tolerates uncertainty, is also an ethical position. Just as science conveys non-epistemic values, I believe that technological ethics can relate to the epistemology of scientific approaches. Technological ethics is thus about epistemology and ethics. What is interesting is the opportunity to fully integrate ethics into a truly interdisciplinary process, and therefore to explore all the dimensions of the GM fish problem ahead of the evaluation, management or decision-making processes, i.e. during the process of knowledge production, hypothesis production, and objectification of issues and risks. Inserted in this process and informed of the many uncertainties, it can shed light on the complex technological trajectory of GM fish. In this process, it will serve to guarantee the *identity-relation* of all disciplines, highlight the links between knowledge drawing from the pluralities involved, while tolerating uncertainties. It will prevent a complex problem from being reduced to a simple cost-benefit analysis. Hence, most importantly, it will enable us to extract the ethical dimension of a complex problem in order to recognise the humanity of man (which in this case is not expressed in accounting terms but in tolerance of uncertainty), which, in my view, is one of the primary ambitions of ethics. It is also connected to ethical evaluation methods, which will benefit from this integrated work. Hence, I hold that technological ethics, through the ability to connect with both the purely scientific interdisciplinary work, and the other layers of ethics, can be considered as a meta-ethics of technological trajectories.

Acknowledgement

All members of the project DOGMATIS: Barrey, S, Boy, L., Doussan, I., Mambrini, M., Mariojouls, C., Schmid, A-F., Bertheau, Y., Sohm, F., Varenne, F.

References

Bertheau, Y. (2009). Détection et traçabilité des poissons génétiquement modifiés. In: The DOGMATIS Workshop, Jouy-en-Josas (Paris), 23 January 2009.

Coutellec, L. (2009). Comment la recherche-action déplace les frontières? In: Schmid, A.F. (ed.) Epistémologie des frontières, Editions Pétra, Paris.

Douglas, H. (2000). Inductive Risk and Values in Science. Philosophy of Science 67(4): 559-579.

Dubois M., Phan D., Schmid, A.F. and Varenne, F. (2007). Epistemology in a nutshell: Theory, Model, Simulation and Experiment. In: Phan, D. and Amblard, F. (ed.) Agent-based Modelling and Simulation in the Social and Human Sciences, The Bardwell Press, Oxford, pp. 357-391.

FAO/WHO (2003). Expert Consultation on Safety Assessment of Foods Derived from Genetically Modified Animals Including Fish, Rome, 17 – 21 November 2003.

Feenberg, A. (1999). Questioning technology. Routledge, London, 243 pp.

Hacking, I. (1999). The Social Construction of What?. Harvard University Press, Cambridge, 272 pp.

Hayes, K.R., Regan, H.M. and Burgman, M.A. (2007). Introduction to the concepts and methods of uncertainty analysis. In: Kapuscinski, A.R., Hayes, K.R., Li, S., Dana, G., Hallerman, E.M. and Schei, P.J. (ed.) Environmental Risk Assessment of Genetically Modified Organisms. CAB International, Oxfordshire and Cambridge, 304 pp.

Lacey, H. (2001). The ways in which the sciences are and are not value free. In: Proceedings of the conference 'Value Free Science: Illusion or Ideal?', Center for Ethics and Values Sciences, University of Alabama at Birmingham, February 23-25.

Le Curieux-Belfond, O., Vandelac, L., Caron, J. and Seralini, G.-E. (2009). Factors to consider before production and commercialization of aquatic genetically modified organisms. Environmental science and policy 12(2): 170-189.

Levasseur C. (2006). Incertitude, pouvoir et résistances. Presses de l'Université Laval, Laval, 450 pp.

Millar, K. and Tomkins, S. (2006). The implications of the use of GM in aquaculture: issues for internationational developpement and trade. In: Kaiser, M. and Lien M. (ed.). Ethics and politics of food. Wageningen Academic Publishers, Wageningen, the Netherlands, pp. 442-445.

Putnam, H. (2001). The Collapse of the Fact Value Dichotomy. Harvard University Press, Cambridge, 208 pp.

Shrader-Frechette, K. (1991). Risk and rationality. University of California Press, Los Angeles, 272 pp.

Van Der Sluijs, J.P. and Kaiser, M. (2008). Vers une compréhension commune du principe de précaution. In: Allard, P., Fox, D. and Picon, B. (eds.) Incertitude et environnement. Edisud, Aix-en-Provence, 479 pp.

Walker, W.E., Harremoes, P., Rotmans, J., Van Der Sluijs, J.P., Van Asselt, M.B.A., Janssen, P. and Krayer von Krauss, M.P. (2003) Defining Uncertainty. A Conceptual Basis for Uncertainty Management in Model-Based Decision Support. Integrated Assessment 4(1): 5-17.

The normative evaluation of the societal interface group

B. Gremmen and L. Hanssen
Centre for Methodical Ethics and Technology Assessment, Wageningen University, Hollandseweg 1, 6768 KN, Wageningen, the Netherlands; bart.gremmen@wur.nl

Abstract

The aim of this paper is to analyse the normative content of the switch from open ended public engagement to genuine public participation (distributive expertise) in the field of plant genomics. In the last few decades (plant) scientists have used a wide array of communication instruments to engage the public in their work and its societal consequences. However, it proved difficult to provide public access to knowledge and to the decisions on the application of that knowledge. The Dutch Centre for BioSystems Genomics (CBSG) is a public/private consortium engaged in genomics research on tomato and potato with the aim of developing new cultivars. In 2005 the CBSG started an experimental series of five Societal Interface Group (SIG) sessions, to involve members of the public in their work. Ten critical and independent people with different relevant societal backgrounds in experience and preparation of food, together with eight members of the Management Team of the CBSG, discussed the research agenda, projects, and societal context of the research. At the end of each session the SIG members formulated a set of recommendations. In 2009 we will evaluate the SIG as a deliberative framework that facilitates upstream engagement in agenda setting and downstream engagement in policy-formulation within the CBSG context. In this paper we will report on the normative aspects involved: transparency and access to knowledge, openness of discussions, pluralism in deliberations, responsibility of the Management Team in addressing the recommendations, and autonomy of the SIG.

Keywords: public engagement, public participation, transmission, transaction

Introduction

The Dutch Centre for BioSystems Genomics (CBSG) is a public/private consortium engaged in genomics research on tomato and potato with the aim of developing new cultivars. In 2003 the management team of CBSG discussed the issue of societal communication on the basis of a stakeholder analysis and a communication plan made by a Dutch consultancy. A strong necessity was felt by the CBSG management team to communicate results of their research to 'the world outside' in a responsible and overt way. After decades full of controversies, e.g. GM crops, scientists recognize the compelling need to become more open and accountable. During the coming years, recovering the public trust will continue to be the most important issue in the process of transferring science and technology to the market (Gaskell *et al.*, 2004; Poortinga and Pidgeon, 2005; Gutteling *et al.*, 2006). To achieve this task, CBSG has to increase its insight into the way science communication works and the effects it achieves and, in that way, provide the groundwork to further interactions between lay people and genomics experts, and also between natural and social scientists (Wilsdon *et al.*, 2005; Hanssen and Van Katwijk, 2007).

The aim of this paper is to analyse the normative content of the switch from open ended public engagement to genuine public participation (distributive expertise) in the field of plant genomics.

From transmission to transaction

Actors with a vested interest in new scientific and technological developments use communication about science and technology to assess the possible support for new developments. The tools they use were developed from the perspective of 'communication as transmission' (Hanssen *et al.*, 2003). Insiders, like scientists, policymakers and corporate managers dominate the discourse with scenarios and policy options of their own. According to the transmission view the actors expect that 'better information' leads not only to more knowledge, but also to more trust and support of scientific developments and technological innovations.

In the transmission view, communication is an asymmetrical process, in which a sender formulates a message for a receiver who has an information shortage. From structured analyses on two public debates on biotechnology and genomics, especially *Food and Genes* and *Genes for your Food, Food for your Genes*, we can place some critical observations to the transmission-mode, commonly used (Van Est *et al.*, 2003; Gutteling et *al.*, 2006; Hanssen and Van Katwijk, 2007; Dijkstra, 2008). First, new risks and scientific uncertainties which accompany new biotechnology and genomics applications call for the associated communication and interaction arrangements:

- to engage all interested parties to suitably and efficiently communication of opportunities and risks;
- to persuade scientists and government officials to be open to public wishes and concerns;
- in cases of high ambiguity, to allow public engagement be an integral part of agenda-setting and policy-making processes (Klinke and Renn, 2002; Pellizoni, 2003; Hanssen, 2006; Pielke, 2007).

Second, in the communication processes, four deficits were identified to:

- public representation of scientific uncertainties;
- imagination of future technological progress;
- moral reflection of implications;
- public and political experience with new products (Nowotny et *al.*, 2001; Hanssen; 2004, Guston; 2008).

In the transaction view, communication is a more or less symmetrical process, and encourages two-way processes, involving 'specialists' and 'laypersons' in the same appraisal process, providing for consistency of framing, mutual interrogation and substantial opportunities for face-to-face deliberations. This leads to experiments in which the relationship between actors and the public are more symmetrical. Recently *societal interface groups* have been developed in which actors and members of the public try to understand each others knowledge, vision, values, and norms (Gremmen, 2007).

The Societal Interface Group: an experiment in public participation

CBSG looked for a new instrument that would enable constructive deliberations between CBSG management and persons from different societal positions. There are many and different possibilities to interact or communicate with society. For example, it is possible to get feedback from all kinds of public outreach activities, from focus groups and public opinion surveys, or even to use an external advisory board. However, CBSG didn't opt for a group of societal advisors at a distance, which may lead to misunderstanding, or to start an experimental information campaign with inherent contingencies. The new CBSG approach was to develop a societal interface that would pick up on and discuss signals from society by integrating 'outsiders', people from different societal backgrounds, and 'insiders', members of the CBSG management team. The intention was to set up an interactive learning process in which on the one hand consumers/citizens concerns and needs could influence scientific developments in the field of genomics, and on the other hand scientists could search for more shared aims, meanings en starting points for their research agenda. In this process not only cognitive matters are important, but also values and belief systems, and issues of mutual trust, responsibility and (corporate) governance are at stake.

In 2004 the Societal Interface Group (SIG) started, and in two years five sessions were held. The aim of the SIG is to involve society in CBSG, more precisely, to translate societal concerns and needs such as sustainability and product quality into research questions, technological applications and, in the end, new products preferred by consumers. In the SIG ten people with different relevant societal backgrounds (a journalist, a chef, a farmer, etc.) together with eight members of the Management Team of the CBSG discussed the research agenda, projects, and societal context of the research.This 'hybrid' character makes the SIG special. The task of the SIG is to identify possibilities and (future) concerns and needs, to develop pro-active intervention strategies, and to integrate their results within CBSG research programme.

The results of the five sessions of the SIG held in CBSG-2007 include:
- a list of societal issues about four relevant themes in the CBSG research programme: product quality, disease resistance, economic and political aspects, and communication aspects;
- recommendations about these four themes in CBSG-2012 policy-formulation;
- recommendations about the research agenda of CBSG-2012, in particular the projects of the societal research programme.

After a general evaluation of these first five sessions the management team of CBSG is positive about the process and the results of the SIG and decided to continue it in CBSG-2012.

The ethics of public engagement

In the SIG communication is embedded in a wider and reflective vision of science's position in society. The meaning of the scientific research of CBSG is clarified on the basis of a dialogue with a broad range of people from the public. Experts of CBSG showed that they are willing to focus more on the social, cultural and moral (re)presentation of scientific knowledge and the social structure of technology.

Interactions between public and experts are often restricted to discussing possible benefits and risks, while the underlying questions about values, visions and interests, which give direction to scientific research and technology development, are hardly mentioned (Stilgoe, 2007; Hanssen and Van Katwijk, 2007). In contrast, the SIG sessions focused on underlying questions about values, norms, etc. In this way, larger questions of the character and direction of scientific and technological change were no longer ignored. The list of issues that resulted from the SIG Sessions contains many societal values.

Another aspect of the interactions between public and experts is the undefined nature of ethical issues, which are often described with reference to the potential aspects or implications of new technologies that need to be studied by special committees of the various projects. These wider issues are therefore framed as other types of risks that have to be dealt with by expert knowledge, and as such, not the object of citizen inquiry (Wilsdon *et al.,* 2005). In the case of the SIG members of the public have been allowed to discuss these issues before the research agenda has been finalised, and before the results left the laboratory. The management team of the CBSG made sure that SIG members got access to all kinds of information they needed to assess plant genomics in open discussions.

An important task in developing and using new instruments is to make use of images that appeal to people, which explain scientific nuances, and which allow for input of 'civic' or 'practical' knowledge (Burgess *et al.,* 2007; Hanssen and Van Katwijk, 2007; Stilgoe, 2007). Despite different value systems and competing interests, the participants of the SIG aimed at shared ambitions and directions for solutions by confronting and eliciting each others perspectives. In this way, participants take account of contesting beliefs and ambitions on the various issues, distributive expertises, and divergent directions of solution.

The two most important ethical principles of the SIG are that all its members ('internal' and 'external') are equal, and that the outcomes of the deliberations are the result of shared responsibility. It was the responsibility of the Management Team to address the recommendations, and autonomy of the SIG. Members of the SIG were allowed to determine the relevant subjects of the CBSG research agenda.

Also, they were allowed to suggest new research topics and to bring in some wider issues. However, within the SIG the 'outside' members were taken out of their environment, had to learn a new idiom, and also had their own expectations about their participation. Also, it takes time and effort to value their contributions compared to the scientific expertise of the 'inside' members.

Conclusion

In her comparative political study of biotechnology in the United Kingdom, Europe and North America, Jasanoff (2005) showed, that where public involvement is insufficiently available formally, it will occur informally, through public protest; market choices, such as consumer rejection of genetically modified foods; or new political structures, such as environmental movements. The SIG as a new instrument for 'participatory deliberation' shows a particular interest in consulting and involving members of the public in the agenda-setting and policy-making processes, at a time when they can still influence research priorities (Rowe and Frewer, 2004; Willis and Wilsdon, 2004). The focus of the SIG sessions on underlying questions about values, norms, etc. shows that larger questions of the character and direction of scientific and technological change were no longer ignored.

The 'external' members of the SIG played their role in a critical and constructive way by showing the importance of public interests and views. Not all members of the SIG had to agree on everything, but deliberations have led to a greater understanding of each other's world views, to a greater joint responsibility, and made them more committed. There was a sense of closure by reaching a joint responsibility and commitment toward 'doing society and genomics' (Gremmen, 2007).

The SIG sessions showed that public expertise is not threatening or irrational, but highly relevant for the development of societal practices in which genomics plays a role. Its input offers the possibility to use the creative potential of future users in the development of genomics applications.

References

Burgess, J., Stirling, A., Clark, J., Davis G., Eames, M., Staley, K. and Williamson, S. (2007). Deliberative mapping: a novel analytic-deliberative methodology to support contested science-policy decisions. Public Understanding of Science 16(3): 299-322.

Dijkstra, A. (2008). Of publics and science. How publics engage with biotechnology and genomics. Thesis. Universtiy of Twente.

Gaskell, G., Allum, N., Wagner, W., Kronberger, N., Torgersen, H., Hampel, J. and Bardes, J. (2004). GM foods and the misperception of risk perception. Risk Analysis 24(1): 185-194.

Gremmen, B. (2007). De Zwakste Schakel. Over maatschappelijk verantwoorde genomics, inaugural address, Wageningen Universiteit.

Guston, D.H. (2001). Boundary organizations in environmental policy and science: an introduction. Science, Technology and Human Values 26(4): 399-408.

Gutteling, J., Hanssen, L., Van der Veer, N. and Seydel, E. (2006). Trust in governance and the acceptance of GM food in the Netherlands. Public Understanding of Science 15(1):103-112.

Hanssen, L., Dijkstra, A., Roeterdink, W. and Stappers, J. (2003). Wetenschapsvoorlichting: profetie of professie. Amsterdam: Stichting Weten.

Hanssen, L. (2004). Verbeelding van Wetenschap. Amsterdam: Stichting Weten.

Hanssen, L. (2006). Governance van biotechnologie. De veranderende rol van wetenschappelijke adviescolleges. COGEM Rapport 2006-1. Bilthoven: COGEM.

Hanssen, L. and Van Katwijk, M. (2007). Paradigmashift in de WTC. Van transmissie- naar transactiedenken. In: Willems, J. (red.) Basisboek Wetenschapscommunicatie, Amsterdam: Boom, pp. 130-149.

Jasanoff, S. (2005). Designs on Nature. Science and Democracy in Europe and the United States. Princeton: University Press.

Klinke, A. and Renn, O. (2002) A new approach to risk evaluation and management: risk-based, precaution-based, and discourse-based strategies. Risk Analysis 22(6): 1071-1094.

Nowotny, H., Scott, P. and Gibbons, M. (2001) Re-Thinking Science: Knowledge and the Public in an Age of Uncertainty. Polity Press, Cambridge.

Pellizoni, L. (2003). Knowledge, uncertainty and the transformation of the public sphere. European Journal of Social Theory 6: 327-355.

Pielke, R.A. (2007). The Honest Brooker. Making Sense of Science in Policy and Politics. Cambridge University Press, Cambridge.

Poortinga, W. and Pidgeon, N. (2005) Trust in risk regulation: cause or consequence of the acceptability of GM food? Risk Analysis 25(1): 199-209.

Rowe, G., Marsh, R. and Frewer, L. (2004). Evaluation of a deliberative conference. Science, Technology & Human Values 29(1): 88-121.

Stilgoe, J. (2007) Nanodialogues. Experiments in public engagement with science. London: DEMOS.

Van Est, R. and Hanssen, L. (2003). Genomics in the agrofood sector. An overview of social questions and dilemmas. Technikfolgenabschätzung. Theorie und Praxis 12(1): 100-105.

Willis, R. and Wilsdon, J. (2004). See-through Science. Why public engagement needs to move upstream. London: Demos.

Wilsdon, J., Wynne, B. and Stilgoe, J. (2005). The Public Value of Science. Or how to ensure that science really matters. London: Demos.

Are consumers' expectations of 'GMO-free' labels answered by production qualities?

M. Henseleit and S. Kubitzki
Institute of Agricultural Policy and Market Research, Justus-Liebig University Giessen, Senckenbergstrasse 3, 35390 Giessen, Germany; meike.henseleit@agrar.uni-giessen.de

Abstract

Consumers, particularly in industrialised countries, are concerned about the application of genetic engineering in food production. However, without appropriate labelling it is impossible for consumers to recognise whether genetically modified organisms (GMOs) have been applied during the production process. Besides labels producers are obliged to put on food products when they contain GMOs, in some countries also labels can be applied on products that are free from GMOs. This type of label is supposed to enable producers to better promote such products. The problem is that requirements for labelling food products as 'GMO-free' can be very different, and therefore it is questionable whether consumers' understanding of 'GMO-free' is consistent with what certain labels actually can guarantee. We conducted a consumer survey in October 2008 in order to explore potential gaps between expectations of 'GMO-free' food and production requirements in the case of the German regulation covering the labelling of foods as 'GMO-free'. Our results indicate significant differences between consumers' views and standards of production.

Keywords: genetic engineering, food labelling, consumer survey

Introduction and background

Genetically modified organisms (GMOs) in food are a growing concern for consumers (Bansal *et al.*, 2007). Although the harmful nature of GMOs has been questioned, especially by commercial seed providers and agricultural producers, and no scientific evidence has been provided yet to suggest that genetic modification of crops could be harmful to humans, many consumers feel a visceral reaction to the thought of eating food that has been genetically modified. Consumers, especially in industrialised countries (e.g. Curtis *et al.*, 2004; Hansen, 2004; Bansal *et al.*, 2007; Lusk and Rozan, 2008), wish to avoid GMOs in food, not only for health reasons but also because of environmental and/or religious, ethical or other non-safety related issues, and, thus, need to be informed whether genetic engineering has been applied during the production process. Hence, the question arises, what kind of labelling is appropriate as viewed by consumers and how should a labelling system be designed to support consumers identifying the products they want.

There are different standards concerning the labelling of GMO-free food products worldwide, if the labels are based on standards at all. They are more or less restrictive and often allow for exemptions, especially in the field of meat and animal products (Gruere and Rao, 2007). This paper tries to shed light on whether the German 'GMO-free' label for animal food products really helps consumers to find the products they want, or if in reality it deceives them. Hence, the following questions emerge: First, what are consumers expecting from a 'GMO-free' label? How large is the gap between what consumers demand from the label 'GMO-free' and what it stands for? Second, does it make a difference in the view of consumers at which stage of the production chain and to what extent genetic engineering has been applied? Third, does the importance of being free from any kind of genetic engineering vary between product categories in the perception of consumers? And finally, how important are different applications of genetic engineering in food production for consumers' purchase decisions?

In order to deal with these questions, we conducted an online survey in October 2008. Our results indicate that consumers strongly refuse genetically modified feed crops, but are rather tolerant of genetic engineering in medication and feed additives.

Survey and methodological approach

Information about which applications of genetic engineering are acceptable for consumers within the process of food production are lacking in the present literature. Therefore, we decided to do an internet-based consumer survey. Online research allows the respondent to see pictures and to participate in complex choice experiments, which was relevant in our case. Furthermore, there is no influence of the interviewer as a person, so there is a smaller incentive to give socially desirable answers. Socially desirable answers are not an unusual phenomenon in interviews considering genetic engineering particularly in Germany, because the debate about the issue is rather emotional. Before we started the survey, we applied several pre-tests of the questionnaire of which the last one was an online survey with about 100 participants in order to improve design and wording. The data collection of the online consumer survey lasted from the 17th until the 28th of October in 2008. The final sample consisted of 1,012 participants recruited from an online-access panel. All participants were living in the federal state of Hesse in Germany, but for the reason that citizens of this region do not diverge significantly in their attitudes towards genetic engineering in food production from at least most European member states (Gaskell *et al.,* 2003, 2006), the results can be seen as transferable to a number of industrialised countries. However, it has to be pointed out that the resistance in Europe against GM foods is somewhat higher than, e.g., in Canada, and significantly higher than in the United States (Curtis *et al.,* 2004; Gaskell *et al.,* 2006; Lusk and Rozan, 2008). The recruitment process for survey participants from the online-access panel included quotas relating to gender, age (only within the range from 16 to 59 years, since elderly people are difficult to recruit via the internet) and place of residence. The average time to complete a questionnaire was about 15 minutes.

According to the literature (e.g. Bredahl, 2001; House *et al.,* 2004; Hartl, 2008) we expected that besides socio-demographic characteristics also consumers' involvement in food products and food neophobia as well as consumption habits, attitudes towards the environment and knowledge about genetic engineering in food production would have an impact on what consumers expect from 'GMO-free' labelled food. Hence, our questionnaire consisted of the following parts: In the first part, people were asked about habits and attitudes related to food and nutrition. Food neophobia and consumer involvement were measured using subsets of items. A reliability analysis of the pre-test data was applied to establish reduced scales of both food neophobia and involvement in food products that were used in the final questionnaire. In the second part we investigated subjective and objective measures of consumers' knowledge and experience with genetic engineering in agriculture, together with questions about image and expectations of a 'GMO-free' label. After that, people were asked in the third part to state preferences regarding the application of genetic engineering in livestock production and to complete a choice experiment in the fourth part, which should help to validate stated preferences. Then, attitudes towards genetic engineering in food production were measured using items from the scale suggested by Bredahl (2001) and Hartl (2008). General attitudes about ethical and environmental issues in the context of genetic engineering were also explored, before participants were asked about socio-demographics and then finally they were given the opportunity to give feedback about the issue and the interview. Altogether, the survey was centred on the labelling of animal-based products as 'GMO-free'. The products used in questions about purchase and consumption as well as in the choice experiment comprised milk, eggs, ground pork and ground beef. In addition to descriptive statistical methods, multivariate methods were applied to analyse the survey data.

Results

The sample is fairly representative of the German population, at least in the age group from 16-59 years. Exactly half of the participants (506 people) are male, which is due to the quota set for the sample. The level of both income and education is higher in the sample than in the population of Germany. This is partly due to the fact that participants of online surveys usually have a higher education level and therefore get higher salaries on average.

Considering knowledge and experience with genetic engineering in food production, most of the participants have already thought about genetic engineering in agriculture, at least once or twice. About 22% stated that they have thought about it often, whereas only about ten percent said that they have never thought about it prior to the survey. When they were shown one of the most common forms of the new 'GMO-free' label, more than seventy percent denied that they had seen such a label before. About 23% of the survey participants ticked the option that they had seen the label already on food products, and less than ten percent gave television or newspapers as the source where they had noticed the label already. Apparently the new label is not yet well-known in Germany.

Considering knowledge about genetic engineering, we first asked participants how well they feel informed about genetically modified food (subjective knowledge). Apparently, the majority felt that they were not very well informed about the issue, and this makes it difficult for people to assess individual risks that are associated with genetic engineering in agriculture. Such a notable degree of uncertainty is partly confirmed by the next question, which is supposed to measure knowledge concerning genetic engineering (objective knowledge). Survey participants were given a list of statements regarding genetic engineering in agriculture and its potential consequences. Only about 8% of respondents of our survey classified all the five of the statements correctly, while the average amount of correct classifications was about 2.6. The Spearman rank correlation coefficient between the perceived level of information and the number of correct answers as an indicator for the objective level of knowledge is highly significant with about 0.3.

As mentioned above, one of the main goals of the survey was to compare consumers' expectations about the new 'GMO-free' label with what it stands for. We asked whether specific applications of genetic engineering would be acceptable for the respondents on products labelled as 'GMO-free'. Process characteristics and whether they are acceptable for labelled products are shown in Table 1. In the last column it is indicated whether products are permitted to carry the 'GMO-free' label based on the German regulation or not.

As can be seen from Table 1, there seems to be a clear difference between consumers' understanding of 'GMO-free' and the legal requirements to carry the label. About 78% did not believe that food products should be allowed to carry the 'GMO-free' label if the fodder has been free from GMOs only for the required time period before product generation. Apart from that, the use of GMOs within the production of drugs and vaccines seems to be more acceptable, as a proportion of more than 23% thinking that products still should be allowed to carry the label shows.

We also investigated whether people differentiated between meat products on the one hand and milk and eggs on the other hand, when considering the acceptance of genetically modified organisms in animal feedstuff. According to our survey data, no difference is made: More than 90% of the respondents gave the same answer for both groups of products. 74% were opposed to the 'GMO-free' label for products in which animals have been fed without genetically modified crops only for a certain period before the raw material was produced. Hence, most consumers expect from 'GMO-free' labelled products that animals have never been fed with genetically modified fodder crops during their lifetime. The majority

Table 1. Expectations and fulfilment of production standards.

	Should be allowed to carry the label	Should **not** be allowed to carry the label	Don't know	Allowed to carry the label based on the regulation
Food product contains genetically modified organisms, e.g. yoghurt cultures, yeasts.	25.1%	59.6%	15.3%	No
Food contains enzymes or has been produced with the aid of enzymes, which were being obtained from genetically modified organisms.	26.9%	56.1%	17%	Yes
Animal feed contains genetically modified organisms.	8.4%	78.3%	13.3%	Yes and no[1]
Animal feedstuff contains additives which have been produced with the aid of genetically modified organisms.	11.9%	73.5%	14.6%	Yes
Drugs and vaccines for the animals have been produced with the aid of genetically modified organisms.	23.4%	61.2%	15.4%	Yes

[1] Until a certain time period before slaughtering / milking / laying eggs.

of consumers are clearly against genetically modified feed crops, whereas the tolerance for genetic engineering in the fields of medicine and food additives is somewhat higher. A further notable aspect is the quota of people choosing the 'Don't know'-option, which is about 15% on average. This again supports the observation that consumers are rather uncertain about the impact of genetic modifications in food production on health, well-being and nature.

In the next part of the questionnaire participants were asked about the importance of certain product characteristics. Before doing so, they were questioned about consumption habits of different animal products (beef products, products containing eggs, dairy products or products made from pork). This screening question made sure that participants were confronted only with a product they regularly buy. Furthermore, a special setting in the online questionnaire provided for equal numbers of the products evaluations, which means that for each of the four product types about 250 participants were consulted. Accordingly, each respondent was asked to rate the importance of characteristics of one type of product only. In Figure 1 the mean level of importance measured on a scale ranging from 1 ('unimportant') to 5 ('very important') is shown.

Apparently, the issue of whether the product is 'GM-free food for the whole lifetime of farm animals' is vitally important to these consumers. Also essential is the avoidance of genetic engineering within the production of both feed additives and medical substances. These aspects are regarded as even more important than the kind of farming (organic) or local origin. Yet asking people in this way about the importance of product qualities, there is no need to make a trade-off and therefore all of the aspects are likely to be rated important. However, the outcome of these stated preferences is verified by the inclusion of a choice experiment in the questionnaire. The experiment was conducted as a choice based conjoint analysis comprising the characteristics of the application of genetic engineering in animal feed production, application of feed additives that have been produced with the aid of GMOs, administration of drugs and vaccines that have been produced with GMOs, and product origin and price. Summarising

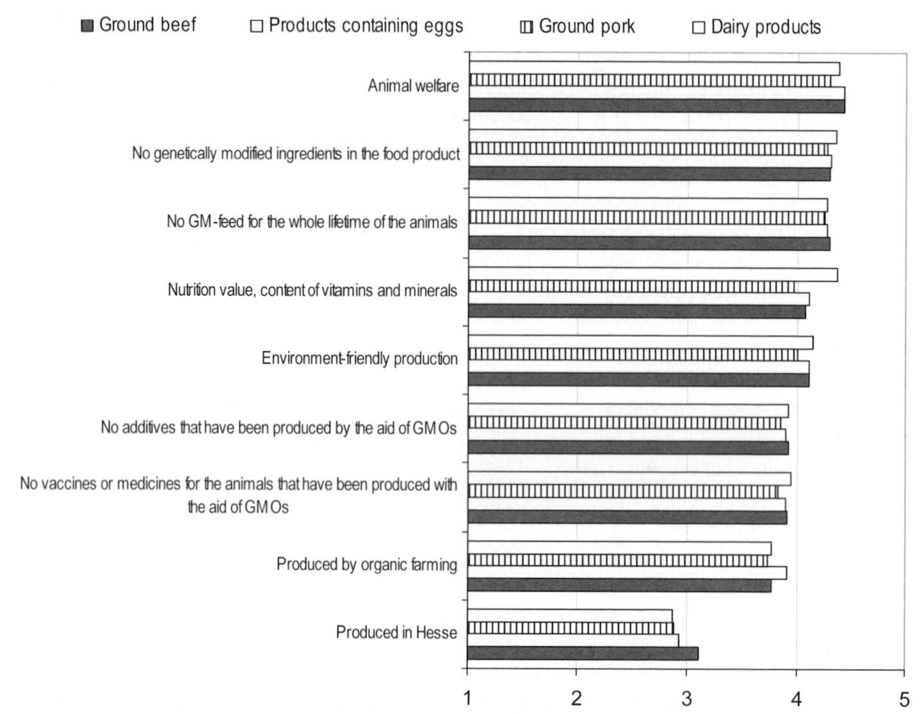

Figure 1. 'Please indicate the meaning of the following product characteristics for you when using ground beef / products containing eggs / ground pork / milk in the kitchen on a scale ranging from '1 = unimportant' to '5 = very important'', n=1,012.

the results of the experiment, highest importance was given to the characteristic of whether genetic engineering had been applied in feed production. This observation confirms the results pictured in Table 1. Furthermore, the question whether animal feed contains additives that have been produced with the aid of GMOs appears to be as important as the price attribute. The relative importance seems to be independent from the type of product, as already mentioned above, so the ranking of attribute importance is almost identical for all the four of the considered products.

Discussion and conclusions

The aim of our research was to identify potential gaps between consumers' expectations of what the GM-free-label in Germany stands for, and what it actually means according to the legal regulation. Furthermore, we wanted to identify relevant factors for the acceptance of genetic engineering techniques within the food production process. We conducted a representative online survey asking more than one thousand consumers about their opinions towards food, environment, genetic engineering and what they would be willing to tolerate in food labelled as 'GMO-free'. We also conducted a short choice experiment which was supposed to validate stated attitudes and behaviour. Overall, our results confirm anticipated relationships between external variables and the preference for GMO-free production chains. In addition, a high impact of attitudinal factors on the aversion towards the application of genetic engineering in the food production was revealed by regression models, presumably built to some degree on missing or biased information,.

More and more producers are starting to use the 'GMO-free' label, hence consumers will soon find a broader range of labelled products on the shelves. However, our survey results demonstrate that consumers expectations of 'GMO-free' labelled food products do not comply with the requirements of the 'GMO-free' label introduced by the German government in 2008. Accordingly, adequate information about the real claims of 'GMO-free' labels is necessary in order to avoid deception of consumers and thus enable them to make conscious buying decisions. If label requirements for animal products remain as they are today in Germany, consumers who wish to avoid any food that has come into contact with genetic engineering, not only for health reasons but also for ethical attitudes, can not rely on the 'GMO-free' label alone.

References

Bansal, S. and Ramaswami, B. (2007). The economics of GM food labels: An evaluation of mandatory labeling proposals in India. Discussion Paper: International Food Policy Research Institute, Washington, DC.

Bredahl, L. (2001). Determinants of Consumer Attitudes and Purchase Intentions With Regard to Genetically Modified Foods – Results of a Cross-National Survey. Journal of Consumer Policy 24: 23-61.

Curtis, K.R., McCluskey, J.J. and Wahl, T.I. (2004). Consumer acceptance of genetically modified food products in the developing world. AgBioForum 7(1,2): 70-75.

Gaskell, G., Stares, S. and Allum, N. (2003). Europeans and Biotechnology in 2002. Eurobarometer 64.3. A report to the European Commission's Directorate-General for Research, Brussels. Available at: http://ec.europa.eu/public_opinion/archives/ebs/ebs_177_en.pdf. Accessed March 2008.

Gaskell, G., Stares, S., Allansdottir, A., Allum, N., Corchero, C., Fischler, C., Hampel, J., Jackson, J., Kronberger, N., Mejgaard, N., Revuelta, G., Schreiner, C., Torgersen, H. and Wagner, W. (2006). Europeans and Biotechnology in 2005: Patterns and Trends. Eurobarometer 64.3. A report to the European Commission's Directorate-General for Research Brussels. Available at: http://ec.europa.eu/research/press/2006/pr1906en.cfm. Accessed March 2008.

Gruere, G. and Rao, S. (2007). A Review of International Labelling Policies of Genetically Modified Food to Evaluate India's Proposed Rule. AgBioForum 10(1): 51-64.

Hansen, K. (2004). Does autonomy count in favour of labelling genetically modified food? In: Journal of Agricultural & Environmental Ethics 17(1): 67-76.

Hartl, J. (2008). Die Nachfrage nach genetisch veränderten Lebensmitteln – Anwendung neuerer Entwicklungen der Discrete-Choice-Analyse zur Bewertung genetisch veränderter Lebensmittel mit Output-Traits. Giessener Schriften zur Agrar- und Ernährungswirtschaft, Heft 34.

House, L., Lusk, J., Jaeger, S., Traill, W.B., Moore, M., Valli, C., Morrow, B. and Yee, W.M.S. (2004). Objective and Subjective Knowledge: Impacts on Consumer Demand for Genetically Modified Foods in the United States and the European Union. AgBioForum 7(3): 113-123.

Lusk, J.L. and Rozan, A. (2008). Public Policy and Endogenous Beliefs: The Case of Genetically Modified Food. Journal of Agricultural and Resource Economics 33(2): 270-289.

Theme 10 – Foundational issues in agriculture and food ethics

The 6 F's of agriculture

S. Aerts[1,2], J. De Tavernier[1], E. Decuypere[1] and D. Lips[1,2]
[1]Boerenbond Chair Agriculture and Society, Centre for Science, Technology and Ethics, Katholieke Universiteit Leuven, Kasteelpark Arenberg 1 -bus 2456, 3001 Leuven, Belgium; Stef.Aerts@kahosl.be
[2]Katholieke Hogeschool Sint-Lieven, Hospitaalstraat 23, 9100 Sint-Niklaas, Belgium

Abstract

Agriculture is an ancient activity with many different functions. These functions can be summarised under 6 F's, the classic 'Food', 'Feed', 'Fuel', 'Fibre', and two additions: 'Flower', and 'Fun'. Historically the production of Fuel and Fibre have lost their importance, but they will regain importance due to climate change policies and fossil fuel shortages. Societal evolutions will increase the importance of Flower and Fun. All six functions will have to be fulfilled, simultaneously, in a framework in which inputs (soil, water, air) are increasingly tight. The debate and the ethical analysis are complicated by oppositions between expected evolutions, moral intuitions and/or practical possibilities, especially when different timescales are considered. In literature, the different functions are often considered independently, although the interactions between them are the main reason for the complexity of the analysis, and for the magnitude of the challenges we face. The multitude of direct and indirect, technical and economical connections make it impossible to separate the different functions. The 6F model shows, by summarising and structuring the future challenges, that production increases will be necessary and input shortages will grow, and therefore that there is no other option than intensification. A well-understood intensification of agriculture is not only unavoidable, but even preferable and necessary. The rest is merely a question of priorities and boundary conditions.

Keywords: future, sustainability, functions, intensification

Introduction

Agriculture has evolved when humans have tried by new means to ensure a good and steady supply of food, at a level that would not be possible by hunting and gathering alone. This basic goal, food production, is still a fundamental goal or a function of agriculture.

Soon, agriculture was supplying more than food alone, it also produced fibres, feed for animals, and energy (e.g. traction). This is still the case, but the relative importance of these four basic functions ('Food', 'Feed', 'Fuel' and 'Fibre') has shifted dramatically, and we are on the verge of yet another shift. But there is more, certainly within a Western context, increasing attention is given to the protection of nature, and an increasing amount of time and money is spent on decoration and recreation. Agriculture will therefore be asked to produce, or give way to 'natural' environments, and there will be an economic pull towards non-food plant production ('Flower'). At the same time recreational use of agricultural areas will increase, especially in the peri-urban areas ('Fun').

In short, the future challenges for agriculture are manyfold, and they can be – somewhat generally – summarised under six Fs: Food, Feed, Fuel, Fibre, Flower and Fun.

The technical challenges

Food

Food production needs are evidently related to the population. Over the last 60 years, the majority of population growth is located in less developed countries. By 2050, there will probably be about 9 billion people, of which 3 billion will live in China and India, and another 5 billion in other less developed countries (US Census Bureau, 2004).

Economical development is another important factor influencing food consumption (and thus production needs). As the majority of the world population is living in areas with low buying power, it can be expected that development will result in (much) higher consumption. Total demand will therefore increase even more than the world population. Depending on eating habits and population size, this may well be a 250% increase by 2050 (Smeets, 2007).

Feed

Economic development usually results in a higher demand for animal products (and other 'high value' products). When these are produced using the current methods, this involves a significant amount of basic commodities such as cereals and soy as these form an important part of the rations that pigs and poultry (but also cattle) are fed on. This will necessitate a further increase in the production of these commodities beyond the population growth.

The United Nations Food and Agriculture Organisation (FAO) estimates a 3-fold increase in cereal demand by 2030, compared to 1960, a 4-fold increase in meat and milk production and a 7.5-fold and 10-fold increase in oil crop and soy production. For the same period the FAO only estimates a 2.6-fold population increase.

Fuel

For centuries, energy for traction was supplied by animals that were fed – partially – on feed produced specifically for that goal. This has diminished with the development of fossil fuel powered engines.

Recently, there is consensus that the global climate is changing (IPCC, 2007), and that we need to take measures to prevent that change or adapt to the consequences thereof. Agriculture is expected to contribute significantly to that challenge, not in the least by supplying commodities that can be transformed into transport and traction energy. These 'biofuels' are mainly biodiesel, bioethanol and biogas. The former two are produced primarily using agricultural commodities, the latter is based on organic waste, supplemented with e.g. maize. Although the high expectations (Anonymous, 2007; Biofuels Research Council, 2006) have diminished much over the last year, it is clear that any significant development of this sector, will have an additional impact on the production volume that needs to be realised by the agricultural sector.

Fibre

The increase in fibre demand is analogous to the increase in food; population size as well as economic development are important. *Ceteris paribus* this means that agriculture will have to increase its production of natural fibres (such as cotton, line, and probably also wood). This is confirmed in the projections of Scanes and Miranowski (2004).

Additionally, at a certain point there will be a decline in the availability of fossil fuels as the basis for the chemical industry. At that time, agriculture will have to step in as a supplier to the 'green chemistry'. Depending on the speed with which the availability decreases (and the prices increase), this will be an important production sector in the shorter or longer term.

Flower

The production of non-edible horticultural products is an important part of the agricultural sector in certain parts of the world (e.g. Belgium, the Netherlands). An increase in buying power beyond that needed for a 'Western style' diet, coupled with an increasing world population, will certainly result in an increasing demand for such products.

Across the world, there is increasing attention to the protection of the remaining wild nature, special habitats, etc. In countries with low population densities, this can often be done by setting aside large areas as national parks or wildlife reserves. In densly populated regions, such as Western Europe, this is difficult, and the local agriculture will be faced with growing pressure, geographically and technically. On the other hand, this may mean that agriculture is asked to assist in the creation of natural or landscape value.

Fun

In densly populated areas recreational activities will spread further into the open space that was traditionally the sole territory of agriculture. This will result in additional pressure on agriculture, but also in new opportunities for diversification of activities.

Another type of 'Fun' may be the drug-related production types. These are not only the drugs in the 'strict' sense (e.g. coca), but also tobacco, and maybe even the cereals used in alcohol production. The demand for this type of products will almost certainly grow as the population and its economic development increases.

Additional challenges

An important way of dealing with growing demand is to select for higher yielding varieties of plants and animals. Great progress can be made through conventional selection; e.g. in 1950 the US produced 56 million tons of maize on 33 million ha and 252 million tons on 29 million ha in 2000 (see Borlaug in Scanes and Miranowski (2004), chapter 2). The question is of course whether there is still enough room for progress, and whether this progress can be made at the speed at which the interactions between the six Fs will influence demand.

Another option would be the increase of the area used for agriculture, but it seems that soil, terrain and water constraints indicate that the vast majority of suitable land is already cultivated (FAO, 2007; Fischer, 2002). As the water use by agriculture already accounts for 70% of the total freshwater withdrawal (FAO, 2006), it is clear that a linear increase of water consumption with the abovementioned production increase will be impossible.

Some thoughts on the ethical challenges

It is tempting to shortcircuit the discussion on agricultural priorities by stating that 'food production is always the first priority'. Rolston (2006) already concluded we do many other things than fighting hunger, and we do not consider this bad. We see no reason why this would (have to) change in the future.

Ensuring access to food is a technical matter that entails structural, political and economical development of vulnerable areas, but it is also a social matter. As Ghandi said: 'the world produces enough for everyone, but not for everyone's greed'. Especially for the world's poor (in the cities of less developed countries, but also in the developed countries) the balance between production and demand is crucial as rising prices have a profound impact on their food consumption potential.

Therefore, we do not consider free trade in food products something to strive for. Lowering means of intervention also means lowering the possibilities for mediation. It is significant that for other 'basic commodities' such as oil, the strategic reserves are not reduced.

Animal production (and its associated feed production) is an ethically ambiguous activity. On the one hand it drains some of the production capacity that could be used for human food products, on the other hand some animal production systems can generate additional production capacity (e.g. ruminants in dry areas). This is true for area and water use, but also for energy and protein efficiencies.

We can at least conclude that within an anthropocentric ethic, animal production becomes more acceptable when it has a positive effect on the world food status. In other words, animal production outside the high-potential crop producing areas and/or animal production using products unsuitable for human consumption is ethically better. This means that in the greater part of Europe, North America, and Asia, we need to ameliorate our production methods. All this relates to a definition of sustainable animal production.

The ethical analysis of the biofuel case is even more complex because of the possible competition between food, feed and fuel production, but also due to the connection with the climate change discussion. In the long term, the limited availability of fossil fuels can be an important element.

Indirect competition (through ground, water or input use) can have just as profound an impact on the availability of food and feed crops as direct competition. Therefore, 'second generation' biofuels may be able to deliver a better energetic and CO_2 balance, but will certainly not only have positive consequences. Biofuel technology needs to be judged, using life-cycle analysis and incorporating all external effects, (1) by the total environmental impact (of which CO_2 is only one element), (2) by its energy gain, and (3) by its impact on agricultural markets and prices.

Our conclusion is that a large focus on biofuels essentially drains efforts from technologies that do have the potential to produce enough energy to 'fuel' the world, something biofuels well never be able to do. The only real solution is to 'decarbonise' energy production, something some European institutions seem to have noticed already (European Commission, 2007), at least in some of their documents. In short, all elements considered, biofuels seem only an ethical long-term option in a very limited amount of very specific situations.

With regard to fibre production and 'green chemistry' the same arguments seem to be relevant, only the timescale will probably be different. Again, direct and indirect competition with food and/or feed crops are relevant, as are the net oil use reduction, the environmental and the social consequences. The difference in ethical analysis of crop production for natural fibres or for green chemistry is probably small. Here too, it seems important to look for synergies in production types and the use of by-products. In the long term, there seems to be no alternative to a combination of natural fibre, green chemistry and better recycling.

It needs little discussion to see that the drug-related elements of 'Flower' can never be an ethical priority, and in some instances be even problematic. In areas where this type of production contributes significantly to the local economy, any reduction, in the short or the long term, should be well managed. The horticultural part seems non-problematic as long as it does not impose unnecessary stress on the environment or on other more essential production (such as food). Up to now, this seems to be the case, certainly in Western Europe, but the expected growing demand of this – and other – production types could change this balance.

Within 'Flower' and 'Fun' the growing demand for recreation and nature preservation in densily populated agricultural areas is a more difficult question. These non-production societal and economic actors will increasingly influence the geographical distribution and technical use of the most important agricultural input, the soil. This is not necessarily bad, but it will mean a paradigm-shift within the agricultural sector, but maybe also within those other 'sectors'. It could be argued that recreation is a less essential need for humans than agriculture, and therefore that agriculture should prevail in case of (potential) conflicts. The competition between agriculture and nature is at another level, as we have stated in the first paragraph of this section, and the choice between them is not as easy to make.

The overall ethical analysis is complex, not least because of the large number of interacting evolutions, the great number of uncertainties, the many very diverse stakeholders, and the fact that most options involve paying short-term costs for long-term benefits. We have tried to illustrate some of the ethical challenges in the paragraphs before this, but it is clear that a proper analysis should be made by evaluating all Fs at the same time. It is indeed the great number of mutual influences that makes the analysis so complex and the challenges so great.

What really complicates the analysis is the multitude of apparent contradictions in future evolutions, moral intuitions and/or pragmatic decisions. This is the type of problem Rolston (2006) sharply translated as 'Feeding People versus Saving Nature?' The moral tension fields Keulartz *et al.* (2005) identified, are equally relevant in this broad context: anthropocentrism versus ecocentrism, mitigation versus adaptation, intergenerational versus intragenerational solidarity, international versus intranational responsibility, process versus product.

The most fundamental conclusion from this discussion is that a production increase will be the main task for the agricultural sector. This production will have to serve much diverser goals, and needs to be realised sustainably while reaching many secondary goals. It seems that, as soon as one attributes any weight to anthropocentric arguments, each discussion about the 6F complex is really a discussion about the cross-compliance between the production increase and these secondary goals and functions. A report published by the FAO ('The ethics of sustainable agricultural intensification', FAO, 2004) literaly states 'given the present and anticipated increases in world population, not to mention current and projected environmental problems and ecological stress from agriculture, further agricultural intensification will be needed'. Thus, efficiency and productivity increases in order to saveguard the future of man and the environment. But, 'there is no route to sustainability on the current path' (Petrella, in Keulartz *et al.*, 2005), so the challenges are great.

Conclusion

The debate, and the ethical analysis of the six main functions of agriculture, the 6Fs ('Food', 'Feed', 'Fuel', 'Fibre', 'Flower', and 'Fun') are complicated by oppositions between expected evolutions, moral intuitions and/or practical possibilities, especially when different timescales are considered. The different functions are often considered independently, although the interactions between them are the main reason for the complexity of the analysis, and for the magnitude of the challenges we face. The multitude of direct and indirect, technical and economical connections make it impossible to separate the different functions.

The 6F model shows, by summarising and structuring the future challenges, that production increases will be necessary and input shortages will grow, and therefore that there is no other option than intensification. A well-understood intensification of agriculture is not only unavoidable, but even preferable and necessary. The rest is merely a question of priorities and boundary conditions.

References

Anonymous (2007). Soy2020. Our vision our destination. Available at: www.soy2020vision.com. Accessed 5 February 2008.

Biofuels Research Council (2006). Biofuels in the European Union – A vision for 2030 and beyond. Office for Official Publications of the European Communities, Luxembourg.

European Commission (2007). Communication from the Commission to the European Parliament, the Council, the European Economic and Social Committee and the Committee of the Regions. A European Strategic Plan for energy technology (SET-Plan). 'Towards a low carbon future'. COM(2007) 723. European Commission, Brussels.

FAOSTAT (2008). FAO Statistics. Food and Agriculture Organisation of the United Nations, Rome.

Fischer, G., Van Velthuizen, H., Shah, M. and Nachtergaele, F. (2002). Global Agro-ecological Assessment for Agriculture in the 21st Century: Methodology and Results. International Institute for Applied Systems Analysis and Food and Agriculture Organization of the United Nations, Laxenburg/Rome. Available at: http://www.iiasa.ac.at/Research/LUC/SAEZ/index.html. Accessed 23 October 2007.

Food and Agriculture Organisation (2003). World agriculture: towards 2015/2030. An FAO perspective. Food and Agriculture Organisation of the United Nations, Rome.

Food and Agriculture Organisation (2004). The ethics of sustainable agricultural intensification. FAO Ethics Series 3. Food and Agriculture Organisation of the United Nations, Rome.

Food and Agriculture Organisation (2006). Livestock's long shadow. Food and Agriculture Organisation of the United Nations, Rome.

Food and Agriculture Organisation (2007). Fao Statistics. Chartroom. Available at: www.fao.org/es/ess/chartroom/default.asp. Accessed 23 October 2007.

Intergovernmental Panel on Climate Change (2007). IPCC Reports. Available at: http://www.ipcc.ch/ipccreports/assessments-reports.htm. Accessed 21 March 2008.

Keulartz, J., Krause, F., Petrella, R., Sachs, W., Vermeersch, E., Weiler, R. and Zwart, H. (2005). Reading the Kyoto protocol. Ethical aspects of the Convention on Climate Change. Eburon Academic Publishers, Delft.

Rolston III, H. (2006). Feeding people versus saving nature? In: Light, A. and Rolston III, H. (eds.) Environmental Ethics. Blackwell Publishing, Malden/Oxford/Carlton, pp. 451-462.

Scanes, C.G. and Miranowski, J.A. (2004). Perspectives in world food and agriculture 2004. The World Food Prize. Iowa State Press, Ames.

Smeets, E.M.W., Faaij, A.P.C., Lewandowski, I.M. and Turkenburg, W.C. (2007). A bottom-up assessment and review of global bio-energy potentials to 2050. Progress in Energy and Combustion Science 33: 56-106.

US Census Bureau (2004). Global Population Profile: 2002. U.S. Government Printing Office, Washington, DC.

Insidious ignorance or burst of biophilia: cultural uses of educational farms in Norway

K. Bjørkdahl
The Ethics Programme and Centre for Development and the Environment, University of Oslo, P.O. Box 1116, Blindern, Norway; kristian.bjorkdahl@sum.uio.no

Abstract

In Oslo, Norway, a number of small farms are designated 'educational farms'. In contrast to profit-oriented farms, these institutions are specifically designed to address visitors, typically kindergartens, school classes, or families. These farms portray an image of farm life which – in Norway as well as in the rest of the Western world – is a thing of the past. While educational farms display a small-scale idyll, Norwegian agriculture has since the 1960s moved steadily towards an increasingly industrialised form of agriculture. This essay aims to interpret the sociological significance of this mismatch and to comment on its implications for the ethics of human-animal relations. I argue that by offering children idealised farms as 'a window to agriculture' we conceal and distort the actual origin of meat and other animal products. But if these farms promote ignorance of animal food production, they also provide the obvious benefit of close interaction with farm animals. I therefore ask: 'Is a bit of self-imposed ignorance really so bad?'

Keywords: human-animal relations, modernity, city farms

Introduction

In a series of short but influential essays from the 1970s, art critic and essayist John Berger offered a distinctive story about human-animal relations in modern societies. He argued that since the 19th century animals have gradually vanished from our lives and that humans now 'belong to a species which has at last been isolated' (Berger, 1980: 26). This estrangement of humans from animals Berger thought the result of a process whereby animals have been physically and culturally marginalised. In food production, he said, animals are 'treated as raw material' and 'processed like manufactured commodities' (*ibid*.: 11), while in the cultural sphere, animals have been 'co-opted into the *family* and into the *spectacle*' (*ibid*.: 13) – made to stand in for human characteristics, human problems, human purposes. In this age of corporate capitalism, argued Berger, the idea that humans and animals live parallel lives is fading and the traditional peasant's familiarity with animals is lost. Animals are thus made *invisible*; they are made into objects we can look at but which do not look back: 'All animals appear like fish seen through the plate glass of an aquarium' (*ibid*.: 14).

Whatever the merits of this narrative as a broader sociological theory of human-animal relations in modernity, Berger's idea that animals are vanishing aptly describes the fate of one particular group of animals, namely livestock. Not much more than a century ago, food animals were still ubiquitous – a constant presence – in city and country alike. In cities, pigs and poultry were kept in backyards, pastures were located near urban settlements, cattle and sheep were routinely driven into town and sold at public markets, slaughterhouses were numerously dispersed around town, and, unless you did your own slaughtering, the butcher's shop was where you bought your meat (Myhre, 1990; Ritvo, 1987; Vialles, 1994). Rural people lived in even greater proximity to animals, at least in Scandinavia, where farm hands always worked, often slept, and sometimes socialised in the buildings where food animals were kept. It was also not uncommon for the peasant to keep pigs, poultry, calves, and lambs within the dwelling house (Frykman and Löfgren, 1979: 151-157). Later on, of course, those very same animals would be slaughtered and eaten by that very same peasant.

With industrialisation, livestock markets moved out of town or disappeared altogether, slaughterhouses moved to the margins of settlements, and butcher's shops were replaced by supermarkets (Vialles, 1994). And as livestock moved out of the city, people moved in. For the first time in history, most people no longer spent time around the animals that were to become their food. The upshot of this development was, just as Berger suggests, a physical marginalisation – a vanishing – of animals.

In this essay I will look at a particular attempt to bring animals back into the life of urban dwellers: the case of educational farms in Oslo, Norway. These are farms, I argue, that allow urban children the chance to experience *bursts of biophilia* – the chance, for a short while, to interact intimately with a range of farm animals. In doing so these farms appear to promote precisely the kind of integrated familiarity with animals that Berger longs for. However, in their outward appearance as well as in the practices they foster, these farms are nothing like the places where food is produced today. While agricultural production in Norway, as elsewhere in the Western world, has moved towards an increasingly industrialised *modus operandi*, educational farms still insist on an intimate, small-scale idyll. As a symbolic exercise, the educational farm experience promotes a pastoral ideal of farmer-livestock familiarity that conceals the economic factors that underlie animal food production in our societies. In place of the standardised mass production of the modern agricultural industry, they provide the intimacy of face-to-face human-animal interaction. So while they may encourage a benign biophilia on the one hand, there is a concern that they promote an insidious ignorance on the other.

Burst of biophilia

Oslo is the capital of Norway and home to approximately 575,000 inhabitants. The city boasts several city farms, most of which are part of the municipality's effort to broaden the range of activities on offer to the city's residents. There appears to be no overall strategy for how these farms are to be run; however, the farms are invariably used by visitors as integral elements of some educational scheme, and they employ teachers who act as guides to visiting groups – mostly kindergartens and primary school classes. Thus their ambition goes beyond being mere animal parks; they aspire to the status of *educational farms*. Postponing an analysis of what that means, exactly, it is in any case evident that these farms are seen by the municipality, by farm employees, and by visitors alike as a service rendered to the city's urban populace. So what do they offer, precisely?

Most obviously, they offer the experience of intimate contact with farm animals, what I call a *burst of biophilia*. The term *biophilia*, as developed by its populariser Edward O. Wilson, refers to the 'innately emotional affiliation of human beings to other living organisms' (Wilson, 1993: 31). Wilson, a biologist, has been particularly concerned to establish biophilia as 'hereditary and hence part of ultimate human nature' (*ibid.*). By highlighting our innate affiliation to other living things, Wilson hopes we can find a cure to the current rates of biodiversity loss. Here, I will not commit to Wilson's broader agenda, but make a rather relaxed use of the concept, that refers to a desire and a willingness to interact closely with other animals.

Informants give ample testament that this motive is central to their activity. A teacher on one of the farms spelled this out: 'The most important thing for me is that children get into close contact with the animals...that they get close and smell them and employ their senses a bit'. She emphasised the fear that many children feel towards big animals and noted that children of immigrant parents in particular tend to be less familiar with farm animals. The primary function of the farm, she explained, was simply to allow children to get used to being around big animals. If she could demonstrate to children that animals are quite comfortable around people, she thought that in turn would have a calming effect on the children. She added that: 'Many people in Oslo live in apartment buildings and are not allowed to have animals at home...I think it is essential that they can have *positive* experiences with animals...to be in the company of animals is therapy in itself'.

Several informants remarked upon the therapeutic value of interacting closely with animals. They tended to highlight this aspect especially when discussing members of staff who were mentally impaired or otherwise challenged. Many of the farms in question employ such people to create an inclusive labour market, and the farm managers thought that the therapy of regular close contact with animals was invaluable for those workers. Said one farm manager: '[In the animals] they have someone who needs them, someone who awaits them in the morning, and not just a stack of paper...someone who *gives back*'. The manager of another farm said about their work training staff: 'For these boys, spending time in the barn is a benign discharge of energy. When they are with the animals they can express an intimacy that they cannot otherwise express in their daily lives'.

When my informants spoke of 'close contact with animals' it was clear they did not just mean being in the same room as an animal or even being close to one. 'Close contact' was a term informants used to convey a more bodily, visceral experience – 'they enter the pens, groom the sheep, pet the rabbits, feed the hens'. To get children into 'close contact' with animals was, in the setting of these farms, simultaneously to take animals out of the invisibility that Berger argues they have receded into. These farms want animals to be something more than 'fish seen through the plate glass of an aquarium'; they want to put humans and animals back into the mode of familiarity that Berger thinks has been lost.

Biophilia is not just a discursive phenomenon at these farms – it is not just talked about – it is actively encouraged as a practice during visits. The educational farm experience is thus in every sense a sensory one. As an illustration, the guide at one farm always asked visiting children to remove their mittens before entering the barn, because, as he said, 'here, we are into close contact'. My visit to one farm demonstrated the range of bodily experiences to be had: the children were let into the sheep's pen to touch and groom the animals, they had their fingers sucked on by a newly fed calf, they got to stroke the pig's bristles, they had rabbits placed in their laps to be stroked and petted, they playfully wrestled the kids in the goats' pen, and were lightly bitten by hens while feeding them. The guides themselves would pet and caress the animals in front of everyone, as if to demonstrate what was meant by 'close contact'. In one instance, the guide – a grown man – lifted up a kid goat in front of the children, caressed it as you would a human baby, and addressed it in 'baby talk'.

At another farm, the guide, as she discovered a newly laid egg in the hen house, stroked the warm egg across the children's cheeks and later did the same thing with a cold egg. Among the more notable experiences I witnessed occurred when a group of children were asked to form a line behind a row of milking cows. Some of the children apparently found the smell rather penetrating, but the guide quickly countered: 'I don't think it smells that bad. It would probably smell a whole lot worse if *we* were the ones standing there'. During the whole session behind the cows, the guide focused on how dung and urine were natural parts of farm life. 'When you work with cows, you can't be too snobbish', he said, shovelling cow dung into the cellar with little concern about getting his bare hands soiled. At one point, one of the cows defecated on his boot, in response to which he shrugged his shoulder's and said: 'now she's pooing on my boot too! Oh well...'. When another cow later urinated – the children still no more than a couple of feet away – the guide responded with enthusiastic applause.

In his classic study *The Civilising Process*, the sociologist Norbert Elias described how Western societies have shifted their 'thresholds of repugnance' towards less tolerance for the animal aspects of ourselves and of our food (Elias 1994 [1939]: 98). In contrast, the farms in question, using a variety of practical means, actively disassemble the idea that humans are elevated above animals and animal functions. There is, as the above examples illustrate, nothing degrading about getting one's boots soiled by cow's dung, nothing uncivilised about staring into the back end of a cow, and perfectly acceptable for a grown man to be seen fondly caressing a young farm animal. The 'close contact' that children get on these farms is not just a furry and cuddly – civilised – experience; it is a mucky and smelly – animal – one as well.

A window to agriculture

The idea and practice of biophilia – in both its furry and mucky aspects – was routinely appealed to as the primary objective of the educational farm, as we have seen. However, informants invariably provided a larger context for the farm's activity. In addition to providing 'close contact' with animals, farms also aimed to function as 'a window to agriculture', as one of my informants expressed it. The ideal educational farm, then, was more than a site of experience; it was also a site of display. Said one farm manager: 'It is our ambition to be a display for Norwegian agriculture. Of course, we have no formal obligations as such; it is rather a task we take upon ourselves. We try to promote positive experiences... to make Norwegian agriculture visible and leave visitors with positive connotations'. On the one hand, the phrase 'window to agriculture' simply expressed the farms' intent to comply with, and even surpass, the rules and regulations issued by agricultural and food authorities. More importantly, however, it constructed the farm as an arena for imparting to urban children some connections that one feared are being lost – not least the connection between animal and meat.

The manager of one farm told me that they wanted to convey to children 'an awareness of animals'. When asked to elaborate on what she meant by that, she turned to a news story she had recently seen about the scalding of a cat – an incident that had clearly disquieted her. She added, in contrast: 'I hope we impart to children some idea of what animals *are*; that they too are living creatures that can feel'. She went on to emphasise, however, that they did not anthropomorphise animals – 'we try not to do that' – and she informed me that they would routinely send their animals to slaughter and 'take back the meat' to use in educational settings, like the salting of lamb's meat. She commented on the difference between Norwegian city farms and their European equivalents, and noted that 'many of those farms are run by vegetarians...and in many of those places, they have sheep that are 15 years old'. That, she added, 'just sounds strange to us Norwegians'. Clearly, the ultimate reason to keep animals on a farm was to kill them for food.

The guide at another farm was particularly straightforward about the connection between animals and meat. During a session where he had gathered all the children around the pig pen for a short lecture, he told the children that: 'Whenever you eat sausages, bacon, or ribs, you are eating pig. This pig is going to be our food'. Then, pointing at one child, he added: 'You may some day eat one of these pigs; that is quite possible'. In a similar session with another group, one child said in response to the guide's explicit account that she thought it was a shame that pigs had to become food, whereupon the guide replied: 'Yes, a lot of people think that is a shame. But then I always say that if you eat meat, you have to accept that the pig gets slaughtered. If you don't want to eat the pig, you have to become a vegetarian'.

John Berger argued that in the traditional relation between humans and animals: 'A peasant becomes fond of his pig and is glad to salt away its pork. What is significant, and is so difficult for the urban stranger to understand, is that the two statements in that sentence are connected by an *and* and not by a *but*' (Berger, 1980: 5). Likewise, the total effect of the educational farm experience is the impression that it is perfectly natural to love the animals that you will later kill and eat. Of course, this assumption can itself be questioned. Jonathan Burt, for instance, asks whether the essential paradox in Berger's essay is not that 'it fails to see...the possibility of the end of relations based on sacrifice' (Burt, 2005: 211).

The point I want to make with regards to educational farms is a bit different. It amounts to saying that, in their endeavour to meet two particular needs – close contact with animals and education about agricultural processes – these farms conceal what ought to be an equally acute concern, namely contemporary practices of agricultural production. As part of their effort to function as 'windows to agriculture', these farms are very eager to make the connection between animal and meat. However, in light of agricultural developments during the last 50 years, it matters a lot *which* agricultural system you are a window *to*.

The mysterious farm 'where only pigs live'

Some of my informants seemed to recognise the discrepancy between farm life as presented on educational farms and what one finds on modern farms. 'We are actually quite old-fashioned compared to Norwegian agriculture', one manager said, 'and there is always a question about what aspect we ought to make visible. Ought we to convey modern Norwegian agriculture or should our approach be more cosy and old-fashioned, where we promote the animals and a nice [farm] environment?'

In actual fact, there is little doubt about what aspect these farms encourage. The abovementioned guide was very intent on making the connection between a particular pig and the children's meat eating – '*you* may some day eat *this* pig' – but in fact, only a tiny minority of pigs are so lucky as to spend their lives on an educational farm. Compared to a conventional modern farm, these farms are much smaller, they employ less machinery, they have a wider range of animals but a much smaller number, the pens are roomier, the ratio of caretaker to animal is immensely greater, and so on. Even more importantly, perhaps, the 'close contact' so eagerly encouraged by educational farms has, for a number of reasons, been made all but impossible on modern farms.

The only mention of conventional food production I ever witnessed during my visits was one guide's comment that the farm's litter of piglets would soon 'move to a farm where only pigs live'. That, I suspect, did not sound all that bad to the children. After all, everyone moves once in a while, and what could be better for the piglets than to spend time with their own kind? The more profound effect of the statement, however, was to contribute to an impression many of us already have, namely that the sites of modern agricultural production are *mysterious places* – places one cannot visit, see, experience, or even learn more about.

I propose that we read the educational farm experience as a symbolic exercise. Taken as such, the first thing to note about them is that they proclaim themselves the primary sites for inculcating children with basic facts about animals, agriculture, and farm life – they are our 'windows to agriculture'. But what kind of agriculture are we asked to see through that window? It is a pastoral scene, to be sure, one wholly undisturbed by the troubles and excesses of urban industrial life. More specifically, the ideal that manifests itself practically during farm visits is Berger's idea that one can be fond of the pig *and* salt away its pork.

An asset greatly valued by these farms is thus the ability to demonstrate acceptance of the animal sacrifice involved in agricultural production. The ability to love and kill animals as parts of the same overall scheme becomes a sign of sophistication, a sign of a worldview that is more grounded and *in-tune* than the detached and alienated urban one. But, as Pierre Bourdieu has pointed out, such 'symbolic capital' makes itself felt only where it 'conceals the fact that it originates in 'material' forms of capital' (Bourdieu, 1977: 183).

The material context in which the agricultural animal sacrifice is situated today is radically different from the one presupposed by educational farms. In short, it is the difference between industry and intimacy. In the traditional state of a near-subsistence economy, the idea that humans and animals live parallel lives – similar but different – did perhaps provide a meaningful context for the fact that animals had to die so that humans could live. But in the current economic system, the proposal to be fond of the pig *and* salt away its pork no longer makes sense. For some, that reaction refers to a principled refusal to salt away the meat of any animal whatsoever. For others, it is simply a result of the fact that there are no longer any animals to 'be fond of', since they have all been reduced to 'raw materials' and 'manufactured commodities'.

The anthropologist Noelie Vialles has argued in her book *Animal to Edible* that the thought of 'one-to-one slaughter, in which the roles of animal and man persist right up until the act of killing, is easier to accept than industrial slaughter' (Vialles, 1994: 31). In order to detach ourselves from the unpleasant thought that our society keeps and slaughters animals on a mass scale, says Vialles, we have put in place a series of disjunctions – 'to keep the mass killing of animals at a reasonable distance' (*ibid.*). My contention is to ask whether the farms in question here are not simply another such disjunction, another effort to draw attention from our practice of treating animals as raw materials, in short, a diversion tactic.

What you don't know can't hurt you, goes the saying. That is true to the extent that every practical mastery involves selective ignorance (Bourdieu, 1977: 19) – every social interaction filters out a lot of things in order to function effectively – but ignorance is no longer bliss when what you are keeping from view is the mass keeping, mass slaughtering, and mass suffering of sentient animals.

Note on sources

This essay is based on visits to five city farms in and around Oslo. Interviews and participatory observation were carried out at three of the five farms in question. Informants' responses were recorded on tape and are rendered here as directly as possible. Three consecutive periods signal either my omission of words or an informant's pause. Italics have been used to represent the informants' actual emphasis.

References

Berger, J. (1980). Why look at animals? In: About looking. Pantheon Books, New York, pp. 1-26. Originally published in three parts in New Society, March/April 1977.

Bourdieu, P. (1977). Outline of a theory of practice. Cambridge University Press, Cambridge, pp. 248.

Burt, J. (2005). John Berger's 'Why look at animals?': a close reading. Worldviews 9: 203-218.

Elias, N. (1994 [1939]). The civilizing process. Blackwell, Oxford, pp. 567.

Frykman, J. and Löfgren, O. (1979). Den kultiverade människan. Gleerups, Malmö, Sweden, pp. 240.

Kean, H. (1998). Animal rights: political and social change in Britain since 1800. Reaktion Books, London, pp. 272.

Myhre, J.E. (1990). Oslo bys historie, bind 3: hovedstaden Christiania. J.W. Cappelens forlag A.S., Oslo, pp. 560.

Ritvo, H. (1987). The animal estate: the English and other creatures in Victorian England. Harvard University Press, Cambridge, USA, pp. 368.

Vialles, N. (1994). Animal to edible. Cambridge University Press, Cambridge, pp. 142.

Wilson, E.O. (1993). Biophilia and the conservation ethic. In: Kellert, S.R. and Wilson, E.O. (eds.) The biophilia hypothesis. Island Press, Washington D.C., pp. 31-41.

Ethics as a concern in agriculture and food research funding programmes of the Agence Nationale de la Recherche

J.M. Chourot[1] and M. de Lattre-Gasquet[1,2]
[1]ANR, 212 rue de Bercy, 75012 Paris, France; jean-marc.chourot@agencerecherche.fr
[2]CIRAD, 42 rue Scheffer, 75116 Paris, France

Abstract

The Agence Nationale de la Recherche (ANR) is a French public organization devoted to competitive project funding in both fundamental and applied research. ANR addresses both public research institutions and industries with a double mission of producing new knowledge and promoting interaction between public and industrial laboratories through the development of partnerships. The research programmes proposed by ANR are translated into publicly available calls for proposals. In the past four years, six programmes related to food, agriculture and sustainable development were launched. All programmes and their subsequent calls for proposals were drafted independently and therefore, ethical issues were not equally dealt with or taken into account in a consistent way. Over time, progress has been made drafting the calls for proposals from specifically 'scientific accuracy and correctness' to a more 'ethical awareness'. But how are the ethical issues of the calls translated in the research proposals and in the research projects? Is the presentation of a call for proposals sufficient to lead to an ethically aware research project? What are the most frequent ethical issues found in the research projects? This paper presents the way ANR has dealt with these issues, and gives examples of ethical issues raised by research projects.

Keywords: ethics, funding agency, programme definition, project selection

The Agence Nationale de la Recherche (ANR), a funding agency

The French national research agency (Agence Nationale de la Recherche – ANR) was created in 2005. Its legal status was modified January 1st 2007 to become an independent public institution for the management of administrative issues. The ANR aims to increase the number of research projects generated from the entire scientific community, and to provide fundings based on calls for proposals and peer-review selection processes. The ANR deals with both public research institutions and industries with a double mission of producing new knowledge and promoting interaction between public and industrial laboratories. It is a mix of a science-based funding agency and a strategic funding agency (Braun, 1998). In 2007, ANR issued 50 calls for proposals, including the trans-national European funding calls. As a result,. 607.4M€ was allocated thorough approximately 1500 research projects. ANR's funding cycle has three steps: i) programming, (2) selection, (3) monitoring and ex post evaluation. Therefore ethical aspects relate to the three ANR steps: the definition of the contents of programmes and calls for proposals, the selection of projects, the follow-up of projects, especially dissemination of projects results.

Ethical issues related to the definition of the programmes of the Ecosystems and Sustainable Development Department

The ANR structure is based on six thematic departments and one non thematic department. The department of Ecosystems and Sustainable Development covers all research related to the ecosystems and their use for human wellbeing. This includes agriculture, genomics and the validation of agricultural product as food. By extension, studies dealing with human nutrition are included. In 2009, the ANR has proposed four programmes translated into five calls for proposals. A short description of each of these programmes is following.

- The programme Systerra deals with ecosystems, living resources, landscapes and agriculture, including the social aspects and the related policy management.
- The programme ALIA (Alimentation et Industries Alimentaires) is the second of the three year food research programme proposed by the ANR. The programme is organised around three research themes: '*For the well being and the better ageing of the population groups*', '*For a more dynamic economy of food production*' and '*For a well balanced society and sustainable development of food productions*'.
- The genomics programme deals with all research related to bacteria, eukaryotic micro-organisms, plants, animals and as a consequence research that involves genetically modified organisms (GMOs) is included.
- The 6th extinction programme deals with the evaluation of the potential loss of biodiversity and the related actions which can be taken.

Compared to the programmes which were published by ANR at its start, the 2009 programmes include much more social dimensions, and therefore ethical concerns. For instance, the first food research programme was very much oriented towards increasing scientific knowledge; the social and economical aspects were one of the seven themes proposed. Within the new programme, the calls for proposals strongly advise researchers to consider and respond to the social (e.g. dealing with specific targeted groups of population), economical (e.g. feasibility and affordability) and environmental (e.g. life cycle assessment for instance) concerns that to any particular proposal. Moreover, the theme entitled 'For a well balanced society and sustainable development of food productions' is strongly oriented towards the development of tools that can assist policy makers.

An important question for a funding agency relates to the correct or best way to encourage ethical awareness within research programmes. It should be noted that, there can be big difference between programmes aiming at fighting what makes life difficult and 'bringing the greatest good to the greatest number' (Sassi *et al.,* 2001) and programmes aiming at improving the wellbeing of the society and individuals. Sen (1973) remarked that 'maximising the sum of individual utilities is supremely unconcerned with the interpersonal distribution of that sum'.

For instance, within the food theme most of the research is oriented towards increasing competitiveness of the industry and aims to better understanding aspects of human nutrition as well as the consumer behaviour. Ethical issues are present within the programme but it is expected that the research proposals will be driven by the aforementioed issues. This also implies that the way to deal with some of these ethical issues is to transfer them from the funding agency to the research teams. However, the responsibility of the funding agency in terms of defining ethics awareness programmes remains intact. During the selection of the proposals, ethical concerns are considered – up to now possibly too loosely – through the peer reviewers guidelines for the evaluation. As part of this process direct questions relating to the ethical dimensions are asked concerning the potential impact of the proposal on consumers or policy makers, and in terms of economical (affordability and availability) aspects and feasibility.

Ethical issues related to the selection and peer review

There are ethical aspects to the selection process, the peer review, the themes of the projects, the protocols, the protection of participants in research, and the nature of the relationships between researchers. To insure that procedures are transparent and that the selection process is impartial, the ANR selection process of research projects has been certified AFAQ ISO 9001 in July 2008. Attention is given to transparency, fair treatment of proposals, and multi-submission of projects. Each project proposal is reviewed by between three to five peer reviewers. To avoid any conflicts of interest and improper use of information provided in the proposals, ANR requires reviewers to acceptance of code of contact when review the proposals (a deontological code).

To insure that protocols consider ethical issues, the ANR has adopted a standard template for the call for proposals. As all research funded by the agency does not take into account ethical issues in the same way, all calls for proposals include a mandatory paragraph which reminds researchers of minimum standards in terms of legal requirements and good practices compliance. ANR relies on existing competent structures, such as national ethic committees, research institute ethical committees, national regulation and laws. However, as ethical aspects do not constitute a clear selection criterion and as all evaluation committees do not have experts on the topic, ANR cannot be certain of the ethical practices of all research activities, e.g. harm to animals or questionable research practices. In its four years of existence, no cases of unethical projects have been identified. However, there may be proposals that do not meet the desired standards, for example, a project was identified that seemed for the sole benefit of a single partner with limited or no feedback to industry, or the civil society; results were not made public available. This raised questions relating to the transfer of ethical concerns from the text of the call to the proposal.

Ethical issues related to the dissemination of projects results

As the projects that were financed after the creation of the agency are starting to end, ANR has yet to be confronted by ethical issues related to the dissemination of projects results, such as fraud, falsification and plagiarism, and the non-disclosure of results. The agency is aware of these issues and has started to consider how to deal with them.

Conclusion

The challenge for a funding agency is to protect both the academic community and stakeholders, so that sustainable, effective, community-based interventions succeed (Giese-Davis, 2008). Since its creation in 2005, ANR has set up procedures to avoid unethical behaviour within the agency, by its partners and in the projects funded. It will continue to review and improve its efforts in the future.

References

Braun, D. (1998). The role of funding agencies in the cognitive development of science. Research Policy 27: 807-821.

Giese-Davis, J. (2008). Community/Research Collaborations: Ethics and Funding. Clinical Psychology: Science and Practice 15(2): 149-152.

Sassi, F., Archard, L. and Le Grand, J. (2001). Equity and the economic evaluation of health care. Health Technol Assessment 5: 1-138.

Sen, A. (1973). On economic inequality. Clarendon Press, Oxford.

Beauchamp and Childress' four principles applied to animal ethics issues

V. Lund[1] and E.M. Forsberg[2]
[1]National Veterinary Institute, P.O. Box 750 Sentrum, 0106 Oslo, Norway; vonne.lund@vetinst.no
[2]Work Research Institute, P.O. Box 6954 St. Olavsplass,0130 Oslo, Norway

Abstract

There is a need for a practical animal ethics that can contribute to a humane keeping of animals, allowing animals to have a function in society while still granting them a good life. The Brambell committee's five freedoms are important principles, but they may be strengthened by including them in a more general framework. The four principles of biomedical ethics, defended by Beauchamp and Childress since 1979, have been widely embraced as an ethical framework among health care professionals. In this paper Beauchamp and Childress' four principles – respect for autonomy, non-maleficence, beneficence, and justice – are discussed and applied to the field of animal ethics. We suggest that although the principled approach in itself may be valuable, the principles proposed for the biomedical domain need to be adapted to the animal ethics domain. Non-maleficence and beneficence seem relevant, but there are problems with applying autonomy and justice straight off. Autonomy can be applied when understood in terms of allowing a species-specific behaviour. Respect for dignity is suggested to complement respect for autonomy, and it is suggested that respect for intrinsic value should be added as an aspect of respect for dignity. Justice may be an applicable principle, but a theory of justice that can incorporate animals first has to be further developed.

Keywords: principlism, ethical principles, animal welfare

Introduction

Animal protection has a long tradition in the Western world. In Norway, cruelty against animals has been prohibited since 1842 and the country has had an Animal Protection Act since 1935. This view of animals has continued to develop and over the years society has gained a morally richer view on the relation between human beings and animals. This is signified by the appearance of the concept of *animal welfare*. Animal welfare is an area of increasing importance for people in the western world and worldwide. For example, an animal welfare action plan was adopted by the European Union in 2006[8]. Since 2001, the World Organisation for Animal Health (OIE), has dealt with animal welfare and has the intention of taking a leading role in the field[9]. In addition, many countries have now revised and sharpened their animal welfare legislation.

When animal welfare became an ethical issue in the 1970's, the ethical theories that became dominating criticised not only taking animal lives, but several ethicists questioned most or all human use of animals. This is true for the utilitarian approach introduced by Peter Singer (1975), as well as the animal rights theories advocated by Tom Regan (1983). However, there are many people who are deeply concerned about animal welfare and who feel that animals should have a good life and be treated with respect, but still do not object to the killing of animals as long as it is performed so that it does not cause suffering. Hence, these individuals accept animal farming, given the above conditions, and they are not vegetarians. Also, sustainable agriculture systems need farm animals, in particular herbivores, since these

[8] http://ec.europa.eu/food/animal/welfare/actionplan/actionplan_en.htm.

[9] http://www.oie.int/Eng/bien_etre/en_introduction.htm.

contribute to a sustainable crop rotation (Lund and Olsson, 2004).This means that there is a need for a practical animal ethics that can contribute to the humane keeping of animals, allowing animals to have a function in society while still granting them a good life. In other words: we need an ethical framework in order to assess the conditions in which animals live. The five freedoms established by the Brambell committee do amount to such a framework. However, these principles are specific to the animal ethics context. Connecting these principles to a more general framework of principles may strengthen the five freedoms and may demonstrate that animal ethics is not a domain isolated from other practical (and theoretical) ethics. In biomedical ethics, Beauchamp and Childress have developed a framework that starts with general *prima facie* ethical principles and specify them for the biomedical domain. The medical profession has widely adopted this practical ethical framework, which has proven useful in ethical evaluations and decisions dealing with the practical treatment of patients, and has been widely accepted as a set of guide principles for health care professionals. It may therefore be useful to see if this framework also can similarly be applied to animal ethics.[10]

The framework in question is the four principles of biomedical ethics, defended by Beauchamp and Childress since 1979, namely respect for autonomy, non-maleficence, beneficence, and justice. These principles are *prima facie*, which means that each of them may justifiably be overridden if they conflict with another moral principle that has stronger power in the particular situation. In several ways there may be similarities between patients and animals in human care, since in both cases there is a situation of care and a power imbalance between the caretaker and the one who needs the care. For example, both human patients and animals in human care are dependent on others for their care. Moreover, in some cases humans, like animals, do not have knowledge to evaluate what is the right decision. However, there are also important differences, the main one obviously being the cognitive abilities of ordinary functioning adult human beings compared to animals. For instance, animals cannot give informed consent, defend or waive a right, etc.

In this paper the application of Beauchamp and Childress' principles to animal ethics issues will be discussed. We will not assume that principles developed to regulate our relations and duties to human beings are suitable to regulate our relation to animals, but we will explore if these principles contain resources that may be used to address animal ethics issues. It should be noted that Beauchamp and Childress' principled approach is the theoretical basis for the ethical matrix method, developed by Mepham (e.g. Mepham, 1995), which has been used for evaluating animal biotechnologies (as well as a number of other issues). The set of principles used in ethical matrix can vary slightly, but the original version proposed by Mepham includes well-being (a combination of beneficence and non-maleficence), autonomy and justice. It should be noted that Mepham's use of Beauchamp and Childress' principles will not be discussed here.

Beneficence and non-maleficence

The most uncontroversial similarity between humans and animals is the fact that both are living, sentient beings. Controversy does indeed reign with regard to which species should be considered sentient, in the sense of 'able to feel pain'. We will not here engage in this discussion, but will discuss the notions of beneficence and non-maleficence with the regard to animals that are non-controversially sentient.

Non-maleficence means to avoid causing harm. To avoid causing harm to animals is the basis for the notion of *animal protection*, which (as noted above) is already deeply entrenched in western societies.

[10] Mullan and Main (2001) have discussed Beauchamp and Childress' principles applied to ethical decision-making in veterinary practice, as have Pullen and Gray (2006) in a book for veterinary nurses. As their context is veterinarian practice, while ours is animal use in general we will not discuss these authors in any detail.

What animal protection means, i.e. how the principle of non-maleficence should be *specified*, is still in dispute, and in most countries actions like castration, dehorning and long transports are still allowed. Moreover, non-maleficence does not seem to cover all our moral obligations towards animals. It is not enough to refrain from harming animals. The principle of non-maleficence does not respond to fulfil animal needs, nor to the increasing public demand for animal welfare. Other principles must therefore be added in an ethical framework of comprehensive ambitions.

The principle of *beneficence* calls for not only avoiding animal suffering but to provide animals with welfare in a positive way. The turn to the concept of *animal welfare* may seem to indicate (in addition to a new focus on the individual animal) an increased importance of the principle of beneficence in animal ethics. The principle of beneficence can be interpreted this way: one should not only avoid harming animals but also take positive steps to promote animal well-being (in some cases some pain must be inflicted in order to secure long term health and welfare). The concept of 'positive welfare' has developed over the last few years. It seems to be oriented towards expression of instinctual behaviour, including both what the animal likes and what the animal wants (Yeates and Main, 2007).

Beauchamp and Childress discuss the line between beneficence as an obligation and as a supererogatory act. This is not clear cut, especially when we do not have a special relation to the person receiving our beneficence. Beneficence makes more sense with regard to doctors' obligations to their patients; their *role* is to provide helpful treatment. This also holds for sick or injured animals. Beneficence is more of an obligation when we stand in a special relationship with an individual, which implies that we should provide healthcare to diseased or injured animals in our custody, but not necessarily to wild animals. However, it also indicates that when we have a caretaker role towards animals, an important aspect of beneficence will be to provide them with appropriate feed and comfortable housing in order for them to physiologically flourish. We can always do more for animals, but at some point these acts start to become supererogatory.

Autonomy

Autonomy means self-legislation, and seems as such not to fit animals well. Animals do not adhere to a behavioural codex other than instinctual behavioural patterns.[11] These instinctual 'laws' are not a form of self-governance that should be respected as an intentional action, but more like natural laws that individual animals are simply the subject of. It is not common to talk about 'respecting' natural laws; we simply acknowledge their existence.

Beauchamp and Childress accept that human beings not always, or even most of the time, are given full autonomy. They also say that 'appropriate criteria of substantial autonomy are best addressed in a particular context' (2001: 60). However, they define autonomous action in terms of 'normal choosers who act (1) intentionally, (2) with understanding, and (3) without controlling influences that determine their action' (p. 59). It is clear that animal behaviour can not be considered to be included in Beauchamp and Childress' understanding of autonomy. However, Beauchamp and Childress mention that autonomy should be respected to the extent it exists. Their example is a mentally disabled person who cannot have autonomy in, for instance, legal affairs, but may express wishes that should be respected; for instance, dietary wishes or wishes to make a telephone call to an acquaintance. Here, there will be a correlation to animals. Animals may express very clear preferences, for instance, a horse may express unwillingness to enter a horse trailer. Respect for autonomy in animals may be interpreted to mean respect for their preferences.[12] However, since animal preferences are governed by their instincts, we suggest that for

[11] Perhaps one can make the case for the existence of human like norms for certain species of primates.

[12] This is in line with Pullen and Gray's (2006) application of the principle of autonomy to animals.

animals autonomy should be interpreted in terms of allowing them to satisfy their behavioural needs. But by interpreting autonomy as 'allowing animals to perform their species-specific behaviour' we can give the obligation of beneficence a practical content, guiding us how this beneficence should be carried out. Although several European countries have sharpened the animal welfare legislation in this direction, most animal rearing systems still do not respect basic ethological needs of those animals they are designed for. This problem also exists in relation to pets – although owners may have the best intentions, since they lack basic knowledge of the animal's needs owners disrespect the autonomy of the animal in the sense that the pet is not given the freedom to live according to its genetic encoding.

In practice this means for example that the environment must be adapted to the animal and not vice versa. A cow that lies down in a meadow does this according to a pattern spelled out by its physiology – first the body is moved forward as the cow gets down on its knees and after that the hind part is placed on the ground. Many tie stalls or cubicle stalls do not allow this forward movement but forces the cow into unnatural laying patterns, resulting in high stress levels and often teat injuries. By learning more about animal behaviour and preferences, we can adjust not only buildings but also animal management to these preferences. For example, ethological studies have shown that during warm summer days, cows prefer to graze during the night and stay in the cowshed during the day. This is the opposite of common management practices today, but it would imply little or no costs to change routines to suit cow preferences. The strength of this prima facie norm will likely vary in different particular contexts. However, since it deals with basic ethological needs it ought to be an obligation not easily overridden by competing moral norms, or simply by human interests.

It seems that there are similarities between positive welfare for animals and what is included as specifications of autonomy in special cases for humans (i.e. humans with reduced cognitive abilities). However, it cannot be denied that taking autonomy to mean expression of instinctual behaviour is a break with the tradition from Kant. Therefore, one may not adopt autonomy as a separate principle here, but include respect for preferences and instinctual behaviour as an aspect of beneficence.

Justice

Although there are several theories of justice the common basis lies in the Aristotelian formal principle of justice, that equals are to be treated equally and unequals to be treated unequally. Beauchamp and Childress discuss different theories of justice as relevant to justice considerations. They do not choose a particular perspective on ethics, but note that different distributive principles may be relevant in different situations. Thus they do not present a simple principle of justice that may (or may not) be applicable in the context of animal ethics. Instead we must generally discuss the applicability of justice to animal issues.

Animal justice is a notion often invoked simply to stress that human beings indeed have moral obligations towards animals; *i.e.*, according animals' moral standing. However, with regard to justice as a moral principle, the issue is not moral standing in itself, but principles for fair treatment, distribution of benefits and burdens, etc. One may identify two general justice concerns relating to animals:

- *Justice among animals:* The same species is treated differently in different contexts of use. For example the pig is used as a pork producer, as prey (wild boar hunting), as pet, and as deliverer of 'spare parts' to humans, and it is treated differently in these contexts. Similarly, the rat is used as a pet and as a research object, but is also regarded as pest.
- *Justice between humans and animals:* Animals are contributing to our good – they should thus have a fair share of what they produce or contribute with.

Feelings of justice (or perhaps especially of *injustice*) run deep in human beings, and these feelings are rooted in our cognitive abilities to compare our own lot with that of others. It is hard to imagine that animals have the cognitive abilities necessary to evaluate the relation between their situation and that of other animals, and thus that they can have feelings of injustice. The urge to promote justice among animals therefore seems not to have its origin in the animals themselves.

The reason why the same animal species is not treated equally across different areas of use is that the balancing of burdens of the animals and benefits to humans will come out differently in the different contexts, and other ethically relevant concerns will also influence the balance. Keeping a pig as a pet will involve developing an emotional bond to the pig, which will influence how it is treated.[13] A wild boar, on the other hand, is a wild animal for which we have weaker responsibilities than animals in our keeping. The reason why pigs providing human spare part are treated better than meat production animals is based on human egoism: these animals must be in good condition and health in order to function in this context. Unequal treatment of individuals of the same species may therefore not be unjust, or even a matter of justice.

The justice in the relation between animals and humans is the justice between unequals. It is thus not reasonable that benefits and burdens are distributed equally. What a fair relation might be is hard to determine, but one might compare the situation to that of an employer and his/her workers. Just as human workers should not be exploited, our animal co-workers should not suffer from exploitation but (as a minimum) have their basic needs fulfilled (Lund *et al.*, 2004). It is difficult to tell what implications such a notion of justice would have in practice over and above the protection and care provided by the principles of beneficence and non-maleficence.

Conclusion

There is one kind of moral argument or concern that does not seem to be addressed by Beauchamp's and Childress' four principles. This is the notion of intrinsic value, inherent value or inner worth. These concepts all signify a concern with something over and above welfare and natural behaviour. Respect for intrinsic value may justify an objection to bestiality even if this activity does not harm the animal at all. It may also justify objections to genetically modify animals simply for entertainment (even if this was possible without suffering to the animal or to experimental animals used in the process). Intrinsic value is most often used when moral protection is to be accorded to non-autonomous moral entities, like the environment, animals, foetuses, etc. Sometimes the notion of intrinsic value is justified by reference to God's creation of the earth, animals and human beings, but intrinsic value can also be defended without such reference. Beauchamp and Childress discuss non-autonomous human beings, but do not pay much attention to intrinsic value. Intrinsic value is not an easily defined concept, but as it does capture a quite widespread intuition[14], it should be included in an animal ethics framework.

Summing up, Beauchamp and Childress' four principles may fit biomedical ethics well, but they need some adjustment to be applied to animal ethics. We do not consider this a flaw of their approach, but as stating some limits to it. Their general approach of prima facie principles that need specifying and balancing can be retained, but their set of principles need to be adapted. A suggestion for such a set of principles is the principles of non-maleficence, beneficence (including allowing animals a species-specific behaviour) and respect for intrinsic value. The question whether or not a principle of justice should be included in this set of principles must be a matter of further discussion.

[13] The special relation to our pets is similar to the fact that we have certain relational obligations to people close to us which we don't have to strangers.

[14] Further, it is included in both Dutch and Norwegian legislation.

References

Beauchamp, T.L. and Childress, J.F. (2001). Principles of Biomedical Ethics, 5th edn. Oxford University Press.
Forsberg, E.-M. (2007). A Deliberative Ethical Matrix Method – Justification of Moral Advice on Genetic Engineering in Food Production. Dr. Art. Dissertation. Oslo, Unipub.
Lund, V., Anthony, R. and Röcklinsberg, H. (2004). The Ethical Contract as a Tool in Organic Animal Husbandry. Journal of Agricultural and Environmental Ethics 17(1): 23-49.
Lund, V. and Olsson, I.A.S. (2004). Animal agriculture: symbiosis, culture or ethical conflict? In De Tavernier and S. Aerts (eds.) Science, Ethics and Society. Preprints. Katholieke Universiteit Leuven.
Mepham, T.B. (1995). Ethical impacts of biotechnology in dairying. In: Phillips, C.J.C. (ed.) Progress in Dairy Science, Wallingford: CAB International, pp. 375-395.
Mullan, S. and Main, D. (2001) Principles of ethical decision-making in veterinary practice. In Practice 23: 394-401.
Pullen, S. and Gray, C. (2006). Ethics, Law and the Veterinary Nurse. Butterworth-Heinemann, Elsevier.
Regan, T. (1983). The Case for Animal Rights. University of California Press, Berkeley, USA.
Singer, P. (1990 [1975]). Animal Liberation, 2nd edn. Avon Books, New York, USA.
Yeates, J.W. and Main, D.C.J. (2007). Assessment of positive welfare: A review. The Veterinary Journal 175(3): 293-300.

The ethical diversity of Europe: from cause for embarrassment to source of enlightenment

G. Meyer and C. Gamborg
Danish Centre for Bioethics and Risk Assessment, Rolighedsvej 25, DK-1958 Frederiksberg, Denmark; gitte@gittemeyer.eu

Abstract

References to different cultural traditions in Europe are frequent in academic literature and in European Union (EU) documents on science, ethics and society. As a rule there is no elaboration of the topic. Rather, it appears to be widespread assumptions that the differences are well known and that they represent an embarrassing fact: the lack of European unity. Thus, the existence of different traditions in Europe is seen as an obstacle to European unity. The current EU approach to Science in Society issues provides illustration of some possible, practical implications of this idea about unity. Originating in open discussion between different understandings of the role of science in society, the field now seems to be marked by closure and the pretence of consensus, linking ethics primarily to rules and regulation, and leaving ethical reflection and discussion behind. This paper argues that the existence of different European traditions and frameworks of thought may be viewed and used as a source of enlightenment, providing access to many different perspectives on reality. This includes different assumptions about the nature of ethical challenges and how they ought to be dealt with. Differences and disagreements, if hidden, may result in embittered conflicts, but if recognized, they may further efforts to deal in a concerted way with science-related ethical challenges. Therefore, different European traditions and frameworks of thought deserve enquiry rather than denial.

Keywords: European unity, pluralism, science-related ethical challenges, science in society

The pretence of consensus

During the last couple of decades Science in Society has become established as an EU field of research and practice in its own right. The development of the field has taken the shape of a series of metamorphoses, and in retrospect each step of development stands out as marked by changes in terminology, beginning with 'Bioethics' and ending up with 'Science in Society'. The process as a whole appears to represent a movement from general conflict and confusion towards consensus, closure and calm efficiency.

As part of this, the kind of concerns that triggered the process in the first place – ethical concerns and disagreement relating to science and technological innovation – has become a minor sub-field, primarily defined as having to do with rules and regulation. In effect, other understandings of ethics – understandings, for instance, that primarily connect ethics to reflection and discussion – have become marginalized.

From one perspective this can be seen as a marker of increasing European unity: A new field has become clearly demarcated and sub-divided; doubt has been left behind, and thereby conditions for concerted action have been secured. From another perspective the same development can be seen as a loss of European diversity that may prevent the realization of European unity.

The latter perspective has informed this paper, making the case that there might be something to gain from exploring, as a possible source of enlightenment, different European traditions as they appear in different understandings of ethics – that is, different assumptions about the nature of ethical challenges and how such challenges should be dealt with. This is also an argument for the re-introduction into the Science in Society field of a wider understanding of ethics, with room not only for rules and regulation, but also for reflection and discussion. The denial of diversity and disagreement, we argue, is unlikely to produce more than the pretence of consensus. The existence of different frameworks of thought – tied to language areas, nation states, religious traditions, professional cultures etc. – is a European fact. The pretence of consensus may bring about much deeper and more embittered conflicts than would have emerged if the presence of different and sometimes directly conflicting points of view had been acknowledged from the outset.

Theoretical background: unity and diversity

Unity is a contested concept (Collier *et al.*, 2006). Interpretations differ according to different basic assumptions. Unity may be connected in a straightforward way to consensus and closure, that is to the absence of differences and disagreement. Thus, unity and diversity are taken to be opposites, and the observation of differences is equalled to the observation of the absence of unity. Currently, this is probably the most widespread understanding of what European unity should be taken to mean. Typically, references to different cultural traditions in Europe are frequent in academic literature and in EU documents on science, ethics and society (European Commission, 2002; Hermerén, 2008), but these differences seem to be treated as embarrassing evidence of the lack of European unity. Taking the differences to be well-known and of no positive significance to a goal of European unity, authors tend to turn their back on the differences rather quickly.

The existence of different European traditions may, on the other hand, also be seen as a basic condition for the realization of European unity. From this perspective, the distinction between unity and uniformity is of utmost importance. European unity is connected to shared responsibilities and collective action; that is, to political life as the living together of men who are all different (Arendt, 1969; Crick, 2005). To this pluralistic understanding, different traditions and frameworks of thought may provide access to many different perspectives on reality and should therefore be used as a source of enlightenment. There is no ideal of closure and complete consensus. On the contrary, there is an ideal of ongoing reflection and discussion between different points of view as a means to dealing with shared challenges in a concerted way.

Both understandings represent European frameworks of thought. Both are committed to an ideal of European unity. The current EU approach to Science in Society issues seems, however, to be dominated by the former understanding and to have marginalized the latter.

The science in society field: a case of closure

In 1991 The European Group on Ethics in Science and the New Technologies was established. This can be seen as a response to public reactions to developments within the broad field of biotechnology in general, and with respect to the possible future production of genetically modified food in particular. Consequently, *Bioethics* acquired significance in the EU. Resources for research under that heading were provided in the research framework programmes FP4 (1994-1998) and FP5 (1998-2002), and the 1990s were marked by economic, intellectual and institutional expansion in this emerging field of EU concern. Thus, during these years *ELSA* was born as an acronym signifying Ethical, Legal and Social Aspects of science and technological innovation. The focus on ethical questions spurred and integrated concern with questions relating to law, to social and economic inequalities and to political decision-making. The field was widening, and a minor project – one person, one year – on philosophical pluralism and bioethics – was actually funded by the FP4 (CORDIS, 2009).

From 2000 onwards a change of focus and direction occurred. The change seems to have been directly linked to the formulation of the Lisbon goal that EU should, by 2010, become the most competitive knowledge-based economy in the world. Bioethical ELSA was substituted by a Science and Society Action Plan (European Commission, 2002), and FP6 provided 88 million euros under the heading of *Science and Society*. In the current research framework programme, FP7, studies regarding Science *in* Society are granted 330 million euros.

Nine sub-fields each carried separate chapters in the final evaluation of the Science and Society Action Plan: Public awareness; Science education and careers; Dialogue with citizens; Involving civil society; Producing gender equality in science; Research and foresight for society; The ethical dimension in science and the new technologies; Risk governance; The use of expertise (The Evaluation Partnership Limited, 2005). The new field as a whole has expanded. Ethics, on the other hand, has been narrowed down to a separate drawer, or, if you like, ethical questions and reflection seem no longer to be taken to be relevant to the eight other sub-fields. The interpretation of ethics has changed.

In FP6, attention to science and society issues had a high priority. Four such areas were identified: Public outreach/dialogue; Education and training; Gender issues; Ethics. The latter area was connected to codes of conduct and respect for ethical standards.

Thus, the financial expansion of the field has continued during the first decade of the new millennium. At the same time, the field has undergone considerable changes with regard to content and focus. The early focus on bioethics has been substituted by a focus on purposes, firstly of achieving the Lisbon goal, secondly of realizing the goals from the White Paper on European Governance, aiming at a more effective enforcement of Community law by way of increased openness and involvement (Commission of the European Communities, 2001).

Gradually, an understanding of ethics as connected primarily to rules, bans, commands, checklists and control has evolved (European Commission, 2009). Concern with ethics – interpreted as the certification that sets of predefined ethical standards have not been violated – has been ascribed significance as a precondition for attending to the purpose of furthering and promoting science as an instrument of technological innovation, economic growth and competitive power.

The integrative approach to ethical, legal and social aspects has been supplanted by a rather strict separation of ethics from science and society issues at large. This separation is not easily taken into practice, as noted in the section on ethics in the final evaluation of the Science and Society Action Plan: 'The nature of the Actions under this theme no doubt accounts for the large number of collaborative activities with communication, dialogue and working group types of projects, focused on enhancing and improving dialogue and cooperation between various actors in this field' (The Evaluation Partnership Limited, 2005).

Still, the isolation of ethics from science and society issues in general has been the overall trend. In 2007 it was even possible to finalize a report on the integration of science in society issues into scientific research with only a footnote on ethics: 'This report was produced before any results of the study in Ethics in FP6 were available, so ethical issues are not covered as a specific theme' (European Commission, 2007: 5).

Closure and the (in)ability to admit new challenges

The delineation of ethics, having emerged from the above process as the official EU understanding to be applied in the Science in Society field, represents one particular idea of ethics. Like the dominant understanding of unity, mentioned in the above, it is connected to ideals of consensus and closure. While perfectly valid, this is neither the only possible idea of ethics, nor is it the only idea with a European birthright. Moreover, it is likely to be inadequate with respect to dealing with new and emerging challenges, characterized by the absence of rules and standards to be respected.

The commercialization of science that has gained momentum during the recent decades, putting time-honoured scientific ideals about disinterestedness and knowledge as a common good (Merton, 1968) into question, can be seen as an example. There does not seem to be room in the current Science in Society field to connect the commercialization process to ethical challenges. European university rectors, meeting in 1988 in Bologna, stated in the Magna Charta of Universities: 'To meet the needs of the world around it, its [the university's] research and teaching must be morally and intellectually independent of all political authority and economic power' (Anonymous, 1988). Efforts would be needed to make this statement compatible with the current process of commercialization, but the Science in Society field appears to be unable to acknowledge the challenge and further such efforts. Within the framework of thought that ties the field unreservedly to a purpose of promoting commercial competitive power, the commercialization of science is seen as a means, not as a development that may at the same time carry ethical challenges to be confronted in the development of a European Research Area (Commission of the European Communities, 2007).

Project experience: unrecognized challenges and conflicts

One particular understanding of ethics appears to have come to prevail in the structure of the Science in Society field. We have argued in the above, that this understanding connects ethics primarily to rules and regulation, that it tends to isolate science in society issues in general from ethics, and that this seems to be taken to represent the universal, rather than a particular, contextual idea about ethics. Understandings of ethics as related to reflection and discussion – and, thus, not to an ideal of closure – appear to have become marginalized.

Now, this is our interpretation, based on the study of official documents. It is nowhere stated in those documents. Their understandings of ethics are not made explicit, and they are silent about the existence of different assumptions and ideas about the nature of ethical challenges and how such challenges should be dealt with. Moreover, their rhetoric may include bits and pieces, picked in a more or less haphazard way from contributions to hearing processes and originating in other frameworks of thought. Consequently, participants in actual or planned projects are left confused: Should scientists be prompted to reflect or instructed to behave according to rules? What is the purpose of providing scientific projects with work packages on ethics?

This confusion, and the consequent impaired ability to facilitate exchanges between different points of view, is likely to be repeated in the actual projects. Experience from a recent, FP6-funded Network of Excellence concerning animal disease genomics may serve as an illustration (EADGENE, 2009). Due to the intensive use of animals in the research, a work package on ethics was obligatory to the project. The purpose of the work package was defined as one of facilitating deliberation on questions about values, social interests and moral disagreement. Thus, the purpose was not based on an understanding of ethics as linked primarily to rules and regulation to be obeyed. In practice, however, the facilitation of deliberation – and the linking of deliberation and actual practice – was hampered by the compartmentalization which is a feature of the Science in Society field. 'Consumer concerns' and 'communication' were placed in separate work packages and not linked to the work package on ethics.

As part of the project, a series of interviews were conducted with scientific researchers linked to the network which included public as well as commercial partners. Dilemmas relating to the commercialization of science – in particular to prospects of patenting and demands that research results be kept confidential

– appeared as pressing issues (Meyer, 2005). These dilemmas were prominent also at a series of internal and external workshops, organizing exchanges within the framework of the Ethical Matrix (Mepham, 2000) between scientists, and between scientists and other, external, stakeholders.

Scientific researchers struggle to somehow reconcile new working conditions with traditional scientific ideals. There is, however, nowhere to go with these concerns. The prevailing, narrow definition of ethics does not recognize such problems as ethical challenges. In so far as ethical reflection and discussion do take place during projects, structural links may not be provided to inform practice – neither within the individual project, nor at a more general level – by concerns and arguments from such deliberative exercises. Moreover, there is a general absence of norms and routines for dealing with the fact that participants in European research projects originate in different cultural traditions and languages and may take very different assumptions of the proper role of such exercises for granted. This also applies to unrecognized interpretational differences with respect to the ideal of openness and other contested concepts which are significant to different understandings of ethics.

In conclusion, it seems possible in practice to go beyond an understanding of ethics as linked primarily to rules and regulation. Deliberation may be introduced, but – due to compartmentalization – it may be of little or no practical avail. And lack of attention to cultural differences and to different interpretations of seemingly shared concepts and ideals may hamper deliberation and, thereby, the development of norms and rules as responses to new challenges.

Exploring the diversity

Exchanges between different points of view and different frameworks of thought are important to a concerted European effort to deal with science-related ethical challenges on the assumption that the acknowledgement of diversity is a condition for concerted action. On this assumption, not only the general recognition of diversity but also the actual understanding of different European frameworks of thought, becomes salient.

Understanding at the level of basic assumptions – making them accessible to discussion – might contribute significantly to the facilitation of exchanges. This includes basic assumptions about ethics as such and about discussion; that is, about disagreement, conflict and consensus, and unity and diversity. It also includes assumptions about the natural and proper relationship between science and production; about the natural and proper role of economic gain as a driver in society; and concerning the significance of universal principles and absolute values to proper argumentation. Attitudes to topics like the legitimate use of animals; academic freedom and commercialization; and openness about scientific uncertainty are at a basic level likely to be informed and to some extent interrelated – not to be confused with determined – by assumptions at that level.

A wide variety of qualitative approaches may be well suited to a goal of increasing the understanding of such assumptions in different European frameworks of thought. One possible approach might be to focus on differences between European language areas, drawing on results, firstly, from comparisons of norms regarding technical and scholarly writing and argumentation (Byrnes, 1986; Clyne, 1993; Galtung, 1981; Pöckl, 1995); secondly, from comparisons of different European approaches to the media and journalism (Hallin and Mancini, 2004; Meyer and Lund, 2008); thirdly, from enquiry into difficulties relating to the translation from one European language to another of key ethical and political notions, like *Öffentlichkeit* (Kleinsteuber, 2001). Suggestions for other possible approaches would be welcomed by the authors.

References:

Anonymous (1988). The Magna Charta of Universities. Available at: http://www3.unibo.it/avl/charta/charta12.htm. Accessed 20 March 2009.

Arendt, H. (1969). The Human Condition. The University of Chicago Press, Chicago and London, 333 pp.

Byrnes, H. (1986). Interactional style in German and American conversations. Text, an interdisciplinary journal for the study of discourse 6(2): 189-206.

Clyne, M. (1993). Pragmatik, Textstruktur und kulturelle Werte. Eine interkulturelle Perspektive. In: Schröder, H. (Hrsg.) Fachtextpragmatik. Gunter Narr Verlag, Tübingen, pp. 3-18.

Collier, D., Hidalgo, F.D. and Maciuceanu, A.O. (2006). Essentially contested concepts: Debates and applications. Journal of Political Ideologies 11(3): 211-246.

Commission of the European Communities (2001). European Governance: A White Paper. COM (2001) 428 final. Available at: http://eur-lex.europa.eu/LexUriServ/site/en/com/2001/com2001_0428en01.pdf. Accessed 20 March 2009.

Commission of the European Communities (2007). Green Paper. The European Research Area: New Perspectives. COM (2007) 161 final. Available at: http://ec.europa.eu/research/era/pdf/era_gp_final_en.pdf. Accessed 20 March 2009.

CORDIS (2009) About projects: Pluralisme des fondements des positions de bioethique et fonctionnement des institutions europeennes. Available at: http://cordis.europa.eu/search/index.cfm?fuseaction=proj.document&CFID=8970353&CFTOKEN=38632554&PJ_RCN=2901345 Accessed 19 March 2009.

Crick, B. (2005) In Defence of Politics. Continuum, London and New York, 245 pp.

EADGENE (2009). European Animal Disease Network of Excellence for Animal Health and Food Safety. Available at: http://www.eadgene.info. Accessed 19 March 2009.

European Commission (2002). Science and Society Action Plan. Office for Official Publications of the European Communities, Luxembourg, 27 pp.

European Commission (2007). Integrating Science in Society Issues in Scientific Research. Main findings of the study on the integration of Science and Society issues in the Sixth Framework Programme. European Commission, Brussels, 22 pp.

European Commission (2009). Research. Available at: http://ec.europa.eu/research/index.cfm?pg=search. Accessed 19 March 2009.

Galtung, J. (1981). Structure, Culture and Intellectual Style. An Essay Comparing Saxonic, Teutonic, Gallic and Nipponic Aproaches. Social Science Information 20: 817-856.

Hallin, D.C. and Mancini, P. (2004). Comparing Media Systems: Three Models of Media and Politics. Cambridge University Press, Cambridge UK, 342 pp.

Hermerén, G. (2008) European Values – and Others. Europe's Shared Values: Towards an ever-closer Union? European Review 16(3): 373-385.

Kleinsteuber, H.J. (2001). Habermas and the Public Sphere: From a German to a European Perspective. Javnost The Public 8(1): 95-108.

Mepham, B. (2000) A Framework for the Ethical Analysis of Novel Foods: The Ethical Matrix. Journal of Agricultural and Environmental Ethics 12(2): 165-76.

Merton, R.K. (1968) Social Theory and Social Structure (Enlarged edition). The Free Press & Collier-Macmillan Limited, New York and London, 701 pp.

Meyer, G. and Lund, A.B. (2008). International language monism and homogenisation of journalism. Javnost The Public 15(4): 73-86.

Meyer, G. (2005). Principles for ethical deliberation in bio-scientific projects. Animal disease genomics: a case study. Project Report 10. Danish Centre for Bioethics and Risk Assessment, Frederiksberg, Denmark, 34 pp.

Pöckl, W. (1995). Nationalstile im Fachtexten? Vom Tabu- zum Modethema. Fachsprache. International Journal of LSP 17(3-4): 98-107.

The Evaluation Partnership Limited (2005). Evaluation of the Science and Society Action Plan. Final report for The European Commission, DG Research – Directorate C, ERA: Science & Society. Available at: http://ec.europa.eu/research/science-society/document_library/pdf_06/sasap-final-report-june-1306_en.pdf. Accessed 20 March 2009.

Ethical food: moral or practical dilemma?

Erik Schmid
6901 Bregenz, Austria; erik.schmid@vorarlberg.at

Abstract

Eurobarometer data indicates that many people are in favour of 'ethical food'. Consumer attitudes to animal welfare strictly depends on clear information and experience of farm visits. If clear labelling was applied to products it would be easy to purchase ethical food and enjoy its consumption. The 'ethical matrix concept' is one of the first promising approaches that can be applied to qualify views of the ethical value of products. Consumer attitudes and the actual shopping decision can differ for various reasons. If the views of various stakeholders are examined, it is claimed that farmers will feel more or less comfortable with the current production situation. Farmers see the setting of higher standards (especially animal welfare or environmental protection) as unnecessary, it should be business as usual with regards to 'Good Manufacturing Practice'. Consumers state that they cannot afford ethical food since it is too expensive. Nevertheless, money is spent on other luxury items, e.g. designer clothing and holidays. In Austria, on average only 12% of the overall cost of living is spent on food. Retailers successfully use their own trade marks to define food quality. These are often clear labels used for specific brands of food that sell well. However, these labels are difficult to compare with each other, they can confuse consumers rather than informing them. Politicians set a clear legal framework for the free market to regulate itself and it is the responsibility of consumers to make their choice within this framework. Hence there is urgent need for a transparent system. At the moment, the 'Animal Welfare Label' takes only one dimension into consideration. Other examples for simple categorisation of complex subjects can be indentified such as The European Energy Award® with its e5. The Ecosocial Forum Europe has created a concept for the ethical assessment of products including five dimensions, these include: Highest product quality for consumers; fair price for the farmers; sustainable care of landscape; good animal husbandry; and efficient use of energy. Applying this concept to animal production systems would ensure a fair competition in production and retail, simple and clear information for consumers to make a fair decision at the point of sale.

Keywords: animal welfare, ethical food, quality labelling, fairness label

Introduction

The attitudes of EU citizens towards animal welfare are well documented in the'Special Eurobarometer 270' published in 2006. This survey highlighted that there is a considerable interest in animal welfare standards across Europe and there is a demand for more extensive and clearer information regarding the welfare conditions of production animals. Within the survey, higher standards were considered to guarantee healthier products and higher quality. For animal friendly products, individuals were willing to pay up to 25% higher prices as well as agreeing that financial compensation for improving welfare standards is needed. However when these attitudes are compared to the point of sale, many consumers still go for the cheapest animal products.

The ethical dilemma

There appears to be an ever increasing moral gap between the articulated attitude of the consumer and actual purchase behaviour. The most common almost reactionary response to the question of why do consumers not buy welfare friendly products is that: They cannot afford it. For instance, consumers often criticise organic food for being too expensive. Considering that the average individual expenditure for nutrition in Austria is only 12% of the overall cost of living, this argument may not be that robust. Comparing the prices of intensive production with those of animal-friendly husbandry, a price difference far above 25% (the threshold identified in the Eurobarometer) can be found for some product. This might be justified for organic products, knowing that nearly 80% of the total costs of production go towards the cost of what is very expensive organic feed. In contrast, higher welfare standards are often self-financed by a farmer through on-going investment to improve the health status and the lifespan of an entire herd.

Consumers display a diversity of often contradictory behaviours when it comes to their animal product choices. As a result many of the big retailing companies have adjusted their strategies to meet this type of consumer schizophrenia. Retailers sell both cheap un-named products or house brands and expensive high quality trade marks, organic or high animal-welfare products. Like in all other businesses, the players in the food sector can increase their earnings if they are provide exclusive or specialised products. Where as the wider market is centred around pricing competitions and crowding out competitors. Organic labelled products, preferably from a local region, are sold at twice the price of non-organic products. Nevertheless, in Austria the proportion of organic farms as well as products sold is steadily increasing, and this appears to not be affected by an economic downturn, as it is the case at the moment.

One of the key issues, when consumers purchase animal products is whether they are aware of the type of production system under which the food is produced. In addition, are they aware of the contradictions of their purchase decisions, such as buying organic produce as well as low value mass produced meat. It can be claimed that the mass market is not only correlated with animal cruelty, but also associated with massive waste of resources. Some of these products really have 'no name', they therefore have no identity, no value. Considering that feeding concentrates to animals, especially ruminants, inefficient or even contradictory in respect to sustainable use of energy, it could be argued that the only moral solution would be to become a vegetarian. If the system of distributing goods in this way is kept up, it could be claimed that it will be impossible to feed the world without cruelty to humans and animals.

A fundamental legal principle in every animal welfare act, such as in Austria, is the need for a justificable reason for killing any animal. Eating meat is still well accepted in Western societies as part of healthy diet or even for spontaneous performance of life finally basing on the ethical argumentation of self-defence. However can the average meat consumption of nearly 80 kg per person and per year be reasonable? Taking into consideration that more than half of the diseases of our modern civilisation are directly correlated with a level of meat consumption, it could be agreed that it is simply unhealthy, as well as socially uneconomical to eat meat in this way. Therefore, killing animals for such a high level of meat consumption can never be ethically justified.

Usually price is the best argument to convince consumers, rather than environmental protection or even ethics. Using a common sense approach, wasting 50% of half price product is as expensive as using 100% of double price product. On top of that, financial support for intensive farming systems causes additional externalised costs due to the severe damage to the environment which is not sustainable. The McKinsey (2009) report on pathways to a low-carbon economy, states that 90% of the total CO_2 emissions are related to land use in forestry and agriculture. If agricultural production changed from intensive to organic systems, 25% of the total CO_2 production could be saved without any other additional measures. In a 21-year study, Mäder systems organic farming is one of the few ways to keep or improve soil fertility and biodiversity. Summarising the economic evidence seems to indicate that the socio-economic balance gives a clear and advice about the harms of intensive farming.

The practical dilemma

There appears to be notable support for animal-friendly farm production. However there does not seem to be an extensive market. Producers, at least on the continent, are traditionally used to a well regulated market, although they are concerned about prices and unnecessary regulations. However, their main problem appears to be overproduction. Fluctuation in the market are immediately balanced between consumers and producers. Retailers are very successful with their own private brands, 'pure nature' 'back to origin' and other sonorous and well-selling labels dominate the market, and this can affect producers drive to improve production systems.

Politicians responsible for a clear legal framework refer back to the free market which is self regulating. Responsibility is delegated to the consumers, closing the circle because the uncountable number of labels is confusing rather than transparently informing their decision-making. The result is claimed to be as clear and unavoidable as the information is unclear: If I cannot be sure, I will take the cheaper product. There is therefore an urgent need for a transparent high quality labelling system.

Quality definition

Product quality traditionally was defined by producers and manufacturers in a very technical way: Percentage of ingredients of the final product, cell counts, bacteriologic status etc. For consumers other aspects of the quality of the production process are far more important, such as the use of fertilisers, the feeding and the rearing conditions of the animals. Accepting the definition of quality should be extend is the first decisive step. The new EU food and feed regulation highlighted animal welfare as an important part of food quality. With 'Welfare Quality', the Commission has not only launched an interdisciplinary scientific project, but the 'Welfare Label' has now become a practical proposal within the EU Animal Welfare Action Plan 2006-2010.

Labelling

Branding and clear labelling can be very successful tools with in any market. Within a saturated society with plenty or almost too much of everything at any time, labels are becoming more important. They are trendy and stylish, they even define a certain life-style as illustrated by LOHAS (Lifestyle of Health and Sustainability), but they also can be very variable and confusing. The current situation within the food-labelling sector is claimed to be far from ideal. However, important preconditions for food labelling are: The message has to be simple and clear-cut, the pictures have to match the reality. The standard(s) has/have to be above the minimum legal requirements. There has to be an independent monitoring body and documentation ensuring reliability.

Strategies

There are two main labelling strategies that can be applied. The first decision is to determine whether a label should be mandatory or voluntary? On example is the 'organic' label which is a standard of label clearly fixed by EU-Regulation, participation is voluntary. This the labelling recommendation are clear: Free membership, legal standard(s). A second more liberal variant could be an official registration scheme, which makes an act of authorisation essential.

The second decision is should a clear-cut single-level or a dynamic multi-level quality system be used? At a first glance, a single-level system seems to be easier. The product only gets the label if the limitations are met. But in the end, this poses the same problem as other legal regulations with minimum standards: It will distinguish strictly between 'good' and 'bad', probably based on one single criterion. However within dynamic and complex systems, like biology, animal welfare and ethology, clear cuts are very difficult to make. Besides, in daily work the pressure for compromise grows, not only in terms of registration but also in monitoring. It is unavoidable to adapt the standards to the situation of the market, but also to scientific evidence. Currently, possible limitations have become the subject of fierce discussions. The danger of losing acceptance within the group of producers as well as the reliability for consumers is increasing and will even grow, especially if there were arbitrary regulations or non-standard members.

The multi-level system has the principal advantage to better reflect complex systems. The best known international example for a tree-level rating system is the 'smiley'. Due to Gaussian distribution, opinion polls that use a grading system will often result in neutral rankings and do not help to differentiate. To avoid this, five-level systems can be used. The big advantage of this approach is that the neutral middle sector gets subdivided in a part with a tendency to good and another to bad. Only with this method, can trends in one or the other direction be estimated. Another benefit from such a multi-level system is that the pressure to get an exact classification is decreased. Entering the system on the first step above legal minimum standards and reaching the next step of quality is much easier, because it is a step by step strategy. Reaching the top is possible in stages, which is also valid for the opposite direction: the closer you get to the minimum standards, the more warning signals you get at every step Finally it would be possible to even combine the multiple-level system with the single-level or other existing system e.g. if 'free range' or 'organic' are taken as defined levels in the system. The labelling of eggs may be a good example of this, but a strategy would be needed to combine the options of the market (quality labelling), as well as a 'stick' (legal minimum standards) and the 'carrot' approach (financial support for compliance with higher standards, e.g. article 40 EC Regulation Nr 1698/2005). The multi-level system would perfectly fit with the instruments of the common market, even with regard of WTO.

The problems of defining 'Animal Welfare' or 'Good Manufacturing Practice' as a one level label are most likely the strongest arguments for a multi-level system. Thinking not only of animal welfare, but about environment, soil management, transport (food miles) and fair prices for the producers as well, not a multi-level, but more likely an index system like the 'Ethical Matrix' could to be generated. The new and convincing approach of this system is the possibility to compare two ore more products with respect to ethical value. But at the point of sale the consumers need a simple labelling system which gives them direct information about the ethical and fairness value of the products that they have to choose between. This will represent a practical dilemma.

European Energy Award, fairea, fairness label

The European Energy Award (eea®) was developed almost eight years ago as a steering and controlling instrument supporting communities to review systematically their energy related activities. It is a Total Quality Management System for communal energy-related services and activities. Certification and awards for energy related achievements and control of success through regular audits are provided. Energy-related activities are summed up in six sectors (i.e. development, buildings, supply and disposal, mobility, organisation and communication). Within the auditing process, the current status of each category is compared to all (100%) possible activities. Thus, the result is not an absolute value, but an overall percentage of reaching the maximum. The certification is 'e5' with one 'e' for reaching 25%, two 'e's for 37.5%, tree 'e's for 50%, etc This type of grading would be applicable for any other certification system. The big advantage of the system is the detailed evaluation procedure within the permanent auditing by an independent body and therefore continuous process of improvement. In Austria the Ecosocial Forum Europe created a concept for the ethical assessment of products including five dimensions. (1) Highest product quality for consumers (2) fair price for the farmers, (3) sustainable care of landscape; (4) good animal husbandry and (5) efficient use of energy including packing and transport. Unfortunately they decided to use the single-level label 'gut so!' (roughly: 'alright!') in combination with fairness and region (faire area). Another competitor within the numerous number of private labels.

Conclusions

Consumers want food to be healthy and they wish to consume with a quiet conscience. Their expectations define quality. Ethical food is not a moral dilemma, as the socio-economic impacts shows there is an urgent need to change farming practice away from intensive farming immediately. In order to take responsibility at the point of sale, consumers need to have clear and simple information about the ethical value of food products. Current labelling of products is more confusing than informing and it is claimed that there is resistance from producers and manufacturers to act. Retailers also prefer their own labels. An independent system is needed that allows consumers to compare the ethical quality of the existing trade labels. The introduction of new Animal Welfare labels could be a first step, which also include aspects of sustainability and fairness. A multi-level system is far more practical to classify complex correlations than single-level systems. It is the only tool that can be used to combine the options of the market with the legal standards and financial support systems. The e5 system of eea® could be a practical model for an European fairness label. This would ensure fair competition in production and retailing, and simple and clear information for consumers allowing then to make fair decisions at the point of sale.

References

Council Regulation (EC) No 1698/2005 of 20 September 2005 on support for rural development by the European Agricultural Fund for Rural Development (EAFRD).

McKinsey and Company (2009). Pathways to a Low-Carbon Economy, Version 2 of the Global Greenhouse Gas Abatement Cost Curve.

Theme 11 – Ethical matrix development

One step beyond the ethical matrix: the sequel on ethical frameworks

V. Beekman and E. de Bakker
Agricultural Economics Research Institute, P.O. Box 29703, 2502 LS The Hague, the Netherlands; erik.debakker@wur.nl

Abstract

At the previous EurSafe conference we presented a paper about developing and working with an ethical framework that combined aspects of the ethical matrix and multi-criteria mapping. We coined this ethical framework as participatory multi-criteria analysis, and told a positive story about our experiences with the framework in the Dutch agricultural and food regulatory context. In comparison with the ethical matrix, we replaced the ethical principles with socio-political values and moved these from the columns to the rows of the matrix, we skipped the actors from the matrix and ensured that impacts on a variety of actors were covered by completing the matrix in an interactive multi-stakeholder setting, and we added scenarios in the columns of the matrix to be able to compare the ethical impacts of different policy options. Although initial experiences with this framework were positive, further experiences with the framework revealed difficulties in incorporating the framework in decision-making processes. We learned that the Dutch Ministry of Agriculture, Nature and Food Quality needed to get a better idea of how socio-political values, ethical principles and legal norms relate to each other. At that point it appeared helpful to return to the principles of the ethical matrix to shed light on the relation between values, principles and norms. The main message of this second report about our experiences with participatory multi-criteria analysis as an ethical framework is that ethical frameworks and competencies are complementary, and need to be developed simultaneously to improve decision-making processes about ethical aspects of agriculture and food production.

Keywords: competencies, ethical principles, multi-criteria mapping, socio-political values

Introduction

At the previous conference of EurSafe in Vienna we presented a paper (Beekman *et al.*, 2007a) about developing and working with an ethical framework that combined aspects of the ethical matrix as developed by Mepham (1996) with aspects of multi-criteria mapping as developed by Stirling and Mayer (2001). We coined this ethical framework as participatory multi-criteria analysis (pMCA), and basically told a very positive story about our experiences with the framework in the Dutch agricultural and food regulatory context. These positive experiences were mainly based on two pilot studies in which we used the framework to evaluate the ethical impacts of different policy options with respect to *Avian Influenza* (Beekman *et al.*, 2007b) and meat and bone meal in animal feed (Stijnen *et al.*, 2008) respectively. However, further experiences with the framework, particularly in a pilot study about the ethical impacts of group housing for pregnant sows, revealed difficulties in incorporating the framework in decision-making processes.

In this second paper about our experiences with pMCA as an ethical framework we will first introduce the basics about the ethical matrix, and summarise the four innovations to this ethical framework that we presented at the previous EurSafe conference. We will then highlight some implementation problems, and reverse one of the four innovations in an attempt to address our learning experience that the Dutch Ministry of Agriculture, Nature & Food Quality (LNV) needed to get a better idea of the relation between socio-political values, ethical principles and legal norms in order to be able to routinely incorporate the framework in decision-making processes. We will conclude with the message that ethical frameworks and competencies are complementary and need to be developed simultaneously.

Ethical matrix

The basic idea of the ethical matrix (Mepham, 2005) is to use a tabular format to map the ethical aspects of socio-political challenges in the domain of agriculture and food production, and to use the ethical principles as developed for biomedical ethics by Beauchamp and Childress (1994) and actors or affected parties as headings for the columns and rows in this table respectively (Table 1).

Standing on the shoulders of a giant

At the previous EurSafe conference we presented a paper (Beekman *et al.*, 2007a), inspired by multi-criteria mapping as developed by Stirling and Mayer (2001), in which we introduced four innovations to the original ethical matrix. The first innovation was to replace the ethical principles with socio-political values, and to move these values from the columns to the rows of the matrix. The second innovation was to skip the actors from the matrix, and to ensure that impacts on a variety of parties were covered by completing the matrix in an interactive multi-stakeholder setting. This latter innovation was in line with earlier arguments by Kaiser and Forsberg (2001) about the importance of developing participatory approaches to the application of ethical frameworks. The third innovation was to add scenarios in the columns of the matrix to be able to compare the ethical impacts of different policy options. The fourth innovation was to introduce some scoring and weighting rules to facilitate balancing the various ethical impacts of different policy options (Table 2).

Implementation problems

Although initial experiences with pMCA as an ethical framework were positive, further experiences with the framework revealed serious difficulties to routinely incorporate it in decision-making processes. Whereas some of these difficulties were of a practical nature – initially required time investment, timing of the application in the policy-making process, communicative willingness of stakeholders – other difficulties related to more foundational issues. One could argue that all ethical frameworks struggle with such practical and foundational issues, and that the experienced difficulties should be understood as an instantiation of these more generic issues. This argument is valid to a certain extent but two striking problems seemed to be specific for pMCA as an ethical framework.

First, the absence of full conceptual clarity about the distinction between socio-political values, ethical principles and legal norms in pMCA caused discomfort on the part of LNV about the ethical and regulatory robustness of the framework. Second, this discomfort revealed that the possession of some

Table 1. Ethical matrix (Mepham et al.*, 2006).*

	Respect for		
	Wellbeing	Autonomy	Fairness
Producers			
Consumers			
Treated organisms			
Biota			

Table 2. Multi-criteria map (Beekman et al., *2007a).*

	Weights	Policy option$_0$	...	Policy option$_N$
Food security				
Economics				
Product quality				
Food safety				
Health				
Environment				
Animal welfare				
Fair trade				
Craftsmanship				
Balance →				

basic ethical expertise is pivotal for a responsible use of the framework. Although many other factors may have hindered the effective implementation of the developed ethical framework, these two problems are significant in foundational terms and should therefore be addressed for a more fruitful application of pMCA. This implies addressing the need for a clear distinction between socio-political values, ethical principles and legal norms, and addressing the need for ethically competent users with an awareness of role, function and limitation of ethical frameworks.

When using pMCA in a pilot study about the ethical aspects of group housing for pregnant sows, we witnessed that some of the participating stakeholders interpreted the framework as nothing more than just another process management tool and therefore questioned its ethical robustness and added value. The next section will present a solution to the aforementioned problems that could strengthen the practical and foundational basis of post ethical matrix frameworks in agricultural and food ethics. The social-political value of animal welfare will be used to illustrate how ethical principles could guide mapping ethical impacts of different policy options.

One step beyond

Autumn 2008 – at that time the first author of this paper temporarily had a different job at the Athena Institute of the Vrije Universiteit Amsterdam, whereas the second author of this paper was about to deliver a finalised version of the ethical framework – we learned that LNV needed to get a better idea of how socio-political values, ethical principles and legal norms relate to each other to be able to work with the framework in their decision-making processes. Further inquiry revealed that three of the four innovations that we made to the original ethical matrix were not affected by the concerns of LNV about the application of the framework. Hence, no revisions were needed with respect to using the framework in an interactive setting to compare ethical impacts of different policy options with some help from scoring and weighting. The one thing that we needed to reconsider was the replacement of ethical principles with socio-political values, since that innovation caused discomfort on part of LNV about the ethical and regulatory robustness of the framework.

Our current understanding is that the introduction of two new columns to the left of the socio-political values could serve as an assurance of the regulatory robustness of the framework. These columns remind users that the nine socio-political values in the framework are not a coincidental selection but rather strongly embedded in the broader guiding vision of sustainable development – with its people, planet and profit value domains – that inspires Dutch agricultural and food policy-making. The reintroduction of the ethical principles in a new column to the right of the socio-political values could serve as an assurance of the ethical robustness of the framework. That column uses the ethical principles to specify four different dimensions that the nine socio-political values might have. The ethical robustness of the framework is thus based on convergence with state-of-the-art (pluralist) ethical theories, ethical frameworks as used in other Dutch policy domains and ethical frameworks as used in agricultural and food policy-making in other European countries (Table 3).

Table 3 also illustrates for the socio-political value of animal welfare what the reintroduction of the ethical principles implies for the first stage in using this most recent version of the ethical framework. The principles of non-maleficence, beneficence, autonomy and justice are used to identify what impacts different policy options could have on negative welfare, positive welfare, natural behaviour and animal rights respectively. Our main reason to reintroduce the principles with the original distinction between non-maleficence and beneficence (Beauchamp and Childress, 1994) is that this is a meaningful distinction within the Dutch regime of animal welfare policy-making. Negative welfare impacts tend to be addressed with the formulation of legal norms, whereas positive welfare – and natural behaviour – impacts tend to be addressed with economic, communicative and new governance instruments (animal rights tend not to be included in the Dutch regime of animal welfare policy-making). A similar distinction in addressing non-maleficence and beneficence impacts holds true for the other socio-political values.

The second stage in using this framework entails scoring the identified ethical impacts and assigning weights to 36 rows in the fourth column, and results in an ethical impacts scorecard (EIS) of the different policy options to address a particular issue in agricultural and food policy-making. This EIS would thus voice a rather strong demand about the ethically preferable policy option with respect to the issue at hand.

Table 3. Ethical impacts scorecard.

Guiding vision	Value domains	Socio-political values	Ethical principles	Weights	Policy option$_0$	...	Policy option$_N$
Sustainable development	People	Food security					
		Product quality					
		Food safety					
		Health					
		Fair trade					
		Craftsmanship					
	Planet	Environment					
		Animal welfare	Non-maleficence		Negative welfare		
			Beneficence		Positive welfare		
			Autonomy		Natural behaviour		
			Justice		Animal rights		
	Profit	Economics					
Balance →							

Afterthought on ethical competency

The main message of this second report about our experiences with pMCA as an ethical framework is that through the development of the ethical matrix into a multi-criteria map and then into the ethical impacts scorecard we hope to have delivered an ethical framework that could be routinely incorporated in Dutch agricultural and food policy-making processes. One caveat remains to be addressed: any ethical framework needs ethically competent users and LNV quite rightly asked the question of who should be these competent users of the developed ethical framework. Three answers seem to be possible to this question: you could outsource application of the framework to professional ethicists at universities, research institutes or consultancies, you could contract a few ethicists to apply the framework as civil servants within the Ministry, or you could give a larger number of civil servants within the Ministry a training to become ethically competent users of the framework. We voiced a preference for the last option and hope that LNV follows that suggestion.

Acknowledgements

We would like to thank Marc Bracke (Animal Sciences Group, Wageningen University and Research Centre) and Tjard de Cock Buning (Athena Institute, Vrije Universiteit Amsterdam) for their always enlightening input in our ongoing discussions about ethical frameworks.

References

Beauchamp, T.L. and Childress, J.F. (1994). Principles of biomedical ethics. Oxford University Press, Oxford, United Kingdom.

Beekman, V., De Bakker, E. and De Graaff, R. (2007a). Standing on the shoulders of a giant: the promise of multi-criteria mapping as a decision-support framework in food ethics. In: Zollitsch, W., Winckler, C., Waiblinger, S. and Haslberger, A. (eds.) Sustainable food production and ethics. Wageningen Academic Publishers, Wageningen, the Netherlands, pp. 95-100.

Beekman, V., De Bakker, E. and De Graaff, R. (2007b). Ethische aspecten dierziektebestrijdingsbeleid: een oefening in participatieve multi-criteria analyse [Ethical aspects of animal disease intervention strategies: a pilot study in participatory multi-criteria analysis]. LEI, The Hague, The Netherlands.

Kaiser, M. and Forsberg, E.M. (2001). Assessing fisheries – Using an ethical matrix in a participatory process. Journal of Agricultural and Environmental Ethics 14: 192-200.

Mepham, B. (1996). Ethical analysis of food biotechnologies: an evaluative framework. In: Mepham, B. (ed.) Food ethics. Routledge, London, United Kingdom, pp. 101-119.

Mepham, B. (2005). Bioethics: an introduction for the biosciences. Oxford University Press, Oxford, United Kingdom.

Stijnen, D., De Bakker, E., Teeuw, J., Van der Spiegel, M., De Graaff, R. and Bracke, M. (2008). Diermeel in diervoeders? Een methodische discussie met stakeholders [Meat and bone meal in animal feed? A methodical discussion with stakeholders]. LEI, The Hague, The Netherlands.

Stirling, A. and Mayer, S. (2001). A novel approach to the appraisal of technological risk: a multi-criteria mapping study of a genetically modified crop. Environment and Planning C: Government and Policy 19: 529-555.

Applying the ethical matrix method: learning from experience

E.-M. Forsberg[1] and K. Millar[2]
[1]Work Research Institute, P.O. Box 6954 St. Olavs plass, 0130 Oslo, Norway; ellen-marie.forsberg@afi-wri.no
[2]Centre for Applied Bioethics, Division of Animal Science, School of Biosciences, University of Nottingham, LE12 5RD, United Kingdom

Abstract

Since the ethical matrix was conceived at the beginning of the 1990s it has been used by an array of organisations for many different purposes. Although the method has not been without it critics, it has been seen by some to be a flexible tool and during the last 10 years different variants have been developed and applied. The theory of the ethical matrix has been quite extensively explored, but the application of the method has not been mapped and reviewed to the same extent. This paper will focus on issues that arise when the ethical matrix is used as a tool for ethical engagement in a group setting. The paper aims to reflect on different application experiences in order to facilitate practical learning and identify research needs.

Keywords: ethical matrix, applied ethics, participatory processes, stakeholder dialogue

Introduction

As technological innovation in many fields, including the agri-food sector, gathered pace in the 1980s and 1990s so did the calls for increased levels of extensive technology assessment and ethical debate regarding the potential impacts of new technologies. One of the results of this growing need for ethical review and assessment was a drive to improve and develop new ethical tools. In the field of applied ethics, a number of methods were developed specifically to facilitate both stakeholder engagement and ethical analysis. The method that will be discussed in this paper is the ethical matrix (EM) method, first proposed by Ben Mepham over 10 years ago (Mepham, 2005).

The EM was originally developed to analyse the ethical issues raised by the application of a proposed strategy (e.g. the use of a dairy biotechnology) through the application of three ethical principles (wellbeing, autonomy and justice as fairness) when applied to a number of interest groups.[22] Since then, the EM has been used by a number of groups as an analytical tool or in stakeholder engagement activities related to food biotechnology. A number of examples include: dairy production, genetic modification of crops, genetic modification of fish, animal disease genomics (Mepham, 1995; Forsberg, 2007a; Kaiser *et al.*, 2007). The approach has also been used in other sectors, for example in assessment of human biotechnologies, bioremediation technology and radiation protection management (Mepham, *et al.* 1996; Kaiser and Forsberg, 2001; Oughton *et al.*, 2004). As an analytical tool, the EM has also been used to structure analyses and reflections within committees, such as the UK Food Ethics Council and the Norwegian Council for Animal Ethics. There has been quite extensive review of the theory of the ethical matrix, for instance with regard to its soundness (Kaiser *et al.*, 2007; Forsberg, 2007b), however the soundness of any particular *application* of the method is dependent on a number of practical, methodological choices and outcomes. This has not been reviewed to same extent and we wish in this paper to look at the practical applications of the method, the factors determining its application and the challenges that face organisers and facilitators of ethical matrix processes. It will be assumed that the reader has some familiarity with the method and therefore the matrix will be discussed in methodological

[22] For further details on the application of the method please refer to Mepham *et al.*, 2007.

detail at the expense of a broader introduction. We believe that it is important at this stage of the life of the ethical matrix to establish a platform for learning, sharing of experience and quality control of practical ethical matrix processes. Only such practical methodological reflection can facilitate learning from experience which will provide the possibility of refining and improving practical applications of the ethical matrix. This methodological exploration by its very nature will also lead to a discussion of further research priorities. This paper is simply meant to be the start of such a learning process.

Application of the ethical matrix: experiences

In order to explore the use of the EM method, this paper will discuss the factors that have varied in the application of the matrix in some of these different uses. In this paper there is unfortunately no room for a comprehensive survey of all applications of the matrix. Moreover, in this paper we shall focus on the ethical matrix as it has been applied in different forms of participatory process. This means that we will not discuss the matrix as an analytic tool that can be used by an individual to assess an issue. A participatory EM process is a process where a number of people (usually a minimum of ten) engage in dialogue using the matrix as a tool to structure the discussion. Different forms of such participatory processes exist; and we will explore some of them. Although they may differ on a number of dimensions, the common element will be that the EM is applied to a given issue. Application choices will be influenced by the purpose of the event, as well as by practical and epistemological factors. These factors can be combined in different ways resulting in different applications of the method. We will now take a closer look at these factors.

Different purposes

The majority of reported uses of the method as an ethical engagement tool have been with multi-participant expert groups or multi-stakeholder groups. However, even within this remit the method has been used in a variety of ways, such as to explore the potential issues raised by a research agenda (Animal Disease Genomics – EADGENE), map the ethical issues raised by the application of a specific biotechnology (GM salmon), or examine possible technology trajectories for an industry (fisheries), to name but a few. In fact there can be many purposes for organising an ethical matrix process:

- Increasing ethical knowledge regarding an issue: An ethical matrix process may be an effective way to draw out an array of issues relevant to a specific topic (as it has been used by the UK Food Ethics Council).
- Making a value based judgement on an issue (e.g. fisheries ethics workshop): An ethical matrix process may be an effective way to reach an ethically informed judgement.
- Increasing value-based dialogue about an issue: In some contentious cases authorities or other agents may simply want to bring parties together in order to facilitate dialogue and new understanding. This is unlikely to be a stated reason for convening a workshop, but it may be part of the reason why such a workshop is organised (an example here may be the Norwegian GM rapeseed workshop).
- Encouraging ethical reflection as individuals or organisations: An EM process may be arranged in order to initiate greater ethical reflection within groups or networks (e.g. EADGENE workshops).
- Facilitating and demonstrating public ethical engagement: An organisation (public or private) may wish to publically facilitate and demonstrate ethical engagement using this method. This is never stated outright, but may be an underlying agenda.

There can also be other purposes and the value of each of these process purposes is often debated amongst users and developers of the method. As can be seen, the purposes are not only related to generating new insights, but may also be related to bringing about practical, process oriented, outcomes. In some cases, there can be more than one purpose for a workshop. In other cases, different stakeholders may have different perceptions of the purpose of a workshop.[23] For instance, an external funder may simply want to increase value-based dialogue on an issue, while the organising institution may want to demonstrate ethical engagement.

Epistemological factors: top down or bottom up approach?

The way the EM is applied in a group process will also depend on the epistemological choices made by the organisers. In the vast majority of these applications the matrix is used as a top-down approach, i.e. the principles and content of the cells is presented to the participants (by the organisers) with reference to specific ethical theories. In these cases, a generic matrix is used as a starting point (Mepham *et al.*, 2007) and then as is necessary, since the standard four 'interest group' categories will naturally vary with each issue, the interest groups will change. For example, the generic 'affected animal' is specified as 'dairy cow'. In addition, in some instances other interest groups are included. In some workshops the organisers have decided to include 'researchers' as a separate interest group, even if researchers strictly speaking might be described as actors rather than those that are simply affected by the decisions. However, in top down versions the principles and most of the content are used without modifications. Other variant uses have used a 'bottom up' approach, anchoring the process in a form of 'common morality' rather than in defined theoretical approaches. This allows the participants greater influence on the categories and content of the matrix. For example, early participant concerns led one workshop group to decide – during the process – to split the principle of wellbeing into two: a principle of increased benefits and reduced harm (Kaiser *et al.*, 2007). In the GM rapeseed workshop the participants were at the start of the process encouraged to revise the proposed content of the matrix, in order to give them more ownership over the matrix that they would be applying to the case.

Practical factors: availability of resources and participants

EM processes are always tightly budgeted and methodological choices will be influenced by this fact. The organisers try to include an adequate number of participants to ensure an appropriate breadth and depth of discussion. However, in most cases this means increased expenditure. Often the application choices are influenced by the fact that there are only resources for a certain number of participants who are available for a fixed amount of time. Moreover, as with all participator processes, it can sometimes be hard to recruit participants to attend these types of workshops. Although most individuals consider ethical discussion to be important it is not always easy to prioritise workshops of this nature in a busy schedule. For instance, for the GM rapeseed workshop approximately 50 people were invited, but only 14 were able to participate.

Group composition and programme design

The three factors identified above will influence the group composition and programme design. As mentioned above, ethical matrix workshops usually include multi-stakeholder groups. For a number of these ethical engagement workshops when specific technologies are being discussed convenors often also invite technical expertise (e.g. scientists who are developing the technology). However, other participation designs have also been used. Lay workshops have been organised, such as a 2004 lay event in Norway which discussed the issues raised by the potential use of GM fish (Kaiser *et al.*, 2007). Furthermore, in 2008 four EM workshops were organised with the aim of encouraging a dialogue on

[23] The Norwegian GM rapeseed project may be an example here. Although the intention of the project leader was to reach a judgement on the issue, this was unlikely to be the main reason why the project was funded. For the funder (the Norwegian Research Council) the most important purpose of the project was to support public discussion of a controversial technological topic.

ethical aspects of animal disease genomics research network. The majority of participants in each of these workshops were members of a single research network, EADGENE, and the workshops were intended to facilitate intra-network ethical reflection.

Ethical matrix workshops have been conducted with groups of varying sizes, from groups of less than 15 participants (e.g. GM rapeseed workshop; livestock genomic research) to approximately 45 participants (i.e. for the 2000 Norwegian fisheries workshop). Increased group size may be correlated with an increasing risk that some participants (usually the more senior participants) may dominate and others may feel intimated or be less interactive in larger groups. In some of these processes – often when the participatory event includes more than approximately 8-10 participants – break out groups have been used where groups of participants are sub-divided into smaller discussion groups. For larger groups, plenary discussions have mainly be used for lectures and group presentations, and it is the authors' view that these sessions have notable limitations if used as the main forum for ethical reflection and the exchange of ideas.

The structure of the programme will vary according to the purpose, group characteristics and the epistemological approach used. If the content of the ethical matrix is not specified, a first group session establishing the structure of the EM will be required. If a top-down matrix is used, then one may start by applying the matrix to the issue at stake. The structure of the programme and the number of sessions will be determined by the objective of the workshop. In terms of practical workshop sessions when the EM is applied to map issues raised, an initial introduction is then followed by sessions that allow the participants to explore (map) potential issues. However, if an evaluation of a particular technology or policy is required, some participatory designs include a weighing session, where the participants indicate which particular values are significant, or of greater importance. If the purpose of the process is to reach a conclusion, a final group session will be devoted to this. This session may include a discussion of the distribution of benefits and costs, weighing of arguments, etc. The only session that is required in all workshops is the session in which the tool is applied to map the issues at hand.

Learning from the experience

As is shown above there is now quite a rich inventory of applications from which to learn. The learning objectives from these experiences must be related to the stated purpose of each workshop. As discussed above, the purpose or objective for using this method can include increasing ethical understanding of an issue or to having some sort of practical impact. Indeed, practical impact is likely to be enhanced if it can be shown that the process contributes to greater insight. Practical impact is likely to depend on whether the participants take ownership of the process and see value in the process and outcomes. But there are some important challenges to achieving this. Some of the practical challenges to achieving a valued and effective process can be summarised as the following:

Time and resource aspects

For some applications a lack of time can be a constraint to effective discussion. This may be considered a minor point, but this can amount to an important practical challenge. The EM in itself may be seen as a complex tool to apply and the issues debated are often complex and hard to differentiate. Many of the issues raised by the application of food biotechnologies have broad ramifications, e.g. for global trade, etc. Moreover, the science is often coloured by uncertainties that can be controversial and difficult to handle.

Group dynamics

Group dynamics may affect the validity of the outcome or the effectiveness of the process in achieving its goals. Limited real heterogeneity is a potential problem especially as participants will be unfamiliar with using such a tool and may be hesitant to object and perhaps reveal a lack of knowledge. Another issue is that some participants may not participate in the group discussions as much as more dominant members. This may compromise the quality of any analysis, the outcomes and the impact of the process.

Transparency about the premises of the workshop

There is a degree of framing of the issue when an EM is used as a reflection tool. This must be clearly communicated both to the participants and in reporting from the workshop. If the process is seen as unduly framed by the organisers and through the use of the matrix itself (although this has not been highlighted as an issue in previous reports of EM use), then by implication, important perspectives may be perceived as absent and the quality of the process will be compromised. The impact of the process may also decrease because participant will not take ownership of the process. These potential challenges must be explicitly addressed by the organisers in order to ensure the quality of the process and to fulfil the objectives of the workshops.

As described above there are a variety of possible applications and motives for use, as well as a variety of process risks. However there are methodological choices that still need further clarification, for instance:
- Soundness of the result: Is the result richer, or more consistent, when a certain type of format is used, for example a multi-stakeholder group? Does richness of deliberation increase with group size?
- Is the value of the process for organisations, politicians or other recipients of results more significant using a multi-stakeholder design, with lay people design or with a homogeneous group?
- Are other facilitator skills required when facilitating decision-making groups rather than reflection groups? In other words, are there additional aspects of group dynamics operating in such groups?
- Do participants actually take more ownership of the matrix in a bottom up process or do they just become more confused and exhausted?

Experience may perhaps indicate some answers to these questions, but not much systematic analysis of the outcomes of these processes has been done and further observation research needs to be conducted. However, one of the few evaluative studies of deliberative richness is being conducted by researchers at the Danish Centre for Bioethics and Risk Assessment, University of Copenhagen. This group is in the process of analysing the use of a participatory EM process with homogeneous groups of scientists (partners from an Animal Disease Genomics Network). This analysis will examine some of the issues raised above and with undoubtedly add to the methodological discussion and highlight further research needs. For most EM workshops participants are asked to fill in feedback forms. This feedback is analysed and often commented upon in the reports from the workshops. However, no systematic study of this data has been carried out. Although there are methodological challenges in carrying out such a comparison, it might indicate some consequences of differences of designs. A condition for carrying out systematic studies is that the workshops are reported in public and that methodological choices are made transparent. Other ways to perform research would be quasi controlled experiments.

With regard to impact, little is known about the effect of these processes on practices, subsequent decisions, or on individuals' or organisations' ethical reflection in the long-term. Participants who have taken part in previous workshops have indicated the value of the process and commented that the ethical matrix workshop experience might be useful for them in their professional situation (e.g. Kaiser *et al.*, 2001). However, no specific impact assessment studies have been carried out beyond the review processes applied by the workshop facilitators. No assessment studies have been carried out to ascertain any direct effects on institutional activities and the authors believe such studies should be conducted. The question of impact opens up a general question relating to the role of ethical reflection process in the wider context of science governance, which unfortunately is not within the remit of this paper. However, further reflection and analysis of the role of these processes in science governance is needed.

Conclusion

This review has briefly highlighted the diverse use of the EM, in particular when applying it as a participatory method. Some of the issues presented above are not unique to EM processes, but in some respects hold for all practical participatory processes. With regard to evaluating public participation exercises, Rowe and Frewer (2004) have identified important research challenges which are also applicable to participatory ethical reflection processes. The value of the applications can be measured using a number of approaches. Although the EM has been evaluated through participant feedback and the reflective reports of users of these reports are dependent on the users' measure of success which can be very different from project to project. It is important for evaluation of the value of this method and its wider applicability that 'success' criteria are clearly stated and measured. Although self reported evaluations of the method are positive, no systematic studies have been conducted. Several groups and institutions have used the EM process and other similar participatory processes, but as yet there has been no comparative assessment of these collective experiences. More work is needed to evaluate how tools of this nature function in practice and what impact they have. There is a need for the development of an evaluation process that will draw on social science knowledge and the social psychology of group processes. Evaluations should include elements such as group dynamics, the different roles of facilitators, ways to anchor these processes in order to extend impact, representation issues, etc. This form of analysis and learning is necessary in order to clearly validate the perceived value of these processes, to ensure effective use of resources, and to avoid stakeholder fatigue and resistance.

References

Beekman, V. *et al.* (2006). Ethical Bio-Technology Assessment Tools for Agriculture and Food Production. Final Report. Ethical Bio-TA Tools (QLG6-CT-2002-02594).

Forsberg, E.-M. (2007a). Report from a Value Workshop on GM Rapeseed. In: Zollitsch, W., Winkler, C., Waiblinger, S. and Haslberger, A. (eds.) Sustainable food production and ethics. Wageningen Academic Publishers, Wageningen, the Netherlands.

Forsberg, E.-M. (2007b). A Deliberative Ethical Matrix Method – Justification of Moral Advice on Genetic Engineering in Food Production. Dr. Art. Dissertation. Faculty of Humanities. Oslo, University of Oslo.

Kaiser, M. and Forsberg, E.-M. (2001). Assessing fisheries – Using an ethical matrix in a participatory process. Journal of Agricultural and Environmental Ethics 14: 191-200.

Kaiser, M. *et al.* (2007). Developing the ethical matrix as a decision support framework: GM fish as a case study. Journal of Agricultural and Environmental Ethics 20(1): 65-80.

Mepham, B. *et al.* (1996). An Ethical Analysis of the Use of Xenografts in Human Transplant Surgery. Bulletin of Medical Ethics 116: 13-18.

Mepham, T.B. (1995). Ethical impacts of biotechnology in dairying. In: Phillips, C.J.C. (ed.) Progress in Dairy Science. Wallingford, CAB International: 375-95.

Mepham, T.B. (2005). Bioethics. An Introduction for the Biosciences. Oxford, Oxford University Press.

Oughton, D. *et al.* (2004). An ethical dimension to sustainable restoration and long-term management of contaminated areas. Journal of Environmental Radioactivity 74(1-3): 171-83.

Rowe, G. and Frewer, L.J. (2004). Evaluating Public Participation Exercises: A Research Agenda. Science, Technology and Human Values 29(4): 512-556.

Natural law and the ethical matrix: putting theory into practice

R.J. Lynch
Department of Philosophy, University of Reading, Whiteknights, Reading, RG6 6AA, United Kindgodm; r.j.lynch@rdg.ac.uk

Abstract

In this paper I will argue that the philosophical justification of the ethical matrix (EM) should come from natural law theory (NL). In order to construct this argument, I assume that NL is the right system of moral philosophy to employ in life generally, not only in the ethical assessment of e.g. agricultural and food biotechnologies. I shall employ a modern representation of NL which has as its central tenets the notions of practical reason and basic goods. The basic goods are self-evidently good and in practical reason these basic goods guide our judgment in distinguishing between acts that are reasonable all-things-considered and those that are not. Something's being reasonable is synonymous with its being morally right. The terms practical reason, basic goods and self-evidently good will all be defined at the outset of the paper. I will argue that the EM operates in accordance with this representation of NL. Well-being, autonomy and fairness are, I shall argue, self-evident goods and the only basic goods we need to guide us in our moral decision making both generally and specifically when applied to the consideration of emerging technologies. As a reasoning process that aims to reach a conclusion concerning what we should do via a consideration of the impact of a specified course of action on different interest groups as assessed through the medium of self evident basic goods, the EM is an instance of NL. A potential problem with using NL to support the EM lies in the anthropocentric nature of much of the work in the NL tradition. Anthropocentricity will make it hard to justify the impartial consideration between human and non-human interest groups that the EM demands. I will argue, however, that the concepts of sustainability and speciesism can be employed (in opposite directions) to suggest that anthropocentricity is unreasonable and that impartial consideration between human and non-human interest groups is not only justified but also demanded by NL.

Keywords: practical reason, biotechnology, speciesism

Natural law theory represented by the ethical matrix

An important aim of the EM is to facilitate ethically-sound decision-making (e.g. on practical issues concerning agricultural and food technologies) by committees of individuals who have rarely had training in ethics. In the following I argue that NL can be represented by the EM and that this representation both facilitates the utility of NL in practical settings and is in accordance with the optimal mode of operation of the EM. In this context, the assumed justifications for the use of NL may also become justifications for the use of the EM.

I will first make clear which version of NL I am employing in this paper. NL is often religious in its derivation but I will employ a secular version and one that aligns what is morally right with what is reasonable all things considered. What is reasonable is determined by the extent to which it is in accordance with, or conducive to, the basic goods. To clarify, I define practical reason/reasonableness as 'any reasoning aiming at a conclusion concerning what to do' (Blackburn, 2002). According to NL, to aid us in this decision making process we have the basic goods – an explication of what we (as a species) do and what we ought to find desirable. As this definition of basic goods implies, the self evidence of such goods is an inherent feature of theirs. It is to the question of the alleged self-evidence of the basic goods that I shall first turn my attention.

Self-evidence

Self evident goods are commonly defined as goods whose goodness can be known without inference (Finnis, 1980). My main motivation for considering first the issue of self evidence is that the way in which the basic goods are often described in NL is entirely anthropocentric and is, therefore, prohibitive to the impartial consideration of non-human species as should be undertaken in ethical decision making that is to avoid the charge of speciesism (Singer, 2002) such as that proposed by the EM.

A major modern proponent of NL, John Finnis (1980), asserts that appreciation of the goodness of the basic goods is underived (from experience, scientific fact etc). The basic goods are alleged to be self evidently good in that they form 'an aspect of authentic human flourishing' (Finnis, 1980). Their goodness 'cannot be demonstrated, but equally it needs no demonstration' (Finnis, 1980). It is essential, however, that I assert that appreciation of the basic goods can be derived. Were this not to be the case, and if our appreciation of what is good for us is totally underived it may be that what is good for other species is knowable only by them, since it is unclear how we would gain access to such knowledge.

Finnis (1980) states 'No value can be deduced or otherwise inferred from a fact or set of facts' – i.e., you cannot get an 'ought' from an 'is'. I do not wish to entangle myself in a discussion of the naturalistic fallacy and nor do I need to. The claim I wish to make against Finnis (1980) is only that he is wrong to stop at self evidence when it comes to ascertaining what the basic goods are. Knowledge of the basic goods can and is gained both initially in virtue of their self evidence but also and subsequently via experience or observation. It could reasonably be retorted that Finnis (1980) in no way contests this claim (in fact as far as I am aware he does not address the issue), but if he doesn't then his and other NL theorists such as Aquinas' failure to extend the notion of basic goods to non-human species is both short sighted and speciesist. If the (human) basic goods are desirable because they are aspects of human flourishing, what reasoning could preclude the possibility of non-human animals having basic goods that are aspects of their flourishing and why would these basic goods not carry the same moral force as ours when it comes to considering the impacts of our activities upon animals? Alternatively, following Midgley (1983), it seems feasible that given the similarities between different species it is likely that the basic goods (i.e. well being) for many species *are the same*.

Aquinas (cited in Clarke and Linzey, 1990), however, believed that there was an important distinction between humans and animals/plants as 'divine providence makes provision for the intellectual creature for its own sake, but for other creatures for the sake of the intellectual creature'. In essence, plants and animals were placed on earth for humans to use and we should only prevent ourselves from acting cruelly towards them 'lest through being cruel to animals one becomes cruel to human beings' (Aquinas cited in Clarke and Linzey, 1990). This is a far from resounding defence of the 'rights' of non-human species and a sentiment that does not seem to have been updated by modern proponents of NL. In direct contradiction to Singer's (2002) assertion that, regardless of species, 'If a being suffers there can be no moral justification for refusing to take that suffering into consideration...the principle of equality requires that its suffering be counted equally with the like suffering...of any other being' and that those who fail to recognise this fact are being 'speciesist', Finnis (1980) states that there is a difference in the quality of experience felt by humans and animals and that even in the case of 'mentally defective persons' we should accord a greater respect for their rights than we would to a 'flourishing, friendly and clever dog' in virtue of the fact that they are human.

Finnis' (1980) argument here seems patently self-contradictory. Firstly, he cites the importance of the quality of human experience but then in the (presumably many) cases where that experience is of an ostensibly lower quality than that of a non-human, he falls back on the humanness of the human as a defence of differential treatment. From this it must be concluded that quality of experience has little to with according humans greater respect than non humans, it is simply our humanness that counts according to Finnis (1980) and this is, for reasons stated above, a difficult claim to defend.

I am following Singer (2002), then, in holding that we should consider like suffering equally and that failing to do this constitutes morally indefensible reasoning. Where, however, does this leave us with regard to things that cannot be said to suffer/are non-sentient, for example plants? It would be difficult to argue against the notion that plants, for example, have instrumental worth to both human and non-human animals as a source of food, shelter and fuel. But what of their intrinsic worth? The debate surrounding the issue of the possible intrinsic worth of plants, ecosystems, etc is ongoing (Mepham, 2008) and is beyond the scope of this paper. In the absence of a discussion of the arguments both for and against ascribing intrinsic worth to non-sentient beings I shall make the assumption that they do have such worth. My justification for this assumption derives from a version of the 'precautionary principle' (Mepham, 2008), i.e. that a lack of scientific evidence as regards its value, should not constitute a reason for degrading our environment.

Through an assessment of the content of the term 'self evident basic goods' I have shown that there is nothing inherently anthropocentric about NL and that it can in principle impartially consider the competing claims of humans, animals and the environment as the EM requires. It remains for me to describe the basic goods that will guide this consideration.

The basic goods

An immediate response to the idea that what is reasonable all things considered is what is morally right could be 'reasonable in whose eyes?' or equally 'how do we determine what is reasonable?' These questions highlight the importance of the idea of basic goods to NL. The notion of basic goods serves as a guide to assist deliberators in deciding which ends to pursue and by what means.

How, though, are we to construct a list of basic goods that will guide decision making that may have to consider the competing claims of humans, plants and non-human animals simultaneously? I contend that the list of basic goods for use in this instance (and in general) could not and should not be similar to the explicit list of goods as employed by NL theorists such as Finnis (1980). My reasoning for this assertion is motivated both by my previously stated objections to the spectre of anthropocentricity in NL but also, and primarily, by considerations of practicality. Constructing lists of basic goods for each individual species would be an onerous task and very much subject to change as knowledge and opinions within scientific communities (and the populace in general) alter.

Different NL theorists have argued for various lists of basic goods. I shall argue for only three, the first of which is, I contend, well-being.

Finnis (1980) argues that there are seven basic goods, namely: (a) life, (b) knowledge, (c) play, (d) aesthetic experience, (e) sociability (friendship), (f) practical reasonableness (of which there are nine sub-principles) and (g) 'religion'. I argue that lists such as Finnis' (1980) seven basic goods merely represent various aspects of human well-being (a term that I shall use interchangeably with flourishing) and could, therefore, simply be replaced by well-being.

The advantages conferred by this move are twofold. In the first instance, well-being seems a less contentious self-evident basic good than i.e. 'religion'. An ethical decision making process that fails to consider the impact on the well-being of affected parties would seem to be a very odd fish whereas one that failed to mention religion would not. Well being is also, unlike Finnis' (1980) list of basic goods, species neutral, and this is very important in the context of ethical consideration of, for example, biotechnology in agriculture as has been discussed. Employing a term as broad as well-being may, however, in spite of its self evident goodness, seem only to shift back one step the requirement to construct more explicit lists of basic goods for each species i.e. onto the decision makers themselves. This indeed is the case but I argue that this is not problematical. It is essential that at some point decision makers do indeed ask, for example, 'What constitutes flourishing for a pig?' but answering this question as part of a decision making process seems to be more acceptable than trying to construct such lists beforehand. This is partly for the reasons of practicality as stated but also because the practice should improve the quality of decision making.

The second basic good that I shall argue for is autonomy. I define autonomy as 'having some capacity for self governance' and in the case of lower animals autonomy may simply refer to such animals' individual preferences. I argue for autonomy as both an instrumental good (in contrast to well being, which is intrinsically good) and an intrinsic good – but primarily as an instrumental one. This is not to say that the intrinsic value of autonomy is not important, it is just that for the use I have in mind for the word (namely as a basic good), its instrumental worth takes precedence.

Autonomy is instrumentally good as it seems that as we usually know ourselves better than others know us, we ourselves are the best judges of what would be conducive to our own well-being. Autonomy is, therefore, instrumentally good as it can serve to further our well-being. It could be argued, based on this assertion, that autonomy should not be listed as a separate basic good as being conducive to well-being it is in a sense part of well being. I refute this assertion, however, by arguing for the intrinsic value of autonomy. It seems possible that one could have well-being without autonomy. I could imagine, for example, flourishing whilst being under the total control of a scientist who has somehow contrived to know me as well as I know myself and makes decisions that are in my best interests accordingly. It seems though that in this example something of real value has been lost. It is not just one's flourishing that matters, but also being the cause of that flourishing. So whilst autonomy seems to be primarily an instrumental good it also appears to have intrinsic value that is separable from its ability to further well-being.

Considering autonomy with reference to animals seems important for all of the reasons listed above but is perhaps better explained with reference to welfare. Stamp Dawkins (2008) defines good welfare as that demonstrated by 'Animals that are healthy and have what they want'. Part of the justification for the use of this definition comes from its overarching nature – there are several ways of measuring animal welfare (which, in this context, I shall use interchangeably with flourishing) but ultimately what we want to know (according to Stamp-Dawkins (2008)) when we consider welfare is are the animals healthy and do they have access to the things that they would choose for themselves, for if the answer to both of these questions is yes, then it is more likely that the animals will be flourishing. Stamp-Dawkins (2008) definition emphasises the value of autonomy to non-human animals.

The third basic good is, I argue, justice. When I refer to justice I am referring to justice as fairness as defined and explained by Rawls (1999). I will not defend the concept of justice as fairness itself as I believe Rawls (1999) has done a better job of this than I could ever hope to approximate. I will though briefly outline what is meant by justice as fairness and state my reasons for arguing that it is a basic good. Rawls (1999) states:

> '...the name 'justice as fairness'...conveys the idea that the principles of justice are agreed to in an initial situation that is fair.'

The initial, fair situation that Rawls (1999) is referring to is one in which the people formulating the principles of justice do so behind a 'veil of ignorance' i.e. they are not aware of their own abilities and position within the world. The principles of justice referred to are as follows:

> 'Social and economic inequalities are to be arranged so that they are both a) to the greatest expected benefit of the least advantaged and b) attached to offices and positions open to all under conditions of fair equality of opportunity' (Rawls, 1999).

Having defined what is meant by justice, it now remains for me to defend my assertion that it is a basic good. Justice, I argue (following Rawls, 1999), is an essential component of a well functioning society. As humans, we are inherently social and so a society without justice would seem to represent a state of affairs that is undesirable as at least suboptimal. With reference to the question of non-human animals, I contend that as humans increasingly take animals' means of independent survival from them (either through destruction of their habitat or by breeding animals incapable of survival without human intervention) in so doing we make them our responsibility and so a part of our society. As a part of our society, animals are equally entitled to justice.

The three basic goods then are well being, autonomy and justice. This assertion concurs with the goods utilised in the EM.

Conclusion

This paper defended the assertion that the ethical justification for the use of the EM could come from NL. In so doing, further moral weight has been added to the EM (given the assumption that NL is the correct method of moral decision making) and NL has been represented in a format that promotes its utility for people who may not be trained ethicists.

References

Aquinas, (1990). Animals are not rational creatures. In: Clarke, P.A.B. and Linzey, A. (eds.) (1990) Political Theory and Animal Rights, Pluto Press, London, pp. 1225-1274.

Blackburn, S. (2002). Oxford Dictionary of Philosophy, Oxford University Press, Oxford.

Mepham, B. (2008). Bioethics: An Introduction for the Biosciences (2nd edition), Oxford University Press; Oxford.

Midgley, M. (1983). Animals and Why They Matter, University of Georgia Press, Athens.

Finnis, J. (1980). Natural Law and Natural Rights, Oxford University Press; Oxford.

Rawls, J. (1999). A Theory of Justice (2nd edition), Oxford University Press, Oxford.

Regan, T. (2004). The Case for Animal Rights (2nd edition), University of California Press, Los Angeles.

Singer, P. (2002). Animal Liberation (2nd edition), Harper Collins, New York.

Stamp Dawkins, M. (2008). What is good welfare and how can we achieve it? in Bonney, R. and Stamp Dawkins, M. (eds.) The Future of Animal Farming: Renewing the Ancient Contract, Blackwell, Oxford.

Use of ethical matrices in formulating policies to address the obesity crisis

Ben Mepham
Centre for Applied Bioethics, University of Nottingham and Department of Policy Studies, University of Lincoln, United Kingdom; ben.mepham@nottingham.ac.uk

Abstract

The paper addresses the question of how governments of democratic states might seek to assist their citizens to reverse the serious trend towards overweight and obesity. Recent reports stress the contributory role of the obesogenic environment, characterising contemporary UK society, and suggest that to be effective policy interventions need to be implemented on several fronts. The paper explores the potential of the ethical matrix as both a procedural and a substantive tool in such programmes, focusing on food production, marketing and consumption.

Keywords: public policy, ethical deliberation, public health

The obesity crisis

Britain, like most western developed democracies, is facing an impending 'obesity epidemic.' Employing this as a case study, equally applicable to these other countries, how should the UK government address the crisis, in the light of the acknowledged rights of individuals to choose their own lifestyles? While democratically-elected governments may presume legitimacy for their policy initiatives, it remains important to ensure that interventionist policies affecting society as a whole are compatible with commonly accepted ethical standards. The specific concern here is the role of the ethical matrix (EM) in providing ethical justifications of relevant public policy decisions. Earlier accounts of the EM (e.g. Mepham, 2000), acknowledged the role in its development of ideas advanced by John Rawls. Here, I stress the perceived relevance of Rawls' thinking by examining the close association between my current ideas on the EM and the revised notion of *justice as fairness*, which Rawls expressed in a restatement of his theory (2002). He identified four roles or objectives of political philosophy, which may be summarised as follows:

1. Provision of '*a focus on deeply disputed questions, to see whether, despite appearances, some underlying basis of philosophical and moral agreement can be found*'.
2. Provision of '*a unified framework within which proposed answers to divisive questions may be made consistent and the insights gained from different kinds of cases can be brought to bear on one another*'.
3. Recognition of the '*profound and irreconcilable differences in citizens' ... philosophical conceptions of the world, and...the moral and aesthetic values to be sought in human life*' thereby attempting to reconcile them '*by showing the reason and indeed the political good and benefits of* (such reconciliation)'.
4. Promotion of the view that '*political philosophy is realistically utopian; that is,* (it probes) *the limits of practicable political possibility*.' (Rawls, 2002: 3-4).

I suggest that the structure, aims and modes of use of the EM may reasonably be said to resonate with these roles and objectives. Thus, in diverse settings, the EM has proved valuable in focusing on contentious issues, by employing a strategy which assesses how far ethical ideals are met by proposed changes and sometimes discovering an 'overlapping consensus' despite marked differences in people's moral values. According to this view, the starting point for ethical analysis is the formulation of the set of *prima facie* principles evident in the *common morality*, but for which no weighting is assigned, and consequently to which prospective users can attach as much significance as is deemed appropriate (which in some cases might be 'none').

The causes of obesity in the UK

'*Being overweight has become a normal condition, and Britain is now becoming an obese society.*' (DIUS, 2007). It does not appear to be a question of people having less willpower than earlier generations, or being more gluttonous, but rather of profound changes having occurred in society over the last 50 years – which have impacted on work patterns, transport, food production and sales, and recreational and leisure activities. Such changes have exposed the underlying tendency for many people to put on weight, so that 25% of adults are now assessed, according to their *body-mass index* (BMI) as 'obese'. The consequences of obesity may be listed as:

- increased risk of a wide range of chronic disease conditions, such as hypertension, cardiovascular disease (including stroke), diabetes and cancer;
- reduced wellbeing due to physical incapacity, social stigma, and low personal esteem;
- the predicted 60% of Britons who will be obese by 2050 will double the National Health Service (NHS) bill attributable to overweight and obesity to £10 billion per annum;
- also by 2050, the wider financial impacts on society and businesses (e.g. in terms of lost productivity) are anticipated to reach (at current prices) £50 billion per annum (DIUS, 2007).

Clearly, at the physiological level, obesity is due to more food energy being consumed than is expended in physical activity. When, as in the immediately post-war period of the 1950s, there was a shortage of food – and physical energy was expended in activities such as manual work, cycling and outdoor sports, most people achieved weight maintenance within the 'normal range.' But now most British people live in an *obesogenic* environment, in which work and leisure activities are largely sedentary, much transport is motorised (while escalators and lifts have replaced stairs), and low food prices and persuasive advertising encourage overeating of *high sugar, salt and fat foods* (HSSFF). It is evident that the *drivers* of obesity are deeply embedded in modern social norms.

An ethical approach to policy decisions

While several government reports on obesity have appeared recently, that of the Nuffield Council on Bioethics (2007) (hereafter 'Nuffield') specifically sought to address obesity from an ethical perspective. Its insights provide some valuable guides to how policy might be shaped, which it is my aim here to augment, amplify and, where appropriate, criticise. Nuffield appeals to a *stewardship model* according to which governments have an obligation to provide conditions that allow their citizens to be healthy – partly though efforts to reduce inequalities in health within the population. While recognising the need to respect personal choice (e.g. by avoiding coercive or intrusive measures), Nuffield considers that to protect the vulnerable (e.g. children and disadvantaged adults) various policy interventions may be ethically justifiable. Depending on the seriousness of the issue, Nuffield considers that the options implemented might be best considered as 'rungs on an intervention ladder,' with the least intrusive entailing no more than monitoring a situation (e.g. the incidence of obesity in different socioeconomic groups) while the most intrusive might entail legal measures to ban a certain foodstuff.

These are useful perspectives, but they omit some important considerations. Use of the EM can clarify the situation by identifying what ethical concerns are at stake at different points in the food chain, and how the needs of different stakeholders would be affected by specific policy interventions. My previous use of the EM has concentrated on its heuristic potential, by means of which users are invited to specify the principles in ways appropriate to their interpretation of the issues and assess anticipated impacts of innovations on the degree of respect these principles are accorded. Others have amended the approach in ways can lead to substantially different approaches, e.g. in which the cells contain questions (Forsberg 2004) instead of specifying principles. Here, I introduce two new forms of matrix, each applicable at a different level of an ethical analysis. These are:

- specified principles matrix (SPM);
- policy objectives matrix (POM).

With reference to these tables, the abbreviations used below identify specific cells (e.g. SPM/PW refers to *producer wellbeing* in the *specified principles matrix*).

The object of the SPM is to identify (groupings of) coherent interest groups and the idealistic objectives to which observing the principles (respect for wellbeing, autonomy and fairness) might reasonably aspire. But since these are *prima facie* principles, some will inevitably take precedence over others, a situation which typically characterises policy decisions, and demands transparent justification. The compositions of the different groups are based on the concept of broadly similar objectives, e.g. in the case of *producers*, all are involved in the growing and manufacture of food, including associated activities, e.g. agrochemical production and butchery.

Clearly, there are wide differences in the circumstances of different groups identified, to which policy makers would need to pay due attention. In the form of EM employed by Kaiser and Forsberg (2001), selected participants in a deliberative exercise were invited to construct an EM by deciding how respect for the different principles (proposed by the organisers) was to be specified. A risk of this procedure is that participants may limit their perceptions of desirable outcomes to what seems 'realistic.' In contrast, reference to idealistic principles aims to guarantee that ethical criteria are prioritised, rather than marginalised by overriding practical limitations. In line with Rawlsian principles, the explicit statement of specified ethical principles facilitates the possibility of discovering whether '*some underlying basis of philosophical and moral agreement can be found.*' Of course, ultimately, practical considerations inevitably play a major role in policy-making, but in this analysis their consideration is appropriately deferred until a later stage. Even so, appealing to 'ideals' allows significant differences of interpretation.

Society members is a term employed to acknowledge that obesity is a condition only affecting some, albeit an increasingly significant proportion, of (the global) society, but that respect for wellbeing, autonomy and fairness demands equal attention for everyone. In the current formulation (intended to be illustrative rather than definitive), the interests of *future generations* of living beings are included (hence the significance of *sustainability* in cell SF). It might also be considered appropriate for this interest group to include ethically-considerable non-human living organisms: but their marked differences from humans might reasonably suggest they should be assigned to a separate category – and they are excluded from Table 1. In summary, the SPM proposes a way in which fundamental ethical principles may be specified in this context. In that the specifications are highly generalised it might be anticipated that they would find support from a substantial majority of people, even acknowledging the '*plurality of reasonable, but to a degree, incompatible doctrines*' characterising modern liberal democracies.

Impacts on policy formulation

Central to Nuffield's analysis was the need to protect vulnerable groups (e.g. infants, senile people, those suffering from addictions) from harm, a motive that is represented by cells CW, CA and CF in one form of POM (Table 2), the primary aim of which is to propose a structured framework by which ethical principles may be translated into policy objectives. Representation in a matrix facilitates assessment of the relative ethical claims of competing interests, and can make explicit the weighting that is deemed appropriate in the subsequent formulation of policies. By bringing ethical considerations to the fore, it acts as a substantive ethical tool; and by requiring policy-makers to articulate their assessments of impacts

Table 1. A proposed specified principles matrix (SPM) pertinent to addressing obesity.

	Respect for		
	Wellbeing	Autonomy	Fairness
Producers of food: farmers and associated workers; food manufacturers; food processors	Satisfactory income and work from producing less obesogenic food	Self-determination	Fair trade laws
Marketers of food: wholesalers; retailers; restauranteurs; advertisers	Satisfactory income and work from selling less obesogenic food	Free market	Fair trade laws
Consumers at risk of obesity	Reduced risks of obesity and associated diseases	Informed food choice	Equality of opportunity e.g. in access to healthy food
Society members	Health and prosperity of global population	Diversity	Sustainability

on each cell of the matrix, it acts as a procedural tool. Table 2 is limited to factors directly affecting food production, marketing and consumption, which represents just one of several forms of POM that would be needed to comprehensively address different aspects of the obesity crisis. For example, the obesogenic environment is susceptible to amelioration through policies relating to provision of sports facilities, cycle tracks, building design, and educational curricula.

Table 2 seeks to define relevant ethical principles in policy-oriented terms that address issues concerning food. But the specifications clearly fall far short of stipulating any actual policy recommendations. Thus it might appear that the analysis has little direct impact on the process of ethical policy formulation. However, that assessment would overlook several important several important considerations. Firstly, such an analysis should provide invaluable stepping stones to the achievement of a Rawlsian 'overlapping consensus' (Rawls, 2002). Secondly, whether or not consensus proves possible, use of the SPM and POM can provide explicit ethical justification for whatever policy decisions are ultimately made. This is because the EM (in the forms described here) is capable of facilitating formulation of judgements that are comprehensive in scope, explicit in articulation, transparent in terms of their justification, and arrived at by rational deliberation. But, thirdly, because ethical considerations necessarily permeate *all* subsequent policy decisions, subsequent use of forms of the EM can have an explicit bearing on how the specified policy proposals in Table 2 might be implemented. This claim will now be examined.

Implementing ethical policies

Nuffield's 'intervention ladder' sought to establish ways governments, acting as stewards of their citizens' interests, might legitimately take increasingly powerful measures (e.g. affecting food producers, or obese individuals) to rectify adverse impacts on public health. Galbraith (1984) identified three ways power is exerted (e.g. by individuals, companies and governments):

- *condign power*, which wins submission by imposing an alternative to the original preferences so that the latter are abandoned (i.e. this entails punishment);
- *compensatory power,* which rewards those who accede to requests to change behaviour;
- *conditioned power,* which entails persuasion and/or education to change behaviour.

Table 2. A policy objectives matrix (POM) indicating some (speculative) proposals relating to food production, marketing and consumption.

	Respect for		
	Wellbeing	Autonomy	Fairness
Producers of food: farmers and associated workers; food manufacturers; food processors	Legislate/regulate to significantly reduce production of obesogenic foods, by diverting production to healthier products	Protect innovative and entrepreneurial practices	Promote fair trade
Marketers of food: wholesalers; retailers; restauranteurs; advertisers	Legislate/regulate to significantly reduce sales of obesogenic foods, by diverting sales to healthier products	Protect innovative and entrepreneurial practices	Promote fair trade
Consumers at risk of obesity	Promote healthier lifestyles to avoid obesity, by encouraging healthy eating	Cultivate informed food choices through education	Ensure equality of access to healthy food and nutrition education, so promoting a healthier lifestyle
Society members	Provide facilities to promote wellbeing	Promote toleration of infirmity and disability through education and non-discriminatory practices	Ensure availability of sustainable supplies of healthy food through long-term planning

Each has its malign aspect (represented e.g. as malicious threats, bribery and brainwashing, respectively), but each also may, at some level, play a benign role in policy formulation. From an ethical perspective, it is important to consider whether recommendations to exert political power by any of these means are justifiable in terms of their impacts on principles specified in the EM. For example, with reference to Table 2, the interests of children assume paramount importance because of their intrinsic naivety and vulnerability to advertising campaigns promoting consumption of HSSFF, coupled with the high risk that early onset obesity will become a permanent condition. Hence, respect for advertisers' *prima facie* rights to exercise 'innovative and entrepreneurial practices' (Table 2, PA and MA) might well be considered justifiably overridden by necessary measures taken to respect CW, CA and CF. Realistically, reducing childhood obesity rates may entail the exercise of (some combination of) condign, compensatory or conditioned power. Can this be achieved while also respecting, e.g., PA and MA (Table 2)?'

Examples of strategies employed to reduce childhood obesity are: (1) requiring schools to introduce low HSSFF menus, (2) banning sale of HSSFF in the vicinity of schools, and (3) restricting TV advertisements for HSSFF. These are all 'command and control' regulations, which imply that professional health experts *know* the answers to the obesity problem. But, in practice, their effectiveness is highly questionable. Thus, a sounder policy, from both ethical and practical perspectives, may be *performance-based regulation*, which e.g. assigns significant responsibility for causing obesity to the large food companies that sell HSSFF (Sugarman and Sandman, 2007). Accordingly, companies would be required to 'put their own house in order' by reducing HSSFF in proportion to the extent to which they are assessed to have contributed to the problem. Internalising cost in this way, by requiring manufacturers to bear the cost of their activities (analogous to the 'polluter-pays' principle) is used by economists to justify industry regulation. It also reduces the burden on taxpayers in a fair way has the advantage that industry's 'innovative and entrepreneurial skills' (POM/PA) will operate on a 'level playing field'.

Conclusions

I have argued that ethically-sound policy interventions aimed at reducing obesity need to prioritise public health. In conformity with Nuffield recommendations, the UK government should adopt a stewardship model to protect the interests of the most vulnerable and seek to achieve equality in society, without overruling responsible personal choice. Use of the EM in the forms outlined here (SPM and POM) could provide a conceptual tool (serving both substantive and procedural roles) to arrive at transparent and ethically justified public policy decisions.

References

DIUS (2007). Foresight, Tackling obesities: future choices- project report. London, Department for Innovation, Universities and Skills.

Forsberg, E.-M. (2004). Ethical assessment of marketing GM Roundup Ready rape seed GT73. In: De Tavernier, J. and Aerts, S. (eds.) Science, Ethics and Society. preprints 5th EURSAFE congress, Leuven, Belgium. pp. 177-180.

Forsberg, E.-M. (2007). A deliberative ethical matrix method – justification of moral advice on genetic engineering in food production [Dr Art dissertation] University of Oslo.

Galbraith, J.K. (1984). The Anatomy of Power. London, Corgi.

Kaiser, M. and Forsberg, E.-M. (2001). Assessing fisheries – using an ethical matrix in a participatory process. Journal of Agricultural and Environmental Ethics 14: 191-200.

Mepham, T.B. (2000). The role of food ethics in food policy. Proceedings of the Nutrition Society 59: 609-618.

Nuffield Council on Bioethics (2007). Public health: ethical issues. London, Nuffield Council on Bioethics.

Rawls, J. (1993). Political Liberalism. New York and Chichester, Sussex, Columbia University Press.

Rawls, J. (2002). Justice as Fairness: a restatement (ed. E Kelly) Cambridge, Mass & London, England, Belknap Press.

Sugarman, S.D. and Sandman, N. (2007). Fighting childhood obesity through performance-based regulation of the food industry. Duke Law Journal 56: 1403-1490.

Theme 12 – Ethical dimensions of organic food and farming

Communication of ethical activities going beyond organic standards in European organic enterprises

K. Gössinger and B. Freyer
University of Natural Resources and Applied Life Sciences, Institute of Organic Farming, Gregor-Mendel-Strasse 33, 1180 Vienna, Austria; Katharina.Goessinger@boku.ac.at

Abstract

In times of rapid growth of the organic market and differentiation processes within the sector, efforts are made on different levels to (re)define and strengthen organic values. The normative part of this process is the reformulation of IFOAM-principles, which form the basis for ethical commitment within the organic movement, without being fully integrated into organic regulations. This article addresses the question of which ethical activities are practised on organic farms and by organic companies that go beyond the organic regulations (referred to as 'organicPlus' in this study), with what arguments these activities are communicated to consumers and how they are related to the dimensions of sustainability. The survey encompasses 100 small and medium-sized enterprises in five European countries (Austria, Germany, Great Britain, Italy and Switzerland) that deal with organic products. In the paper various 'organicPlus' approaches are analysed with a particular focus on communication arguments. Innovative communication strategies are seen as a valuable tool for the strategic positioning of organic companies. The mapping of companies resulted in a total of 72 different generic communication arguments relating to 'organicPlus' activities. The paper offers a categorisation of these communication arguments on the basis of the economic, ecological/environmental and the social pillar of sustainability extended by a fourth cultural dimension. Furthermore, the relations between the companies' 'organicPlus' approaches, business size, company type and product category are analysed. Finally, the paper reflects on status quo and the potential of integrating 'organicPlus' approaches into organic enterprises and their communication to consumers.

Keywords: communication strategies, differentiation process, marketing, values

Background and research questions

Ongoing differentiation processes within the organic sector (Freyer, 2008) form the context of this study. Parts of this process are professionalisation strategies, increases in inputs at farm level or the integration of organic products in conventional trade structures. The growth process is not necessarily contradictory to organic farming, the question is rather how it is expressed. The organic movement has particular roots which are related to a holistic based thinking and acting. This is why efforts are made on different levels to (re)define and strengthen the so called organic values and not to use sustainability as a marketing gag. According to the four IFOAM principles, ecological/environmental and social matters should be practiced on a high level in organic farming. In addition to the IFOAM principles, the concept of sustainability, ideas of the concept Corporate Social Responsibility (Hiß, 2005) and other ethical oriented concepts, are also reflected in this study.

The overall framework of the study is the CORE Organic project 'Farmer Consumer Partnerships', that aims to strengthen the market position of organic companies sticking to higher standards than the organic guidelines. The term organicPlus was introduced for all economic, ecological/environmental, social and cultural activities that go beyond European organic regulations (EEC/2092/91, to be replaced by EC/834/2007 and implementing rules) and the organic standards of private associations (Padel and Gössinger, 2008). In this article we focus on the following research questions:

- Which organicPlus activities are practiced in organic farms/enterprises and which arguments are used to transport these activities to the consumers?
- Are there any relations between the communicated organicPlus activities of a company and the company size, the company type (100% organic, organic & conventional) as well as the product categories?

Data collection and analysis

The empirical research included the mapping of organicPlus activities of approximately 100 organic SMEs in five different European countries (Austria, Germany, Great Britain, Italy, Switzerland). Apart from company size, the selection criteria included the existence of written documentation of the companies' organicPlus activities. In addition, the company had to be involved in the production or processing of organic food. Data collection comprised a semi-standardised questionnaire as a basis for the mapping and written material of the enterprises (websites, product labels, folder etc.). In many cases telephone interviews with the company owners/ personnel were also conducted. For data structure and interpretation we used a set of categories (see results).

Results

We identified 72 different communication arguments about organicPlus activities that companies use. The first grouping of arguments followed the categories 'internally' (e.g. fair prices for farmers) or 'externally' (e.g. contribution to rural development). Secondly, the arguments were classified according to the group of actors in the supply chain they are referring to: one group of actors (e.g. farmers), two groups of actors (e.g. farmers and consumers) or the whole supply chain. Finally, the 72 arguments were grouped on the basis of the economic, ecological/environmental and social pillar of sustainability, extended by a cultural dimension (Brocchi, 2007). Several communication arguments were cross-cutting, therefore clear categorisation according to each sustainability dimension was difficult in some cases (Table 1).

Table 1. OrganicPlus dimensions, sub-dimensions and number of companies (Padel and Gössinger, 2008)[1].

Dimension	Sub-dimensions	No. of companies
I. Economic	Fair price – farmer oriented	14
	Fair price- farmers and consumers	10
	Fair price - additional value	1
	Added value in the region	23
II. Ecological / environmental	Biodiversity	25
	Landscape	17
	Resources	26
	Animal welfare	17
III. Social	Family farms & farmers in need	16
	Care farming	25
	Social projects & charity	22
	Working conditions of employees	12
IV. Cultural	Local & regional	41
	Communication & information	18
	Rural customs/traditions & originality	13
	Individual attitudes	14

[1] Number of companies using communication arguments of the specific sub-dimension.

In particular, criteria for activities associated with regional supply chains and local foods were challenging to categorise because they contain a cultural, social and economic dimension: Local food is considered to benefit the consumers in terms of fresher food and traceability; the farmers in terms of better prices; and the environment in terms of reduced transport. Due to the difficulty of precisely categorising the communication arguments according to the dimensions of sustainability, we decided to create 16 sub-dimensions and to focus on them in the further analysis.

The majority (65%) of the companies communicated arguments of several sub-dimensions. Hardly any arguments were found which we could categorise as country-specific, except the argument of an Italian care farm that is cultivating land confiscated from the Mafia. Arguments related to the sub-dimension of 'local & regional' were the single, most frequently issue mentioned within company mapping, covering issues of local food provision and the preservation of small structures.

In the following section trends concerning the relations between the communication arguments (categorised in sub-dimensions) and company size, company type and the product categories are illustrated. The empirical findings show a relationship between the company size and the content of communication arguments (Table 2). Micro companies tend to communicate arguments related to 'rural customs & traditions' and 'family farms & farmers in need' more often. 'Working conditions' is mainly communicated by medium-sized companies. 'Fair price' arguments are important for companies of all size.

Furthermore, frequent combinations between the sub-dimensions of communication arguments and the company type (100% organic or organic *and* conventional) can be seen in Table 3. Mainly companies that deal with organic *and* conventional products use arguments related to fair prices for the farmer. Also cultural arguments are used by this group, whereas arguments related to 'biodiversity' and 'care farming' are mainly communicated in purely organic companies.

Arguments referring to a fair price are mainly used by dairy farmers, whereas animal welfare is rarely communicated by them (Table 4). Social activities as well as ecological/environmental activities do not seem to be related to any specific product category, while 'traceability and transparency' issues (grouped under the sub-dimension 'communication & information') are frequently communicated by vegetables and fruit farmers.

Table 2. Company size and sub-dimensions of communication arguments (frequent combinations) (Padel and Gössinger, 2008).

Company size	Dimensions of sustainability / sub-dimensions of communication arguments			
	Economy	Ecology / environment	Social	Cultural
Micro	Fair price-farmer oriented	Biodiversity	Family farms & farmers in need	Rural customs & traditions
Small	Fair price-farmers and consumers; added value in the region		Care farming; social projects & charity	Local & regional; communication & information
Medium-sized	Fair price-farmers and consumers	Landscape	Working conditions of employees	

Table 3. Company type and sub-dimensions of communication arguments (frequent combinations) (Padel and Gössinger, 2008).

Company type	Dimensions of sustainability / sub-dimensions of communication arguments			
	Economy	Ecology / environment	Social	Cultural
Organic & conventional	Fair price-farmer oriented; added value in the region		Working conditions of employees	Local & regional; rural customs & traditions
Pure organic		Biodiversity	Care farming	Communication & information

Table 4. Product categories and sub-dimensions of communication arguments (frequent combinations) (Padel and Gössinger, 2008).

Product categories	Dimensions of sustainability / sub-dimensions of communication arguments			
	Economy	Ecology / environment	Social	Cultural
Dairy products	Fair price - farmer oriented	Landscape; resources	Family farms & farmers in need	
Vegetables & fruits		Resources	Social projects & charity	Communication & information
Meat		Biodiversity; animal welfare	Care farming	
Cereals		Resources; biodiversity	Social projects & charity; care farming; support of family farms & farmers in need	

The empirical findings also show that the knowledge and awareness of the Corporate Social Responsibility (CSR) concept and of the IFOAM principles among organic SME companies is limited. Besides, the study illustrates that there is a group of companies that do not (fully) integrate their ethical concepts in their marketing. Some organic farmers or small company-owners follow higher standards because of their own beliefs and commitment to certain 'organic' values, but they do not necessarily wish to market these personal perspectives. Some companies and organic certification bodies are likely to develop and introduce auditing and certification schemes for the verification of organicPlus claims, other refuse additional audit schemes for different reasons as e.g. additional bureaucracy and costs.

Conclusion

Based on 100 cases in five different European countries, we identified diverse organicPlus approaches. Arguments relating to regional development issues, regional supply chains or food miles appeared to be of particular relevance in all participating countries. In addition to the so called 'conventionalisation process', we found the integration of organicPlus approaches as another pathway of differentiation in organic agriculture. It is obvious that some companies use this approach for improving the strategic positioning of organic companies in the market place, while others argue that ethical motivated activities should not be part of any marketing activities.

In addition, hardly any comprehensive organicPlus approach exists that systematically integrates ethically driven activities going beyond the organic standards according to specific frameworks, as e.g. the four IFOAM Principles. There seems to be therefore a need to investigate the accessibility of these concepts for small and medium-sized organic companies and to consider a broader information strategy on IFOAM Principles in the organic movement.

In today's business world, many companies – both within and outside the food industry – implement ethical activities, the most prominent example within the food sector being Fairtrade. In this context, the organic movement is under pressure to intensify its own efforts to maintain its leading position with regard to the sustainability of agri-food business. However, ethical acting cannot be reduced to criteria alone as it will always be driven by experiences, knowledge, beliefs, responsibility and openness. Consequently, the implementation of ethical concepts and ethical values in the market place represents a considerable challenge for all stakeholders involved.

Acknowledgements

The authors gratefully acknowledge funding from the Federal Ministry of Agriculture, Forestry, Environment and Water Management of Austria for the financial support for the CORE Organic Project 'Farmer Consumer Partnerships'.

References

Brocchi, D. (2007). Die Umweltkrise – eine Krise der Kultur. In: Altner, G., Leitschuh-Fecht, H., Michelsen, G., Simonis, U.E. and Weizsäcker, E.U.V. (eds.) Jahrbuch der Ökologie 2008. München, C.H. Beck, pp. 115-126.

Freyer, B. (2008). The Differentiation Process in Organic Agriculture (OA) – between Capitalistic Market System and IFOAM Principles. In: IFOAM, ISOFAR (eds.), 16th IFOAM Organic World Congress; Cultivating the Future Based on Science, Vol. 2 Livestock, Socio-economy and Cross disciplinary Research in Organic Agriculture, 16th IFOAM Organic World Congress; June 16-20, 2008. Modena, Italy, pp. 374-377.

Hiß, S. (2005). Warum übernehmen Unternehmen gesellschaftliche Verantwortung? Ein soziologischer Erklärungsversuch, Frankfurt New York, Campus Verlag, 339 pp.

Padel, S. and Gössinger, K. (eds.) (2008). Farmer Consumer Partnerships. Communicating ethical values: a conceptual framework. CORE Organic Project Series Report. http://orgprints.org/11028/01/CORE_FCP_Vol1_Final_31_July.pdf.

A conceptual framework for the communication of ethical values going beyond standards in organic farming

S. Padel[1] *and H. Röcklinsberg*[2]
[1]*Institute of Biological, Environmental and Rural Sciences, Aberystwyth University, SY23 3AL, United Kingdom; sxp@aber.ac.uk*
[2]*Centre for Theology and Religious Studies, Lund University, Allhelgona Kyrkog. 8, 223 62 Lund, Sweden*

Abstract

Values are part of the core concept of organic farming but the term organic has different meaning to different people. Organic producers may practise organic farming in ways that go beyond European standard requirements (labelled as organicPlus values), but communication about such activities can be lost, especially in long supply chains and more complex trading structures. This represents an opportunity for differentiation by communicating a better ethical quality in an increasingly competitive market, but could also have negative impact for the credibility of all organic labels. In an approach guided by the stages of the Corporate Moral Responsibility Manual the paper summarises organicPlus concerns under principal headings according to the impact on the environment, on animals, and people in terms of economic and social impact. Concerns referring to the whole system or supply chains – such as preferences for local food and concerns about the integrity of the organic supply chain – have a likely impact in several areas and are more difficult to categorise in such a way. Both private standards and public regulations have responded with more detailed standard requirements to concerns about environmental impact and animal welfare in recent years, whereas activities related to economic and social concerns are largely voluntary. Several concerns can be argued from at least two, if not three, different moral traditions. We explore whether the moral reasoning behind organicPlus concerns influences the nature of activities of the organic sector in response to them. It appears that attempts to develop verification and certification tools largely focus on concerns that are based on utilitarian reasoning. Other ways will need to be found for a credible communication of aspirational organic values (e.g. respect for animals and for nature) that are more difficult to operationalise in the form of the pass/fail criteria required by organic certification.

Keywords: organic standards, stakeholder concerns, certification

Introduction

The global market for organic food has tripled in value in the last eight years with the vast majority of sales concentrated in Europe and in North America. The European market was estimated to be worth €16.2 billion in 2007. It has grown substantially since the middle of the 1980s as a result of growing consumer demand and increased policy support as part of agri-environmental and rural development programmes, increased supply and a European Regulation defining organic production (Willer and Kilcher, 2009). In Europe, the marketing of organic food is governed by the regulations setting out the rules for labelling a food product as 'organic' or the equivalent terms 'biological' or 'ecological' in other languages and stating the requirements for organic operators (producers, processors and traders) (EC, 2007).

Values are part of the core concept of organic farming but the term organic has different meaning to different people. Organic producers may act on values that are not part of the European standard requirements (labelled as organicPlus values. In 2005, four principles of organic agriculture were formulated by IFOAM (International Federation of Organic Agricultural Movements) (IFOAM,

2005b; Luttikholt, 2007). These principles reflect the values associated with organic farming, but not all values are part of organic standards and certification (Padel *et al.*, 2009) and communication about related activities can be lost, especially in long supply chains and more complex trading structures.

Companies report on their corporate social responsibility within and outside the food sector and the organic sector is challenged to maintain its leading position. The paper explores whether the moral reasoning behind organicPlus concerns expressed by organic farmers, consumers or other stakeholders appears to influence the nature of activities of the organic sector. Clear standards and certification are important for the credibility of the organic concept with consumers (e.g. Zanoli, 2004). This could be undermined by the communication of organicPlus claims that cannot be verified. It appears that most attempts to develop verification tools for such claims have focussed on concerns that are argued from a utilitarian perspective, such as aspects of environmental impact and animal welfare whereas to communicate respect for the integrity of animals or nature through certification appears challenging.

The work presented in this paper is part of the CORE funded project Farmer Consumer Partnerships (FCP) aiming to develop innovative generic communication arguments that can strengthen the partnership between producers and consumers through better communication of ethical attributes of organic food.

Methods

The first stage was to map the various stakeholders of organic food systems. These include consumers/ citizens, farmers and growers, processors, traders, as well as researchers, standard-setting bodies, policy-makers and regulators. Their ethical concerns and values reported in the literature together with values expressed in four principles formulated by IFOAM (health, ecology, fairness and care) were summarised and categorised according to impact. Concerns were then contrasted with the European Regulations for organic food to identify so called organicPlus values going beyond minimal organic standards requirements. We then explored which concerns private standards have responded and where other certification schemes exist. Finally, the moral reasoning behind the concerns in terms of important traditions was analysed in a similar way to the Ethical Matrix Tool[24]. The approach was guided by the stages of the Corporate Moral Responsibility Manual (Brom *et al.,* 2006).

Results

Values and concerns of consumers and producers are reported in a number of studies, but less attention has been paid to the concerns of traders, processors, certification bodies and other actors (for details see Padel and Gössinger, 2008). The organicPlus concerns were summarised under headings according to impact.

Environmental concerns

These are mentioned by most stakeholders, relating to the question of sustainability of resource use and the protection of biodiversity. These are additional to the EU Regulations which cover input use and minimising pollution. Concerns for sustainability of resource use (such as fossil energy, land, water) can be related to the ethical norm of justice/fairness in terms of similar or fair access to resources for current and future generations. The protection of biodiversity could be related to a deontological reasoning that the rights of various species are protected, or a human duty to protect and care for nature and /or future

[24] http://www.ethicalmatrix.net/ developed by Ben Mepham and Sandra Tomkins, Centre for Applied Bioethics, University of Nottingham.

generations. The protection of bio-diversity can also be argued from an anthropocentric utilitarian perspective, stressing that there may be unknown future needs that require protection of a wide genome for medical research and future breeding. Alroe *et al.* (2006) refer to 'ecological justice' described as fairness to other living organisms with regard to a common environment. Therefore, producers, consumers and future generations have also to be seen as beneficiaries of any action. In addition to the environment and respect for nature and responsible use of resources, the new EU Regulation indicates that standard-setters are aware of these concerns. Additional requirements in the IFOAM basic norms (IFOAM, 2005a) and in several private standards illustrate the potential for market differentiation.

Concerns for animal welfare

Animal welfare concerns are expressed mostly by consumers and other stakeholders (in particular researchers) rather than producers. Animal welfare can be argued from utilitarian thinking of wellbeing as the core interest of animals and an acceptance that humans have an obligation to protect the wellbeing of domestic animals. The deontological and fairness-based reasoning that animals have rights remains more controversial and is less widely accepted. This is a likely consequence that follows from using and viewing animals as production units. There are potential conflicts between animal health and welfare and, for example, a fair economic return and environmental protection, although high welfare standards also pay off in terms of higher quality (Blokhuis *et al.*, 2008). Coverage of animal health and welfare in organic standards is changing. Current standards address the conditions on which good animal health and welfare should be based (feeding, breeding, housing, treatment). The IFOAM basic norms (IFOAM, 2005a) emphasise the rights of animals to express natural behaviour and some national and private standards contain more detailed requirements about housing rules with likely positive implications for welfare (Schmid *et al.*, 2007). The new European organic regulation (EC) 834/2007 has a strong emphasis on animal welfare in its principles, but the details of implementing rules are similar to previous provisions. There is an ongoing debate about whether the organic inspection systems should monitor more animal-based indicators (e.g. Vaarst *et al.*, 2008). Animal welfare is also subject to national legislation and a number of private certification schemes for welfare-friendly production (e.g. Freedom Food) exist. The EU-funded project Welfare Quality®[25] is developing auditing tools for animal welfare parameters. Concerns about respecting the integrity and intrinsic value of animals will nevertheless remain difficult to cover in organic standards and certification.

Economic concerns

Concerns about fair prices for producers or consumers are mainly although not exclusively raised by those stakeholders that are likely to benefit from action. The ethical reasoning is partly utilitarian – everybody involved should be provided with a satisfactory income – but also contains an element of deontological reasoning of treating all farmers with respect and not just as commodity producers, and justice-based reasoning in terms of fair/equitable returns to all. Similarly, the underlying ethical reasoning of availability and affordability to consumers could be related to the wellbeing of consumers, assuming that eating organic food is considered beneficial but the concern is mainly about fairness and social justice. European organic standards do not directly address fairness issues, but an increasing number of organic products from developing countries are also Fairtrade certified. Several fairness related examples in the dairy sector are reported in the literature (e.g. Thiele and Burchardi, 2006) and some voluntary ethical organic trade schemes and initiatives throughout Europe indicate that this is a potential area for market differentiation and organicPlus communication.

[25] http://www.welfarequality.net.

Social concerns

Concerns relating to the workplace and to transparency were raised by all stakeholder groups. This includes workplaces being safe and fair for employees and for farmers. A considerable proportion of care farms providing work opportunities for disadvantaged members of society are organic farms. The ethical reasoning can be both utilitarian, related to the wellbeing of those who work on farms, and deontological, treating each person in his/her own right and with respect. Activities in this area would benefit employees and farmers and indirectly also animal welfare. Organic producers are also required to meet national labour laws but the EU regulation and private standards are largely silent on these issues (with some exceptions) indicating that this is not a prime area of activity within the organic sector.

Other concerns

There is concern that as a consequence of market growth, organic farmers have lost ownership of the process of defining what organic farming stands for (e.g. Vogl *et al.*, 2005). However, there is a strong tradition of democratic and bottom-up processes in the development of organic ideas and standards, exemplified by the process of developing the IFOAM Principles (Luttikholt, 2007). The main ethical reasoning for transparency is deontological and relates to the rights and responsibilities of all actors. Producers and certifiers are in a situation of power, and have a duty to behave in a fair way with 'open cards' not allowing themselves anything they would not allow anybody else. However, there also is a utilitarian rationale that all operators in organic supply chains – producers, intermediaries and consumers – would benefit from greater openness in relation to the further development of organic standards and wider communication about those values that go beyond minimal requirements.

Concerns about the integrity of the organic supply chain, system health or the preference for local food cannot easily be categorised according to impact. For example, the preference for local supply chains can be argued as achieving greater wellbeing (economic prosperity) of the local community but also social justice/fairness for both farmers and consumers. Proponents of local food also refer to environmental benefits, such reduced energy for transport. The association of organic food with 'local' takes on the character of an ethical argument in itself, through the expectation that local structures will bring benefits in terms of ecology and the environment, social relationships for producers and consumers, care and resistance to globalisation. 'Buying local food' can almost be considered a rule in itself in the sense of deontological ethics. However, it is less certain whether local food delivers on all the expectations that rest on it. The ethical values attached to local may be internally contradictory and prioritise a particular one, such as ecology over social justice (Clarke *et al.*, 2008). Also animal welfare seems to be less vital to reasoning about the supply chain, although short transport distances for storing animals and to slaughter would promote both decreased transport as well as employment, i.e. social aspects.

Conclusions

Organic farming is practised in ways that go beyond the minimum requirements of the European Organic Regulations, in line with the broad range of values and sustainability goals associated with the concept of organic farming. These so called 'organicPlus' values represent an opportunity for differentiation by communicating a better ethical quality in an increasingly competitive market. However, the legal framework of standards and inspection/certification is only one trust builder for 'organics' (e.g. Zanoli, 2004). Communicating other activities could have negative implications for the overall credibility of the organic concept so there is a need for caution and for considering the possibilities how ethical claims could be verified.

OrganicPlus values can be summarised under headings of impact (or likely beneficiaries), such as impact on the environment, animals, and impact on people in terms of social and economic impact. Concerns referring to the whole system or supply chains – such as preferences for local food and concerns about the integrity of the organic supply chain have potential impacts in several areas and are more difficult to categorise in such a way.

Both private standards and public regulations have responded to some concerns, further developing certification as a trust builder for their label. Both the IFOAM basic norms (IFOAM, 2005a) and several private standards contain additional coverage of environmental and animal welfare issues which are also mentioned in the principles of the new European organic regulation (EC) 834/2007 on organic food. Certification activities related to socio-economic concerns are largely exploratory or voluntary and some organic products carry additional certification for animal welfare or Fairtrade. To our knowledge standard setters have not widely covered supply chain and integrity issues, but Lautermann *et al.* (2005) explored the feasibility of an ethical management system for the German natural food sector (*Naturkost*).

Several concerns can be argued from at least two, if not three, different traditions. The main reasoning in relation to environmental impact is utilitarian but the concept of ecological justice extends the notion of fairness to the environment and future generations. The utilitarian reasoning of responsibility for the wellbeing of animals is widely accepted, but the idea of animals having rights in parallel to human rights (i.e. respect for the being itself) remains controversial. Economic concerns relate both to the concept of wellbeing and, more strongly, to reasoning based on fairness/justice. This raises the difficult but important question of how a fair distribution of benefits should be attempted. For social concerns, utilitarian, deontological and fairness-based reasoning can be argued, while some values such as transparency and civic responsibility could also be seen to describe virtues.

Preference for one way of reasoning over the other could lead to different outcomes in terms of grouping or categorising concerns but also in terms of what action should be taken. It appears that the organic sector has been more active in developing standards and certification procedures in response to concerns that are based on utilitarian reasoning than for others. Aspirational organic values (e.g. deontological respect for animal integrity and for nature) will remain difficult to operationalise in the form of the pass/fail criteria that are needed for certification. Other ways to credibly communicate such values is necessary. More work would be needed to explore further the impact of ethical reasoning on explication of a broader range of values in organic standards and certification.

Acknowledgements

The authors gratefully acknowledge funding from the Department for Environment, Food and Rural Affairs, London for the financial support as part of the CORE organic partnership of national funding bodies for CORE Organic Project 'Farmer Consumer Partnerships'.

References

Alroe, H.F., Byrne, J. and Glover, L. (2006). Organic Agriculture and Ecological Justice: ethics and practise. In: Halberg, N., Alroe, H.F., Knudsen, M.T. and Kristensen, E.S. (eds.) Global Development of Organic Agriculture: Challenges and Prospects CAB International, Wallingford, pp. 75-112.

Blokhuis, H.J., Keeling, L.J., Gavinelli, A. and Serratosa, J. (2008). Animal welfare's impact on the food chain. Trends in Food Science and Technology 19: S79-S87.

Brom, F., De Bakker, E., Deblonde, M. and De Graaff, R. (2006). Corporate Moral Responsibility Manual. LEI, The Hague.

Clarke, N., Cloke, P., Barnett, C. and Malpass, A. (2008). The spaces and ethics of organic food. Journal of Rural Studies 24: 219-230.

EC (2007). Council Regulation (EC) No 834/2007 of 28 June 2007 on organic production and labelling of organic products and repealing Regulation (EEC) No 2092/91. Official Journal of the European Union L 189, 1-23.

IFOAM (2005a). The IFOAM Norms for Organic Production and Processing. Version 2005. IFOAM.

IFOAM (2005b). Principles of Organic Agriculture International Federation of Organic Agriculture Movements, Bonn.

Lautermann, C., Pfriem, R., Wieland, J., Fürst, M. and Pforr, S. (2005). Ethikmanagement in der Naturkostbranche – Eine Machbarkeitsstudie [Business Ethics and Sustainability in the Organic food Sector] Lehrstuhl für Allgemeine Betriebswirtschaftslehre, Unternehmensführung und Betriebliche Umweltpolitik, Carl von Ossietzky Universität, Oldenburg.

Luttikholt, L.W.M. (2007). Principles of organic agriculture as formulated by the International Federation of Organic Agriculture Movements NJAS Wageningen Journal of Life Sciences 54: 347-360.

Padel, S. and Gössinger, K. (eds.) (2008). Farmer Consumer Partnerships Communicating Ethical Values: a conceptual framework. Aberystwyth University and University of Natural Resources and Applied Life Sciences Aberystwyth and Vienna.

Padel, S., Schmid, O. and Röcklinsberg, H. (2009). The implementation of organic principles and values in the European Regulation for organic food. Food Policy, in press.

Schmid, O., Huber, B., Ziegler, K., Jespersen, L.M., Jens Gronbech Hansen, Plakolm, G., Gilbert, J., Lomann, S., Micheloni, C. and Padel, S. (2007). Comparison of the EEC Reg.2092/91 and selected national and international organic standards as regards compliance and identification of specific areas where harmonisation, regionalisation or simplification may be implemented in EEC 2092/91. EEC 2092/91 (Organic) Revision: Project report D 3.2. Forschungsinstitut Biologischer Landbau (FIBL), Frick.

Thiele, H. and Burchardi, H. (2006). Preispolitische Spielräume für regional erzeugte ökologische Produkte: Analyse und Umsetzung einer regionalen Marketingstrategie bei Biomilchprodukten Bundesanstalt für Landwirtschaft und Ernährung (BLE), Bonn.

Vaarst, M., Padel, S., Younie, D., Hovi, M., Sundrum, A. and Rymer, C. (2008). Animal Health Challenges and Veterinary Aspects of Organic Livestock Farming Identified Through a 3 Year EU Network Project. The Open Veterinary Science Journal 2: 111-116.

Vogl, C.R., Kilcher, L. and Schmidt, H. (2005). Are standards and regulations of organic farming moving away from small farmers' knowledge? Journal of Sustainable Agriculture 26: 5-26.

Willer, H. and Kilcher, L. (eds.) (2009). The World of organic Agriculture, Statistics and Emerging Trends 2009. IFOAM, FiBL, ITC, Bonn, Frick, Geneva.

Zanoli, R. (ed.) (2004). The European Consumer and Organic Food. School of Management and Business, University of Wales, Aberystwyth.

Consumers' preferences for ethical values of organic food

K. Zander and U. Hamm
Agricultural and Food Marketing, University of Kassel, Steinstrasse 19, 37213 Witzenhausen, Germany; k.zander@uni-kassel.de

Abstract

Consumers increasingly criticise food that is produced under unsatisfactory social and environmental conditions. From the very beginning, organic food production included ethical aspects and did not only care for the environment and animal welfare but also for social aspects of people affected by the organic supply chain. With the organic sector gaining additional market share, products from organic mass production become more and more important and competition is predominantly a question of price. Ethical values going beyond the standards of the EU regulation on organic farming are no longer features of large parts of organic production. Against this background, the question arises of whether there is a demand and an increased willingness to pay for 'ethical' organic food. This would create possibilities of product and market differentiation with respect to ethical products. This paper investigates various ethical communication arguments from consumers' perspective by means of an Information-Display-Matrix (IDM) in five European countries. The IDM is a process tracing method aiming at monitoring the information acquisition and purchase behaviour of consumers. Seven different ethical attributes and the product price were tested with about 1200 consumers. The most important ethical attributes turned out to be 'animal welfare', 'regional production' and 'fair prices for farmers'. These attributes were followed by 'product price', indicating that consumers tend to prefer cheaper products over ethical products with attributes like 'care farming', 'social criteria of production', 'protection of biodiversity' and 'cultural aspects'. There are only minor differences between the countries regarding the order of the most important attributes. Altogether, the results allow the conclusion that a large share of consumers of organic food would be willing to pay a price premium for 'ethical' products.

Keywords: consumer behaviour, market research, information-display-matrix

Introduction

Globalisation and anonymity of trade with organic products are seen as a problem by many organic farmers and consumers in Europe. European organic farmers often fear competition with producers from countries, in which production costs are much lower due to climatic conditions, lower costs of labour or land, lower production standards, etc. Against this background, organic farmers are under pressure to find a way to decrease their production costs in order to keep up with world-wide competition, either by realising economies of scale through increased production or by lowering their production standards and serving the organic mass market. Organic farmers and other suppliers such as manufacturers or traders, who do not make use of all possibilities to lower their production costs but offer ethical values, face competitive disadvantages and will finally disappear from the market.

From the very beginning, organic farming was based on ethical aspects and did not only care for the environment and animal welfare but also for social aspects of people affected by the organic supply chain (Lautermann *et al.,* 2005). With the organic sector gaining additional market share, products from organic mass production become more and more important and competition is predominantly a question of price. Ethical values going beyond the standards of the EU regulation on organic farming are no longer part of large sections of organic production.

On the other hand, various publications indicate that consumers are interested in ethical values. Ethical consumerism is a growing trend worldwide and moral responsibility is a relevant buying motivation among various consumer groups (Carrigan *et al.,* 2004; Shaw and Shiu, 2003). 'Organic consumers are widely perceived to be ethical, although their motivations for buying organic are said to be based on health and environmental criteria rather than on workers' welfare' (Browne *et al.,* 2000: 87). Several examples illustrate that consumers of organic food are willing to pay a price premium if ethical values are added to organic products and are well communicated, such as fair trade products from developing countries or the direct support of small farmers' initiatives in disadvantaged (mountainous) areas (Zanoli *et al.,* 2004; Schmid *et al.,* 2004). Recently, very successful 'fair milk price' projects were initiated by organic dairy farmers in Germany and Austria (Sobczak and Burchardi, 2006; IG Milch, 2006).

Based on literature review and on an analysis of 'ethical' arguments used by organic farmers and farmers' initiatives, a wide range of arguments going beyond organic standards according to the EU regulation was identified (Padel and Gössinger, 2008). Existing arguments were categorised according to economic, social and environmental concerns as well as cultural issues. Within this research, consumers of organic food were confronted with these ethical concerns and with the product price in order to identify the ethical aspects that are most important for consumers in their decision to buy organic food.

Methodological approach

The Information-Display-Matrix (IDM) is one among several methods for analysing information acquisition behaviour. The IDM is a process tracing method aiming at monitoring information search, judgement and choice. The two-dimensional matrix lists alternative product stimuli in columns, while product attributes are listed in rows. Each cell contains concealed information about a product-related attribute, which has to be accessed one after another by the subject in order to obtain the information (Jacoby *et al.,* 1987, Mühlbacher and Kirchler, 2003). The method enables a detailed analysis of cognitive processes within decision making. A variety of measures exist on the kind, sequence and amount of

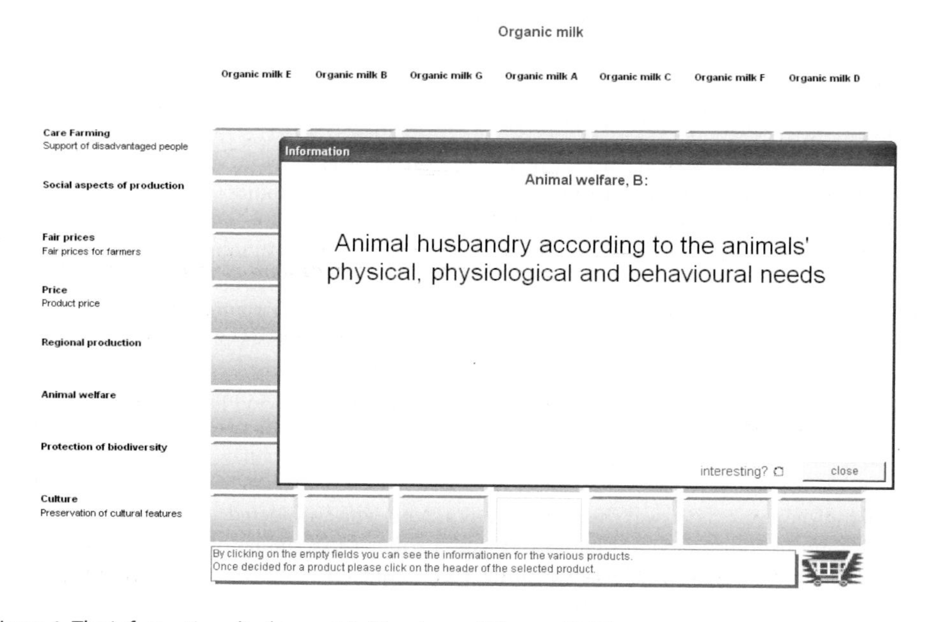

Figure 1. The information-display-matrix (Zander and Hamm, 2009).

information sought as well as on the duration and structure of the information acquisition preceding choice and decision. That way, relevant product-related criteria and their relative importance for the purchase decision can be identified (Jasper and Shapiro, 2002).

In this research, the IDM was used to track the information search regarding ethical values of organic food and to identify the ethical attributes most relevant for decision making. Based on the outcome of the research of Padel and Gössinger (2008), 7 ethical attributes were selected for further testing: 'biodiversity', 'animal welfare', 'regional production', 'fair prices for farmers', 'care farming', 'social aspects of production' (e.g. support for family farms and good working conditions for farm workers), and the 'preservation of cultural features' (traditional processing methods and landscape). The corresponding ethical arguments were assigned at random on six hypothetical products (1 litre of organic milk each). For the use within the IDM, the product price was added as a reference to the list of attributes. Thus, the seventh product was defined without any additional ethical value but at a lower price, namely 1.00€ per litre of organic milk, compared to 1.20€ for all other products with ethical attributes.

The respondents were asked to open various information fields one after the other according to their individual preferences and interests. The number of fields to be opened was not limited and repeated opening of the same field was also possible. Fields once opened changed colour after closing. In order to help the test persons not to get lost in the rather large matrix they were offered the possibility to mark interesting information fields. These fields got another (third) colour after closing. Once the respondents had come to a purchase decision they had to click the header of the selected product. The software then asked the respondents to confirm their choice. With this virtual final purchase decision, the IDM exercise stopped and the decision time measurement ended. The IDM software saved the information of all accession incidents, the order of accession, the time spent on each information field, the total time needed from starting the information acquisition until deciding for a product and the chosen product.

The IDM was accompanied by a questionnaire aiming at the validation of the results obtained by the IDM, at an explanation of the information acquisition behaviour and at giving answers on 'real life' information behaviour concerning organic food. About 1200 organic milk buyers were interviewed face-to-face in a computer assisted manner between Mai and July 2008 in the five study countries Austria (AT), Germany (DE), Italy (IT), Switzerland (CH) and United Kingdom (UK).

Results

The main aim of the IDM was to identify the most important ethical attributes of organic food from the consumers' point of view. According to economic theory, the most valuable information will be asked for first, since its marginal utility is highest (Foscht and Swoboda, 2004; Solomon *et al.*, 2006). The indicator provided by the IDM is the share of each attribute in all first accessions (Table 1). It turns out that 'animal welfare' and 'regional production' are the most important attributes followed by 'product price'. 'Cultural aspects', 'biodiversity' and 'social criteria' are the attributes least frequently accessed first. There are some differences between the countries: 'product price' was most frequently accessed first in Italy and least frequently in Switzerland. 'Cultural aspects' seem to be least important in Germany. The results indicate that many consumers prefer cheaper products over products with additional social or cultural values.

The second indicator calculated was the share of respondents who considered an attribute at least once within their information search. This indicator places attributes in a similar order. More than 86% of the respondents considered 'animal welfare', followed by 'regional production' (84%), 'fair prices for farmers' (81%) and the product price (80%). These results are quite similar in all study countries. Only Austrian and Italian consumers considered the product price more frequently than the attribute 'fair

Table 1. Share of attributes in all first accession incidents (%) (Zander and Hamm, 2009).

Attribute	All	AT	CH	DE	IT	UK
Animal welfare	21.4	21.3	27.6	22.1	18.0	17.9
Regional production	21.2	19.2	25.1	22.9	21.9	17.1
Fair prices for farmers	13.8	17.1	13.4	15.4	8.2	14.6
Product price	13.3	13.8	6.7	11.3	20.6	14.6
Care farming	8.2	9.6	4.6	7.9	9.4	9.6
Social criteria of production	7.8	6.3	5.9	10.8	9.4	6.7
Biodiversity	7.3	5.0	9.2	5.8	6.9	9.6
Cultural features	7.0	7.9	7.5	3.8	5.6	10.0
Total	100.0	100.0	100.0	100.0	100.0	100.0

prices for farmers'. In Austria, information on the product price (87%) was even more frequently asked for than 'regional production' (83%).

Socio-demographic factors affecting the respondents' perception of the most important attribute were found to be gender, age and education: Women are more interested in 'fair prices for farmers' than men; younger and older respondents assess 'animal welfare' to be more important than respondents belonging to other age groups and test persons with a higher educational level seem to value 'animal welfare' less than others.

The fact that on average 80% of the consumers considered the product price before deciding for a product implies that about 20% of the test persons made their purchase decision without having asked for information on prices. The product price thus seems to be less important for the purchase decision of organic consumers than commonly assumed.

This finding is supported by the results on the purchase decision of test persons. As explained before, one product without any additional ethical value was offered at a lower price. This product was preferred by only 6% of the respondents on average of all countries. Thus, the vast majority of consumers appears to be willing to pay a price premium for additional ethical values of organic food.

For the aim of contrasting the results of the IDM with results from direct inquiry within the survey part, the questionnaire included a question asking for the relevance of different criteria for the purchase decision on organic food. In direct inquiry, social production criteria like 'good working conditions for farm workers' and 'support for family farms' as well as 'preservation of cultural landscape' ranked higher than 'regional production' and 'fair prices for farmers'. These were the most important arguments in the IDM following 'animal welfare', which is most important according to both methodological approaches. The most outstanding example for differences in the preference structure is the relative importance of the product price. When asked directly, consumers ranked the price lowest (AT respondents second lowest) of all arguments, while in the IDM it was in close competition with the 'fair price' argument at the third or fourth position, respectively. Similar results with respect to revealed preferences for prices were reported by Aschemann and Hamm (2008), who also compared results of an IDM with those obtained by direct questioning in a single source approach. It can therefore be concluded that the IDM is an adequate instrument to reduce socially desired answers in consumer surveys.

Conclusions

The overall conclusion is that a considerable group of consumers seem to be willing to pay a price premium for some 'ethical' attributes of organic food. Accordingly, organic food with additional ethical values offers an opportunity for product differentiation – given that these ethical qualities are effectively communicated in an increasingly competitive market. Communication concepts should focus on attributes that are most important to consumers, such as 'animal welfare', 'regional production' and 'fair prices for farmers' in order to be successful.

The results were produced in a test environment, so that there might be some differences compared to real life behaviour. However, the aim of this research was to elicit relative preferences of consumers. There is no reason for assuming that relative preferences obtained in an experimental setting differ considerably from real behaviour, given that similar information is provided to consumers. Moreover, contrasting the results of the IDM with those from direct inquiry within the survey part, the IDM proved to be a valuable tool to elicit consumer preferences by reducing biases due to social desirability of answers. However, there are clear limitations to the use of the IDM. In order not to cause information overload, the number of products, attributes and arguments used in an IDM has to be restricted. Nevertheless, the IDM turned out to be suitable to rank different ethical values according to consumers' point of view within this research.

Acknowledgements

The authors gratefully acknowledge the financial support for this report provided by the members of the CORE Organic Funding Body Network, being former partners of the FP6 ERA-NET project, CORE Organic (Coordination of European Transnational Research in Organic Food and Farming, EU FP6 Project no. 011716), which was finalised in September 2007. The text in this report is the sole responsibility of the authors and does not necessarily reflect the views of the national funding bodies having financed this project.

References

Aschemann, J. and Hamm, U. (2008). Information acquisition behaviour of Fair-Trade-Coffee consumers – a survey by means of an Information-Display-Matrix. In: Neuhoff, D. *et al.* (eds.) Cultivating the future based on science. Proceedings of the 2nd Scientific conference of ISOFAR, June 18 – 2008. Modena, 338-341.

Browne, A.W., Harris, P.J.C., Hofny-Collins, A.H., Pasiecznik, N. and Wallace, R.R. (2000). Organic production and ethical trade: definition, practice and links. Food Policy 25: 69-89.

Carrigan, M., Szmigin, I. and Wright, J. (2004). Shopping for a better world? An interpretative study of the potential for ethical consumption within the older market. Journal of Consumer Marketing 21(6): 401-417.

Foscht, T. and Swoboda, B. (2004). Käuferverhalten. Wiesbaden, Germany.

IG-Milch – Interessengemeinschaft Milch (2006). A faire Milch. http://www.afairemilch.at/start.htm (Accessed 2007-06-27).

Jacoby, J., Jaccard, J., Kuss, A., Troutman, T. and Mazursky, D. (1987). New directions in behavioural process research: implications for social psychology. Journal of Experimental Social Psychology 23: 146-175.

Jasper, J.D. and Shapiro, J. (2002). MouseTrace: A better mousetrap for catching decision processes. Behaviour Research Methods, Instruments & Computers 34(3): 375-382.

Lautermann, C., Pfriem, R., Wieland, J., Fürst, M. and Pforr, S. (2005). Ethikmanagement in der Naturkostbranche. Metropolis-Verlag, Marburg.

Mühlbacher, S. and Kirchler, E. (2003). Informations-Display-Matrix. Einsatz- und Analysemöglichkeiten. Der Markt, 42, Nr. 166/167:147-152.

Padel, S. and Gössinger, K. (2008). Farmer Consumer Partnerships. Communicating ethical values: A conceptual framework. CORE Organic project FCP, Project report, Aberystwyth and Vienna. http://orgprints.org/12821/.
Schmid, O., Hamm, U., Richter, T. and Dahlke, A. (2004). A guide to successful marketing initiatives. Research Institute of Organic Agriculture, Frick.
Shaw, D. and Shiu, E. (2003). Ethics in consumer choice: a multivariate modelling approach. European Journal of Marketing 37(10): 1485-1498.
Sobczak, A. and Burchardi, H. (2006). Erzeuger Fair Milch. Faire Preise für heimische Biobäuerinnen und Biobauern. In: Agrarbündnis (ed.) Der kritische Agrarbericht 2006. AbL Verlag, Hamm, 264-268.
Solomon, M., Bamossy, G., Askegaard, S. and Hogg, M.K. (2006). Consumer behaviour. A European perspective. Prentice Hall, Harlow.
Zander, K. and Hamm, U. (2009). Farmer Consumer Partnerships: Information search and decision making – the case of ethical values of organic products. CORE Organic project FCP, Project report, Kassel-Witzenhausen. http://orgprints.org/15199/.
Zanoli, R., Bähr, M., Botschen, M., Laberenz, H., Naspetti, S. and Thelen, E. (2004). The European consumer and organic food. – Organic marketing initiatives and rural development Vol. 4, University of Wales, Aberystwyth.

Testing of communication tools on ethical values with focus groups

Raffaele Zanoli and Simona Naspetti
Dipartimento di Ingegneria Informatica,Gestionale e dell'Automazione, Facoltà di Ingegneria, Università Politecnica delle Marche, Italy; zanoli@agrecon.univpm.it

Abstract

Organic farmers and processors are often engaged in activities with ethical content that are well beyond European Organic standards, but communication about such activities is still problematic and largely unexplored. A public call to advertising companies was carried out for the production of communication tools (e.g. product labels) using the most promising arguments to communicate such ethical values, often referred to as organicPlus values. These arguments were identified through a specially devised survey in a previous stage of the research project, and consisted of: animal welfare, regional/local products, fair prices to farmers. Two 'creative ideas' for each argument were developed and will be tested in each country by Focus Group discussions (three focus groups per country) to find out the most promising for each country. A recall-test by telephone with the participants to the Focus Group discussions ten days after the discussions will identify participants' recollection of the main ethical arguments and therefore help to choose the most promising proposals. In this paper we discuss the methodological problems derived from implementing such approach in a multi-cultural, international context.

Keywords: organic egg labels, advertising, focus group

Introduction

The work presented in this paper is part of the CORE funded project Farmer Consumer Partnerships (FCP) which is aiming to develop innovative generic communication arguments that can strengthen the partnership between producers and consumers through better communication of ethical attributes of organic food. The aim of this paper is to present on-going research work on communication tools to convey additional ethical claims on organic products to consumers. By using a qualitative iterative research technique (focus groups combined with paper-and-pencil questionnaires and a telephone interview administered to the same subjects), the aim of the work is to select the preferred communication concepts to express selected arguments regarding an organic benchmark product (eggs) in each country.

Material and methods

Theoretical background

An effective advertising message strategy should include the message idea, a headline, the body copy and the creative format. The message idea (or argument) is the main topic or benefit to be communicated in the message. The argument may be substantiated in one or more claims, i.e. statements that are used in the body text that addresses some specific benefit to be gained from using the product; for example: free from OGM feed, free-range, locally produced, etc. The body copy (or copy text) is a written statement that fully describes the message idea; it constitutes the main copy block of an advertisement as distinguished from headlines, subheadings, logo, illustrations, and the like. The headline is a sentence, phrase, word, or group of words set in large, bold type in a printed advertisement (in our case, the label). The purpose of a headline is to attract attention and usually to encourage the reading of the following copy. In print advertising, the headline is considered to be the most important element, because it invites the reader into the advertisement. The creative format is how to communicate the message idea to the target. It is a mix of graphical elements (illustrations), design, typeset, headline and body copy that convey the message idea or argument.

All the elements of the advertising message are intended to conform to three principal communication goals: Reminding, Informing, Persuading (RIP). The communication mix is planned in order to inform consumers about the product. A persuasive communication increases consumers' loyalty and preference for the product but also reduces substitution strategies. Finally, the communication has to remind consumers about product characteristics that are unique and that are strongly connected with their final goals and values (Zanoli, 2004).

In an international market, target consumers are often subject to different cultural influences and they reply to communication messages in a different way according to many variables. Culture and subculture are surely relevant environmental variables, though it is well known that 'measuring the content of culture is actually a tricky matter' (Peter *et al.,* 1999).

The reason for this is that our own culture is often used as a frame of reference and often this may lead to misinterpretation of other cultures. This is very relevant when working in a multi-cultural, international study and has implications for both the researcher and the communication specialist. There are two basic approaches in consumer research when culture is involved, which are labelled 'emic' and 'etic'.

Emic research emphasizes the uniqueness of each culture, and allows insight into a particular culture, but cannot be used for comparisons across cultures. Emic approaches involve using culture-specific symbolism, concepts and terms.

Etic research, on the other hand, aims to compare different cultural settings, and therefore tries to employ terms, concepts and symbols which will be common across the cultures to be investigated. Etic research can therefore be used for cross-cultural studies (Peter *et al.,* 1999).

Focus groups and other qualitative approaches, by having the advantage that they provide rich and redundant information, reduce the danger of misinterpretation in cross-culture context and allow for a full account of cross-cultural differences in consumer perception of communication devices. At the same time, by using quantitative survey instruments like questionnaires, even in a qualitative study, allows for tests of cross-cultural validity of these instruments as tools for measuring consumer values across cultures.

We have therefore used both focus groups and questionnaires in order to capture consumer variability of reactions to a communication tool (product label) in 5 different EU countries.

Research methodology & data collection

This study builds on the results of previous research carried by the CORE ORGANIC FCP partnership by means of an Information Display Matrix (IDM) survey of consumers of 5 countries (Austria, Germany, Italy, Switzerland, United Kingdom). This involved studying different 'ethical' arguments selected from those actually used by organic farmers and processors, in order to find the three most promising arguments to communicate organicPlus values. These arguments were identified as: animal welfare, regional/local products, fair prices to farmers (Zander and Hamm, 2009).

Since this is an 'etic' study allowing for cross-country comparisons, eggs were chosen as the product to advertise, as other products (e.g. milk or pasta), have different connotations and are perceived quite differently in the various EU countries. In order to test the communication tool (product label), carrying the deployment of the message idea (3 different organicPlus arguments) on organic food products, a two-stage process was devised.

During the first stage, an advertising company – selected via a public international call – was asked to prepare proposals (creative formats) for a portfolio of six printed labels (2 for each argument) in colours, composed of a headline, body copy and symbolic images. In order not to influence the consumers in selecting the preferred creative format for each argument, a common design and background colour was selected for all the six labels, and the same symbolism was used for each of the two competing creative formats for each argument. The cross-cultural project team selected the adopted homogeneous design and a green background colour by democratic vote from the five different combinations of design and colour proposed by the advertising company.

The design was based on various 'heart' images, symbolising care, love and respect, as well as 'deeply felt' ethical values in all cultures involved in the study[26]. This imaging is 'reflected' in three graphical element/illustrations for each argument:

- the 'hearty hen' for animal welfare;
- the 'hearty farm/region/earth' for regional/local product;
- a 'hearty farmer' for fair prices.

The advertising company, after a briefing with the research team, proposed various communication concepts (headlines and body copy) reflecting the three organicPlus arguments. The claims substantiating the arguments were based on previous project results (Padel and Gössinger, 2008) and the literature (Zanoli, 2004). After a discussion with partners the combinations of headline and body copy were set out in order to have comparable concepts and claims despite cross-cultural and language differences among the five countries involved in the survey (see Appendix A). Labels were also planned to be consistent according to the different country legal requirements, in order to have reliable and trustworthy labelled packages.

At a second stage, to measure the effectiveness of the labels, a total amount of 30 Focus Group (FG) discussions will be carried out in five different countries – Austria, Germany, Italy, Switzerland, United Kingdom – in order to investigate consumers' attitudes and preferences towards the advertising labels. In each country three FG repetitions will be conducted in capital cities or large metropolitan areas. Only organic, either regular or occasional, egg consumers and buyers will be asked to participate in the discussions. Selected participants need to have exclusive or shared responsibility for household food purchases.

The focus groups will explore consumer attitudes in three steps. First, 'top-of-mind' statements on the ads will be elicited, in order to explore the recognition of the communication arguments and respective claims in each of the six labels. Second, the labels will be shown paired per argument (two at a time), in order to explore the liking and preference of the communication concepts proposed. Finally, the effectiveness of the communication will be explored by asking participants which label may influence them in buying the product the most.

After the focus group, the participants will complete a questionnaire to measure their general attitude towards advertising (Purvis and Mehta, 1995) and, specifically, their Emotional Quotient Scale towards

[26] 'The heart has long been used as a symbol to refer to the spiritual, emotional, moral, and in the past also intellectual core of a human being. As the heart was once widely believed to be the seat of the human mind, the word heart continues to be used poetically to refer to the soul, and stylized depictions of hearts are extremely prevalent symbols representing love.' (Viswiki, 2009).

each label (Wells, 1964), as well as the label believability (Beltramini, 1982). Ten days after the discussions the participant will participate to a recall survey by telephone, in order to test which argument and concept was related to 'value messaging', that is the communication of the claims[27].

The discussion guide, questionnaires as well as a complete moderator guide (research package) were developed and tested in a pre-test focus group session, carried out in Italy. After further discussions with partners the final qualitative 'research package' was released in English and translated in Italian and German.

Discussion

In advertising, some companies are able to convey messages that are common to many countries in the world. This is the case e.g. of Coca Cola whose ads are often international in nature. But other companies (e.g. McDonalds) in some countries need to change the approach. For example, McDonalds had little in Italian appeal for some targets, so they needed to make more 'Mediterranean' style dishes and advertisements. Multicultural marketing research is relevant since it is quite difficult to find an advertisement successful in many countries.

This research will explore the extent to which ethical claims of organicPlus farmers and processors are able to be communicated homogenously in the five countries under investigation. Cross-cultural validity of arguments, claims and communication concepts will be explored, albeit by qualitative research instruments.

It is well known that cultural barriers could prevent the identification of a commonly preferred concept, and that some or all concepts could be totally refused in some countries, due to 'local' cultural conditions. In this respect the limit of our research is embedded in the limited funding that does not allow for multi-stage testing of various communication concepts should the proposed concept fail in one or more countries.

Acknowledgements

The authors gratefully acknowledge funding from the Ministero delle Politiche Agricole e Forestali (MIPAF) and the Italian Government for the financial support as part of the CORE organic partnership of national funding bodies for CORE Organic Project 'Farmer Consumer Partnerships'.

References

Beltramini, R.F. (1982). Avertising Perceived Believability Scale. In D.R. Corrigan, F.B. Kraft, and R.H. Ross (eds.). Proceedings of the Southwestern Marketimg Association. Wichita (KS). 1-3.

Mehta, A. and Purvis, S.C. (1995). When attitudes towards advertising in general influence advertising success. Paper presented at the 1995 Conference of the Aamerican Advertising Academy, Norfolk (VA). 1-18. http://www.gallup-robinson.com/reprints/whenattitudestowardsadvertising.pdf.

Padel, S. and Gössinger, K. (eds.) (2008). Farmer Consumer Partnerships Communicating Ethical Values: a conceptual framework. Aberystwyth University and University of Natural Resources and Applied Life Sciences Aberystwyth and Vienna.

Viswiki (2009). The heart symbol, http://www.viswiki.com/en/Heart_(symbol).

[27] We are well aware that, traditionally, recall questions are more connected with functional benefits than values, and in our research the labels aim to communicate claims that substantiate ethical values. Therefore, we use recall as an aim to check the recall of values more than benefits. Am emotional bond can be assumed to be stronger if the recall of such values is correct.

Wells, W.D. (1964). EQ, Son of EQ, and the Reaction Profile. Journal of Marketing, 28: 45-52.
Zander, K. and Hamm, U. (2009). Farmer Consumer Partnerships: Information search and decision making – the case of ethical values of organic products. University of Kassel, Agricultural and food marketing.
Zanoli, R. (ed.) (2004). The European Consumer and Organic Food. School of Management and Business, University of Wales, Aberystwyth.

Appendix A. Label texts in the 3 different languages, per argument and claim

Arguments	German (DE/AT/CH)	English	Italian	Claims
AW1	**Die Wahl des Herzens** Die Hennen werden mit Liebe und Respekt gehalten. Sie bekommen gentechnikfreies Futter und können im Freien herumlaufen.	**The heart's choice** The hens are looked after with love and care, fed organic feed free from GMOs and are free to live and roam outdoors!	**La scelta del cuore** Le galline sono allevate con amore e rispetto, libere da mangimi OGM, libere di crescere e di razzolare all'aperto!	Love & respect Freedom GMO-free
AW2	**Mit dem Herz erzeugt!** Das Wohlbefinden unserer Hennen liegt uns am Herzen. Sie können im Freien herumlaufen und bekommen natürliches, gentechnikfreies Futter. Für sie haben wir ein 100prozentiges Bio-Leben ausgesucht.	**Produced with the heart!** The welfare of our hens is close to our heart! They have access to the outdoors where they are free to roam, and they are fed on natural, GMO-free feed. For them we have chosen a 100% ORGANIC healthy life!	**Prodotte con il cuore!** Ci sta a cuore il benessere delle nostre galline! Sono allevate libere di razzolare all'aperto ed alimentate naturalmente e senza OGM. Per loro abbiamo scelto una vita sana 100% BIO!	Welfare & care Freedom GMO-free
Local 1	**Aus dem Herzen unserer Region** Diese Bio-Eier stammen aus der Gegend, in der ich wohne. Sie kommen auf kurzen Transportwegen und mit geringer Umweltbelastung auf meinen Tisch.	**From the heart of our region** These organic eggs are produced close to where I live and are brought to my table with minimum transport and less pollution.	**Dal cuore della nostra regione** Queste uova bio sono prodotte a due passi da casa mia e arrivano sulla mia tavola senza compiere lunghi e inquinanti tragitti.	Local and near Food miles Environment
Local 2	**Das Herz der Tradition!** Unsere Region liegt uns am Herzen. Dieses regionale Produkt trägt zum Erhalt bäuerlicher Kultur und Traditionen bei.	**The heart of tradition!** Our region is close to our heart. This regional product safeguards our rural values and traditions.	**Il cuore della tradizione!** Ci sta a cuore la nostra regione. Questo prodotto tutela i valori e le tradizioni rurali del nostro territorio.	Regional Rural values & traditions
Fair1	**Ich unterstütze die, denen unsere Welt am Herzen liegt!** Der Kauf dieser Eier honoriert die Arbeit der Bio-Bäuerinnen und Bio-Bauern, die unsere Mutter Erde pflegen und schützen.	**I support those who have our world at heart!** Buying these eggs rewards the work of organic farmers who safeguard and preserve our mother Earth!	**Io sostengo chi ha a cuore il mio mondo!** Comprando queste uova bio premio il lavoro degli agricoltori biologici che tutelano e custodiscono la nostra madre Terra!	Fair prices/reward for stewardship
Fair2	**Das Wohl unserer Bauern liegt uns am Herzen!** Ein faires Geschäft: Der Kauf dieser Eier honoriert die schwere Arbeit von Bio-Bäuerinnen und Bio-Bauern und ihren Familien und sichert ihr Überleben.	**The wellbeing of our farmers is close to our heart!** A fair deal: buying these eggs rewards the hard work of organic farmers and their families and secures their survival!	**Ci sta a cuore il benessere dei nostri agricoltori!** Un affare equo: l'acquisto di queste uova premia il duro lavoro degli allevatori biologici e delle loro famiglie e assicura la loro sopravvivenza!	Fair prices/reward for family farms

Theme 13 – Other contributions

Functional bodies, functional foods: how will health be re-defined?

R. Ibanez Martin
STS Department, Philosophy Institute, Centre for the Humanities and Social Sciences (CCHS) Spanish National Research Council, (CSIC), C/ Albasanz, 26-28, 28037 Madrid, Spain; rebecaibanezm@gmail.com

Abstract

In this paper, I argue that functional foods research, marketing, and consumerism respond to a -not that novel- idea of perfection and high performance of the body. In such an ideal, functionality and hope for better health play a crucial role. The scheme of functionality refers to a set of ideas about control, risk management, and health-related ideas and concepts. In this particular case, early prevention through body interventions in the present and the achievement of enhanced body functions are key concepts in understanding functional foods.

Keywords: functional foods, biomedicalization, economies of hope

In this text, I will focus in order to illustrate my argument, on one of the multiple characteristics of bodies. Bodies as healthy, fully functional, and more real than real, as Balsamo puts it in the introduction to her book *Technologies of the Gendered Body*. My goal is to cover and put in conversation three issues that I consider crucial for an accurate account of what is going on in our techno-cultures. These are, in the first place, bodies and biotechnologies, in short, bodies as the locus or the target of so many current biotechnologies and future 'yet to be' technologies. Secondly, there are perfection ideas of the body and its reflection and negotiation with those very technologies. In this scheme, I take up and describe the biomedicalization of bodies as encompassing these technologies and lastly, functional foods and functionality, in which illness gets elicited as the new paradigm of moral perfection.

Functional foods as a material artefact have entered our lives, our guts and to a large extent, have influenced consumers' hopes for a better and long-lasting health. The term future foods has often been used to refer to functional foods. The term future foods also designates 'yet to be' foods that will be potentially engineered by nanotechnology, among others. The standard definition of functional foods that I extract from the food sciences literature refers to those products with a health benefit claim that goes beyond the inherent nutritional value of the product. Generally, we find two types of health claims. One kind typically refers to an enhanced body function. The second refers to a reduced risk of disease, such as high levels of cholesterol or blood pressure (Wieringa *et al.,* 2008).

Functional foods, as we all know, is a phrase that contains a concept that holds so many wishes and promises for a better health. Functional foods also symbolize certain promises of the future of foods that will cure, that will endure, and that will be, so to speak 'more than foods'. In that respect, it is remarkable that after an analysis of the marketing of the most widely accepted functional foods in Spain, the results clearly indicate that the target population are females[15]. A conclusion I extract is that the commercialization of functional foods or 'more than foods' is targeted to women, and bought by 'more than women' in the sense that they are enacting an enhanced idea of the role of femininity in relation

[15] Some of the sources reviewed are: Vidal-Guevara, M.: 'Is Spain the New Japan of Functional Foods?' *Nutraingredients*, Availible at http://www.nutraingridients.com accessed October 15, 2008). Quintana, Yolanda (2008)'¿Alimentos que todo lo pueden? La realidad de las declaraciones nutricionales y de salud en el etiquetado' CEACCU, Available at: http://www.ceaccu.org/index.php?option=com_content&view=article&id=190&Itemid=93 Accessed March 08, 2009. And several retailers' websites such as Danone available at (http://www.danone.es/Danone/default.jsp).

with health and nourishment. I will argue that this kind of foodstuff portrays, and embodies promises of a better, longer, healthier future for the humans that will consume them thanks to food science and innovative technologies. In that sense, functional foods mobilize regimes of hope[16].

When it comes to functional foods, we find many promises of personal transformation within those kinds of products that specially pinpoint and target some gender anxieties and roles. For example, we find these kinds of allegations: (low fat that 'makes you feel better' 'or it reduces the levels of cholesterol in blood' (targeted at men, but we see in the commercial a woman telling the male character he should try it) or 'makes your immune system stronger' (we see a family gulping down a shiny liquid with a woman in the centre of the image explaining the beneficial effects) or, 'with bacteria that is good for your intestines', and many, many more). Of course, since those products represent and communicate promises of personal transformation, we can find some ethical, and more saliently, legal problems.

I situate and locate functional foods within the more widespread trend of technologies that have the transformation of the body as their fundamental target. We are living the time of the life sciences, which seem to have taken the first place in the position to reveal the innermost truths of human nature, which is to say the truths of Western Civilization. New body biotechnologies are often promoted and rationalized as life-enhancing and even lifesaving technologies (Fukuyama, 2002; Franklin and Lock, 2003). I am thinking of reproductive technologies, cosmetic and plastic repairing technologies, therapeutic, and sex assignment and re-assignment technologies. All these technologies respond to the current biomedicalization of the bodies as a complex phenomenon that has as locus and target in the body. Therefore, inscribing the body with metaphors and ideals of culture. Within this trend, I also localize and situate the rise and commercialization of functional foods as technologies of enhancement of the body. For biomedicalization I understand, following the work of Adele Clarke and her collaborators (Clarke *et al.,* 2003) all 'the increasingly complex, multi-sited, multidirectional processes of medicalization that today are being both extended and reconstituted through the emergent social forms and practices' (Clarke *et al.,* 2003: 45). The extension of medical jurisdiction over health itself -and not only to treat disease-, but the commodification and commercialization of health are fundamental in this scheme of biomedicalization (*ibid.*). Health itself and the management of it are becoming individual moral responsibilities to be fulfilled through access to, not only to knowledge and expertise, but also by the appropriation, consumption and mobilization of expectations on health (ibid).

In the ground-breaking article co-authored by Adele Clarke, the authors identified 5 key processes of biomedicalization. These are: (1) major political economic shifts; (2) a new focus on health and risk and surveillance biomedicines; (3) the technoscientization of biomedicine; (4) transformations of the production, distribution, and consumption of biomedical knowledge's; and (5) transformations of bodies and identities (*ibid.*).

I argue that with the rise of functional foods, we are moving towards a self-surveillance or what can be labelled as 'know your body obsession'. The absence of illness becomes a process when it comes to functional foods, since most of them are designed to prevent -intervene in the present to treat the future, manage the risk of- specific illnesses or enhance a body function. As a result, we have to pay attention to the biotechnopolitics[17] of these technologies in which health becomes a constant process of realization,

---

[16] The explanation of the notion of regimes of hope can be looked up in Brown, 2005. The main idea is that there is a tension that takes place between the regimes of truth (facts) and the regimes of hope (promises, hopes, expectations). In this scheme, research and marketing of some biotechnologies require the mobilization of regimes of hopes in future oriented technologies.

[17] Beatriz Preciado (2008) coined this term following the work of Foucault (1988) (biopolitics) and Haraway (1997) (technocultures), see the bibliography.

not a state characterized by the lack of illness. Within this biomedicalization scheme, the focus on health, risk and surveillance is radical (Clarke *et al.*, 2003), in the way that health becomes a commodity. In that sense, health becomes not only a moral state of being, but a moral obligation. Health becomes extremely individualized since it becomes an individual goal (through individual consumption) and not a public goal (through public health policies). Functional foods can be understood as technologies of susceptibility because a mobilization of pre-symptomatic processes take place (Rose, 2007). An intervention in the present to treat the unknown future is necessary. In this scheme, subjects may become long time pre-patients in an ethical and political scheme of intervention.

Consequently, the consumer becomes responsible for her own health. What is more is that she is asked to be aware of the potential health risks out there. If the consumer becomes embedded in those processes of knowing, controlling, and monitoring her own health, in a process where the consumer looks up and looks for scientific information regarding food science, nutrition, and health, then, she becomes and an expert in the sense of expertise conceived as knowledge resource management (Rozin, 1997; Crossley, 2002). It is crucial then to look at the places where the consumer goes to find this kind of information, what are the sources of 'legitimate expertise', and the sources of information that the consumer gives credit to.

References

Balsamo, A. (1996). Technologies of the Gendered Body. Reading Cyborg Women. Durham University Press, Durham.

Brown, N. (2005). Shifting Tenses: Reconnecting Regimes of Truth and Hope. Configurations 13: 331-355.

Clarke, A.E., Shim, J.K., Mamo, L., Fosket, J.R. and Fishman, J.R. (2003). Biomedicalization: Technoscientific Transformations of Health, Illness, and U.S. Biomedicine. American Sociological Review 68(2): 161-94.

Foucault, M. (1988). Technologies of the Self. In H. Gutman L. H. Martin, and P. H. Hutton (eds.) Technologies of the Self: A Seminar with Michel Foucault. University of Massachusetts Press, Amherst, MA.

Franklin, S. and Lock, M. (eds.) (2003). Remaking Life & Death. Toward an Anthropology of the Biosciences. School of American Research, Santa Fe.

Fukuyama, F. (2002). Our Posthuman Future. Consequences of the Biotechnology Revolution. Farrar, New York.

Haraway, D. (1997). Modest_Witness@Second_Millenium. FemaleManã_Meets_OncoMouseÔ. Routledge, New York and London.

Preciado, B. (2008). Testo Yonqui. Espasa, Madrid. There is a French and English version of the book availible.

Rabinow, P. (1996). Making PCR: A Story of Biotechnology. University of Chicago Press, Chicago.

Rose, N. (2007). Molecular Biopolitics, Somatic Ethics and the Spirit of Biocapital. Social Theory and Health 5: 3-29.

Rozin, 1997 quoted in Urala, N. (2005). Functional Foods in Finland. Consumer's Views, attitudes and willingness to use. Doctoral Dissertation, University of Helsinki, Helsinki.

Wieringa, N.F., Van Der Windt, H.J., Zuiker, R.R.M., Dijkhuizen, L., Verkerk, M.A., Vonk, R.J. and Swart, J.A.A. (2008). Positioning Functional Foods in an Ecological Approach to the Prevention of Overweight and Obesity. Obesity Reviews 9: 464-473.

Efforts of the international community to address the issue of food security against the background of global environmental change

A.E.J. McGill
International Union of Food Science and Technology (IUFoST): Future for Food, 89 Melvins Road, Riddells Creek, Victoria 3431, Australia; albert.mcgill@futureforfood.com

Abstract

Food security has been a major theme for the International Union of Food Science and Technology (IUFoST) since its Budapest Declaration in 1995. In reviewing its actions, a proposal was initiated by IUFoST in 2007 that allowed the evaluation of related projects by similar global organisations. Participant observation was arranged for a sample of conferences that were scheduled by such organisations within the available timeframe of the proposal. Reports were made on each conference, its relevance and its possible involvement in the IUFoST strategic plan. A summary of the relevant points from each conference programme in relation to food security and the actions of food professionals is included here, as well as suggestions as to the value of collaboration or networking between IUFoST and the organisations cited. The lack of involvement of food professionals and representation of the food industry in many of the conference agendas or discussions is of concern.

Keywords: food professionals, food supply chain, urban health/wellbeing

Introduction

The International Union of Food Science and Technology (IUFoST) is the sole existing global professional food science and technology organisation. It is a voluntary, non-profit association of national food science organisations that links the world's best food scientists and technologists. Its present membership comprises 67 countries, each of which is represented through its Adhering Body (AB). The food professions of each member country contribute to their national AB as best fits their individual circumstances. This allows for wide differences between groups of countries, depending on their types of professional organisations and stages of economic development.

In 1995 the delegates to the IUFoST Congress in Budapest approved a declaration that consolidated their intentions and drew upon the outcomes of the International Conference on Nutrition (ICN, 1992). The main element of the declaration was '. . . the determination to work for the elimination of hunger and the reduction of all forms of malnutrition throughout the world . . . recognising that access to nutritionally adequate and safe food is the right of each individual' (IUFoST,1995). In 1996 the Food and Agriculture Organization of the United Nations (FAO) adopted its definition of Food Security and IUFoST accepted that definition and theme in place of hunger, in the intentions of their declaration.

At IUFoST Congresses in 2001 and 2003 concern was expressed that nothing seemed to have emanated from the organisation that could be seen to have impacted in any way on the state of global Food Security. In 2005-2006 a study of activities directed towards the reduction of food insecurity, and conveyed to the Governing Council of IUFoST, indicated that many global institutions had programmes directed at the study and even the alleviation of the problems of food insecurity, which indeed formed the basis of the first of the Millennium Goals. The World Health Organization (WHO), the Food and Agriculture Organization (FAO), the World Food Programme (WFP), the Special Programme for Food Security (SPFS) and many charity groups such as, OXFAM, World Vision, Save the Children Fund and others, are larger and more complex organisations with significantly greater funding than IUFoST. The best way to promote the spirit of the Budapest Declaration was considered to be for IUFoST to deploy its limited resources and available personnel to collaborate with, support and enhance existing programmes of similar intent, where possible. Some funding was made available through the International Relations Task Force of IUFoST to evaluate the work of a number of diverse organisations whose activities were being showcased through their conferences scheduled during 2007 – 2008. The objective was to identify, through participant observation, how the programmes being presented, paralleled or incorporated the work of food professionals and how IUFoST might make any form of contribution to their programmes. The observational focus at each conference was directed at the elements of it that could be identified as related to the work of food professionals or upon which they might have some input or from which they might derive some intelligence. Consequently, some parts of the conference agendas lay outside this scope and were not highlighted. The selection of organisations and conferences to observe was based on relevant contents, access and timing, in relation to the IUFoST funding timescale. Some diversity in themes of conferences allowed the observations to be considered under headings of Policy, Products, Actions and Impacts.

Sources of observation

Policy

The International Food & Agriculture Trade Policy Council (IPC), 'Sustainability in the Food & Agricultural Sector: The Role of the Private Sector and Government', Stratford – upon – Avon, United Kingdom, 12-13 October, 2007.

This conference was attended by a wide range of international representatives from both governments and the private sector. The direct food sector was represented by two multi-national companies, but through their agricultural and economic advisers, and two supermarket retailers, concerned mainly in showcasing their 'green' credentials. The main messages were:
- Global food production must double by 2050 to meet projected demand and the remaining available arable land is <18%.
- Significant increases in funding are required urgently for agricultural R & D, especially in the USA, remembering that there is usually a 15 years lag time between funding and outcomes.
- Agriculture on a global basis is not sustainable as currently practised, with for example, UK dairy farmers receiving 20% less for milk than production costs, and farmers in developing countries having their exports blocked by divisive safety legislation in Europe and the USA.
- There is a breakdown in perception between populations and their knowledge of their own food supply fragility. Global distribution practices makes the concept of seasonality of food supply irrelevant until a natural disaster occurs.
- The general public views food safety failures as they would an air disaster or a road accident.

A number of attempts were made during discussion sessions to encourage the involvement of the food industry or food professionals of participant companies without success.

Products

The Healthy Foods European Summit, 'Policy, regulatory and national issues: Markets, trends, innovations and futures', The Royal Horseguards Hotel, London, United Kingdom, 23-24 October, 2007.

Although the broad theme of this conference was healthy foods, it acted as an opportunity for many European companies to present their latest 'miracle' foods and ingredients to prospective clients, hence the 'Products' heading. The major relevant issues were:

- All governments' concerns for healthy diets targets, particularly to target obesity and type 2 diabetes. Revisions of dietary guidelines were still to be shown to be effective.
- New EU policies and regulations on allowable health claims on foods were implemented in July, 2007, but the European Food Safety Authority (EFSA) was struggling to deal with three times as many claims lodged as were anticipated.
- In terms of dietary guidance, there was confrontation between the UK government 'traffic lights' labelling system and others, especially implemented by major supermarket groups, that were considered to be more useful to consumers. The cost of re-labelling was considered to be an excessive impost and unnecessary bureaucracy.
- The market had estimated that major consumer concerns were; tiredness, overweight and stress. Any products targeting these concerns had strong market response.
- The strongest market drivers were the attributes: Fresh, Tasty, Free from.
- 'Superfoods' were seen as those which exploited the 'health halo'. An example was the perceived high antioxidant contents of pomegranates – anything with added pomegranates was seen as 'cool'.

Some effort was made to question the impacts of the demands for the unique raw materials being presented as new exotic elixirs of health on their countries of origin or their agricultural communities. Some companies explained their 'fair trade' policies, others were more reticent. The potential effects on the food security status of the developing countries in question were not regarded as of primary interest.

Action

The International Council for Science (ICSU), Workshop to advise the Science for Health and Wellbeing in the Changing Urban Environment (SHWB) Planning Committee, International Institute for Applied Systems Analysis (IIASA), Vienna, Austria, 24-25 January,2008.

IUFoST is a full member of ICSU and was invited to nominate a representative to participate in this Workshop, prior to the Planning Committee meeting. As events developed, no invitation was offered subsequently to attend the Planning meeting and only the Workshop provided any sense of the directions that the project might take. From a food science viewpoint, the key outcomes were:

- The urban environment was selected because as from 23 May, 2007, more people live in urban than in rural environments.
- The focus was on an Applied Systems approach as the best scientific means to study a complex system, develop appropriate mathematical models and enable scenario projections which can be tested and refined experimentally. The location of the meeting was indicative of the future format of the work of the project.
- The major themes of the discussions related to infrastructure and pollution reduction, with the effects of chronic poverty in the larger cities of equal concern.
- The only food related topic was nutrition based and related wholly to the performance of dietary guidelines and expansion of chronic obesity and type 2 diabetes.
- There was no interest in discussing the role of the food industry, the importance of the food supply chain or food safety and hygiene as factors in urban health.

Given the role that the food industry, particularly in the ubiquitous supply of 'fast food' to the diets of low income consumers in inner city areas (McGill, 2006), it was surprising that no input was sought from IUFoST in the subsequent deliberations of the Planning Committee.

Impact

The International Council for Science Project on Global Environmental Change and Food Systems (ICSU-GECAFS), 'Food Security and Environmental Change – Linking Science, Development and Policy for Adaptation', St Catherine's College, University of Oxford, United Kingdom, 2-4 April, 2008.

The ICSU-GECAFS project was relocated to the Environmental Change Institute (ECI) at Oxford University in 2007 and this was an opportune location for this Conference bringing together both those interested in the subject but, more importantly, interim results from the project's four field locations in Sub-Saharan Africa, the Indo-Gangetic Plain, the Caribbean and Europe. The 'Food Systems' aspect of the project gave some hope of a more food science or industry context than had been experienced elsewhere. The main points from the conference were:
- Food Systems cause environmental change! They contribute 30% of the global greenhouse gas (GHG) output.
- The primary effects of environmental change were observed on land, water, vegetation and local temperature variations.
- Food was dealt with, almost exclusively, as agriculture. No aspect of storage, value-added processing, transport or retail was included.
- Only one presenter, from the University of Amsterdam, saw the relevance of food processing and the supply chain as important parts of food systems.
- Adaptation processes to cope with climate change were active in many areas, but limited and variable especially in developing countries.

The presentations to the conference were generally of very high standard and covered a very wide range both in content and geography. The scope for more consideration of issues such as the food supply chain and the role of the global food industry may have been limited. However, if the project is intended to include 'food systems' in its main theme then the lack of inclusion of any adequate discussion is to be regretted.

Discussion

Initial investigations into progress on the improvement of the state of world food security had indicated just how many organisations were actively involved with diverse projects whose goals were seldom integrated. Few of these could be seen to include or involve food professionals, as would be defined by IUFoST. Although the conferences cited here comprised a very limited sample, as dictated by both the timescale and available funding, enough intelligence was accumulated to give strong indications to IUFoST both as to its status as perceived by other organisations, and to areas of future action that would be feasible in terms of its strategic plan for food security and would fill gaps in existing programmes. Opportunities have been identified for developing better and more creative relationships with organisations having similar goals. Following a seminar session at the 2008 IUFoST Congress in Shanghai, a new executive has been established with a more active approach to the very intentions as were expressed originally in Budapest in 1995.

Conclusions

Attendance at the four conferences listed gave a limited perspective on the efforts of some sectors of the international community to address climate change issues and food security as they appeared relevant to each of their core businesses. Primary interests were focussed on agriculture rather than food and where food was the theme, labelling health claims and new 'magic' ingredients were key factors and food security appeared only through a more marketable diet. Urban infrastructure, social welfare and nutrition (through dietary guidelines) were critical to consumer health but the food choice and the inputs of the food industry were ignored. Overall there was little interest in the importance of the food supply chain and even less knowledge of the role of IUFoST. In revising its approach to food security and global hunger, there are many suggestions for new approaches to networking and collaborative enterprise by IUFoST. The number of food insecure people in the world continues to increase (Food and Agriculture Organization, 2008).

References

Food and Agriculture Organization (FAO) (2008). The State of Food Insecurity in the World. High food prices and food security – threats and opportunities. Rome, Italy. ISBN 978-92-5-106049-0, 56pp.

Healthy Foods European Summit (2007). Report available at http://www.healthyfoodssummit.com/. Accessed 31 March 2009.

International Conference on Nutrition (ICN) (1992). Final Report of the Conference on Nutrition. ICN/92/3 and PREPCOM2.ICN/92/FINAL REPORT. FAO and WHO, Rome, Italy.

International Council for Science (ICSU) (2008). Workshop to advise the Science for Health and Wellbeing (SHWB) Planning Committee. http://www.icsu.org/5_abouticsu/STRUCT_comm_Adhoc_health.html. Accessed 31 March 2009.

International Council for Science-Global Environmental Change and Food Systems (ICSU-GECAFS) (2008). Food Security and Environmental Change – Linking Science, Development and Policy for Adaptation. http://www.gecafs.org/gecafs_conference_2008/index.html. Accessed 31 March 2009.

International Food & Agriculture Trade Policy Council (IPC) (2007). Sustainability in the Food & Agricultural Sector: The Role of the Private Sector and Government. http://www.agritrade.org/pressroom/documents/Sustainability_conference_October_2007_000.pdf. Accessed 31 March 2009.

McGill, A.E.J. (2006). Urban challenges and solutions for ethical eating. In: Kaiser, M. and Lien, M. (eds.) Ethics and the Politics of Food, Wageningen Academic Publishers, Wageningen, the Netherlands, pp 368-375.

The International Union of Food Science and Technology (IUFoST) (1995). World Food Congress, Budapest, Hungary. http://www.iufost.org/. Accessed 5 January 2005.

Farmers and professional autonomy: from human right to civil duty

F.R. Stafleu and F.L.B. Meijboom
Ethics Institute, Utrecht University, Heidelberglaan 8, 3508 TC Utrecht, the Netherlands;
F.R.Stafleu@uu.nl

Abstract

At the previous EurSafe conference we suggested that the concept of 'professional autonomy' for farmers could be a tool for farmers in the public debate. The reason we gave was that farmers are members of a profession and that farmers have a distinct 'farmers' morality' that may have an important input in the public discussion. In our ongoing research we have elaborated further on the reasons for attributing professional autonomy to professionals. First we redefine 'professional autonomy' as 'professional moral autonomy (PMA)'. Secondly we introduce the term Moral Acting Professional (MAP) for those professionals who deal professionally with moral questions. We found two reasons for attributing PMA to MAPs: First for individual MAPs, having certain autonomy in their work is a human right and secondly, for MAPs as members of the profession, autonomy is a necessary prerequisite for fulfilling the civil duty of contributing to the public moral debate. Our research has also showed that there is not one single farmer's morality but that at least three groups can be recognized. The differences between these groups may lead to differences in the sort of MPA that can be attributed to them. Furthermore we found that the conditions needed to speak of 'a profession' have not been fully met and so farming is a developing profession. However, being a profession is necessary for getting attributed PMA, because the organisational structure of a profession is necessary to regulate the equilibrium between societal demands and PMA.

Keywords: farmers' morality, professional autonomy, professionalism, public debate

Introduction

Farmers have lost a great deal of the autonomy they once had. Traditionally, farmers were entrepreneurs or tenants who were relatively independent in their choices how to run their farms. There was little interference from 'outside'. This situation, however, changed after World War Two. After WWII, food security became an important public goal. All developments in agriculture started to focus on the production of sufficient safe food for all citizens of Europe. To pursue this, a common agricultural policy was considered necessary (Brom *et al.*, 2007). This development had direct influence on the independence of individual farmers. During this war much of the infrastructure of Europe was damaged and the economic part of the society had to be rebuilt. Farmers did not escape from this development. The resulting rationalisation as it was called in the Netherlands meant that farmers had to scale up their production and had to strive for maximum efficiency, which in the end resulted in 'intensive farming'. Many (small) farmers had to quit. This development was the first infringement of the farmer's autonomy. This infringement mainly affected choices concerning production methods. A few decades later, when food security was obtained, the intensification became subjected to criticism by society. This time the technical and economical aspects of farming were not at stake, but the moral dimensions of intensive farming were targeted, resulting in demands and rules on issues like animal welfare and the environment. These demands and rules represented a further loss of autonomy for farmers, this time not only concerning technical matters but also concerning moral matters. In this paper we want to focus on this last aspect of autonomy. As it concerns the autonomy of farmers in their professional role, we label this autonomy as a version of professional autonomy. The term 'professional autonomy' however includes more aspects than the moral dimension that we discuss in this paper. In the case of medical doctors for example professional autonomy also includes a certain extent of freedom with respect to medical technical questions like treatment methods and socio-economical factors like controlling who, under what circumstances, is to be admitted to the profession. But also in this group autonomy on moral matters is part of professional autonomy. In this paper we want to focus on the freedom of *moral* choice that someone has in his profession. We will call this: Professional Moral Autonomy (PMA)

Professional moral autonomy: from fact to contested

A profession may possess PMA as a fact, meaning they are to a certain extent free to act on moral matters as they wish. In this case the professional is accountable only to himself. Before WWII, farmers dealt with issues of animal welfare, but farm animal welfare was not an issue in the public debate. If a moral question came up, it was dealt with by the person himself or within the family of the farmer. So PMA as a fact means that there is a moral issue in the profession, but that it is undisputed in the public debate and is, as a fact and not intentionally, left to the professionals to deal with. PMA may also be challenged. This happens when the moral issue at stake is part of the public debate. For example in the last decades, (farm) animal welfare became a subject in the public debate. It follows from the nature of morality, that the public debate has its consequences for the factual farmer's PMA on this issue. This is because what is 'right' in a certain issue in the public debate will also be claimed to be right in professional practice. If in the public debate, the principle of 'not harming animals' is thought of as being relevant, then this principle is also relevant in the professional practice (ceteris paribus). In such an example professional autonomy might be challenged and it is, by definition, not a fact anymore. What happens next depends on whether the moral opinions in the debate and the practice differ. If they do not differ, a situation may develop in which the public entrusts the moral practice to the professional practice. In this case PMA is not (only) a fact any more but it is 'uncontested PMA'. When the moral opinions differ however a dialectical process starts. This is what happened with the publications of 'Animal Machines' (Harrison, 1964) and Animal Liberation (Singer, 1990) etc. Sharp criticism was aimed at intensive farming. This was not criticism aimed at the technical properties of intensive farming, but at the ethical properties: Farmers were attacked because, practicing intensive farming, they were accused of harming animal welfare. The resulting discussion focused on the factual side of the matter. It was not so much discussed whether, or to what extent, harming animal welfare could be justified by economic benefits. The discussion focused mainly on the question of whether the welfare of the farm animals was indeed harmed. Farmers kept on denying the harm; voices from the public kept on affirming it. A full dialectical process developed in which farmers, politicians, NGO's, philosophers etc discussed or took (sometimes radical) action. Elsewhere (Stafleu, 2004) we have discussed the fact that farmers were not well equipped for a moral discussion. In the countryside morality existed but critically reflecting on it was not custom. It was in the city, where many different cultures exist, that reflection on morality was more common. The discussion on intensive farming was and is a discussion between the countryside and the city, so it was a moral discussion between parties unequally equipped for moral discussion. The (urban) opponents spoke the moral language much better and won. Farmers lost moral 'voice' in the discussion and were literarily 'overruled' i.e. many rules were issued on animal welfare. The rules are so many that that the farmers' PMA was narrowed considerably or was virtually lost. In this paper we will argue that such a loss is not fruitful and that reallocation of PMA to farmers is necessary from a human right point of view and because it will enhance and enrich the quality of the public debate. We will defend this view by giving arguments that concern the relationship between society and the professional in general and will then return to the special case of the farmer.

Professionals and morally acting professionals

Before we scrutinize the concept of PMA in more detail, we have to clarify the concept of a professional. The dictionary meaning of a profession is: occupation requiring training and intellectual abilities, practiced so as to earn a living (Penguin English Dictionary, 1996). When we discuss PMA we are interested in a subgroup of professionals, i.e. those that in their practice are involved in ethical matters which concern more than their persons or their profession, but that touch societal problems. For example doctors deal with moral problems as euthanasia and abortion, lawyers deal with problems concerning freedom of speech. The action of these professionals directly influences or shapes the public discussion. We will call this subgroup 'Morally Acting Professionals' (MAP's).

Arguments for attributing PMA to MAP's

There are two main arguments for attributing PMA to MAP's: the ' human rights argument' and the 'civil duty' argument.

The human rights argument

A MAP is both a professional and also a person. The society imposes certain ethical standards upon the profession and so upon the individual MAP as a person. The ethical standards of society and those of the individual MAP as a person may conflict. Every human has the right of conscience (has moral autonomy). When a MAP wants to act according to this right and this concerns his professional action, a conflict may arise with the ethical standards imposed by society. A clear cut hierarchy does not exist, so an equilibrium has to be found between the moral autonomy of professional as a person and the ethical standards of the society. This means that the MAP must have a certain moral autonomy concerning his professional acting: i.e. a certain professional autonomy.

The civil duty argument

The human rights argument mentioned above concerns a moral conflict in the relation between the professional as an individual civilian and the society as a whole. In the civil duty argument the same relationship is at stake. The basic assumption in this argument is that society needs the practical moral expertise of the MAP in the public debate and that the MAP has a moral duty to advance this public debate. By definition the MAP is in his professional work confronted with moral issues. The MAP is, so to say, the fieldworker in the field where the moral issues actually are at stake. For example, the doctor is the one who is confronted with 'end of life decisions', the farmer has direct influence on the welfare of animals etc. This means that the MAP has practical knowledge about these moral issues. This practical moral knowledge is a mixture of technical knowledge, moral considerations and experience of how these two combine in reality. This kind of knowledge creates new moral insights and is indispensable for a meaningful public debate. This indispensability gives rise to a moral duty for the MAP to share this knowledge in the public debate. When, however, the MAP is forced to act only on the moral prescriptions of the society, the advancement of practical moral knowledge is hampered and the MAP cannot fulfill its duty. So a certain amount of PMA is necessary to advance the public debate.

Necessary conditions for PMA

As described above there is a need for a *certain amount* of PMA. When moral considerations clash between the MAP and the society, an equilibrium has to be found. As there is a conflict situation, attributing PMA to the MAPs in such an equilibrium will not be unconditional We interpret the classical features ascribed to 'professions' (as opposed to 'occupations') as the necessary conditions for the development of such an equilibrium (Carr, 1999). First of all the MAPs must be organized so there is a party to negotiate with. It must also be clear what basic norms and values the profession has and on which the society can count. Some kind of professional code will fulfill this requirement. The MAPs must have moral competence. This means they know the moral principles at stake and know how to pursue an ethical discussion. In practice this means special education.

PMA and farmers

Are farmers MAPs? As mentioned farmers deal with moral questions concerning e.g. animal welfare and environmental questions, so the moral action is there. But are they professionals? If we look at the requirements, i.e. 'a certain' training and intellectual abilities, then the situation is unclear. There is no official farmer's license and to be a farmer a special education and certain intellectual capabilities are not required (compare e.g. the situation with doctors). On the other hand there are agricultural schools and universities and most farmers visit these schools before becoming a farmer. It seems that farmers as a group are in a process of professionalization. If we look at the necessary conditions for PMA, then we see that farmers are getting organized, but that there is no clear professional ethics and no ethical code. Also ethical competence is not always shown. So farmers as a group are at the moment not clear cut professionals.

But when we look at the arguments for attributing PMA to professionals we see that these arguments are fully valid in the case of farmers: farmers do have strong moral convictions concerning their job. These strong moral convictions are to a certain extent typical for farmers and for that reason may play an important role in the public debate. So the basic reasons are there, but concerning the conditions needed to give PMA to farmers, they fall short as a group.

Different farmer's moralities

An important reason to give Farmers PMA is that they have a strong moral conviction, rooted in tradition and practice, which may enrich the public debate on e.g. animal welfare and other 'green' issues. In our in-depth interviews with a large number of farmers we found three groups of farmers. (De Rooij & De Lauwere, in press; see also Grimm, 2006) The first group is strongly focused on production and does not see any moral issues concerning this production. They are convinced that public concern is rooted in ignorance, and will be solved with better information. The second group is strongly convinced of the moral dimension of their work and knows how to deal with it. Most organic farmers fall in this group. So this group sees the moral problems, but claims they know the answers. The third and largest group lingers in between. They know that farming raises ethical questions, but they do not know exactly how to handle these questions. These three groups may have a different contribution to the public debate. The possible contribution of the first group seems to be small. New moral insights are not to be expected. The second group surely has a specific attribution to make but it seems to be aimed at a special section of the public. The third group may have an important contribution, but only insofar they succeed in translating their moral questions into terms usable in the discussion. As the quality of the contribution for the public debate is one of the main reasons to give PMA to professionals, it seems to be clear that PMA for farmers as one group is not feasible. What seams to be feasible is that different forms of PMA, with different range and content, are given to the different groups of farmers. The first group, which has little to add to the moral discussion, because they see no moral problems, may get a very limited form of PMA. The second group may have a very tailor made form of PMA that mirrors their special view on their profession and of the special views of their customers. The third group seems to be the most promising group, because they are still open for debate and may experiment with new ideas, which could be the basis for new insights and further development of morality.

References

Brom, F.W.A., Visak, T. and Meijboom, F. (2007). Food, citizens and market: the quest for responsible consuming. In: Frewer, L. and Van Trijp, H. (eds.) Understanding consumers of food products, Cambridge: Woodhead Publishing, pp. 610-623.

Carr, D. (1999). Professional Education and Professional Ethics. Journal of Applied Philosophy 16(1): 33-46.

Grimm H. (2006). Animal Welfare in Animal Husbandry – How to Put Moral Responsibility for Livestock into Practice. In: Kaiser, M. and Lien, M. (eds.) Ethics and the Politics of Food, Wageningen Aacademic Publishers, Wageningen, the Netherlands.

Harrison, R. (1964). Animal Machines: The New Factory Farming Industry. Vincent Stuart Ltd. London.

The Penguin English Dictionary, Revised Edition.(1969). Allen Lane the Penguin Press. London.

Singer, P. (1990). Animal Liberation. New Revised Edition. Avon Book. New York.

Stafleu, F.R., De Lauwere, CC, De Greef K.H. Sollie,P. and Dudink, S. (2004). Boerenethiek, eigenwaarden als basis voor een 'nieuwe ethiek', een inventarisatie. Verkende studie. NWO's Gravenhagen. the Netherlands.

Short supply chains as a criterion for sustainable food production and consumption

L. Voget
Institute for Botany and Landscape Ecology, University of Greifswald, Grimmer Str. 88, D-17487 Greifswald, Germany; umweltet@uni-greifswald.de

Abstract

This paper aims to analyse the notion of short supply chains as a fruitful concept for the description of sustainable ways of living with regard to food production and consumption. Most concepts of sustainability give a prominent place to three strategies for sustainable development: efficiency, consistency and sufficiency. Sufficiency asks for ways of living that promote sustainable development. Since the effect of efficiency strategies is limited because of the rebound effect and since consistency will take time to achieve, sufficiency can be regarded as a necessary complementary strategy. Advocates of sufficiency assume that the standard of living, i.e. the kind and amount of goods consumed can be lowered without negative impacts on quality of life. With recourse to the décroissance (degrowth) movement this paper aims to show that such efforts towards sustainable living do not only imply the consumption of greener products but also the substitution of mere consumption for processes which are characterized by a lesser degree of product character. Décroissance thinkers argue for the central importance of reciprocity and relational good. Along this line of thought the ideas of local food networks as a possibility of re-embedding the food market into social and environmental systems via a kind of 'food citizenship' or 'civic agriculture' are analysed. The shortness of supply chains is identified as a joint characteristic of certain types of food production and consumption such as local and fair trade food. With regard to organic food, this concept allows for a distinction between small scale organic production on mixed farms, marketed through farmers markets or specialized dealers and big scale, 'conventionalized' organic food. Furthermore the concept allows for a strong argument in favour of producing a certain amount of food either at home or collectively.

Keywords: sufficiency, local food networks, organic agriculture

Introduction: sustainability, sufficiency, décroissance

This paper aims to analyse the notion of short supply chains as a fruitful concept for the description of sustainable ways of living with regard to food production and consumption.

In accordance with the World Commission on Environment and Development (WCED) definition I conceive sustainable development as 'development that meets the needs of the present without compromising the ability of future generations to meet their own needs.' (World Commission on Environment and Development, 1987). In this definition sustainability comprises not only justice towards future generations (intergenerational justice) but also justice towards contemporary humans (intragenerational justice). Without going into detail I will further presuppose that these claims for justice do imply the necessity to conserve natural capital as well as obligations to advocate and work for living conditions that allow for all contemporary and future humans to live a good human life.

Most concepts of sustainability give a prominent place to three strategies for sustainable development: efficiency, consistency and sufficiency. Efficiency aims at either providing the same amount of products or services while drawing on a smaller amount of energy and resources or drawing on the same amount of resources and energy while providing a larger amount of products or services. Consistency aims to reform the economic system leading to a state in which the economic system does not harm or disrupt natural systems. Sufficiency asks for ways of living that promote sustainable development (Ott and Voget, 2007; Linz, 2004). Since the effect of efficiency strategies is limited because of the rebound effect and since consistency will take time to achieve (Linz, 2004; Sachs, 2002), sufficiency can be regarded as a necessary complementary strategy.

In terms of a narrow understanding, sufficiency asks for a reduction of consumption. There are two underlying notions:

- On the one hand it should be noted that consumption presupposes production and that production relies on the use of natural resources. Thus a reduction in consumption is seen as a way to reduce pressure on natural resources and thus to allow for a reallocation of unjustly distributed goods (intragenerational justice) and to maintain the foundations for human life on earth for future generations (intergenerational justice) (Linz, 2004).
- On the other hand the distinction between standard of living and quality of life is emphasized. Standard of living draws on the notion that anything valuable for a human life can be bought by monetary means. Therefore the standard of living as the amount of money a person can spent is conceived as a reasonable measure of a person's welfare. In contrast, quality of life is a broader concept which acknowledges that when asking the question does a person lead a good human life, this cannot be answered by looking at her income alone. Advocates of sufficiency claim that above a certain threshold, standard of living and quality of life are no longer linear or positively correlated. Instead the correlation may be unclear or indeed negative. (Max-Neff, 1995, Easterlin, 2005; Bruni *et al.,* 2005). Thus it is stated that above a threshold it is possible to have a low standard of living without diminishing the quality of life.

However, the description of sufficiency solely as an issue of consumption levels seems too narrow. To begin with sufficiency as a strategy for sustainable development must not be conceived as a question hinging solely on individual consumption decisions. It should be clear that such an 'individualization of responsibility' (Maniates, 2002) overburdens consumers. Secondly it should be highlighted that the level of the threshold above which a lowering of the standard of living does not lead to a lower quality of life does depend on social and political circumstances. Thus sufficiency is not only an individual strategy or question, but it should be conceived as a social and political challenge.[18] Among other things this challenge comprises a need to develop circumstances in which individuals cannot only consume less but can substitute consumption for other processes.

The Décroissance- or Degrowth-Movement is composed of different voices criticizing economic growth from a broad range of perspectives. Their common denominator is a critical attitude towards not only growth *sensu strictu,* but the theoretical presuppositions and practical consequences of neoclassical standard economics in general. With regard to sufficiency the examination of the anthropology underlying neoclassical standard economics is especially interesting. Instead of viewing humans as mutually disinterested utility-maximisers décroissance-thinkers represent a concept of human beings as depending on the recognition of others to flourish. This concept stresses the importance of so called relational goods. Relational goods are goods whose production and consumption can only be accomplished collectively. An illustrative example is friendship but the term would also comprise goods like personal social networks, neighbourly help or a fruitful working atmosphere.

The important link to sufficiency as discussed above is a link between the claim for substitution of consumption by other processes and the concept of relational goods. On the one hand the production and consumption of relational goods can in itself constitute possible substitutes for classical consumption. For

[18] For a similar arguments see Seyfang (2006: 391).

example this is the case if somebody is frustrated and instead of going shopping visits a friend. On the other hand relational goods can allow access to good lives that do not rely on consumption (Muraca, 2009).

Localisation

At first view it may be conceived as somewhat strange to choose food as an example to illustrate the importance and opportunities deriving from substituting consumption by other processes. After all food consumption is one of the processes that satisfy basic human needs. Therefore food consumption itself cannot be substituted by other processes. Nevertheless it is possible to link food purchase and consumption with the simultaneous production and consumption of relational goods. Local food networks are often claimed to offer this possibility. In this context '[l]ocalisation [...] means simply that food should be consumed as close to the point of origin as possible' (Seyfang, 2006: 386). The engagement in a local food network is said to have positive impacts on food-consumers and -producers and thus seems to be a promising approach on the way to sustainable food production and consumption. Seyfang names two main positive aspects of localisation (2006: 386). With regard to environmental impacts, she stresses the reduction of 'food miles', e.g. of the distance food travels between production and consumption. Since food transport is linked to the emission of greenhouse gases and other pollution as well as a loss in freshness and taste, the reduction of food miles can have positive impacts on both the environment and product quality. With regard to social impacts, Seyfang points to the embeddedness of food markets into social relations. Social embeddedness 'ensues from the possibility of face-to-face interactions and mutual knowledge.' (Hinrichs, 2003: 36) and is characterized by 'social connection, reciprocity and trust' (Hinrichs, 2000: 296).[19] It contrasts the sharp division which conventional food retailing creates between producers and consumers (Hinrichs, 2003: 41). Social embededdness can be related to the following positive aspects:

- A strengthening of local economies: While the increasing amount of locally circulating money (Seyfang, 2006: 386) can be seen as an independent effect of localisation, the embededdnes is displayed in 'the potential to mobilise new forms of association which might resist the conventional price-squeeze' (Seyfang, 2006: 386). Such forms of association are build on regard or even sympathy for farmers (e.g. Winter, 2003: 29).[20]
- Consumer-education: 'Local [...] food networks have an influence on consumers in developing informed, educated communities around food – through education, outreach, literature, farm visits, web sites, etc. [...]'(Seyfang, 2006: 393).
- Traceability and accountability: In line with the feeling of sympathy or responsibility for 'their' farmer, people tend to trust suppliers they have met face-to-face (Seyfang, 2006: 391; Winter, 2003: 29-30; Holloway and Kneafsy, 2000).
- Environmental management: Hinrichs (2000: 35-36) points out that 'pursuing local direct market opportunities would seem to encourage attention to environmental management practices by farmers who anticipate surveillance by concerned consumers.' However, this holds only with regard to management practices whose impact on the environment is evident for lay consumers.

In summary the social embeddedness allows for 'feedback-mechanisms which are absent when food comes from distant origins' (Seyfang, 2006: 386).

[19] There appears to be a broad intersection between the concepts of social embededdness and relational capital even though the two concepts emanate from different starting points.

[20] Thus for example in Community Supported Agriculture (CSA) customers pay an annual share in return for weekly supplies of fresh vegetables. The annual share supplies the CSA-farmer with a guaranteed base income while the weekly customer's supplies vary in size depending on time and the conditions of the growing season. Thus the consumers bear part of the risk of farming (Hinrichs 2000: 299-301).

The discussion of the non-food mile-environmental impacts of localism forms a link to the negative aspects of localisation:

- Local is not organic: As mentioned above, if at all, farmers marketing their produce locally will foremost be motivated to change those practises that might evoke attention with consumers. Thus it is mistaken to equate localism with organic and ecological products (Winter, 2003: 29-31; Holloway and Kneafsy, 2000: 286-287).
- The defensive stance of localism: 'localisation can be reactionary and defensive stance against a perceived external threat from globalisation and 'others' and this exclusiveness can hinder the acceptance of diversity and difference [...]. Furthermore the local can be a site of inequality and hegemonic domination, not at all conducive to the environmental and social sustainability often automatically attributed to processes of localisation by activists.' (Seyfang, 2006: 386, see also Allen, 1999).
- Non-availability: A wide range of products cannot be produced locally, e.g. coffee, tea, chocolate, bananas (from a European perspective). To hope that embeddement will lead to an uncomplaining renunciation of these goods by consumers seems to overextend the concept of embeddement.
- Lacking accessibility for low-income people: some forms of local supply chains involve higher costs (monetary or time) for consumers (Allen, 1999: 125, Macias, 2008: 1088).

Short food supply chains

Regarding these negative aspects, localisation seems to be a somewhat ambivalent concept. Nevertheless, the motion of positive impacts of an embeddedness of food markets into social relation seem a fruitful concept with regard to the ideas of sufficiency and relational goods and their importance for sustainability as outlined above. A concept that is able to encompass these positive aspects while avoiding some flaws of localisation is the concept of short food supply chains (SFSC). SFSCs can be defined as 'giving clear signals on the provenance and quality attributes of food' (Renting *et al.,* 2003: 398). They can be further differentiated by their extension in time and space (Renting *et al.,* 2003: 399-401). This does imply that food networks characterized by SFSCs are not congruent with local food networks: While local food networks failing to be characterized by short supply chains seem hardly imaginable, several short food supply chains do not fall into the local realm. Examples are fair trade products like coffee, chocolate or honey (Renting *et al.,* 2003: 400). These products originate from far away regions, but they are traded through SFSCs and this leads to strong accountability. On the other hand, guarantees regarding purchase quantities and prices allow for a certain security for primary producers.

The concept of SFSC allows these products to keep the notion of embeddedness. By enabling worldwide connections via SFSCs it seems possible to avoid the defensive stance of localism. Ultimately these worldwide connections facilitate the purchase of products that cannot be produced locally. Thus a first proposal for sustainable food consumption could be to purchase food locally where possible and to purchase it via short supply chains where it is not locally available.

On the other hand the abandoning of localisation is accompanied by the loss of food miles savings. And similar to local food networks, SFSCs alone can neither guarantee any particular positive impact with regard to environmentally friendly farming methods will occur, e.g. organic agriculture, nor do they resolve the problem of accessibility for low income groups. Having not enough money to spend on high quality food or having not enough time to purchase food and prepare meals is not specific to the sustainability of a *food* system but constitutes a general problem of justice. In contrast being able to handle the environmental impacts of food production and consumption should count as an integral criterion for the sustainability of a food system. Therefore SFSC can not be used as the sole criterion for sustainable food production, but neither can organic agriculture. This is illustrated by Smith and Marsden (2004) who show that organic agriculture is increasingly retailed by large supermarkets (p.

346) which leads to organic farmers becoming caught like conventional farmers in a 'farm-gate price-squeeze' (p. 349/50). Seyfang points in a similar direction when she claims that '[...] we can see that the mainstream has superficially adopted the niche consumption market for organic food, but has done so in a way which keeps the technical point (not using pesticides or fertilisers in growing) but discards the essence of the project – namely to promote a different relationship between people and food and build alternative provisioning systems.' (Seyfang, 2007: 118)[21].

Embeddedness, relational goods, sufficiency

To come back to the opening question, what is the relevance of the embeddedness resulting from SFSCs for relational goods and in turn for sufficiency when examining sustainable life styles? Embeddedness was described as resulting from face-to-face interactions and mutual knowledge. Thus embeddedness is closely linked to personal relations between and among producers and consumers. How does such a relational wealth affect the possibilities for sufficient lifestyles with regard to food? The crucial point is how the personal relation affects the purchase, processing and attitudes towards food. Several aspects can be distinguished

- Community: Through personal contact with the producer or with a small retailer the purchase itself acquires a relational aspect. If the purchase is accompanied by a small talk with the producer or retailer, this can be seen as the simultaneous production and consumption of a relational good. Furthermore, by joining into cooperatives (producers and consumers) or by 'sharing' the same producers new relational networks may be established and information exchanged among and between producers and consumers.
- Cooking: In buying produce sold in SFSC-contexts consumers frequently agree to purchase food that is not processed or processed only to a certain degree. This forces consumers to acquaint themselves with how to process such foods themselves. Furthermore in Community Supported Agriculture (CSA) and box schemes, consumers agree to accept the food provided by the farmer instead of choosing what food to purchase. This reflects another attitude towards the purchase. King customer is substituted for a more cooperative transaction.
- Working: In CSA as well as in some consumer cooperatives, consumers agree to contribute a part of the personal share as work. Thus consumers actively participate in the production process.
- Learning: Both cooking and work constitute genuine learning experiences. Besides learning how to process certain foods or how certain foods are grown and reared this may lead to a change in attitude towards the food itself.

To conclude, my hypothesis is that via personal relations between producers and consumers SFSC can contribute to an alternative attitude towards consumption that is characterized by regard or even sympathy for the producer, by interest in the production methods, by the willingness to learn and to contribute ones lot. I assume that such an attitude could be conducive to sufficiency not only as sustainable consumption but as a sustainable way of life.

Acknowledgement

I gratefully acknowledge funding for my research via a PhD scholarship by the Deutsche Bundesstiftung Umwelt (DBU).

[21] The claim that the promotion of a different relationship between people and food constitutes the essence of organics is naturally debatable. However, I personally adhere this opinion.

References

Allen, P. (1999). Reweaving the food security safety net: mediating entitlement and entrepreneurship. Agriculture and human values 16: 117-129.

Bruni, L. and Porta, P.L. (ed.) (2005). Economics and Happiness. Framing the analysis. Oxford University Press. United Kingdom.

Easterlin, R.A. (2005). Building a better theory of well-being. In: Bruni, Luigino; Porta, Pier Luigi (ed.): Economics and Happiness. Framing the analysis. Oxford University Press. United Kingdom, pp. 29-64.

Hinrichs, C.C. (2000). Embeddedness and local food systems: notes on two types of direct agricultural market. Journal of rural studies 16: 295-303.

Hinrichs, C.C. (2003). The practice and politics of food system localization. Journal of rural studies 19: 33-45.

Holloway, L. and Kneafsy, M. (2000). Reading the space of the farmers' market: a preliminary investigation from the UK. Sociologia ruralis 40(3): 285-299.

Linz, M. (2004). Weder Mangel noch Übermaß. Über Suffizienz und Suffizienzforschung. (Wuppertal Paper, 145). Available at: http://www.wupperinst.org/uploads/tx_wibeitrag/WP145.pdf Accessed 16 March 2009.

Macias, T. (2008). Working Toward a Just, Equitable, and Local Food System. The Social Impact of Community-Based Agriculture. Social science quarterly 89(5): 1086-1101.

Maniates, M. (2002). Individualization: Plant a tree, buy a bike, save the world. In: Princen, T., Maniates, M. and Conca, K. (ed.) Confronting Consumption: MIT Press. USA: 43-66.

Max-Neff, M. (1995). Economic growth and quality of life: a threshold hypotheses. Ecological Economics 15: 115-118.

Muraca, B. (2009). Nachhaltigkeit ohne Wachstum? Auf dem Weg zur Décroissance Theoretische Ansätze für eine konviviale Post-Wachstum-Gesellschaft. In: Egan-Krieger, T.V., Schultz, J., Thapa, P. and Voget, L. (eds.) Sammelband zur Theorie und Praxis starker Nachhaltigkeit. In Press.

Ott, K. and Voget, L. (2007). Ethical dimensions of education for sustainable development. Bildung für Nachhaltige Entwicklung 2. Available at: http://www.bne-portal.de/coremedia/generator/pm/en/Issue__001/01__Contributions/Ott__Voget_3A_20Ethical_20Dimensions.html Acessed 16 March 2009.

Renting, H., Marsden, T.K.and Banks, J. (2003): Understanding alternative food networks: exploring the role of short food supply chains in rural development. Environment and Planning 35: 393-411.

Sachs, W. (2002). Die zwei Gesichter der Ressourcenproduktivität. In: Wuppertal Institut (ed.): Von nichts zu viel. Suffizienz gehört zur Zukunftsfähigkeit (Wuppertal Paper, 125): 49-56. Available at: http://www.wupperinst.org/uploads/tx_wibeitrag/WP125.pdf Accessed: 16 March 2009.

Seyfang, G. (2006). Ecological citizenship and sustainable consumption: examining local organic food networks. Journal of rural studies 22: 383-395.

Seyfang, G. (2007). Cultivating carrots and community: local organic food and sustainable consumption. Environmental values 16: 105-123.

Smith, E. and Marsden, T.K. (2004). Exploring the 'limits to growth' in UK organics: beyond the statistical image. In: Journal of rural studies 20: 345-357

Winter, M. (2003). Embeddedness, the new food economy and defensive localism. Journal of rural studies 19: 23-32.

World Commission on Environment and Development (WCED) (1987). Our common future. New York. USA.

Moral consideration of plants for their own sake

Ariane Willemsen
Executive Secretary, Federal Ethics Committee on Non-Human Biotechnology ECNH, c/o Federal Office for the Environment FOEN, CH-3003 Bern, Switzerland; ariane.willemsen@bafu.admin.ch

Abstract

The Swiss Constitution knows three forms of protection for plants: the protection of biodiversity, species protection, and the duty to take the dignity of living beings into consideration. The constitutional term 'living beings' encompasses animals, plants, and other organisms. Previous discussion relates the term dignity of living beings to the value of the individual organism for its own sake. Given the legal interpretation, the general ethical questions remain unanswered: do we have moral obligations towards plants and why? The concept of dignity of living beings clearly refers to the protection for their own sake, not to the protection for the sake of others. The Federal Ethics Committee on Non-Human Biotechnology ECNH was mandated by the Swiss administration to clarify the meaning of this concept for the handling of plants in research. Since for plants – unlike animals – it is almost impossible to refer to moral intuition, a theoretical approach was followed. Which fundamental ethical positions actually are open to the moral consideration of plants *for their own sake*? What is the individual entity of a plant? What is the weight of a moral obligation towards plants compared to moral obligations towards other living beings? In April 2008, the ECNH presented a report that analyses these questions. The report not only caused some controversy in Switzerland, but also internationally. This paper intends to present the contents of the ECNH-report as well as to critically discuss the committee's majority opinion.

Keywords: plants, dignity of living beings, inherent value, regulation

Introduction

Since 1992 the Swiss Constitution requires that the dignity of living beings (*Würde der Kreatur*) be taken into consideration. The German term *Würde* builds primarily on a theological tradition with two different roots: the *bonitas-* and the *dignitas-Würde*. *Bonitas* refers to all beings created by God, while only the human being as creation in the image of God has *dignitas*. The French version of the Swiss constitution speaks of *intégrité des organismes vivants* and makes this distinction from human dignity transparent. In the Italian version (*dignità della creatura*) as well as e.g. in the English translation the different etymological traditions are lost. The term 'living beings' (*Kreatur*) encompasses animals, plants, and other organisms. In contrast to human dignity the concept of dignity of living beings does not provide an absolute protection. The obligation to take the dignity of living beings into account when handling animals, plants and other organisms is fulfilled when the result of a weighing of goods shows that their goods weigh less than the interests obtained from inflicting impairments on them.

In what context did this concept find its way into the Constitution? In the late sixties came the breakthrough of genetic engineering. During the seventies and early eighties several initiatives by Swiss parliamentarians called for the legal regulation of genetic engineering. Finally, in 1987, a popular initiative (a change to the Constitution proposed by citizens) requested the introduction of a new article prohibiting the misuse of reproductive and genetic engineering applied to humans. The Federal Council (Swiss Government) reacted by proposing an alternative amendment in form of the new constitutional Article 24^{novies} (now Article 120), which aimed to protect not only humans but also the environment from the misuse of gene technologies and to regulate the handling of genetic material of animals, plants and other organisms. Additionally, it introduced the concept of dignity of living beings that is to be respected with regard to animals, plants and other organisms. Following this alternative suggestion by the government, the popular initiative was withdrawn. In the 1992 referendum vote, the Swiss people and the cantons approved the new constitutional article. After further years of controversial debates on the regulation of gene technologies, the Swiss Parliament adopted the Gene Technology Act (GTA) in 2003. Article 8 GTA confirms the concept of dignity of living beings. However, in the Act, the scope of the term was limited to animals and plants, omitting the 'other organisms'.

The respect of the dignity of living beings restricts other constitutional rights such as economic freedom and freedom of research. Since its establishment in 1998 the ECNH has been charged with proposing ways of clarifying and concretising this concept from an ethical perspective, not only for animals but also for the handling of plants in research. With regard to the concretisation for animals it was possible to refer to a large consensus of moral intuitions, at least as a starting point of the discussion. In the case of plants, although the authority of intuition in the ethical discourse is contested, it was still hoped to draw on the discussion on typical concrete examples so as to achieve a consensus on general criteria for dealing with plants. However, where plants are concerned, our moral intuitions leave us at loss. Even within the committee, the intuitions about the extent and justification of moral responsibilities towards plants were highly heterogeneous. Since the intuitive approach did not lead any further, a theoretical procedure was followed. In April 2008, the ECNH presented its report 'The dignity of living beings with regard to plants – Moral consideration of plants for their own sake', where it analysed and discussed the following questions.

Do we have moral obligations towards plants for their own sake?

There are three possible reasons to justify the protection of plants: (1) Plants should be protected because of their instrumental value to others as long as they are of benefit to humans or other living beings, e.g. as crops or as part of biodiversity. (2) They should be protected because of their relational value, i.e. because someone considers them to be worthy of protection. Their worthiness of protection is in relation to a value ascribed to them because of particular properties. (3) Plants should be protected for their own sake because they possess inherent worth. Previous legal interpretation relates the term 'dignity of living beings' to the protection of living beings for their own sake, not for the sake of others. Given this legal interpretation, the general ethical questions remain whether and why we do have moral obligations towards plants and how such obligations can be ethically justified.

Ethical positions open to a moral consideration of plants for their own sake

To solve the issue of whether organisms ought to be morally considered for their own sake, the ethical positions were examined in terms of two questions: (1) who is the moral object and (2) can an individual being itself be harmed? The committee members examined the following positions: Theocentrism, Ratiocentrism, Pathocentrism, and Biocentrism. The basis of a theocentric position is the idea of a God who is creator, and therefore also the creative origin of all living organisms. What counts for its own sake is God. All other organisms including plants count because of their relationship to God. In a ratiocentric position the issue of whether beings count for their own sake depends on their (potential) capacity for reason and their capacity for abstract speech. Plants therefore are merely of instrumental and relational value. The position of pathocentrism is based on the sentience of living organisms. They count morally for their own sake if they are sentient and are therefore able to experience something, in some way, as good or bad. According to this position, plants possess inherent value and have their own interests *if* they are sentient. Otherwise, they are of instrumental and relational value only. In a biocentric position, all living organisms count morally for their own sake because they are alive. The fact that they are alive is a good of their own that must be considered. The ethical positions open to a moral consideration of plants for their own sake are therefore pathocentrism (if plants are sentient) and biocentrism.

No committee member defends a theocentric position. Ratiocentric and pathocentric positions are only represented by small minorities. The clear majority takes a biocentric position.

Can plants be benefited or harmed?

If a pathocentric position is represented, the question of whether a plant can be benefited or harmed is linked to the question of whether a plant has some form of internal experience. It must be able to experience harm or benefit as good or bad. The condition for an independent positive or negative experience is sentience. An organism which satisfies this prerequisite has its own interests. An act which can be experienced by the organism as harm is therefore morally relevant. If, however, it is unable to experience harm as negative, such an act is of no moral significance.

If we know that plants are *not* sentient, they have no interests of their own and there is no sense in referring to plants as moral objects. If we know that plants *are* sentient, they are part of the moral community. If we *do not know* whether plants are sentient or not, we can either speculate, or investigate whether scientific findings exist that give an indication of sentience. For animals, we are in possession of clear indications that they are sentient. For vertebrates, decapods and cephalopods there is even a socially broadly supported agreement that they are sentient. This has been embedded in the Swiss Animal Protection Act: These animals are protected from pain, suffering, fear and stress. Interventions that cause such kinds of harm to an animal require justification. For plants, on the other hand, we lack comparable evidence that would indicate some kind of inner experience. Internal experiences are linked to a kind of consciousness. For plants, we do not have any evidence that they posses such a consciousness.

Although plants do not have a central nervous system, they may nevertheless fulfil the necessary conditions for a kind of sentience. The question arises of whether sentience necessarily depends on a central nervous system. Since we do not have the kind of access to plants that would enable us to know whether plants perceive disturbances consciously, we simply cannot answer. It is nevertheless imaginable that plants have other possibilities of experiencing harm or benefit. Studies in cell biology show that plants and animals, which share a developmental history lasting three billion years, have many processes and reactions that do not differ fundamentally at the cellular level. Plants can choose between various ways of behaving and can change their behaviour. For example, plants undergo complex interactions with their environment, just as animals do. While animals move and respond to external stimuli with flight or fight, plants react by modifying their developmental processes and adapting their growth. They thus express great plasticity of behaviour. Plants also have a differentiated hormonal system for internal 'communication'. The action potential of cellular communication also shows similarities to the signals of nerve fibres in animals. Plants react to touch and stress or defend themselves against predators and pathogens in highly differentiated ways.

Based on the results of such investigations, we may ask whether the moral consideration of plants can be discarded with the argument that plants lack the conditions of negative or positive experience. It is not clear that plants have sentience, but neither is it clear that this is not the case. It can therefore be argued that the reasons for excluding plants from the circle of moral objects have been eliminated. Where we do not know whether plants are sentient, we must decide what consequences this has. If it is morally irrelevant that we do not know, plants are not part of the moral community. If, on the other hand, it is morally relevant, the consideration of plants for their own sake is not excluded. Thus, the pathocentric position is open to the moral consideration of plants for their own sake whether we know that plants are sentient beings or we do not know, but consider this lack of knowledge as morally relevant.

What weight do our obligations have towards plants?

The *egalitarian position* takes the principle that for all living organism, like should be evaluated and treated as like and unlike should be evaluated and treated as unlike. It concedes the possibility that plants genuinely could have the same interests as other organisms, and that these should then be treated equally. According to the *hierarchical position*, all living organisms deserve moral respect, but not all of them equally. Either what counts is the species to which it belongs and human interests are weighted more than the same interests of plants (or animals), or it is the complexity of properties. The more similar the properties are in terms of their complexity to those of humans, the higher their moral significance.

The majority of the committee members take a hierarchical position: Less strong reasons are needed to justify the use of plants than is required to justify the use of (vertebrate) animals. Nevertheless, use requires justification. However, the committee seems to accept in an intuitive approach that any reason whatsoever can serve as a justification for harming plants. Thus, the only criterion that morally prohibits harming plants is having no reason at all to do so, i.e. arbitrariness.

What is the individual entity of a plant?

Legal interpretation relates the term 'dignity of living beings' to the *individual* organism. The moral consideration of plant individuals assumes that we know what plant unit it is, which may be considered to have a good of its own, or, in the case of sentience, interests of its own. There are several possible individual plant entities that may correspond to this definition: plant collectives, species, individual plants, or independent survival-competent components.

It can be argued that the decentral organisation of plants and their dependence on a location makes them more diffuse and permeable to their environment than organisms that are centrally controlled by nervous systems and can move to new locations. For this reason, the collective plays a special role for plants. As moral objects that should be protected for their own sake, plant collectives can be viewed from various perspectives, e.g. plant networks, plant populations, reproductive communities, biotopic communities, plant communities including their interaction with microorganisms or even a more comprehensive definition of the collective that goes beyond the community of plants.

For some committee members it seems clear that the term of collective must be defined even more broadly than plant populations or reproductive communities, since all organisms stand in mutual relationship with one another. However, all positions that assume any kind of plant collectives to have inherent worth are confronted with the problem of distinguishing this community, as an entity, from others. These positions have to formulate the boundaries of these entities and therefore to produce plausible reasons why, and at which point, one collective is distinguished from another.

Those who take the position that it is the species which should be considered morally for its own sake assume that the concept 'species' exists in real terms. Such a definition assumes that all members of a species necessarily have particular essential characteristics. This position is defended by the argument that the species is defined biologically as a reproductive community. Another concept assumes that species are nominal definitions. The selected properties we use to name and classify organisms as a species are empirically observable phenomena. However, there are no clearly delimited and unchangeable characteristics, i.e. no essence that necessarily appertains to all organisms of one species. Instead, organisms are attributed to a species on the basis of a broad spectrum of properties, because processes of change in organisms are always gradual. If a 'species' is understood to be a nominal definition it cannot count for its own sake.

Plants are not constructed centrally but in components. We must, therefore, additionally examine whether plant components that can survive independently of the mother plant could be the object of moral consideration. It remains uncontested that plants do have an overall coordination of their individual components. Experiments with cloned plants growing next to one another show that their roots are able to differentiate between themselves and the Other. It could thus be concluded that the option of considering plant components as objects of moral consideration for their own sake can be excluded, as long as this components are still part of the mother plant.

The majority of the committee members take the position that the object of moral consideration for its own sake is the individual plant. A smaller majority additionally considers plant networks as objects of moral consideration. A minority assumes that we do not know enough about plants and therefore this question can only be answered on a case by case basis.

Conclusions

The concept of dignity of living beings does not provide a core of absolute protection but merely requires a weighing of goods. The ECNH has adopted this legal interpretation of the concept for its ethical analysis. According to the committee members the only morally impermissible harm to plants is arbitrarily inflicted harm. Arbitrary acts cannot be justified by definition because there is no good reason, and therefore nothing to be weighed on the side of the interests in favour of harming the plant. All other harms inflicted when handling plants can be justified by weighing the interests at stake.

The hierarchical position that is taken by the majority of the committee members either ranks the species of humans on the highest level and animals higher than plants, or it ascribes human properties a privileged weight and classifies other living beings according to their degree of similarity of their properties with human ones. Such a position is nothing more than a pragmatic anthropocentrism: humans always rank highest and plants always rank lowest and there is no example within sight where harming plants as a result of a weighing of interests would be morally impermissible. In other words, where plants are concerned, dignity of living beings remains an empty construct, except in the case of arbitrary harm inflicted on plants.

The discussion of the dignity of living beings with regard to plants originates in the legal context of biotechnological plant research. From a legal point of view, this is the only instance where it is required to take the dignity of living beings into account. Such research projects have to be approved by or at least notified to the authorities. In their applications or notifications the researchers must describe their projects and give explicit or implicit reasons for undertaking their research. Considering the low weight that the goods of plants possess in a hierarchical position and considering that the committee members intuitively seem to accept that any reason whatsoever can serve as a moral justification, it can almost be taken for granted that no research project is considered arbitrary. Then, however, the constitutional and legal concept, when applied to plants, is reduced to a moral appeal to the researchers to be aware that they are handling living beings, and has no further consequences.

Following the biocentric position of the majority of the committee members, the morally relevant aspect is that plants are living beings. Such a position has to demonstrate what distinguishes living beings from the non living. This question cannot be avoided and is raised again even more pressingly with regard to the next steps of the discussions within the ECNH: the ethical aspects of synthetic biology and, even more so, the moral status of artificial life. In the context of the discussion on plants, the majority of the members disputed the fact that species count for their own sake on the grounds that a 'species' is understood to be a nominal definition because species develop gradually. If this is so, it would be consistent to assume that not only the development of species but also the development of life itself is a gradual process. 'Life' must then be understood as a nominal definition as well and 'being alive' cannot be morally relevant for its own sake. It therefore ensues that a biocentric position can no longer be defended. Other morally relevant criteria have to be defined.

References

Baranzke, H. (2002). Würde der Kreatur?, Die Idee der Würde im Horizont der Bioethik. Königshausen & Neumann, Würzburg, Germany.

Brenner, A. (2007). Leben, Eine philosophische Untersuchung. Beiträge zur Ethik und Biotechnologie, Vol. 3, Bern, Switzerland.

Federal Ethics Committee on Non-Human Biotechnology, Federal Committee on Animal Experiments and Willemsen, A. (2001/2008). The Dignity of Animals. Joint report by the Federal Ethics Committee on Non-Human Biotechnology and the Federal Committee on Animal Experiments, Bern, Switzerland.

Federal Ethics Committee on Non-Human Biotechnology and Willemsen, A. (2008). The dignity of living beings with regard to plants, Moral consideration of plants for their own sake. Report by the Federal Ethics Committee on Non-Human Biotechnology, Bern, Switzerland.

Rippe, K.P. (2008). Ethik im außerhumanen Bereich. Mentis, Paderborn, Germany.

Stöcklin, J. (2007). Die Pflanze, Moderne Konzepte der Biologie. Beiträge zur Ethik und Biotechnologie, Vol 2, Bern, Switzerland.

Authors index

Keyword index